QUANTUM INFRARED PHYSICS

Proceedings of the Workshop on

QUANTUM INFRARED PHYSICS

The American University of Paris
6 – 10 June 1994

Editors

H M Fried
Brown University

B Müller
Duke University

World Scientific
Singapore • New Jersey • London • Hong Kong

Published by

World Scientific Publishing Co. Pte. Ltd.
P O Box 128, Farrer Road, Singapore 9128
USA office: Suite 1B, 1060 Main Street, River Edge, NJ 07661
UK office: 57 Shelton Street, Covent Garden, London WC2H 9HE

QUANTUM INFRARED PHYSICS

ISBN 981-02-2173-8

Printed in Singapore by Uto-Print

Preface

These Proceedings mark the second Workshop devoted to fundamental Physics held at the American University of Paris (AUP), during the week of 6-10 June 1994. Eight half-day sessions, each dealing with a distinct and different theme, were organized and chaired by eight Session Chairmen, and comprised the principal part of the first four days' Program.* A fifth day plus three brief evening sessions dealing with other, more general topics completed the Workshop's Program. In addition to the the Scientific program, Prof. Maurice Lévy (Université Pierre et Marie Curie) and Prof. David Saxon (UCLA) presented after-dinner speeches on the Future of Physics, at the Workshop banquet.

Many, many people are owed thanks by the undersigned, and indeed by all the participants, for their kind and thoughtful performance of the varied tasks needed for a successful Workshop: To Mary Ann Rotondo, of Brown's High-Energy staff, for welcoming and assisting participants as they arrived at AUP, and for supervising all relevant housing, lecture schedules, and manuscript preparation for these Proceedings, prior to, during, and after the Workshop; to Karen Wagstaff of AUP, for her continuing and good-humored assistance in all matters of organization; to David Masters and his helpful computer assistants at AUP, for setting up temporary e-mail accounts for all participants; and to Karen Archer, for her aid in Workshop material preparation at AUP.

Finally, we must acknowledge our continuing debt to AUP's Dean Bill Cipolla, for his warm hospitality and enthusiasm; never have so many theoretical physicists been so courteously received. To all of the above, and to AUP's President Glenn Ferguson and all of our other friends at AUP, we express our warmest thanks and deep appreciation.

H. M. Fried, Brown University

B. Müller, Duke University

---

* Papers were presented by authors listed in the Table of Contents; all co-authors are indicated on the manuscript.

CONTENTS

Gribov Horizons in QCD

THE UNRELIABILIY OF THE PERTURBATIVE METHOD IN NON-ABELIAN MODELS*

ADRIAN PATRASCIOIU
Physics Department, University of Arizona
Tucson, AZ 85721, U.S.A.

and

ERHARD SEILER
Max-Planck-Institut für Physik
– Werner-Heisenberg-Institut –
Föhringer Ring 6, 80805 Munich, Germany

Abstract

We show that both in $2D$ classical Heisenberg ferromagnets and in lattice Yang-Mills theory in dimension $D \geq 2$ the perturbative method fails by producing results that depend on the boundary conditions chosen to approximate the models in a finite volume. This failure occurs only in non-Abelian models, starting at $O(1/\beta^2)$. It is related to the existence of a hitherto unnoticed type of excitation, dubbed super-instanton, that dominates the low temperature (weak coupling) regime of those models.

1.Introduction

It has been recognized for about two decades that the $2D$ $O(N)$ spin models (non-linear σ-models) and $4D$ lattice Yang-Mills models have many features in common. Commonly it is said that they share the features of asymptotic feedom and dynamical mass generation, provided the symmetry group is non-Abelian. In this talk I want to point out that those models indeed share something else: they both have problems with perturbation theory (PT), on which the above claims are based. It is our belief that actually the commonly made claims of asymptotic freedom and dynamical mass generation are incorrect; for the case of the $2D$ $O(N)$ models we have given a detailed argument, based on percolation theoretic considerations, leading to the

*Talk given by E.Seiler

conclusion that the models in fact have a phase transition to a massless spin wave phase, whether the symmetry group is Abelian or not[1,2].

The $O(N)$ models are described in terms of the action (classical Hamiltonian)

$$S = -\sum_{\langle ij \rangle} s_i \cdot s_j \tag{1}$$

where the sum is over nearest neighbor pairs $\langle ij \rangle$ and the spins s_i are unit vectors in $\mathbb{R}^{\mathrm{N}}$. The partition function corresponding to this action is

$$Z = \int \prod_{\langle ij \rangle} \exp(\beta s_i \cdot s_j) \prod_i \delta(s_i^2 - 1) ds_i \tag{2}$$

and the corresponding Gibbs measure is given by $Z^{-1} e^{\beta S} \prod_i \delta(s_i^2 - 1) ds_i$

Similarly the lattice Yang-Mills (LYM) model has the partition function

$$Z = \int e^{\beta S} \prod_{\langle ij \rangle} dU_{ij} \tag{3}$$

where U_{ij} is an element of a compact Lie group G associated to the 'link' $\langle ij \rangle$, dU is the Haar measure of G and the action S is given by Wilson's expression[3]

$$S = \mathcal{R}e \sum_P \mathrm{tr} U_{\partial P} \tag{4}$$

with the sum runing over the plaquettes of the lattice.

As mentioned already, the conventional belief is that in the non-Abelian versions of these models, i.e. for $N > 2$, respectively a non-Abelian group G, there is exponential clustering (i.e. a mass gap) for all values of β and that the mass gap m varies with β asymptotically for large β as follows:

$$\frac{d\beta}{d \ln m} = -c_1 + \frac{c_2}{\beta} + O(\frac{1}{\beta^2}) \tag{5}$$

with $c_1 > 0$ in the non-Abelian models. On the other hand, it has long been known[4,5] that in the Abelian versions of those models, i.e. $N = 2$, respectively $G = U(1)$, there is a value β_c such that for $\beta \geq \beta_c$ there is no exponential clustering, or, in other words, the mass gap and the β-function vanish identically.

The reason for the belief concerning the non-Abelian models is PT, or more specifically the perturbative Callan-Symanzik β-function β_{CS}, which is related to the left hand side of Eq.(5) by $\beta_{CS}(g) = \frac{1}{2} g^3 \frac{d\beta}{d \ln m}$. The question that arises here is obvious: Can we trust PT in these models? Our answer is *no*, and we will explain why.

2. What are the Low-Lying Excitations?

In order to understand better what is going on in these models at low temperature (large β) we have to find out what are the low-lying excitations above the completely ordered ground state. It has been known since the seventies that some of the models possess topological excitations known respectively as vortices or instantons. These excitations have, however, either finite (instantons) or even logarithmically divergent (vortices) energy, and will therefore be very rare at low temperature. But there is a different type of excitation that seems to have been overlooked; since it has arbitrarily low energy and still roughly as much entropy as the instanton, it will clearly dominate at low temperature; for this reason we dubbed it 'super-instanton'.

Let us describe the super-instanton for the case of the $2D$ $O(N)$ models: Consider a disc of radius R, fix the spin in the middle to point in a given direction e. Fix the spins at the boundary of the disc to be all equal to a unit vector e'. Try to find a configuration interpolating between those values with the lowest possible energy. A little consideration shows that this minimizing configuration will turn smoothly from e to e'; the turning angle will be a harmonic function (i.e. it will be annihilated by the lattice Laplacean) and the energy will be

$$E_{min} = O(\frac{1}{\ln R}) \tag{6}$$

If instead we consider the analogous problem in dimension $D > 2$ we find that $E_{min} \geq E_o > 0$.

The super-instanton configurations are the 'enforcers'of the Mermin-Wagner theorem[6,7] that states that in $2D$ there will be no long range order in continuous spin systems; in fact Pfister's proof[7] is based on something like those configurations. Intuitively it quite clear that the existence of configurations that move the spin arbitrarily far from its value at a reference point at arbitrarily low cost in energy will destroy any possible long range order.

Let us now turn to lattice Yang-Mills theories. Again there are configurations that disorder the system at arbitraily low cost in energy. They can be described as the configurations of minimal energy that force the parallel transporter around a long, thin loop (length L, width 1) to assume a certain group value $U_o \in G$. Another, possibly more intuitive, description uses the maximal temporal tree gauge, in which all link variables in a certain ('time') direction are gauge fixed to 1. In this gauge the system can be considered as a collection of $1D$ spin chains, coupled among each other by the 'spatial' plaquettes. The super-instanton in this picture looks like the super-instanton for one of those spin chains. The minimal energy of such configurations can be seen to be

$$E_{min} = O(\frac{1}{L}) \tag{7}$$

for all $D \geq 2$ and they have again entropy of position and orientation. So again excitations of this type will be present abundantly at large β and will cause large fluctuations in the spatial link variables in the temporal gauge. These fluctuations are the reason for the well-known infrared problems encountered in PT in that gauge, but their true importance has not been recognized so far.

The common conclusion for both spin models in $2D$ and gauge models in $D \geq 2$ is: PT is a priori suspect, because it is an attempt to expand the models in the deviations from an ordered ground state, in spite of the fact that large fluctuations away from the ordered ground state are present.

Somebody might object to this argument by saying that in LYM there is a gauge specifically designed to minimize those fluctuations: the Landau gauge. It appears to be free of such large fluctuations at least in $D \leq 4$. While this may be true, the Landau gauge for the non-Abelian models has a well known problem, related to the existence of Gribov copies[8], which again puts PT into doubt: The integration has to be cut off at the boundary of the 'fundamental modular domain', which comes arbitrarily close to the trivial vacuum[9]; so the fluctuations can no longer be treated as approximately Gaussian.

3.What Could Go Wrong with Perturbation Theory?

Consider the $2D$ $O(N)$ model in a square of side L and let β get large. Clearly the Gibbs factor $\exp(\beta s_i \cdot s_j)$ will make sure that the system is very well ordered. So a saddle point expansion around a completely ordered ground state will be justified (this is nothing but Laplace's method for obtaining an aymptotic expansion for the integral of a sharply peaked integrand). The result will be PT in a finite volume, which will be an asymptotic expansion in powers of $1/\beta$. Specifically there will be PT coefficients $c_j(L)$ such that

$$|\langle s(0) \cdot s(1) \rangle - \sum_{j=0}^{k} \frac{c_j(L)}{\beta^j}| < \frac{d_{k+1}(L)}{\beta^{k+1}} \tag{8}$$

It is a remarkable result, due to Elitzur[10] and David[11] that the limits as $L \to \infty$ of the coefficients $c_j(L)$ exist. But it is a completely different question whether the remainder terms $d_k(L)$ have limits or are at least bounded. That this is not an academic question can be seen in $1D$, where, using periodic boundary conditions (b.c.), one finds that $c_2(L)$ has a finite limit, but $d_2(L)$ is unbounded. In fact, we can take the $L \to \infty$ limit of $\langle s(0) \cdot s(1) \rangle$ and compute its asymptotic expansion in powers of $1/\beta$. But this expansion has nothing to do with the termwise limit of Eq.(8) for periodic b.c. (it is a special feature of $1D$ that with free b.c. one can avoid this problem).

We will show below that also in $2D$ the limit of c_2 will depend on the b.c. used in the finite volume approximation. This unphysical behavior of PT shows that it cannot be trusted. There is no rational argument why any specific choice of b.c. should produce the right answer in the thermodynamic limit. The problem in lattice Yang-Mills is similar.

4.Something *is* Wrong with Perturbation Theory

Since the super-instantons discussed in section 2 become degenerate with the completely ordered ground state for $R \to \infty$, it is legitimate to use them as reference states for PT. In fact it can be shown that expectation values of local observables will in the thermodynamic limit become independent of any such choices of b.c.. This is a consequence of the absence of spontaneous symmetry breaking which is assured by the Mermin-Wagner theorem for the $2D$ spin models and by an old result for LYM in the axial gauge[12,13]. We showed by explicit calculation (numerical evaluation of Feynman graphs) that the thermodynamic limit for the PT coefficient of order $1/\beta^2$ is in fact different for periodic and super-instanton b.c.[14,15].

For reasons of computational simplicity we consider only the 'trivial super-instanton'. This means for the spin model that we fix a spin in the middle of the lattice to be equal to all the spins on the boundary ($e = e'$); for LYM it means that we freeze the parallel transporter around a loop of length L and width 1 to the identity.

For the $2D$ $O(N)$ model we get with 'trivial super-instanton' b.c.

$$\begin{aligned} \langle S(0)\cdot S(1)\rangle &= 1 - \frac{(N-1)}{2\beta}G(1,1) \\ &- \frac{1}{\beta^2}\Big\{(N-1)^2\Big[\frac{(G(1,1)^2}{8} - \frac{t}{4}\Big] \\ &+ (N-1)\Big[-\frac{G(1,1)^2}{4} + \frac{v}{2}\Big]\Big\} + O(\frac{1}{\beta^2}) \end{aligned} \tag{9}$$

where

$$\begin{aligned} t &= \sum_{<i,j>} [G(1,i)^2 - G(1,j)^2][G(i,i) - G(j,j)], \\ v &= \sum_{<i,j>} [G(1,i) - G(1,j)]\Big\{G(1,j)[G(i,j) - G(j,j)] \\ &- G(1,i)[G(i,j) - G(i,i)]\Big\}; \end{aligned} \tag{10}$$

and $G(i,j)$ is the Green's function of the lattice Laplacean with the Dirichlet b.c. at the origin and on the boundary.

This is to be compared with the formula for periodic b.c. which was written down by Hasenfratz[16]. It turns out that

(1) $\lim_{L\to\infty} G(1,1) = \frac{1}{2}$

(2) $\lim_{L\to\infty}(v - \frac{t}{2}) \approx \frac{1}{8}$

(3) $\lim_{L\to\infty} t \approx 3 \times \frac{1}{8}$

(1) implies that to order $1/\beta$ (Gaussian approximation) PT 'forgets' the b.c. and agrees with the result obtained using periodic b.c., (2) implies that for the Abelian case $N = 1$ PT becomes independent of b.c. (and agrees with Hasenfratz's result) to order $1/\beta^2$. (3) implies that for the non-Abelian case $N > 2$ the limits remain dependent of b.c.; agreement with Hasenfratz's result would require the limit of t to be $\frac{1}{8}$. So the perturbative method fails. For more details see[14].

In the case of LYM we compared the Abelian group $U(1)$ with the non-Abelian group $SU(2)$, and for reasons of limited computing power we worked in $3D$, but the phenomenon is clearly dimension independent. Again we found that to order $1/\beta^2$, PT becomes independent of b.c. only in the Abelian case, whereas for the non-Abelian group $SU(2)$ the answers for 'trivial superinstanton' and periodic b.c. are different. Details can be found in[15].

5. Conclusions

We have confirmed that PT 'works' in Abelian models to order $1/\beta^2$. This is not a new result, but rather a check that we calculated correctly. The validity of PT for the Abelian models had in fact be proven long ago by Bricmont et al[17].

On the other hand we found that PT to the same order $1/\beta^2$ produces results dependent on b.c., which are therefore not trustworthy. It is easy to understand why the phenomenon first occurs at this order: this is the lowest order at which PT involves a loop containing summation over the whole lattice, so that the b.c. have a chance being felt, even in a local quantity like the one considered here.

What we found should be contrasted to the conventional claim that PT is good for short distance quantities. We found in fact that this is not so, the reason being clearly that the models in question do not allow a neat separation of short and long distances. In the loop graphs all distances are being entangled and we have found no regime in which conventional PT can be trusted.

Since we found trouble with PT already for short-distance quantities, it should not come as a surprise that the same trouble appears in long range quantitities such as the Callan-Symanzik β-function. In the $2D$ $O(N)$ models we have checked that the super-instanton b.c. give a different coefficient of the $(\ln x)^2$ term of the two-point function $\langle s(0) \cdot s(x) \rangle$ than periodic b.c., signalling a change in the leading coefficient of the PT β-function. Elsewhere[1,2] we have given arguments for the existence of a soft phase at large β in all $2D$ $O(N)$ models, which implies that the true β-function vanishes identically in that regime.

References

1. A.Patrascioiu, *Existence of Algebraic Decay in non-Abelian Ferromagnets*, University of Arizona preprint AZPH-TH/91-49.
2. A.Patrascioiu and E.Seiler, *Percolation Theory and the Existence of a Soft Phase in 2D Spin Models, Nucl.Phys.B.(Proc. Suppl.)* **30** (1993) 184.
3. K.Wilson, *Phys.Rev.* **D10**, (1974) 2445.
4. J.M.Kosterlitz and D.J.Thouless, *J. Phys. (Paris)* **32** (1975) 581.
5. J.Fröhlich and T.Spencer, *Commun.Math.Phys.* **81** (1981) 527.
6. N.D.Mermin and H.Wagner, *Phys. Rev. Lett.* **17** (1966) 1133; N.D.Mermin, *J. Math. Phys.* **8** (1967) 1061.
7. C.Pfister, *Commun.Math.Phys.* **79** (1981) 181.
8. V.N.Gribov, *Nucl.Phys.* **B139** (1978) 1. (1981) 527.
9. D.Zwanziger, *Nucl.Phys.* **B 378** (1992) 525 and these proceedings.
10. S.Elitzur, *Nucl.Phys.* **B212** (1983) 501.
11. F.David, *Commun.Math.Phys.* **81** (1981) 149.
12. B.Simon and L.G.Yaffe, *Phys.Lett.* **115B** (1982) 145.
13. M. Lüscher, *Absence of Spontaneous Gauge Symmetry Breaking in Hamiltonian Lattice Gauge Theories*, preprint DESY-77/16.
14. A.Patrascioiu and E.Seiler, *Super-Instantons and the Reliability of Perturbation Theory in Non-Abelian Models*, preprint MPI-Ph/93-87, AZPH-TH/93-33, ⟨hep-lat 9311019⟩.
15. A.Patrascioiu and E.Seiler, *Super-Instantons in Gauge Theories and Troubles with Perturbation Theory*, preprint MPI-PhT/94-07, AZPH-TH/94-03, ⟨hep-lat 9402003⟩.
16. P.Hasenfratz, *Phys.Lett.***B141** (1984) 385.
17. J.Bricmont, J.-R.Fontaine, J.L.Lebowitz, E.H.Lieb and T.Spencer, *Commun. Math. Phys.* **78** (1981) 545.

HORIZON CONDITION HOLDS POINTWISE ON FINITE LATTICE WITH FREE BOUNDARY CONDITIONS

MARTIN SCHADEN* and DANIEL ZWANZIGER†
Physics Department, New York University
New York, N. Y. 10003, U.S.A.

ABSTRACT

It is proven that the "horizon condition", which was found to characterize the fundamental modular region in continuum theory and the thermodynamic limit of gauge theory on a periodic lattice, holds for every (transverse) configuration on a finite lattice with free boundary conditions.

1. Introduction

In his seminal work, Gribov[1] pointed out that the non-triviality of the fundamental modular region in Coulomb or Landau gauge has dynamical consequences. [The fundamental modular region is defined explicitly in Eq. (4) below.] Recently[2], the fundamental modular region was studied in the infinite-volume or thermodynamic limit of a periodic lattice. It was found that the fundamental modular region is characterized by a "horizon condition" in the sense that the Euclidean probability gets concentrated where the horizon function $H(U)$, a bulk quantity of order V, defined in Eq. (41) below, vanishes. More precisely, it was shown that at large volumes the vacuum expectation value of $H(U)$ is of order unity, $\langle H(U)\rangle = O(1)$, (instead of V) that its variance is of order V , $\langle H^2(U)\rangle = O(V)$, instead of V^2. It was also found[2] that the horizon condition may be implemented by a Boltzmann factor $\exp[-\alpha H(U)]$, where α is a thermodynamic parameter determined by the constraint $\langle H(U)\rangle = 0$. Recently a calculation of glueball masses has been carried out[3] in which the mass scale is set by α. Previous calculations of glueball masses from the fundamental modular region of gauge theory have been done using a mode expansion by Cutkosky and co-workers[4] and by Van Baal and co-workers[5].

The argument whereby the horizon condition is established in the *infinite-volume* limit of the *periodic* lattice, relies on two technical hypotheses[2]. In the present contribution, we shall prove that, remarkably, the horizon condition holds *point-wise* for every transverse configuration on the *finite* lattice with *free boundary conditions*, $H(U) = 0$.

*Research supported by Deutsche Forschungsgemeinschaft under grant no. Scha/1-1
†Research supported by the National Science Foundation under grant no. PHY93-18781

2. Lattice with Free Boundary Conditions in Minimal Landau Gauge

We consider a finite hypercubic lattice without periodicity conditions, in D Euclidean space-time dimensions. Along each principal axis of the lattice there are L links and $L+1$ sites. Lattice configurations U are described by link variables $U_{xy} = U^\dagger_{yx}$, and local gauge transformations g by the site variables g_x, where U_{xy} and g_x are both elements of $SU(n)$. The gauge-transform U^g of the configuration U by a local gauge transformation g, is given explicitly by

$$(U^g)_{xy} = g_x^\dagger U_{xy} g_y \,. \tag{1}$$

For the purposes of analytic calculations, it is convenient to minimize gauge fluctuations by choosing a gauge which makes the link variables U_{xy}, on each link (xy) of the lattice, as close to unity as possible, in an equitable way over the whole lattice. For this purpose, we take as a measure of the deviation of the link variables from unity, the quantity

$$I(U) = \sum_{(xy)} n^{-1} \text{Re tr}(1 - U_{xy})\,, \tag{2}$$

where the sum extends over all links (xy) of the lattice. It is positive, $I(U) \geq 0$, and vanishes, $I(U) = 0$, if and only if $U_{xy} = 1$ on every link (xy). [The continuum analog of this expression is the Hilbert norm of the connection A, $I[A] = \int \mathrm{d}^D x |A(x)|^2$.] The restriction of this quantity to the gauge orbit through an arbitrary configuration U is given by the Morse function

$$F_U(g) \equiv I(U^g)\,, \tag{3}$$

regarded as a function on the local gauge group G, for fixed U. Gauge fluctuations are minimized by the gauge-fixing which consists in choosing as the representative, on each gauge orbit, that configuration U which yields the absolute minimum of $F_U(g)$ at $g_x = 1$. The set Λ of all configurations U such that this function is an absolute minimum on each gauge orbit at $g_x = 1$, constitutes the fundamental modular region,

$$\Lambda \equiv \{U : F_U(1) \leq F_U(g) \text{ for all } g\}\,. \tag{4}$$

Degenerate absolute minima occur only on the boundary of the fundamental modular region, and are identified topologically.

At any local or absolute minimum, the minimizing function is stationary, which gives the local gauge condition. The second variation is non-negative. To obtain the explicit form of these quantities, we write the local gauge transformation $g_x = \exp(\omega_x)$, where $\omega_x = \omega_x^a t^a$. Here the t^a form an anti-hermitian basis of the defining representation of the Lie algebra of $SU(n)$, $[t^a, t^b] = f^{abc} t^c$, normalized to $\text{tr}(t^a t^b) =$

$-\frac{1}{2}\delta^{ab}$. We have, to second order in ω,

$$\begin{aligned} F_U(g) &= F_U(1) - (2n)^{-1} \sum_{(xy)} [\mathrm{tr}\{[U_{xy} - U^\dagger_{xy}](\omega_y - \omega_x + \tfrac{1}{2}[\omega_x, \omega_y])\} \\ &\qquad + \tfrac{1}{2}\mathrm{tr}\{[U_{xy} + U^\dagger_{xy}](\omega_y - \omega_x)^2\}] \end{aligned}$$

$$\begin{aligned} F_U(g) &= F_U(1) + (2n)^{-1} \sum_{(xy)} [A^c_{xy}(\omega^c_y - \omega^c_x + \tfrac{1}{2} f^{abc} w^a w^b) \\ &\qquad + \tfrac{1}{2}(\omega^a_y - \omega^a_x) G^{ac}_{xy}(\omega^c_y - \omega^c_x)] \,. \end{aligned} \tag{5}$$

Here we have introduced the real link variables

$$A^c_{xy} \equiv -\mathrm{tr}[(U_{xy} - U^\dagger_{xy})t^c] \,, \tag{6}$$

which approach the classical connection in the continuum limit, and

$$G^{ab}_{xy} \equiv -\tfrac{1}{2}\mathrm{tr}\{(t^a t^b + t^b t^a)[U_{xy} + U^\dagger_{xy}]\} \,. \tag{7}$$

We introduce some elementary geometry of this lattice that will be useful. Let $\phi_x \equiv \phi(x)$ be a site variable, and let (xy) denote a link of a lattice. Corresponding link variables are defined by

$$(\nabla\phi)_{xy} \equiv \phi_x - \phi_y \tag{8}$$

$$(a\phi)_{xy} \equiv \tfrac{1}{2}(\phi_x + \phi_y) \,. \tag{9}$$

These are defined for all links, including those for which x or y may lie on a face, edge or vertex. (For $\nabla\phi$ we could alternatively use the Cartan notation $\mathrm{d}\phi$). We may also uniquely designate links by $(xy) = (x, \mu)$, for $y = x + e_\mu$, where e_μ points in the positive μ-direction, and we shall also use, as convenient, the notation for link variables

$$\nabla_\mu \phi(x) \equiv (\nabla\phi)_{yx} \tag{10}$$

$$a_\mu \phi(x) \equiv (a\phi)_{xy} \tag{11}$$

(These expressions are <u>undefined</u> for sites x where $y = x + e_\mu$ is not a site of the lattice.) Given two link variables V_{xy} and W_{xy}, we may form the link variable $P_{xy} = V_{xy} W_{xy}$, etc.

We also introduce the lattice divergence of a link variable $A_\mu(x)$, which is a site variable $(\nabla \cdot A)_x$, defined by the dual

$$(\nabla \cdot A, \phi) \equiv -(A, \nabla\phi) = -\sum_{(xy)} A_{xy}(\nabla\phi)_{xy} = \sum_x (\nabla \cdot A)_x \phi_x \,, \tag{12}$$

where the sums extend over all links and sites of the lattice. Similarly the site variable $(a \cdot A)_x$ is defined by

$$(a \cdot A, \phi) \equiv (A, a\phi) = \sum_{(xy)} A_{xy}(a\phi)_{xy} = \sum_x (a \cdot A)_x \phi_x \,. \tag{13}$$

The lattice divergence is defined for all sites x, and represents the sum of all link variables leaving the site x. (It is *not* formed simply of differences of link variables when x is a boundary point of the lattice.)

With the help of these definitions, we rewrite Eq. (5) in the form

$$F_U(g) = F_U(1) + (2n)^{-1}[-(A, \nabla\omega) + (\nabla\omega, D(U)\omega)]\,. \tag{14}$$

Here $[D(U)\omega]_{yx}$ is a well-defined link variable which we call the lattice gauge-covariant derivative of the site variable ω_x,

$$D_\mu^{ac}\omega^c(x) = G_\mu^{ac}(x)\nabla_\mu\omega^c(x) + f^{abc}A_\mu^b(x)a_\mu\omega^c(x)\,. \tag{15}$$

It is the infinitesimal change in $A_\mu(x)$ under an infinitesimal gauge transformation ω_x. We also define the lattice Faddeev-Popov matrix $M(U)$ by

$$(\omega, M(U)\phi) = (\nabla\omega, D(U)\phi)\,. \tag{16}$$

At a minimum, the minimizing function $F_U(g)$ is stationary, namely $(A, \nabla\omega) = -(\nabla\cdot A, \omega) = 0$. This holds for all ω, so the gauge condition is expressed by the transversality of A,

$$\nabla\cdot A = 0\,. \tag{17}$$

Because of transversality, the gauge just defined falls into the class of lattice Landau gauges, and we call it the "minimal Landau gauge". At a stationary point of the minimizing function, the matrix of second derivatives is symmetric, so when A is transverse, the Faddeev-Popov matrix is symmetric,

$$M \equiv -\nabla\cdot D(U) = -D(U)\cdot\nabla \qquad (\text{for } \nabla\cdot A = 0)\,. \tag{18}$$

This matrix is non-negative at a minimum, and the two conditions together define the Gribov region Ω,

$$\Omega \equiv \{U : \nabla\cdot A(U) = 0 \text{ and } M(U) \geq 0\}\,. \tag{19}$$

Because the set Λ of absolute minima is contained in the set of relative mimima, we have the inclusion

$$\Lambda \subseteq \Omega\,, \tag{20}$$

where Λ is the fundamental modular region.

3. Vanishing of the Horizon Function on the Finite Lattice with Free Boundary Conditions

Because there are $L+1$ sites, but only L links on each row of the lattice with free boundary conditions, the transversality condition is more restrictive for free boundary conditions than on a periodic lattice, so various additional constraints hold, as we shall now show.

Choose a particular direction, $\mu = 0$, on the hypercubic lattice, and call the corresponding (Euclidean) coordinate t and the other coordinates $\mathbf{x}$. Consider the sum over the hyperplane labelled by t, for a transverse configuration, $\nabla \cdot A = 0$,

$$\sum_{\mathbf{x}} (\nabla \cdot A)(t, \mathbf{x}) = 0 . \tag{21}$$

Contributions to this sum from links that lie within the hyperplane vanish, because all such links are connected to 2 sites within the hyperplane and these 2 contributions cancel. There remain the contributions from perpendicular links. For all t that label interior hyperplanes $1 \leq t < L-1$, they give

$$Q_0(t) - Q_0(t-1) = 0 , \tag{22}$$

whereas on the boundary hyperplanes they give

$$Q_0(L-1) = Q_0(0) = 0 . \tag{23}$$

where

$$Q_0(t) \equiv \sum_{\mathbf{x}} A_0(t, \mathbf{x}) . \tag{24}$$

We thus have proved the lemma:
Let $A_\mu(x)$ be a transverse configuration, $\nabla \cdot A(x) = 0$, on a lattice with free boundary conditions. Then the "charges" vanish,

$$Q_0(t) \equiv \sum_{\mathbf{x}} A_0(t, \mathbf{x}) = 0 \qquad (t = 0, 1, \ldots L-1) . \tag{25}$$

[For a lattice with periodic boundary conditions, only the weaker condition $Q_0(t) =$ const. holds.] If we sum the last equation over t, for generic μ, we obtain:
Corollary. Let $A_\mu(x)$ be a transverse configuration $\nabla \cdot A(x) = 0$ on a lattice with free boundary conditions. Then the zero-momentum component of A vanishes,

$$\sum_{x} A_\mu(x) = 0 . \tag{26}$$

Note that for the periodic lattice, transversality imposes no condition whatsoever on the constant component of $A_\mu(x)$. On the other hand, for the periodic lattice, the core of the fundamental modular region was found[2] to satisfy Eq. (26).

We shall now prove that the horizon condition is satisfied point-wise for a finite lattice with free boundary conditions. Consider the integral

$$\begin{aligned} I &\equiv \int \mathrm{d}\phi \mathrm{d}\phi^* \exp[(\phi_i^*, M(U)\phi_i)] , \\ &= \int \mathrm{d}\phi \mathrm{d}\phi^* \exp[-(\nabla_\lambda \phi_i^{*a}, D_\lambda^{ac}(U)\phi_i^c)] , \end{aligned} \tag{27}$$

where the $\phi_i^a(x)$ variables are integrated over the real axis, and the $\phi_i^{*a}(x)$ are independent variables that are integrated over the imaginary axis. The lattice gauge-covariant derivative $D_\lambda^{ac}(U)$ acts on the upper index, of $\phi_i^c(x)$, and $i = 1, \dots f$ is a dummy index. For transverse $A = A(U)$, as we assume, the symmetric matrix $M(U) \equiv -\nabla \cdot D(U) = -D(U) \cdot \nabla$ has a null-space, H_0, consisting of constant functions, so the integral (27) would diverge if $\phi(x)$ and $\phi^*(x)$ were independent integration variables for every x. The integral is made finite by the constraint at a fixed point y of the lattice

$$\phi_i^c(y) = \phi_i^{*c}(y) = 0\,, \tag{28}$$

and the definition

$$\mathrm{d}\phi\mathrm{d}\phi^* \equiv \prod_{x \neq y, i, a} \mathrm{d}\phi_i^a(x)\mathrm{d}\phi_i^{*a}(x)\,. \tag{29}$$

With this specification, the integral has the value

$$I = \int \mathrm{d}\phi \prod_{i=1}^{f} \delta(M\phi_i) = \det{}^{-f} M_y\,, \tag{30}$$

where M_y is the matrix obtained from M by deleting the rows and columns that bear the label y.

Because $M(U)$ has a null space, H_0, consisting of constant functions $\nabla\omega = 0$, an alternative expression for I is

$$I = \mathrm{d}\phi\mathrm{d}\phi^* \exp[(\phi_{i\perp}^*, M(U)\phi_{i\perp})]\,, \tag{31}$$

where $\phi_\perp$ and $\phi_\perp^*$ are the projections of ϕ onto the orthogonal subspace $H_\perp$,

$$\phi_\perp(x) = \phi(x) - S^{-1} \sum_x \phi(x)\,, \tag{32}$$

and S is the total number of sites in the lattice. The variables $\phi(x)$ and $\phi^*(x)$ for $x \neq y$ constitute a complete set of (linear) coordinates on $H_\perp$, and the change of basis (32) is configuration independent. We conclude that

$$I = \det{}^{-f} M_y = \det{}^{-f} M_\perp\,, \tag{33}$$

which shows that I is independent of y. (The equation $\det M_y = \det M_\perp$,holds for any symmetric matrix M with the property that the sum of each row and each column vanish.)

Now let the dummy index i represent the pair $i = (\mu, a)$, so $\phi_{\mu,a}^c(x)$ is a site variable with a preferred direction μ, and make the shift on ϕ and ϕ^* given by

$$\begin{aligned} \phi_{\mu,a}^c(x) &= \phi_{\mu,a}^{\prime c}(x) + x_\mu \delta_a^c \\ \phi_{\mu,a}^{*c}(x) &= \phi_{\mu,a}^{*\prime c}(x) + x_\mu \delta_a^c \end{aligned} \tag{34}$$

which are the lattice analogs of shifts previously introduced in continuum theory[6]. This shift makes no sense on a (finite) periodic lattice, but it is well defined on a finite lattice with free boundary conditions. (Note that the constraint at the lattice point y for the shifted fields differs from (28), unless it is imposed at the "origin" $y = 0$.) For $\nabla \cdot A = 0$, one finds after a simple calculation

$$\begin{aligned} I \equiv & \int d\phi d\phi^* \exp[(\nabla_\lambda \phi^{*b}_{\mu,a}, D^{bc}_\lambda(U)\phi^c_{\mu,a}) + (D^{ac}_\lambda(U)\phi^c_{\lambda,a}) \\ & + (D^{ac}_\lambda(U)\phi^{*c}_{\lambda,a}) + \sum_{x,\mu} G^{aa}_\mu(x)] \,. \end{aligned} \tag{35}$$

It is convenient to introduce a field variable $B^c_{\lambda,a}(x)$, by duality

$$(B^c_{\lambda,a}, \phi^c_{\lambda,a}) \equiv -(D^{ac}_\lambda(U)\phi^c_{\lambda,a}) \,. \tag{36}$$

Like $\phi^c_{\lambda,a}$, the new variable $B^c_{\lambda,a}$ is a site variable with a preferred direction μ. We have

$$B^c_{\lambda,a}(x) = (\nabla \cdot G)^{ac}_\lambda(x) - f^{abc}(a \cdot A)^b_\lambda(x) \,, \tag{37}$$

where $(\nabla \cdot G)^{ac}_\lambda(x)$ represents the contribution to the lattice divergence $\nabla \cdot G^{ac}(x)$ associated with the λ axis, and similarly for $(a \cdot A)^b_\lambda(x)$. With this definition,

$$\begin{aligned} I \equiv & \int d\phi d\phi^* \exp[(\nabla_\lambda \phi^{*b}_{\mu,a}, D^{bc}_\lambda(U)\phi^c_{\mu,a}) - (B^c_{\lambda,a}, \phi^c_{\lambda,a}) \\ & - (\phi^{*c}_{\lambda,a}, B^c_{\lambda,a}) + \sum_{x,\mu} G^{aa}_\mu(x)] \,. \end{aligned} \tag{38}$$

We may now effect the ϕ and ϕ^* integrations by making the shifts

$$\begin{aligned} \phi^b_{\lambda,a} &= \phi'^b_{\lambda,a} + (M^{-1})^{bc} B^c_{\lambda,a} \\ \phi^{*b}_{\lambda,a} &= \phi^{*\prime b}_{\lambda,a} + (M^{-1})^{bc} B^c_{\lambda,a} \,. \end{aligned} \tag{39}$$

This expression is well-defined because $B^c_{\lambda,a}$ is orthogonal to the null-space of M,

$$\sum_x B^c_{\lambda,a}(x) = 0 \,, \tag{40}$$

and we may choose coordinates adapted to these subspaces. To see this, observe that $\sum_x (\nabla \cdot G)^{ac}_\lambda(x)$ vanishes because each link is connected to two sites and gives opposite contributions from each. Moreover $\sum_x (a \cdot A)^b_\lambda(x) = \sum_x A^b_\lambda(x)$ vanishes by the preceding corollary. This gives $I = I \exp[-H(U)]$, where $H(U)$ is given by Eq. (41). We have proven:

Theorem. Let U be a transverse configuration $\nabla \cdot A(U) = 0$, and let the horizon function $H(U)$ on a finite lattice with free boundary conditions be defined by

$$H(U) = (B^b_{\lambda,a}, (M^{-1})^{bc} B^c_{\lambda,a}) - \sum_{x,\mu} G^{aa}_\mu(x) \,, \tag{41}$$

where B is defined in (37) and $G_\mu^{ab}(x)$ in (7). Then $H(U)$ vanishes

$$H(U) = 0\,. \tag{42}$$

Remark. On a periodic lattice, the horizon condition does not hold point-wise. For example, the vacuum configuration $U_\mu(x) = 1$ on a *periodic* lattice gives $B^c_{\lambda,a}(x) = 0$, so $H(U) = -(n^2-1)DV$. For the vacuum configuration on the finite lattice with *free* boundary conditions, $B^c_{\lambda,a}(x)$ is entirely supported on the boundary. Thus boundary contributions remain important at arbitrary large volume. Long range boundary effects have also been found by Patrascioiu and Seiler[7].

Discussion. The result (42) is quite remarkable and unexpected. Recall that as the *volume of the periodic lattice approaches infinity*, the probability gets concentrated where the horizon condition holds[2] $H(U) = 0$, and where $\sum_x A(x) = 0$. Here we have just proven that these conditions are satisfied for all transverse configurations U on the finite lattice with free boundary conditions.

4. References

[1] V. N. Gribov, Nucl. Phys. B139 (1978) 1.

[2] D. Zwanziger, Nucl. Phys. B412 (1994) 657.

[3] M. Schaden and D. Zwanziger, *Glueball Masses from the Gribov Horizon: Basic Equations and Numerical Estimates*, NYU preprint ThPhSZ94-1.

[4] R. E. Cutkosky, J. Math. Phys. 25 (1984) 939; R. E. Cutkosky and K. Wang, Phys. Rev. D37 (1988) 3024; R. E. Cutkosky, Czech J. Phys. 40 (1990) 252.

[5] P. van Baal and N. D. Hari Dass, Nucl. Phys B385 (1992) 185; J. Koller and P. van Baal, Nucl. Phys. B302 (1991) 1; P. van Baal, Acta Physica Pol. B20 (1989) 295; P. Van Baal and B. van den Heuvel, *Zooming in on the SU(2) Fundamental Domain*, (preprint) University of Leiden, INLO-PUB-12/93.

[6] N. Maggiore and M. Schaden, *Landau Gauge within the Gribov Horizon*, NYU preprint, October 1993, Phys. Rev. D (to be published).

[7] A. Patrascioiu and E. Seiler, *Super-Instantons in Gauge Theories and Troubles with Perturbation Theory*, hep-lat 9402003, MPI-PhT/94-07, AZPH-TH/94-03, Phys. Rev. Lett (to be published).

CAN AN INFRARED-VANISHING GLUON PROPAGATOR CONFINE QUARKS?

CRAIG D. ROBERTS
Physics Division, Bldg. 203, Argonne National Laboratory
Argonne, IL 60439-4843, USA

ABSTRACT

It is shown that the solution of the quark Dyson-Schwinger equation in QCD obtained with a gluon propagator of the form $D(q) \sim q^2/[q^4 + b^4]$ and a quark-gluon vertex that is free of kinematic singularities does not describe a confined particle and that there is always a value of $b^2 = b_c^2$ such that chiral symmetry is not dynamically broken for $b^2 > b_c^2$.

1. Introduction

Dynamical Chiral Symmetry Breaking (DχSB) and confinement are two crucial features of quantum chromodynamics (QCD). They are responsible for the nature of the hadronic spectrum; DχSB ensuring the absence of low mass scalar partners of the pion and confinement ensuring the absence of free quarks, for example. A natural method for studying both DχSB and confinement in QCD is the complex of Dyson-Schwinger Equations (DSEs).[1]

One goal of DSE studies is to develop this nonperturbative approach to the point where it is as firmly founded as lattice QCD and calculationally competitive. Although more needs to be done in order to achieve this goal there has been a good deal of progress, especially in the study of Abelian gauge theories where direct and meaningful comparisons can be made, and agreement obtained, between the results of lattice and DSE studies.[2]

Herein a recent study[3] of the fermion DSE in which the gluon propagator vanishes at $q^2 = 0$, aimed at determining whether such a gluon propagator can support DχSB and/or generate a confining quark propagator, is described. There has been renewed interest in such a form of the gluon propagator, which was argued in Ref. [4] to be associated with the elimination of Gribov copies, because of the recent work of Häbel *et. al.*[5] and Zwanziger.[6]

In Sec. 2 the DSE for the quark propagator is described in detail. There is also a discussion of what is known about the dressed gluon propagator and quark-gluon vertex in QCD. The analysis of DχSB is reported in Sec. 3 and quark confinement is discussed in Sec. 4. The results are summarised in Sec. 5.

2. Dyson-Schwinger Equation for the Fermion Self Energy

In Minkowski space, with metric $g_{\mu\nu} = \text{diag}(1, -1, -1, -1)$ and in a general covariant gauge, the inverse of the dressed quark propagator can be written as

$$S^{-1}(p) = \not p - m - \Sigma(p) \equiv Z^{-1}(p^2)\left(\not p - M(p^2)\right) \equiv A(p^2)\,\not p - B(p^2), \qquad (1)$$

with: m the renormalised, explicit chiral symmetry breaking mass (if present); $\Sigma(p)$ the self-energy; $M(p^2) = B(p^2)/A(p^2)$ the dynamical quark mass function; and $Z(p^2) = A^{-1}(p^2)$ the momentum-dependent renormalisation of the quark wavefunction. The unrenormalised DSE for the inverse propagator is

$$S^{-1}(p) = \not p - m^{\text{bare}} - i\frac{4}{3}g^2 \int \frac{d^4k}{(2\pi)^4}\gamma^\mu S(k)\Gamma^\nu(k,p)D_{\mu\nu}((p-k)^2), \qquad (2)$$

where $D_{\mu\nu}(q^2)$ is the dressed gluon propagator and Γ^ν is the proper quark-gluon vertex, which is illustrated in Fig. 1.

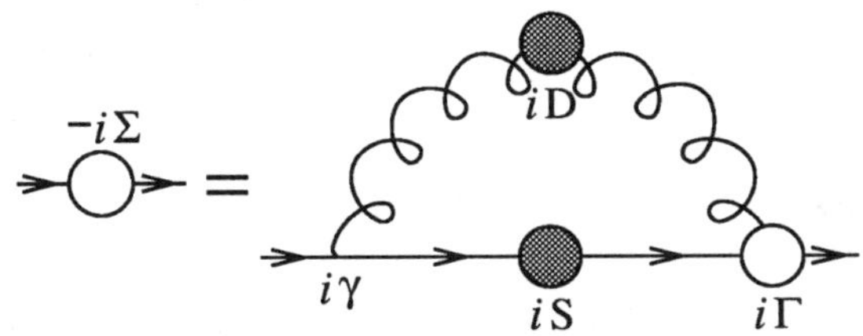

Fig. 1. The Dyson-Schwinger equation for the quark self-energy.

The renormalised, massless ($m = 0$) DSE can be written as

$$\Sigma_R(p) = (1 - Z_S)\,\not p + iZ_\Gamma\frac{4}{3}g^2 \int^\Lambda \frac{d^4k}{(2\pi)^4}\gamma^\mu S_R(k)\Gamma_R^\nu(k,p)D_{\mu\nu}^R((p-k)^2), \qquad (3)$$

where Z_S and Z_Γ are quark-propagator and quark-gluon-vertex renormalisation constants, respectively, which depend on the renormalisation scale, μ, and ultraviolet cutoff, Λ. Hereafter we suppress the label R.

The solution of this equation provides information about DχSB . The quark condensate, $\langle\bar{q}q\rangle \propto \text{tr}[S(x=0)]$, is a chiral symmetry order parameter. If there is a solution of Eq. (3) with $B \neq 0$ then the quark has generated a mass via interaction with its own gluon field and the chiral symmetry is therefore dynamically broken. The solution also provides information about quark confinement, as discussed in Sec. 4.

2.1a *Gluon Propagator*

In a general covariant gauge the dressed gluon propagator, which is diagonal in colour space, can be written:

$$D^{\mu\nu}(q^2) = \left[\left(g^{\mu\nu} - \frac{q^\mu q^\nu}{q^2}\right)\frac{1}{1-\Pi(q^2)} + \xi\frac{q^\mu q^\nu}{q^2}\right]\frac{1}{q^2}, \qquad (4)$$

where $\Pi(q^2)$ is the gluon vacuum polarisation and ξ is the gauge parameter. In covariant gauges the longitudinal piece of this propagator is not modified by interactions, which follows from the Slavnov-Taylor identities in QCD.

The Dyson-Schwinger equation for the gluon propagator is given diagrammatically in Fig. 2. The symmetrisation factors of 1/2 and 1/6 arise from the usual Feynman

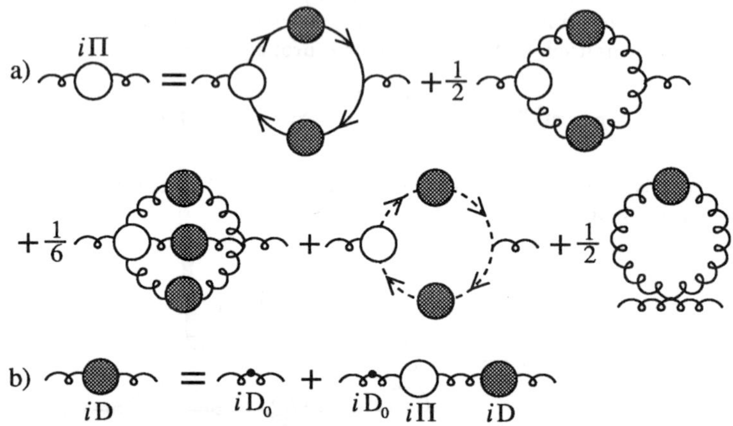

Fig. 2. The Dyson-Schwinger equation for the gluon propagator. [Here and below the broken line represents the propagator for the ghost field.]

rules, which also require a negative sign [unshown] to be included for every fermion and ghost loop. This equation has been studied extensively.[5,7–9] There have also been attempts to determine the gluon propagator from numerical simulations of lattice-QCD.[10,11]

The results of the DSE and lattice studies are summarised in Sec. 5.1 of Ref. [1] and are represented in Fig. 3. This figure illustrates that for spacelike-$q^2 > 1$ GeV^2 the gluon propagator is given by the two-loop, QCD renormalisation group result, with the next order correction being <10%. For spacelike-$q^2 < 1$ GeV^2, however, the form of the propagator is not known. The DSE studies of Refs. [7–9] suggest a regularised infrared singularity, represented by $1/q^4$ in the figure. That of Ref. [5], which differs mainly in that the Ansatz used for the triple-gluon vertex has kinematic singularities, suggests an infrared vanishing form, characterised by $q^2/(q^4+b^4)$, which has also been argued[4,6] to be the form necessary to completely eliminate Gribov copies.

The lattice Landau-gauge studies of Ref. [10] favour the massive vector boson form, $1/(q^2+m_g^2)$, which is broadly consistent with the improved studies of Ref. [11]. On $16^3 \times 40$ and $24^3 \times 40$ lattices at $\beta = 6.0$ these numerical results allowed a fit of the form $q^2/(q^4+b^4)$, with $b \sim 340$ MeV, but a fit using a standard massive particle propagator could not be ruled out. On a lattice of dimension $16^3 \times 24$ at $\beta = 5.7$ it was found that the gluon propagator was best fit by a standard massive vector boson propagator with mass $\sim$ 600 MeV. There is a problem with these studies,

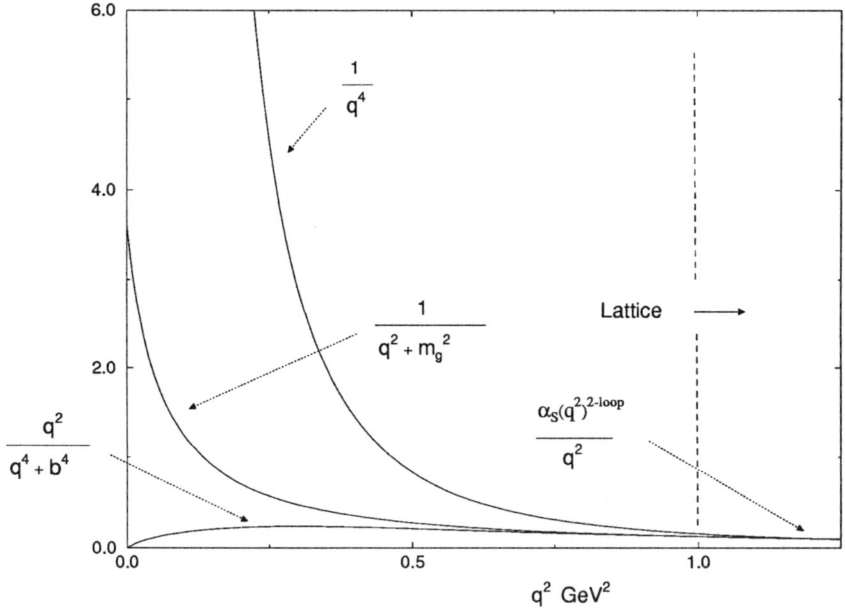

Fig. 3. Results of studies of the gluon propagator in QCD. Typical values of $m_g = 0.5$ GeV and $b = 0.7$ GeV have been used.

however, which is indicated by the dashed vertical line at the right of Fig. 3. With present technology, the domain of spacelike-$q^2 < 1$ GeV^2 is actually inaccessible in lattice studies and, since all forms of the propagator are very nearly the same outside this domain, it is clear that these results are both qualitatively and quantitatively unreliable.

This brief discussion indicates that at present one can only say that

$$D(q^2) \equiv \frac{1}{q^2\left[1 - \Pi(q^2)\right]} = \frac{q^2}{q^4 + b^4} \tag{5}$$

is not implausible in QCD, at least at small q^2. For this reason it is of interest to determine whether such a form of $D(q^2)$ can lead to DχSB and a confining quark propagator, which will provide further insight into the validity of this form of gluon propagator. In keeping with the ultraviolet behaviour illustrated in Fig. 3, the results obtained with the "ultraviolet-improved" form:

$$\frac{g^2}{4\pi}\, D(q^2) = \alpha(\tau; q^2)\, \frac{q^2}{q^4 + b^4}\,, \tag{6}$$

with $\alpha(\tau;q^2) = (d\pi)/\left(\ln\left[\tau + q^2/\Lambda_{QCD}^2\right]\right)$, $d = 12/[33 - 2N_f]$, $N_f = 4$, where $\tau \geq 1$

is an infrared regularising parameter, are described herein.

2.1b *Gluon Condensate*

The gluon condensate can be calculated from the nonperturbative part of the gluon propagator.[12] Using Eq. (6) one obtains

$$\langle \alpha_S GG \rangle_\mu = -\frac{3b^4}{\pi^2} \ln\left(\ln\left[\frac{\mu^2}{\Lambda^2_{QCD}}\right]\right) . \tag{7}$$

For $b = 340$ MeV, $\Lambda_{QCD} = 200MeV$ and $\mu = 1$ GeV this yields $\langle \alpha_S GG \rangle_\mu = -0.0047$ GeV^4, which should be compared with the value inferred from QCD sum rules: $\langle \alpha_S GG \rangle_\mu \sim 0.04$ GeV^4. The "wrong sign" is due to the fact that Eq. (6) is weaker than perturbative gluon exchange for all spacelike-q^2 and is an harbinger of the results to follow.

2.2 *Quark-Gluon Vertex*

The quark-gluon vertex satisfies its own DSE but hitherto there have been no studies of this equation in QCD. Making use of the "Abelian approximation",[8] the Slavnov-Taylor identity for this vertex reduces to the Ward-Takahashi identity familiar from QED:

$$k_\mu \Gamma^\mu(p,q) = S^{-1}(p) - S^{-1}(q) , \tag{8}$$

$k = (p - q)$, and the quark-gluon vertex can be written in the general form[13]

$$\Gamma_\mu(p,q) = \Gamma^{\rm BC}_\mu(p,q) + \sum_{i=1}^{8} T^i_\mu(p,q)\, g^i(p^2, p\cdot q, q^2) , \tag{9}$$

where the eight tensors, T^i_μ, are transverse, $(p-q)^\mu T^i_\mu(p,q) = 0$ and

$$\Gamma^{\rm BC}_\mu(p,q) = \Sigma_A(p,q)\gamma_\mu + (p+q)_\mu \left\{\Delta_A(p,q)\frac{1}{2}\left[\gamma\cdot p + \gamma\cdot q\right] - i\Delta_B(p,q)\right\} \tag{10}$$

with $\Sigma_F(p,q) \equiv [F(p^2)+F(q^2)]/2$ and $\Delta_F(p,q) \equiv [F(p^2)-F(q^2)]/[p^2-q^2]$, for $F = A$ or B. In this approximation QED studies[14] can be used to place constraints on the functions g^i. Taking these constraints into account leads to the Ansatz:[15] $T^i_\mu = 0$, $i \neq 6$, and

$$T^6_\mu(p,q)\, g^6(p^2, p\cdot q, q^2) = \frac{\gamma^\mu(k^2-p^2) - (k+p)^\mu(\not k - \not p)}{2d(k,p)}\left[A(k^2) - A(p^2)\right] , \tag{11}$$

$$\text{with } d(k,p) = \frac{1}{(k^2+p^2)}\left((k^2-p^2)^2 + \left[\frac{B^2(k^2)}{A^2(k^2)} + \frac{B^2(p^2)}{A^2(p^2)}\right]^2\right) . \tag{12}$$

This Ansatz, Eq. (9) with Eq. (11) and $T^i_\mu = 0$ for $i \neq 6$, satisfies the Ward-Takahashi Identity, is free of kinematic singularities (i.e., has a well defined limit as

$(p \to q)$, reduces to the bare vertex in the free field limit in the manner prescribed by perturbation theory, transforms correctly under charge conjugation and Lorentz transformations and preserves multiplicative renormalisability in the quark DSE. (Of these properties the vertex $\Gamma_\mu^{\rm BC}$ satisfies all but the last and, for the most part, the calculations described herein were performed with the Ansatz $\Gamma_\mu = \Gamma_\mu^{\rm BC}$.)

It should be noted that the absence of kinematic singularities is an important and physically reasonable constraint. To understand this one can simply consider an analogy with PCAC. In the case of the axial-current vertex, Γ_μ^5, there is a kinematic singularity in the chiral limit, which is identified with the massless pion excitation. A kinematic singularity in the quark-gluon vertex would therefore entail the existence of an hitherto unknown, massless excitation in QCD.

2.3 *Quark DSE in Euclidean Space*

In the Abelian approximation one has $Z_S = 1 = Z_\Gamma$ at one-loop in Landau gauge. Using this result here considerably simplifies Eq. (3), which can be written in Euclidean space, with metric $\delta_{\mu\nu} = {\rm diag}(1,1,1,1)$ and γ_μ hermitian,

$$\Sigma(p) = \frac{4}{3} g^2 \int^\Lambda \frac{d^4k}{(2\pi)^4} \gamma_\mu S(k) \Gamma_\nu(k,p) D_{\mu\nu}((p-k)^2) \tag{13}$$

where

$$S^{-1}(p) = i\gamma\cdot p + \Sigma(p) = i\gamma\cdot p A(p^2) + B(p^2) \tag{14}$$

and all the other elements in this equation are taken to be specified by the expressions given above evaluated at Euclidean (spacelike) values of their arguments. The important, subtle considerations associated with the transformation between Minkowski and Euclidean space are discussed in Sec. 2.3 of Ref. [1].

Equation (13) is actually a pair of coupled, nonlinear integral equations for $A(p^2)$ and $B(p^2)$. With $\Gamma_\mu = \Gamma_\mu^{\rm BC}$ one obtains

$$\begin{aligned} B(p^2) = \frac{16\pi}{3} \int^\Lambda \frac{d^4k}{(2\pi)^4} \alpha(\tau;(p-k)^2) \frac{(p-k)^2}{(p-k)^4 + b^4} \frac{1}{A^2(k^2)k^2 + B^2(k^2)} \times \\ \left\{ 3B(k^2) \frac{A(k^2) + A(p^2)}{2} + \left[B(k^2)\Delta A(k^2,p^2) - A(k^2)\Delta B(k^2,p^2) \right] h(p,k) \right\}, \end{aligned} \tag{15}$$

with $h(p,k) = 2\left[k^2p^2 - (k\cdot p)^2\right]/(p-k)^2$, and a similar but more complicated equation for $A(p^2)$, while with $\Gamma_\mu = (\Gamma_\mu^{\rm BC} + T_\mu^6\, g^6)$, from Eq. (11), one has

$$\begin{aligned} B(p^2) &= \text{RHS of (15)} \\ &+ \frac{16\pi}{3} \int^\Lambda \frac{d^4k}{(2\pi)^4} \alpha(\tau;(p-k)^2) \frac{(p-k)^2}{(p-k)^4 + b^4} \frac{B(k^2)\Delta A(k^2,p^2)}{A^2(k^2)k^2 + B^2(k^2)} \frac{(k^2-p^2)}{2d(k,p)} 3(k^2 - p^2), \end{aligned} \tag{16}$$

with, again, a similar but more complicated equation for $A(p^2)$.

The results discussed below were obtained from an iterative, numerical solution of these equations on a logarithmic grid of $x = p^2/\Lambda_{\rm QCD}^2$ and $y = k^2/\Lambda_{\rm QCD}^2$ points.

The solutions were independent of the seed-solution and grid choice and also of the UV cutoff, which was $\Lambda^2 = 5 \times 10^8 \ \Lambda^2_{\rm QCD}$.

3. Dynamical Chiral Symmetry Breaking?

The gauge-invariant quark condensate is an order parameter for DχSB and is obtained from the trace of the quark propagator:

$$\langle \bar{q}q \rangle_\mu = -\frac{3}{4\pi^2} \ln\left(\frac{\mu^2}{\Lambda^2_{\rm QCD}}\right)^d \lim_{\Lambda^2\to\infty} \left(\ln\left(\frac{\Lambda^2}{\Lambda^2_{\rm QCD}}\right)^{-d} \int_0^{\Lambda^2} ds\, s\, \frac{B(s)}{sA(s)^2 + B(s)^2} \right) , \quad (17)$$

where μ is the renormalisation point for the condensate, which is usually fixed at 1 GeV. This is the parameter that is used to study DχSB in lattice QCD. A nonzero value signals DχSB.

In Fig. 4 the condensate obtained from the numerical solution of Eq. (15) and the associated equation for $A(p^2)$ for values of $\ln \tau$ in the domain $[0.0, 0.7]$ and b^2 in $[0.1, 1.0]$ is plotted. This figure shows regions of unbroken and dynamically broken chiral symmetry. Notably there is no DχSB for the value of $b^2 \sim 3\ \Lambda^2_{\rm QCD}$ inferred from lattice simulations.

The numerical results suggest that $\langle \bar{q}q \rangle_\mu$ rises continuously from the transition boundary and hence that the transition is second order. Assuming therefore that, for a given value of $\ln \tau$, the order parameter behaves as $\langle \bar{q}q \rangle_\mu(b^2) \approx C\,(1 - b^2/b_c^2)^\beta$ one obtains $\beta = 0.572$ with $\sigma_\beta = 0.020$. Including the T^6_μ term in the vertex only leads to a small quantitative change in the results. For example, with $\ln \tau = 0.6$ one finds $\beta_{T^6} = 0.579$, $\sigma_{\beta_{T^6}} = 0.015$. This suggests that for any vertex that is free of kinematic singularities there is a critical value of $b^2 = b_c^2$ such that there is no DχSB for $b^2 > b_c^2$.

4. Quark Confinement?

In order to determine whether the quark propagator obtained as a solution to Eq. (13) with an infrared-vanishing gluon propagator can represent a confined particle we follow Ref. [16] and adapt a method commonly used in lattice QCD to estimate bound state masses. Writing $\sigma_S(s) = B(s)/[sA(s)^2 + B(s)^2]$, $s = p^2$, defining

$$\Delta_S(T) = \int_{-\infty}^{\infty} \frac{dy}{2\pi}\, \sigma_S(y^2)\, e^{i\,y\,T} \quad (18)$$

and, for notational convenience, $E(T) = -\ln[\Delta_S(T)]$, it follows from the axioms of field theory that if there is a *stable asymptotic state* with the quantum number of the quark then

$$\lim_{T\to\infty} \frac{dE(T)}{dT} = m\ ; \quad (19)$$

where $m \geq 0$ is the mass of this excitation; i.e, this limit yields the dynamically generated quark mass.

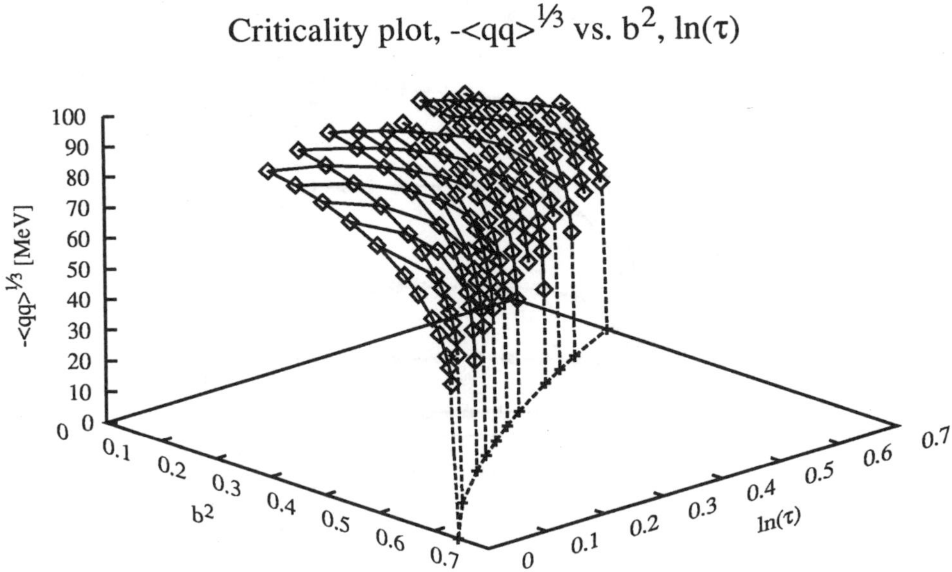

Fig. 4. Criticality plot for $(-\langle\bar{q}q\rangle_\mu)^{\frac{1}{3}}$ as a function of $\ln\tau$ and b^2. The condensate, $(-\langle\bar{q}q\rangle_\mu)^{\frac{1}{3}}$, is in units of MeV, scaled to $\mu^2 = 1\text{GeV}$, and b^2 is in units Λ^2_{QCD} [$b^2 = 0.49 \Rightarrow b \sim 140$ MeV]; the gluon regulator τ is dimensionless.

As a simple example one can consider the Nambu-Jona-Lasinio model[17] in which the dressed quark propagator is $S(p) = 1/[i\gamma\cdot p + M]$ and hence $\sigma_S(s) = M/[s+M^2]$. In this case $\Delta_S(T) = \exp(-M\,T)/2$, which, from Eq. (19), yields $m = M$, as one would expect.

This confinement test was applied to the numerical solutions in the following cases: 1) The propagators obtained with $\ln\tau = 0.1$ and b^2 in the range $[0.1, 1.0]$; 2) The propagator obtained with $\ln\tau = 0$ and $b^2 = 0.35$, which yields the largest value of $(-\langle\bar{q}q\rangle_\mu)$ on the $(b^2, \ln\tau)$ domain considered; 3) Two propagators obtained with $(b^2, \ln\tau) = (0.1, 0.6)$ - one using $\Gamma_\mu = \Gamma^{\rm BC}_\mu$ and the other using $\Gamma_\mu = (\Gamma^{\rm BC}_\mu + T^6_\mu\, g^6)$. Plots of $E'(T)$ for the family of propagators obtained with $\ln\tau = 0.1$ are presented in Fig. 5. (It should be noted that a quark propagator with complex conjugate poles[5] leads to a form of $\Delta_S(T)$ with zeros and hence to $E'(T)$ with zeros and poles; a strong signal of which there is no sign in Fig. 5.) Since the behaviour of all the other solutions was qualitatively the same as that described by the results presented in Fig. 5 one concludes that, independent of b, $D(q) \sim q^2/[q^4 + b^4]$ does not yield a

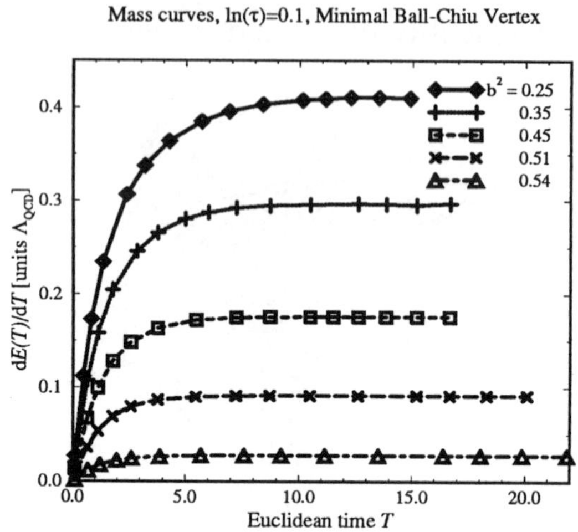

Fig. 5. Dressed-quark-mass curves for the family of propagators with the minimal Ball-Chiu vertex and $\ln\tau = 0.1$. The masses are in units of $\Lambda_{\rm QCD}$.

confining quark propagator.

5. Summary

A study of the quark Dyson-Schwinger equation in QCD using a gluon propagator that vanishes as $q^2 \to 0$, Eq. (6), and a dressed quark-gluon vertex, $\Gamma_\mu(p,q)$, that has no kinematic singularities; i.e., has a well defined limit as $p \to q$, is described. The results indicate that such a gluon propagator can only support DχSB for values of $\ln\tau$ and b^2 less than certain critical values and does not confine quarks. The results are qualitatively independent of the model dependent elements in this study. One is therefore lead to conclude that the dressed gluon propagator in QCD does not vanish in the infrared.

Acknowledgments.

I would like to thank and congratulate the organisers, Berndt Müller and Herb Fried, for bringing about this timely and stimulating meeting, and Dan Zwanziger for giving me the opportunity to make this presentation in his session. This work was supported by the Department of Energy, Nuclear Physics Division under contract number W-31-109-ENG-38.

References

1. C. D. Roberts and A. G. Williams, *Dyson-Schwinger Equations and their Application to Hadronic Physics*, in *Progress in Particle and Nuclear Physics*, Vol. 33, pp. 477-575, ed. by A. Fäßler (Pergamon Press, Oxford, 1994).

2. C. J. Burden, *How can QED_3 help us understand QCD_4*, in Proceedings of "The Workshop on QCD Vacuum Structure", American University of Paris, 1-5 June 1992, Editors H. M. Fried and B. Müller (World Scientific, New York, 1993).

3. F. T. Hawes, C. D. Roberts and A. G. Williams, Phys. Rev. D **49** (1994) 4683.

4. V. N. Gribov, Nucl. Phys. B **139** (1979) 1.

5. U. Häbel, *et al*, *Z. Phys.* **A336** (1990) 423, 435.

6. D. Zwanziger, Nucl. Phys. **B364** (1991) 127.

7. S. Mandelstam, *Phys. Rev.* **D20** (1979) 3223.

8. U. Bar-Gadda, *Nucl. Phys.* **B163** (1980) 312.

9. M. Baker, J. S. Ball and F. Zachariasen, *Nucl. Phys.* **B186** (1981) 531,560; **B226** (1983) 455; N. Brown and M. R. Pennington, *Phys. Lett.* **B202** (1988) 257; *Phys. Rev.* **D38** (1988) 2266.

10. J. E. Mandula and M. Ogilvie, *Nucl. Phys.* **B1A** (Proc. Suppl.) (1987) 117; *Phys. Lett.* **B185**, (1987) 127.

11. C. Bernard, C. Parrinello and A. Soni, Phys. Rev. D **49** (1994) 1585.

12. C. D. Roberts, A. G. Williams and G. Krein, Int. J. Mod. Phys. A **7** (1992) 5607.

13. J. S. Ball and T.-W. Chiu, Phys. Rev. D **22** (1980) 2542.

14. Z. Dong, H. J. Munczek and C. D. Roberts, Phys. Lett. B **333** (1994) 536, and references therein.

15. D. C. Curtis and M. R. Pennington, Phys. Rev. D **46** (1992) 2663, and references therein.

16. L. C. L. Hollenberg, C. D. Roberts, and B. H. J. McKellar, Phys. Rev. C **46** (1992) 2057.

17. See, for example, S. P. Klevansky, Rev. Mod. Phys. **64** (1992) 649.

IR Models and QCD

RENORMALIZATION OF SCALE INVARIANT FIELD THEORIES

VINCENZO BRANCHINA

Department of Physics,Louis Pasteur University,Strasbourg, France

ABSTRACT

In the semiclassical expansion of classically scale invariant field theories the instanton gas generates *new relevant coupling constants* in addition to the perturbatively relevant one. This clearly shows a *non-perturbative breakdown of the concept of universality* which is based on low dimensional operators and has far reaching phenomenological consequences.

1. Introduction

The fundamental difference between a quantum field and a classical one is that the former is never smooth at any scale [1]. This intrinsic roughness is at the very heart of the origin of divergences in a quantum field theory. In the perturbative regime it manifests itself through the appearance of intermediate states with arbitrarily high values of energies and momenta. A quantum field theory is then necessarily defined by the help of a cut-off which is simply a reflection of our ignorance concerning the physics of too high energies. A cut-off Λ in momentum space means that we do not consider fluctuations of the fields with wavelengths smaller than Λ^{-1}. The high energy degrees of freedom which are invisible in the theory defined in this way manifest their presence through an explicit dependence on the cut-off of the parameters in the theory. For instance in the case of Quantum Electrodynamics we write

$$\mathcal{L}_\Lambda = -\frac{1}{4}F^{\mu\nu}F_{\mu\nu} + \bar{\psi}\Big(i\partial_\mu\gamma^\mu - e_B(\Lambda)A_\mu\gamma^\mu - m_B(\Lambda)\Big)\psi. \tag{1.1}$$

The energy scale E of the physical phenomena described by this lagrangian is much lower than the cutoff scale :

$$E << \Lambda. \tag{1.2}$$

Under the condition (1.2) we usually expect that the ultraviolet energy regime, i.e. the physics at the scale of the ultraviolet cutoff Λ, is totally decoupled by the physics in the observation region, or in other words that the short range details of the interaction are irrelevant to the physical predictions of the theory at the "infrared" scale E.

Actually the anomalies provide an example of intimate connection between the infrared and the ultraviolet regime [2]. As another example we may consider the "Strong Coupling QED" [3]. Going back to the lagrangian (1.1) we note that there are actually other terms which respect the same Lorentz and gauge symmetries. The Pauli and the Fermi interaction terms, for instance, between many others :

$$\frac{g}{\Lambda}\bar{\psi}F^{\mu\nu}\sigma_{\mu\nu}\psi \qquad , \qquad \frac{G}{\Lambda^2}\bar{\psi}\psi\bar{\psi}\psi. \tag{1.3}$$

According to Renormalization Group arguments based on the (perturbative) power counting theorem these operators are irrelevant to the low energy physics. By studying the problem of chiral symmetry breaking in "massless QED" through the Schwinger-Dyson equation in the planar approximation [4], it has been actually shown that the quadrifermionic interaction term (which has a naive dimension $d = 6$ and is then perturbatively irrelevant) acquires an anomalous dimension $d = 4$ at the critical point $\alpha = \alpha_c$ where the symmetry is broken and then it has to be taken consistently into account together with the term (1.1). The short distance fluctuations cancel the Λ^{-2} suppression factor which appears at the tree level order in the quadrifermionic term and give rise to an O(1) interaction.

I want to present here another cutoff effect [5] which occurs in the semiclassical expansion of classically scale invariant theories which possesses non trivial saddle points as is the case for the non-linear sigma model in two dimensions, the CP^N models and the $SU(N)$ theory. The specific example which is worked out is that of the two dimensional non linear sigma model. The physical implications of the result, however, are valid for any scale invariant theory.

2. Semiclassical Approximation : Tree Level

The two dimensional $O(3)$ non linear sigma model with the constraint

$$\vec{\phi}\cdot\vec{\phi} = \sum_{a=1}^{3}\phi^a\phi^a = 1, \tag{2.1}$$

is commonly defined through the lagrangian

$$S[\vec{\phi}] = \frac{1}{2f}\int d^2x \; \partial_\mu\vec{\phi}\cdot\partial_\mu\vec{\phi}. \tag{2.2}$$

There are actually other terms, as for instance

$$\left(\frac{C_n}{\Lambda}\right)^{2n-2}\left(\partial_\mu\vec{\phi}\cdot\partial_\mu\vec{\phi}\right)^n, \tag{2.3}$$

which respect the same $O(3)$ symmetry. They are not considered because according to the power counting argument they should be irrelevant to the low energy (infrared) physics.

We'll show that this is not the case and that the model is sensitive to the choice of "would be" irrelevant coupling constants. For our purposes we add to the lagrangian the two "irrelevant" terms of the type (2.3) with $n = 2,4$ and consider the semi-classical expansion of this model. We look first for the saddle points of the path integral

$$Z = \int D[\vec{\phi}] \prod_x \delta(\vec{\phi}^2(x) - 1) e^{-S[\vec{\phi}]}, \tag{2.4}$$

where the action $S[\vec{\phi}]$ is

$$S[\vec{\phi}] = \frac{1}{2f} \int d^2x \left(1 - \frac{3}{4} \frac{C_2}{\Lambda^2} \partial_\mu \vec{\phi} \partial_\mu \vec{\phi} + \frac{5}{64} \frac{C_4}{\Lambda^4} (\partial_\mu \vec{\phi} \partial_\mu \vec{\phi})^2 \right) (\partial_\mu \vec{\phi} \partial_\mu \vec{\phi}). \tag{2.5}$$

The numerical coefficients have been chosen for later convenience and the relative sign between the constants C_2 and C_4, both positive, anticipates the case for which stable saddle points exist.

The saddle point equation is

$$\left(1 - \frac{3}{2} \frac{C_2}{\Lambda^2} (\partial_\mu \vec{\phi} \partial_\mu \vec{\phi}) + \frac{15}{64} \frac{C_4}{\Lambda^4} (\partial_\mu \vec{\phi} \partial_\mu \vec{\phi})^2 \right) \left(-\partial_\mu \partial_\mu - \partial_\mu \vec{\phi} \partial_\mu \vec{\phi} \right) \vec{\phi} = 0. \tag{2.6}$$

This equation actually contains two factors. Solutions of the eq. (2.6) are field configurations for which either the first factor is zero all over R^2 , or the second factor is zero all over R^2 , or the first factor is zero in a certain domain D of R^2 and the second factor is zero in the complementary domain.

To obtain the partition function (2.4) in the semiclassical approximation we need to sum up the various contributions coming from the quantum fluctuations around solutions of the (2.6) in the topologically disconnected sectors of the configuration space characterized each by a different value of the winding number N. Let's consider the $N = 1$ sector. By elementary energy entropy considerations we may convince ourselves that the relevant solution to eq. (2.6) which has to be taken into account in given by the second factor equal to zero all over R^2 , i.e. it is the well-known Belavin Polyakov [6] one-instanton solution. In the semiclassical approximation to (2.4) then the contribution from the $N = 1$ sector is obtained by considering the quantum fluctuations around the one-instanton. We only consider now the contribution to Z coming from this sector. In the dilute instantons gas approximation this turns out to be all what we need to compute the partition function.

Let's call $\vec{\phi}_0(x)$ the one-instanton solution to the (2.6). It can be parametrized in the simple following way:

$$\begin{cases} \phi_0^{(1)} = \dfrac{2\rho(x-x_0)}{\rho^2+|z-z_0|^2} \\ \phi_0^{(2)} = \dfrac{2\rho(y-y_0)}{\rho^2+|z-z_0|^2} \\ \phi_0^{(3)} = \dfrac{|z-z_0|^2-\rho^2}{\rho^2+|z-z_0|^2}. \end{cases} \tag{2.7}$$

where x_0 and y_0 are the instanton center coordinates (z and z_0 are simply the complexified notation for the coordinates in the plane : $z = x + iy$) and ρ is the instanton size.

Inserting the (2.7) in the action we get for the one instanton configuration:

$$S[\vec{\phi}_0] \equiv S_0(\rho) = \frac{4\pi}{f}\Big(1 - \frac{2C_2}{\rho^2\Lambda^2} + \frac{C_4}{\rho^4\Lambda^4}\Big). \tag{2.8}$$

The irrelevant terms remove the action degeneracy with respect to the scale transformations already at the tree level. Deriving the (2.8) with respect to ρ we find that this action has a minimum for

$$\overline{\rho}^2 = \frac{C_4}{C_2}\frac{1}{\Lambda^2}, \tag{2.9}$$

i.e. there is only one stable instanton with dimension given by the (2.9). For such an instanton the action is

$$S_0(\overline{\rho}) = \frac{4\pi}{f}\Big(1 - \frac{2C_2^2}{C_4}\Big). \tag{2.10}$$

It should be noticed that the existence of such a stable instanton is related to the particular choice we have done for the coupling constants. Would we have chosen the two irrelevant coupling constants with the same sign, there would have been no stable instanton. Such a dependence from the choice of the coupling constants is already unexpected.

3. Quantum Fluctuations

Let's call Z_1 the contribution to the partition function (2.4) from the N=1 sector. To compute the quantum fluctuation around the one-instanton solution to eq. (2.6) we expand the field $\vec{\phi}$ around the instanton configuration $\vec{\phi}_0$,

$$\vec{\phi} = \vec{\phi}_0 + \vec{\eta} \tag{3.1}$$

and retain terms up to the quadratic order in $\vec{\eta}$. Inserting (3.1) in (2.4) we get

$$Z_1 = \int D[\vec{\eta}\,] \prod_x \delta(\vec{\eta}(x)\cdot\vec{\phi}_0(x)) \;\; e^{-S_0(\rho)} \times$$
$$\exp\left\{-\frac{1}{2f}\int d^2x\; \vec{\eta}\Big(1 - \frac{3}{2}\frac{C_2}{\Lambda^2}(\partial_\mu\vec{\phi}_0)^2 + \frac{15}{64}\frac{C_4}{\Lambda^4}(\partial_\mu\vec{\phi}_0)^4\Big)\Big(-\partial_\mu\partial_\mu - (\partial_\mu\vec{\phi}_0)^2\Big)\vec{\eta}\right\}. \tag{3.2}$$

Roughly speaking we have to compute the determinant of the product of the two operators

$$A(x,y) = \delta(x-y)\left(1 - \frac{3}{2}\frac{C_2}{\Lambda^2}(\partial_\mu\vec{\phi}_0(x))^2 + \frac{15}{64}\frac{C_4}{\Lambda^4}(\partial_\mu\vec{\phi}_0(x))^4\right), \tag{3.3}$$

and

$$B(x,y) = -\frac{1}{2f}\delta(x-y)\left(\partial_\mu\partial_\mu - (\partial_\mu\vec{\phi}_0(x))^2\right). \tag{3.4}$$

The operator (3.4) is well known [7]. It has six zero modes in the N=1 sector and no negative eigenvalues. An instability to the instanton could come from the (3.3) operator. But as far as we restrict ourselves to the case

$$C_n << 1, \tag{3.5}$$

the operator (3.3) is positive definite and there is no place for instability. Under this conditions the determinant of $A \times B$ can be calculated as the product of the determinant of the ultralocal operator (3.3) times the well known determinant of the operator (3.4) [7], [8], [9], [10]. The zero modes of the operator (3.4) are treated in the standard way by the collective coordinates method [11].

The contributions from the various topological sectors are usually normalized to the contribution from the $N = 0$ sector, which we call Z_0. In the $N = 0$ sector only the operator

$$B_0(x,y) = -\frac{1}{2f}\delta(x-y)\Big(\partial_\mu\partial_\mu\Big) \tag{3.6}$$

appears while $A(x,y) = \delta(x-y)$. The determinant of the operator B is actually well known, as well as B_0. By considering the stereographic projection of the (x,y) plane onto the sphere, the problem is traced back to the evaluation of the eigenvalues of the laplacian on the sphere [9]. The result for the ratio Z_1/Z_0 is then

$$\frac{Z_1}{Z_0} = (2\pi)(\frac{16}{3})^2 \frac{1}{f^2} \int_{-\infty}^{+\infty} dx_0 \int_{-\infty}^{+\infty} dy_0 \int_0^{\infty} \frac{d\rho}{\rho^3} e^{-\frac{4\pi}{f}\left(1-\frac{2C_2}{\rho^2\Lambda^2}+\frac{C_4}{\rho^4\Lambda^4}\right)} \times \left[\frac{\prod_{j>0}\left[j(j+1)\right]^{(2j+1)}}{\prod_{j>1}\left[j(j+1)-2\right]^{(2j+1)}} \right] \det{}^{-\frac{1}{2}} A. \tag{3.7}$$

In (3.7) the integrations come from the three collective coordinates, (x_0, y_0) (the center of the instanton) and ρ (the instanton size), the argument of the exponential is the classical action while the ratio in the second line is the ratio between the determinants of the operators B_0 and B , i.e. the ratio between the product of eigenvalues of the laplacian for the $N=0$ and $N=1$ case where the zero modes have already been excluded. We have left indicated the determinant of the operator A because it is of no particular importance for our later discussion (see ref. [5]).

4. Regularization and Renormalization Flow

The ratio of the eigenvalues product is divergent and has to be regularized. Having developed the fluctuation in spherical harmonics over a sphere of radius ρ , the natural choice is to regularize such a product by demanding a dimensionless cut-off in the eigenvalues related to the original momentum cut-off by the relation

$$j(j+1) - 2 \leq \rho^2\Lambda^2. \tag{4.1}$$

The functional space is truncated in such a manner that in the expansion of a function on the sphere we retain only those spherical harmonics for values of j up to the maximal value

$$j_{max} = [\rho^2\Lambda^2]. \tag{4.2}$$

We now introduce a convenient notation. Let's call $D(\rho\Lambda)$ the regularized ratio of the eigenvalues and similarly $C(\rho\Lambda)$ the $\det^{-\frac{1}{2}} A$. Neglecting terms inessential for the following considerations we write then

$$\frac{Z_1}{Z_0} = \int_0^{\infty} \frac{d\rho}{\rho^3} e^{-\frac{4\pi}{f}\left(1-\frac{2C_2}{\rho^2\Lambda^2}+\frac{C_4}{\rho^4\Lambda^4}\right)} D(\rho\Lambda)C(\rho\Lambda). \tag{4.3}$$

It is interesting to notice that if we only consider large instantons,

$$\rho^2\Lambda^2 >> 1 \tag{4.4}$$

we have (see [9] and [5])

$$D(\rho\Lambda) \to e^{\ln\rho^2\Lambda^2} \quad , \qquad C(\rho\Lambda) \to 1 \quad , \qquad e^{-\frac{4\pi}{f}\left(1-\frac{2C_2}{\rho^2\Lambda^2}+\frac{C_4}{\rho^4\Lambda^4}\right)} \to e^{-\frac{4\pi}{f}}. \tag{4.5}$$

Then the integrand of the (4.3) become

$$\exp\left\{-\frac{4\pi}{f}+\ln\rho^2\Lambda^2\right\}. \tag{4.6}$$

By requiring that (4.6) is cut-off independent we get immediately

$$\frac{1}{f(\rho)}=\frac{1}{f(\Lambda)}-\frac{1}{2\pi}\ln\rho\Lambda\ , \tag{4.7}$$

from which we are led to the popular statement [12], [8], that the renormalization of the theory in the instanton (N = 1) sector doesn't require additional counterterms; operators which are irrelevant in the $N=0$ sector are also irrelevant in the $N=1$ sector. It doesn't come as a surprise that for such a "slowly varying" background field the renormalization properties of the theory are not different from the case of a constant background field.

Let's now go back to our model. It is by no means obvious that the path integral should be dominated by large instantons. We have already seen at the tree level that for certain values of the irrelevant coupling constants we have mini-instanton solutions. As for the one loop level now, let's go back to the eq. (4.3) and change the variable $\rho \to \frac{1}{\rho^2}$. We get :

$$\frac{Z_1}{Z_0}=e^{-\frac{4\pi}{f}\left(1-\frac{2C_2^2}{C_4}\right)}\int\limits_0^\infty d\left(\frac{1}{\rho^2}\right)e^{-\frac{4\pi}{f}\frac{C_4}{\Lambda^4}\left(\frac{1}{\rho^2}-\frac{1}{\bar\rho^2}\right)^2}D(\rho\Lambda)C(\rho\Lambda), \tag{4.8}$$

where $\bar\rho$ is given by (2.9) and is the dimension of the stable mini-instanton. The interesting thing about eq. (4.8) is that the integral is actually gaussian in the $\frac{1}{\rho^2}$ variable. In this expression we have a competition between the gaussian exponent and the one loop term $D(\rho\Lambda)C(\rho\Lambda)$. This integral actually receives two big contributions, one coming from the peak at the mini-instanton location the other coming from the infrared region. The peak width is actually $O(\sqrt{f})$ while the distance between these two regions is $O(1)$. We have then a sufficiently large interval which separates the mini-instanton from the "dangerous" infrared region. As far as we don't enter this "profound infrared region", delimited by Λ_σ , i.e. if we put an infrared cut-off in the (4.8) ($0 \to \Lambda_\sigma$), we can perform the integration in the saddle point approximation. In the mini-instanton region the functions $D(\rho\Lambda)$ and $C(\rho\Lambda)$ may be well approximated by their value at $\bar{\rho}$. We then get

$$\frac{Z_1}{Z_0} = \frac{1}{4f^2} D(\overline{\rho}\Lambda) C(\overline{\rho}\Lambda) e^{-\frac{4\pi}{f}\left(1-\frac{2C_2^2}{C_4}\right)} \Lambda^2 \sqrt{\frac{f}{C_4}}. \tag{4.9}$$

This is actually our central result. In any physical quantity defined at the infrared scale μ, according to the power counting theorem, the irrelevant coupling constants C_n should always appear in combination with a suppression factor $(\mu/\Lambda)^n$:

$$C_n \left(\frac{\mu}{\Lambda}\right)^n. \tag{4.10}$$

Being Z_1/Z_0 an extensive (i.e. $\mu \sim 0$) dimensionless quantity, we would then expect

$$\frac{Z_1}{Z_0} = F(\Lambda_\sigma/\Lambda). \tag{4.11}$$

Actually the "would be" irrelevant coupling constants C_2 and C_4 appear in eq. (4.9) without any suppression factor.

We see here the similarity with the case of strong coupling QED which we have mentioned earlier. In that case the strong short distance interactions generate little bound states which populate the vacuum and produce long distance effects : the strong quantum fluctuations cancel the $O(\Lambda^{-2})$ tree level suppression factor which appears in the quadrifermionic term.

We may have now a similar scenario. The vacuum may be populated by "small" instantons, i.e. inverse cutoff size instantons, which provide the mechanism to enhance the contribution of operators that appear as irrelevant in the $N = 0$ sector.

As we have already said the situation changes drastically if we chose the two coupling constant with the same sign. In this case no stable saddle point exists. The large size saddle points are almost stable giving rise to a renormalization structure similar to the one obtained in the $N = 0$ sector. In this case the two irrelevant coupling constants are still irrelevant. The model has then a richer phase structure then is usually believed : it is sensible to the choice of irrelevant coupling constants.

In the usual renormalization prescription we don't need to fine tune the irrelevant the coupling constants: a need for a fine tuning of those coupling constants would signal a departure from the established power counting theorem.

If we then keep C_2 and C_4 constant, we may determine the flow of the coupling constant f from the cutoff independence of (4.9),

$$\Lambda \frac{d}{d\Lambda} \left\{ \frac{1}{f^2} e^{-\frac{4\pi}{f}\left(1-\frac{2C_2^2}{C_4}\right)} \Lambda^2 \sqrt{\frac{f}{C_4}} \right\} = 0. \tag{4.12}$$

Up to second order terms in f we get the flow :

$$\Lambda\frac{df}{d\Lambda} = \beta(f) = -\frac{f^2}{2\pi}\frac{1}{1-\frac{2C_2^2}{C_4}}.$$

Obviously in the dilute instanton gas approximation the bulk of the contribution comes from the zero winding number sector where we have the same density of instantons and antiinstantons. Each of these instantons is a mini-instanton and generates the beta function (4.13). Even though the constant background belongs to such a sector and gives the well known beta function (independent on the C_n) such a configuration is suppressed by the instanton entropy. We can then summarize all the previous considerations by saying the dilute instanton gas generates new relevant coupling constants and that the first coefficient of the beta function is non universal.

5. Conclusions

We have seen that the semiclassical expansion of a classically scale invariant theory is actually more involved than it is usually believed. Due to the peculiarity of the quantum theory, which necessarily requires the presence of irrelevant operators in the quantum action, in the semiclassical expansion of the theory we are forced to explore a larger subspace in parameter space than that spanned by those coupling constants which are proven to be relevant in the zero winding number sector. In the dilute instanton gas approximation there are regions in the parameter space where "would be" irrelevant coupling constants are promoted to the rang of relevant ones. A scenario which could could be responsible for the enhancement mechanism is that a gas of mini-instantons populate the vacuum giving rise to long distance effects.

If this is a sensible mechanism we may expect relevant effects in the application to QCD. Let's consider for instance the evolution equations for the structure functions whose momentum dependence is obtained in the framework of the operator product expansion [13] as a power series in $\frac{\Lambda^2}{Q^2}$. At sufficiently high energies we just need to take into account the leading order, i.e. the contributions coming from the lower twist. As we consider lower momentum scales we have to include subdominant terms in this series.

If new coupling constants are relevant in the asymptotic scaling region naturally the fixing of the theory at high energies becomes more complicated. However some of the contributions which in the operator product expansion should come from higher twist could actually be produced at a lower order due to the presence of these additional coupling constants. It should then be possible, by the way, to follow the evolution equations until lower energies. This problem is presently under investigation.

There is another intriguing question associated to this phenomenon. We may expect a similar effect in the soliton sector of a system like the Yang-Mills-Higgs which possess 't'Hooft-Polyakov monopole solutions. Even though the static soliton energy as a function of the monopole size has a well pronounced minimum determined by the physical scale of the model, the presence of irrelevant higher order derivative terms may create another minimum in this function for mini-monopoles, i.e. for monopoles of inverse cutoff size. It remains to be seen if these mini-monopoles play a role in the low energy phenomena of the model. In this case we should have an effect for instance on the monopole induced proton decay [14]. Work is in progress in this direction.

ACKNOWLEDGEMENTS

I wish to thank Janos Polonyi whose deep physical insight in the problem has made this work possible.

REFERENCES

[1] G.P. Lepage, "What is renormalization?", preprint,CLNS,89/970. Newman Lab. of Nuclear Studies, Cornell University.

[2] S.L. Adler, *Phys. Rev.***177** (1969) 2426 ; J.S. Bell and R. Jackiw, *Nuovo Cimento* **A 60**(1969) 47.

[3] P.I. Fomin, V.P. Gusynin, V.A. Miranski and YU.A. Sytenko, *Riv. Nuovo Cimento*, **6**, No. 5 (1983).

[4] C.N. Leung, S.T. Love , William A. Bardeen, *Nucl. Phys.***B 323** (1989) 493 ; C.N. Leung, S.T. Love , William A. Bardeen , *Nucl. Phys.***B 273** (1986) 649.

[5] V.Branchina, J.Polonyi, Instantons in cut-off theories, hep-th 9401079, to appear in *Nucl. Phys.***B**.

[6] A.Belavin and A.Polyakov, JEPT Letters **22** (1975) 245.

[7] A.Polyakov, JEPT **41** (1976) 988.

[8] A.Jevicki, *Nucl. Phys.***B127** (1977) 125.

[9] D.Forster, *Nucl. Phys.***B130** (1977) 38.

[10] V.A. Fateev, I.V. Frolov and A.S. Schwarz, *Nucl. Phys.***B 154** (1979) 1

[11] D. Bohm and D. Pines, *Phys. Rev.***92** (1953) 609.

[12] G. 't Hooft, *Phys. Rev.***D 14** (1976) 3432.

[13] K. Wilson, *Phys. Rev.***179** (1969) 1499.

[14] V.A. Rubakov, *Nucl. Phys.***B203** (1982) 311.

DEPENDENCE OF HOLOMORPHICITY OF THE GAUGE COUPLING CONSTANT ON THE MASS MATRIX IN SUSY THEORIES

H.S. LI* AND K.T. MAHANTHAPPA†

Department of Physics, Campus Box 390,

University of Colorado, Boulder, CO 80309

ABSTRACT

We investigate the dependence of holomorphicity of the gauge coupling constant function on the mass matrix at one- and two-loop levels in supersymmetric theories. Gauge invariance puts constraints on the mass matrix. These constraints at one-loop level lead us to three cases of mass matrix that require different ways of regulating infrared contributions: massive, pseudo massive and intrinsically massless. The first two give rise to a holomorphic gauge coupling constant function whereas the last one does not. Two-loop dependence on the mass matrix is found to fall under the same three cases as at the one-loop level. Making use of our two-loop results we also verify the holomorphicity of the Wilson coupling based on general arguments of Shifman and Vainshtein.

1. Introduction

The effective low energy supersymmetric lagrangian derived from superstrings or supergravity is constrained to obey the original supersymmetries as well as the holomorphic properties of certain functions including the gauge coupling constant function f. Recent works [1, 2] find that in their investigations this function has non-holomorphic one-loop contribution. This has its origin in the infrared divergence due to zero mass modes of the matter field. In this talk we analyze the dependence of the holomorphic property of f in terms of the mass matrix. The details are contained in Ref. [3].

⋆ Current address: MCI Telecommunications Corp., 707 17th St., Suite 4100, Denver, CO 80202

† Speaker

2. Holomorphicity

Supergravity is a local realization of SUSY. This theory is invariant under both the general coordinate transformation and the local SUSY transformation. Models based on supergravity have many attractive features including a nicer ultraviolate behavior. The action for $N = 1$ supergravity has the following form in the superspace [4]:

$$\begin{aligned}\mathcal{A} = &\int d^4x d^4\theta \; E\big[\Phi(S, \bar{S}e^{2V}) + \mathrm{Re}(\frac{1}{\mathcal{R}}P(S))\big] \\ &+ \int d^4x d^4\theta \; E \; \mathrm{Re}(\frac{1}{\mathcal{R}}f_{ab}(S)W^aW^b).\end{aligned} \tag{2.1}$$

Where a and b indicate different group sectors, E is the superspace determinant, $\mathcal{R}$ is the super curvature, S is the chiral superfield, V is the super gauge field of the group G, W is the the super field strength, Φ is an arbitrary real function, $P(S)$ is the superpotential and $f_{ab}(S)$ are the gauge coupling functions. $f_{ab}(S)$ should be analytic functions of S (but not $\bar{S}$). This analytical dependence is called *holomorphicity* of $f_{ab}(S)$.

The bosonic part of the Lagrangian density is given by

$$\begin{aligned}e^{-1}\mathcal{L}_B = &R + \Big(\frac{1}{4g^2(\varphi,\bar{\varphi})}\Big)_{ab} F^a_{\mu\nu}F^{b\mu\nu} + i\frac{\Theta_{ab}(\varphi,\bar{\varphi})}{32\pi^2}F^a_{\mu\nu}\tilde{F}^{b\mu\nu} \\ &+ \frac{1}{2}G_{i\bar{j}}(\varphi,\bar{\varphi})D^\mu\varphi^i D_\mu\bar{\varphi}^{\bar{j}} + V(\varphi,\bar{\varphi}),\end{aligned} \tag{2.2}$$

where R is the canonical gravitational curvature, e is the determinant of the space-time metric, D is the covariant derivative, $G_{i\bar{j}}(\varphi,\bar{\varphi})$ is the metric for the scalar fields φ^i, $V(\varphi,\bar{\varphi})$ is the potential for them, F is the gauge field strength and $\tilde{F}$ is its dual. We have

$$\Big(\frac{1}{g^2(\varphi,\bar{\varphi})}\Big)_{ab} = \mathrm{Re}f_{ab}(\varphi), \qquad \frac{\Theta_{ab}(\varphi,\bar{\varphi})}{8\pi^2} = \mathrm{Im}f_{ab}(\varphi). \tag{2.3}$$

Holomorphicity of $f(\varphi)$ is a tree level result. It is desirable to have holomorphicity even with quantum corrections, one reason being that Θ_{ab} cannot be obtained directly from Feynman diagrams, but can be obtained from derivatives. We need the following integrability conditions to have a well-defined Θ_{ab}:

$$\frac{\partial}{\partial\langle\varphi^j\rangle}\Big\{\frac{\partial\Theta_{ab}}{\partial\varphi^i}\Big\}_{\text{effective}} = \frac{\partial}{\partial\langle\varphi^i\rangle}\Big\{\frac{\partial\Theta_{ab}}{\partial\varphi^j}\Big\}_{\text{effective}}. \tag{2.4}$$

These conditions are only true if holomorphicity holds.

Considering the global super gauge theory (SGT) as a derivative of this local supergravity, the couplings in the SGT should also be holomorphic functions of $\langle\varphi^i\rangle$. For example, considering a mass matrix $M = M(\langle\varphi\rangle)$, the dependence of the gauge couplings on M should be holomorphic. Recently, Dixon *et al* [1] considered a SGT based on the E_6 group with two 27-dimensional families of chiral fermions and one $\bar{27}$-dimensional antifamily. The matter mass matrix is

$$M(\langle\varphi\rangle) = \begin{pmatrix} 0 & 0 & \langle\varphi_1\rangle \\ 0 & 0 & \langle\varphi_2\rangle \\ \langle\varphi_1\rangle & \langle\varphi_2\rangle & 0 \end{pmatrix} \otimes I_{27}. \tag{2.5}$$

They calculated one-loop correction to $1/g^2$ and got

$$\Delta\frac{1}{g^2} = -\frac{6}{16\pi^2}\ln\left[|\langle\varphi^1\rangle|^2 + |\langle\varphi^2\rangle|^2 + \mathrm{O}(p_0^2)\right] + \text{const.}. \tag{2.6}$$

This is not the real part of a holomorphic function.

The infrared divergence is the cause of the non-holomorphic dependence. We investigate holomorphicity in the SGT at both one- and two-loop levels, and examine the conditions for holomorphicity based on the mass matrix M of the matter fields.

3. Explicit One-Loop Calculation

Our calculation of the one-loop dependence of $1/g^2$ on the mass matrix M follows the way β-function is calculated, except that one can ignore M-independence terms. The SUSY action for gauge field coupled to matter is

$$\begin{aligned} \mathcal{A} = &-\frac{1}{4g^2}\operatorname{Tr}\int d^4x d^4\theta\left[(e^{-V}D^\alpha e^V)\bar{D}^2(e^{-V}D_\alpha e^V)\right] \\ &+\int d^4x d^4\theta\; \bar{\phi}^{\mathrm{T}} e^V \phi + \int d^4x\left[d^2\theta\; P(\phi) + \text{h.c.}\right], \end{aligned} \tag{3.1}$$

where V is the gauge vector super field, ϕ is the matter chiral super field, $P(\phi)$ is the super potential, D_α and $\bar{D}_{\dot\alpha}$ are covariant spinor derivatives and g is the tree-level gauge coupling. The gauge fixing term is

$$\mathcal{A}_{\mathrm{gf}} = -\frac{1}{2\alpha g^2} \operatorname{Tr} \int d^4x d^4\theta D^2 V \bar{D}^2 V.$$

This leads to Feynman propagators

$$\begin{aligned} \langle V^\sigma(1) V^\rho(2) \rangle &= -\frac{ig^2}{\partial^2} \delta_{12} \delta^{\sigma\rho} \\ \langle \phi(1) \phi^{\mathrm{T}}(2) \rangle &= M^\dagger \frac{i}{\partial^2 - MM^\dagger} \bar{D}_1^2 \delta_{12} \\ \langle \bar{\phi}(1) \bar{\phi}^{\mathrm{T}}(2) \rangle &= \frac{i}{\partial^2 - MM^\dagger} M \, D_1^2 \delta_{12} \\ \langle \phi(1) \bar{\phi}^{\mathrm{T}}(2) \rangle &= \frac{i}{\partial^2 - M^\dagger M} \bar{D}_1^2 D_1^2 \delta_{12} \\ \langle \bar{\phi}(1) \phi^{\mathrm{T}}(2) \rangle &= \frac{i}{\partial^2 - MM^\dagger} D_1^2 \bar{D}_1^2 \delta_{12}, \end{aligned} \tag{3.2}$$

where $\delta_{12} \equiv \delta^4(\theta_1 - \theta_2)\delta^4(x_1 - x_2)$.

By calculating Z_g, we can get one-loop correction to $1/g^2$. Z_g denotes the r.c. for g. Z_g can be obtained from the relation $Z_g Z_V^{\frac{1}{2}} Z_\phi = Z_\lambda$; Z_λ denotes the r.c. for vertex $\bar{\phi}^{\mathrm{T}} V \phi$; Z_ϕ and Z_V denote the r.c. for ϕ- and V-wave functions. In our work $Z_\lambda = Z_\phi$ and hence $Z_g = Z_V^{-\frac{1}{2}}$. For wave function r.c. Z_V, we have eight diagrams

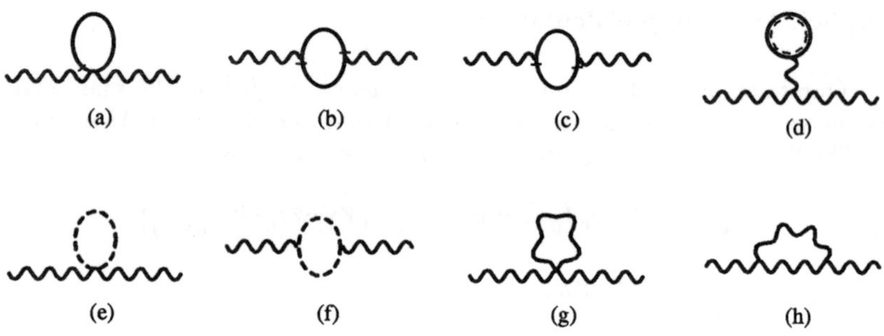

FIG. 1. Self-energy of the gauge field

Here the dashed line stands for the ghost propagator and the dashed-solid line stands for either a super matter, super gauge or super ghost propagator. We then have

$$Z_V = 1 - ig^2 \frac{T_R}{2d_R} \operatorname{Tr} \int \frac{d^4q}{(2\pi)^4} \frac{1}{(q^2 + M^\dagger M)[(q+p)^2 + M^\dagger M]}, \tag{3.3}$$

where T_R is given by $\operatorname{Tr}(T^\sigma T^\rho) = T_R \delta^{\sigma\rho}$. This means

$$\Delta \frac{1}{g^2} = -i \frac{T_R}{2d_R} \operatorname{Tr} \int \frac{d^4q}{(2\pi)^4} \frac{1}{(q^2 + M^\dagger M)[(q+p)^2 + M^\dagger M]}, \tag{3.4}$$

as external momentum $p \to 0$.

The last integral is ultraviolet divergent. We cannot use the dimensional regularization scheme since holomorphicity is a four-dimensional phenomenon and every calculation has to be performed in four dimensional space-time. Instead, we use Pauli-Villars regularization scheme

$$\frac{1}{k^2+M^\dagger M} \to \left(\frac{1}{k^2+M^\dagger M} - \frac{1}{k^2+\Lambda^\dagger\Lambda}\right), \tag{3.5}$$

where Λ is the matrix of ultraviolet momentum cutoff. In this scheme

$$\Delta\frac{1}{g^2} = -i\frac{T_R}{2d_R}\,\mathrm{Tr}\int\frac{d^4q}{(2\pi)^4}\frac{(\Lambda^\dagger\Lambda)^2}{(q^2+M^\dagger M)^2(q^2+\Lambda^\dagger\Lambda)^2}. \tag{3.6}$$

Whether this integral is infrared divergent or not depends on the mass matrix M. M is not arbitrary. The invariance of the mass term, $-\frac{1}{2}\phi^{\mathrm{T}}M\phi$, under gauge transformation requires

$$T^{\sigma\mathrm{T}}M + MT^\sigma = 0. \tag{3.7}$$

This puts constraints on the mass matrix.

4. Representations of the Mass Matrix

We now solve the constraints Eq. (3.7) for the form of M. For the (reducible) real or pseudo-real representation of the group, we write

$$G(R) = \mathrm{diag}\{\overbrace{G_r,\ G_r,\ \ldots,\ G_r}^{l}\}, \tag{4.1}$$

where the submatrix G_r is an irreducible $n\times n$ real or pseudo-real representation. We assume a total of l G_r's. Then we have

$$M = \begin{pmatrix} a_{11} & a_{12} & \cdots & a_{1l} \\ a_{21} & a_{22} & \cdots & a_{2l} \\ \vdots & \vdots & \ddots & \vdots \\ a_{l1} & a_{l2} & \cdots & a_{ll} \end{pmatrix} \otimes J.$$

Here J is $n\times n$ dimensional matrix. For the real representation $J = I$ and for the pseudo-real representation $J^2 = -1$.

For the (reducible) complex representation of the group, we have

$$G(R) = \mathrm{diag}\{\overbrace{G_c,\ G_c,\ \ldots,\ G_c}^{l},\ \overbrace{G_c^*,\ G_c^*,\ \ldots,\ G_c^*}^{\bar{l}}\}, \tag{4.2}$$

where G_c is an irreducible $n\times n$ complex representation and G_c^* is its complex conjugate. We assume a total of l G_c's and $\bar{l}$ G_c^*'s, and we call l as the number of

n-dimensional families of chiral fields and $\bar{l}$ as the number of $\bar{n}$-dimensional antifamilies of chiral fields. Then

$$M = \begin{pmatrix} & 0 & & b_{1l+1} & \cdots & b_{1l+\bar{l}} \\ & & & \vdots & \ddots & \vdots \\ & & & b_{ll+1} & \cdots & b_{ll+\bar{l}} \\ b_{l+11} & \cdots & b_{l+1l} & & & \\ \vdots & \ddots & \vdots & & 0 & \\ b_{l+\bar{l}1} & \cdots & b_{l+\bar{l}l} & & & \end{pmatrix} \otimes I.$$

5. One-Loop Holomorphicity

a) For the real and pseudo-real representations, the mass matrix can be massive (no zero eigen-value) or massless (at least one zero eigen-value). b) For the complex representation with balanced numbers of families and anti-families ($l = \bar{l}$), the mass matrix can be massive or massless. c) For the complex representation with unbalanced numbers of families and anti-families ($l \neq \bar{l}$), the mass matrix is always massless. We call the massless cases in both a) and b) *pseudo massive* and call case c) *intrinsically massless.*

Massive Mass Matrix: Now the mass matrix itself is an infrared regulator, we have

$$\Delta\frac{1}{g^2} = -\frac{T_R}{2d_R}\,\mathrm{Tr}\left(\frac{1}{16\pi^2}\right)\ln\left(\frac{M^\dagger M}{\Lambda^\dagger\Lambda}\right), \tag{5.1}$$

which is the real part of the holomorphic function

$$-\frac{T_R}{d_R}\,\mathrm{Tr}\left(\frac{1}{16\pi^2}\right)\ln\frac{M}{\Lambda}.$$

Thus, for this case, holomorphicity holds. *Pseudo Massive Mass Matrix*: We can add a perturbative matrix ϵ to M by defining $M_{\mathrm{p}} = M + \epsilon$. We choose M_p to be massive and adopt M_{p} as the infrared regulator while keeping the perturbed mass term gauge invariant:

$$\begin{aligned}\Delta\frac{1}{g^2} &= -\frac{T_R}{2d_R}\,\mathrm{Tr}\left(\frac{1}{16\pi^2}\right)\ln\left(\frac{M_{\mathrm{p}}^\dagger M_{\mathrm{p}}}{\Lambda^\dagger\Lambda}\right). \\ &= -\frac{T_R}{2d_R}\lim_{\epsilon\to 0}\mathrm{Tr}\left(\frac{1}{16\pi^2}\right)\ln\frac{(M^\dagger+\epsilon^\dagger)(M+\epsilon)}{\Lambda^\dagger\Lambda},\end{aligned}$$

which is, again, the real part of a holomorphic function. Recently, Dine and Shirman [5] have modified the model of Dixon *et al* [1] by adding a $\overline{27}$ leading to holomorphicity. This result follows from our analysis as the modified model has equal number of families and anti-families. *Intrinsically Massless Matrix*: In this case, the group is a complex representation with $l \neq \bar{l}$. We cannot perturb the mass matrix M so that it becomes massive while the mass term is still gauge invariant. The perturbed mass matrix cannot be used as the infrared regulator. We need to put in an infrared momentum cutoff, p_0:

$$\Delta\frac{1}{g^2} = -\frac{T_R}{2d_R}\lim_{p_0\to 0}\mathrm{Tr}\left(\frac{1}{16\pi^2}\right)\ln\left(p_0^2 + M^\dagger M\right). \tag{5.2}$$

Since we cannot factorize $p_0^2 + M^\dagger M$, the above expression is not the real part of a holomorphic function.

We thus conclude that the cause of non-holomorphicity is the unbalanced numbers of families and anti-families, combined with the infrared divergence. The models with an intrinsically massless mass matrix have non-holomorphic one-loop corrections. But the ones with a massive or pseudo massive mass matrix have holomorphic f.

6. Two-Loop Calculations

We use the super background field method to calculate two-loop corrections. We have three vacuum Feynman diagrams at two-loop level

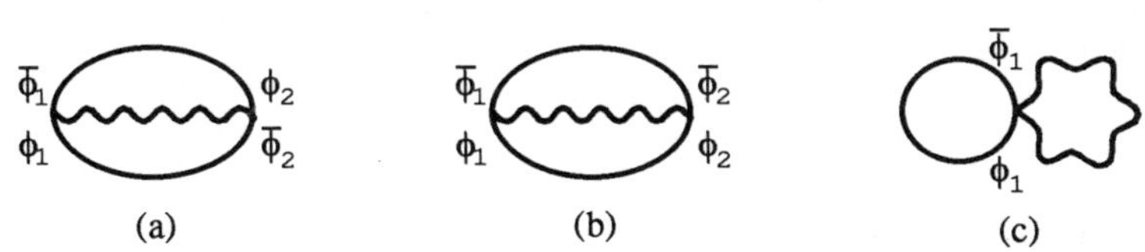

FIG. 2. Two-loop vacuum Feynman diagrams in non-Abelian gauge theory

The background gauge fields are hidden inside the quantum propagators. We need to expand the quantum propagators $\frac{1}{\Box_+ - M^\dagger M}$, $\frac{1}{\hat{\Box}_A}$ and $\frac{1}{\hat{\Box}_A}$ around the external background gauge fields. From the expansions, Figs. 2(b) and 2(c) do not contribute, up to the second order in W and Γ. This contributions from Figs. 2(a) can be represented by six Feynman diagrams

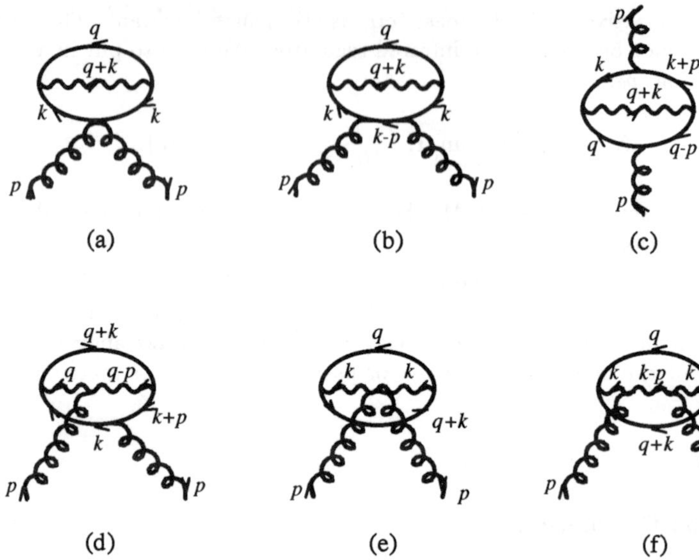

FIG. 3. Expansions in non-Abelian gauge theory

Curled lines stand for the external W and Γ. We have a total of eighteen non-vanishing contributions.

7. Two-Loop Holomorphicity

Calculations show that the second order logarithmic terms, $[\ln(M^\dagger M)]^2$, cancel out. For massive and pseudo massive cases we get

$$\Delta\frac{1}{g^2} = -\frac{T_R}{d_R}\left(\frac{g}{16\pi^2}\right)^2\Big[(C_R + C_A)\,\mathrm{Tr}\ln\left(\frac{M^\dagger}{\Lambda^\dagger}\cdot\frac{M}{\Lambda}\right) + \text{const.}\Big], \tag{7.1}$$

where $\sum_\sigma T^\sigma T^\sigma = C_R I$ and $C_A \equiv C_R$ for adjoint R. For intrinsically massless case, we get

$$\Delta\frac{1}{g^2} = -\frac{T_R}{d_R}\left(\frac{g}{16\pi^2}\right)^2\Big[(C_R + C_A)\,\mathrm{Tr}\ln\left(p_0^2 + M^\dagger M\right) + \text{const.}\Big]. \tag{7.2}$$

Apart from the coefficient difference, these results are the same as one-loop's. Hence, one- and two-loop holomorphic properties are identical: a) holomorphic for the massive and pseudo massive matrices and b) non-holomorphic for the intrinsically massless matrix.

8. Holomorphicity of "Wilson Coupling Constant"

Shifman and Vainshtein have argued that $1/g_W^2$, where g_W is the coupling constant in Wilsonian effective action, is always holomorphic. Because $1/g_W^2$ and $1/g^2$ are related by a term having the wave function r.c. of the matter field, Z_ϕ, we can verify this at the two-loop level. In SQED the relation is

$$\frac{1}{g_{\rm W}^2} = \frac{1}{g^2} + \frac{T_R}{8\pi^2 d_R} \operatorname{Tr} \ln Z_\phi .$$

Substituting on the r.h.s. with

$$\Delta \frac{1}{g^2} = -\frac{T_R}{d_R}\left(\frac{g}{16\pi^2}\right)^2 \left[\operatorname{Tr} \ln \left(p_0^2 + M^\dagger M\right) + \text{const.}\right]$$

and the one-loop result

$$Z_\phi = 1 + \frac{g^2}{32\pi^2} \ln(M^\dagger M + p_0^2),$$

we see that the mass dependent term disappears making $1/g_W^2$ holomorphic. In super Yang-Mills theory, we have

$$\frac{1}{g_{\rm W}^2} = \frac{1}{g^2} - \frac{C_A}{8\pi^2} \ln \frac{1}{g^2} + \frac{T_R}{8\pi^2 d_R} \operatorname{Tr} \ln Z_\phi$$

and the one-loop Z_ϕ

$$Z_\phi = 1 + \frac{g^2}{32\pi^2} \ln(M^\dagger M + p_0^2).$$

Again we see that the mass matrix dependence of the second and third terms at one-loop level cancel the mass-matrix dependence of $1/g^2$ at two-loop level, making $1/g_W^2$ holomorphic.

9. Conclusions

a) We have calculated the dependence of $1/g^2$ on the mass matrix and tried to express the dependence as the real part of a holomorphic function of the mass matrix. In the cases of massive and pseudo massive matrices, it is holomorphic. In the case of intrinsically massless mass matrix which arises for the complex representation with unbalanced number of families and anti-families, the infrared divergence leads to non-holomorphicity. b) Our calculations show that the holomorphic properties of one-loop and two-loop corrections are identical. It is likely that this is true at all loops. c) Using our results we have been able to explicitly verify at the two-loop level the assertion of Shifman and Vainshtein that $1/g_W^2$ is independent of mass matrix.

REFERENCES

1. L. Dixon, V.S. Kaplunovsky and J. Louis, Nucl. Phys. **B 355**, 649 (1991); J. Louis, in the Proceedings of the Second International Symposium on Particles, Strings and Cosmology, edited by P. Nath and S. Reucroft (World Scientific, Singapore, 1992), p. 751.

2. M.A. Shifman and A.I. Vainshtein, Nucl. Phys. **B 359**, 571 (1991).

3. H.S. Li and K.T. Mahanthappa, Phys. Lett. **B 319** , 152 (1993) and Phys. Rev. D **49**, 5532 (1994).

4. See, for example, E. Cremmer, S. Ferrara, L. Girardello and A. Van Proeyen, Nucl. Phys. **B 212**, 413 (1983).

5. M. Dine and Y. Shirman, University of California at Santa Cruz Report No. SCIPP94/11.

INFRARED TECHNOLOGY IN PERTURBATIVE QCD CALCULATIONS

DAVID A. KOSOWER
Service de Physique Théorique, Centre d'Etudes de Saclay
F–91191 Gif-sur-Yvette cedex, France

ABSTRACT

I review the formalism developed by Giele and Glover,[1] along with the author,[2] for handling infrared cancellations in a general-purpose numerical perturbative QCD calculation.

1. Introduction

Hadron colliders, that is the presently-operating Tevatron and the planned LHC, are likely to play an important role in elucidating the physics underlying the standard model. Within the context of the QCD-corrected parton model, which describes hadronic collisions in the high-energy limit of interest, the initial states are colored, and the hard-scattering event is associated with the production of gluonic radiation. The hadronization products of this radiation form narrow clusters of particles in the detector, that is, jets. Because of their ubiquity, a detailed understanding of the distributions of jets is important to searches for new physics. In particular, if we wish to use short-distance 'final' states containing quarks or gluons, corresponding to detector final states containing jets, to search for new physics, we must understand possible QCD backgrounds. The ongoing search for the top quark utilizing, among other modes, the W+4 jets mode, requires a detailed understanding of ordinary QCD production of the same signal.[3]

The basic estimates of cross sections and distributions for a process with n jets are given by leading-order calculations, using tree-level calculations of the $(n+2)$-parton short-distance matrix element. While such calculations have proven reasonable in general, they do suffer from a number of drawbacks: (a) renormalization-scale uncertainty (b) incorrect sensitivity to experimental cuts and jet algorithm. For example, a leading-order inclusive jet calculation will not predict the dependence of the cross-section on the cone size. Next-to-leading order calculations can alleviate some of these problems. The dependence on the renormalization scale, present in fixed-order perturbative calculations, decreases at next-to-leading order, as the explicit dependence in the matrix element compensates the dependence in the coupling constant. Furthermore, the presence of real radiation beyond the basic number of partons required introduces a dependence on experimental cuts and on the experimental jet algorithm, bringing theory closer to what is observed in experiments.[4]

2. Infrared Cancellations

In order to provide theoretical modelling of all distributions analyzed by experiments, incorporating all experimental cuts, such as cuts on rapidity, and allowing for the different choices of the jet algorithm, one would ideally want a completely differential numerical calculation, where all cuts and jet algorithms are implemented in a computer program rather than through an analytic calculation. On the other hand, next-to-leading order calculations involve two separate classes of contributions, arising respectively from virtual corrections and real-emission corrections to the matrix elements.

For example, the construction of a numerical next-to-leading order computer program for the calculation of two-jet distributions involves both the one-loop corrections to the basic $2 \to 2$ matrix elements, and the integration over those configurations of the $2 \to 3$ matrix elements that the jet algorithm classifies as having two jets. If two partons are sufficiently nearby, the parton-level version of the jet algorithm will merge the two into a single jet. Each of these contributions has infrared divergences.

The technical complexity of such programs arises from the simultaneous requirements of regulating the infrared divergences, while allowing a completely numerical determination of the boundary in phase space between the two- and three-jet regions. For practical purposes, in gauge theories the only available regulator is a dimensional one, $D = 4 - 2\epsilon$; on the other hand, the expressions in computer programs must be four-dimensional ones, since computers cannot do arithmetic in $D = 4 - 2\epsilon$.

To perform a calculation within these seemingly contradictory requirements, we proceed along lines originally suggested by van Neerven and Vermaseren[5] in the context of e^+e^- physics, variants of which were used by later authors in QCD calculations as well.[6] The idea behind this 'slicing' technique is the following: divide the integrations over the real-emission phase space into two pieces, using a theoretical parameter much smaller than all characteristic energy scales or invariants in the problem. For our purposes, it will be convenient to choose this parameter to be an invariant mass, which we will call $s_{\rm min}$. We must split the phase space integrations in such a way that all singular parts of the matrix element lie inside the region of small $s_{\rm min}$. We must also choose $s_{\rm min}$ small enough that the inner region lies completely inside the experimental jet cuts. In the example of the two-jet calculation, we must ensure that no three-jet configuration lies inside the small-$s_{\rm min}$ region; this will be the case so long as $s_{\rm min}$ is smaller than the minimal invariant mass of any two partons in a leading-order three-jet configuration.

Inside the small-$s_{\rm min}$ region, we perform the integrations over the singular parts of real phase space analytically, in $4 - 2\epsilon$ dimensions, but leaving the hard phase to be done numerically. Outside, all integrations be performed numerically. We then add the pieces computed inside the small-$s_{\rm min}$ region to the (renormalized) virtual corrections; the $1/\epsilon^2$ and $1/\epsilon$ poles will cancel if all colored particles are in the final state. (Mass factorization will remove the remaining singularities for colored initial-state particles.) We can then take the limit $\epsilon \to 0$, to obtain expressions which can be evaluated numerically. For these virtual+singular pieces, we must perform the hard phase-space integrals numerically; here, each jet will correspond to one *resolved* parton ($s_{\rm min}$ being the parton resolution parameter). For the remaining real-emission phase space, the expressions are of course finite in four dimensions, so we can perform the integrations numerically. We must integrate both over the 'hard' phase space and the region outside the theoretical dividing parameter $s_{\min}$, but inside the two-jet region as defined by the jet algorithm. The analytic integration of the singular pieces will leave behind $\ln^2(s_{\min})$ and $\ln(s_{\min})$ terms, which of course reproduce the required infrared-singular pieces in the limit $s_{\min} \to 0$. The numerical integration of the remaining real-emission pieces produces similar logarithms, but with opposite sign, so that (up to statistical errors in the Monte Carlo integration) the final results for cross sections and distributions will not depend on $s_{\min}$, but will contain a logarithmic dependence only on the jet cuts (y_{min}, for example, in the e^+e^- JADE algorithm, E_{Tmin} or ΔR in Snowmass-type algorithms used in hadron-collider experiments).

In principle, we want to compute the limit of the sum of these two pieces as $s_{\min} \to 0$. In performing the analytic integrations inside the $s_{\min}$ boundary, we will want to use soft or collinear approximations for the matrix element, and will want to drop terms that are of $\mathcal{O}(s_{\min})$, which are not universal. This introduces errors of that order, which may be minimized by taking $s_{\min}$ as small as possible. In contrast, the numerical integrations for the region outside the $s_{\min}$ boundary will most typically be done by Monte Carlo integration, which introduces statistical errors into the calculation. These get worse as $s_{\min}$ gets smaller, because one gets a cancellation of larger and larger numbers (in absolute value) producing the final answer, and thus are minimized by choosing $s_{\min}$ to be not too small. In practice, one searches for a value of $s_{\min}$ where the statistical errors are not too large, but the net answer is numerically nearly independent of $s_{\min}$ (as it should be in the limit).

3. Singular Parts of the Matrix Element

In order to perform the analytic integration over the singular phase space inside the $s_{\min}$ boundary, we must first extract the singular parts of the matrix element. If one simply takes the squared matrix element for the desired process, and attempts to do this, the exercise is quite laborious even for the simplest processes, and yields poles in many different corners of phase space. Integrating all these poles, and stitching together the different pieces is also quite complicated.[7]

Fortunately, it turns out that it is possible to summarize the singular behavior, integrated over singular phase space, of all tree-level QCD matrix elements, in a set of process-independent functions. This is possible due to the factorization (in the precise sense to be explained below) of the QCD matrix element in the relevant limits, the soft-gluon and collinear-parton limits.

Consider first the collinear region, $k_i \parallel k_j$ for (say) two final-state momenta. In this region, the tree-level QCD squared matrix element as a whole factorizes,

$$|\mathcal{M}_{n+1}|^2 \stackrel{s_{ij}\to 0}{\longrightarrow} g^2 \frac{1}{s_{ij}} P(z)\, |\mathcal{M}_n|^2 + \text{non-singular}\,, \tag{1}$$

where $P(z)$ is the Altarelli-Parisi splitting function. In QED, the same would also be true in the region where one vector boson becomes soft, $E_i \to 0$; but in QCD the structure is more complicated. It can be understood in terms of colored currents,[8] but it is much better to proceed by first performing a color decomposition.

At tree level, one can write the full tree-level amplitude $\mathcal{A}_n$ for a pure-gluon process in terms of color-ordered partial amplitudes A_n,

$$\mathcal{A}_n\left(\{k_i,\lambda_i,a_i\}\right) = g^{n-2} \sum_{\sigma\in S_n/Z_n} \mathrm{Tr}\left(T^{a_{\sigma(1)}}\cdots T^{a_{\sigma(1)}}\right) A_n\left(k_{\sigma(1)},\lambda_{\sigma(1)};\ldots;k_{\sigma(n)},\lambda_{\sigma(n)}\right), \tag{2}$$

where the T^a are color charge matrices, and S_n/Z_n is the set of non-cyclic permutations. Similar formulæ hold for amplitudes with external quarks; indeed, the partial amplitudes for the two kinds of processes are related by supersymmetry, given that the color factors are stripped off.

The partial amplitudes A_n are not fully Bose-symmetric; the Bose symmetry is recovered only after summing over permutations. The utility of this decomposition becomes apparent when we consider the factorization properties of the partial amplitudes. In the soft limit, they factorize in a simple, QED-like manner,

$$A_n(k_1,\lambda_1;\ldots;k_n,\lambda_n) \stackrel{k_n\text{ soft}}{\longrightarrow} s(k_{n-1},k_n,k_1)\,A_{n-1}(k_1,\lambda_1;\ldots;k_{n-1},\lambda_{n-1}) \tag{3}$$

where ignoring phases,

$$|s(k_{n-1},k_n,k_1)| = \sqrt{\frac{k_{n-1}\cdot k_1}{2\,k_{n-1}\cdot k_n\,k_n\cdot k_1}}\,. \tag{4}$$

The A_n also factorize in the collinear limit, of course.

To understand how universal functions arise in this formalism, it is easiest to consider the leading-color parts of the squared matrix element; the structure is more complicated for the subleading-color terms, but the same universal functions can be used in them as well. (See ref. 2 for explicit examples.) If we square the matrix element, and sum over colors, then

$$\sum_{\text{colors }N_c} |\mathcal{A}_{n+1}(k_1\cdots k_{n+1})|^2 = g^{n-1}N_c^{n-1}(N_c^2-1) \sum_{\sigma\in S_{n+1}/Z_{n+1}} \left|A_{n+1}(k_{\sigma(1)}\cdots k_{\sigma(n+1)})\right|^2 + \text{subleading color}\,; \tag{5}$$

If we take one gluon (say the last) to be soft, each permutation factorizes,

$$\sum_{\text{colors }N_c} |\mathcal{A}_n(k_1\cdots k_n)|^2 \stackrel{k_n\text{ soft}}{\longrightarrow} g^{n-1}N_c^{n-1}(N_c^2-1) \sum_{\sigma\in S_n/Z_n} s_F(\sigma)\,|A_n(\sigma)|^2 + \text{subleading}\,; \tag{6}$$

where the permutation soft factor is just the cyclic sum of the square of factors given in equation (4). The color decomposition also guides us in slicing up phase space; it turns out to be convenient to define soft and collinear phase space in terms of color-adjacent invariants, two color-adjacent invariants smaller than $s_{\rm min}$ corresponding to the 'soft' region, and a lone invariant smaller than $s_{\rm min}$ the 'collinear' region. (The boundary between the two is arbitrary.) Integrating the permutation soft factor over the soft phase space then produces the universal function summarizing the soft behavior of QCD amplitudes,

$$\frac{\alpha_s N_c}{2\pi}\frac{1}{\Gamma(1-\epsilon)}\frac{1}{\epsilon^2}\left(\frac{4\pi\mu^2}{s_{\rm min}}\right)^{\epsilon}\sum_i\left(\frac{s_{i,i+1}}{s_{\rm min}}\right)^{\epsilon}. \tag{7}$$

Analogous functions can be constructed for the 'collinear' regions.

The virtual corrections can be separated into singular and non-singular pieces; the singular pieces have a structure similar to that of the tree matrix element. We can then combine all the singular pieces, soft S_F, collinear C_F and virtual V,

$$\sum_{\sigma\in S_n/Z_n} \left(S_F(\sigma)+C_F(\sigma)+V(\sigma)\right)\left|A^{\rm tree}(\sigma)\right|^2 + \text{ finite virtual } + \text{ subleading color}\,. \quad (8)$$

Because of the KLN theorem on cancellation of infrared divergences, the combination inside the parentheses is finite as $\epsilon \to 0$; we may thus take the limit to obtain a quantity which can be evaluated numerically. Note that the tree amplitude multiplying it does *not* need to be evaluated in $4-2\epsilon$ dimensions, but only in four dimensions.

To incorporate initial-state radiation, it is convenient to proceed in a crossing-invariant fashion. One first treats all colored partons as outgoing, and cancels all singularities as in equation (8). Doing so, however, treats the collinear momentum fraction $z = E_i/(E_i+E_j)$ incorrectly for the initial state legs, the fused leg rather than the (external) initial-state leg carrying the sum of the momenta. We must subtract the error from our result; this introduces a new singular piece, of precisely the correct form to be removed by mass factorization into the initial-state parton distributions; and new finite terms, the crossing functions, which are essentially convolutions of the parton distributions with functions of z. The crossing functions are again universal objects, and need be computed only once for a given set of parton distributions.

References

1. W. T. Giele and E. W. N. Glover, Phys. Rev. D46:1980 (1992).
2. W. T. Giele, E. W. N. Glover, and D. A. Kosower, Nucl. Phys. B403:633 (1993).
3. CDF Collaboration, preprints Fermilab–Pub–94/097–E and Fermilab–Conf–94/152–E.
4. F. Abe et al., CDF Collab., Phys. Rev. Letts. **68**, 1104 (1992).
5. W. L. van Neerven and J. A. M. Vermaseren, Nucl. Phys. B238:73 (1984).
6. W. L. van Neerven, J. Smith, and J. A. M. Vermaseren, Z. Phys. C30:621 (1986); H. Baer, J. Ohnemus, and J. Owens, Phys. Rev. D40:2844 (1989); D42:61 (1990); Phys. Lett. B234:127 (1990).
7. S.D. Ellis, Z. Kunszt, and D.E. Soper, Phys. Rev. **D40**, 2188 (1989).
8. F. Fiorani, G. Marchesini, and L. Reina, Nucl. Phys. B309:439 (1988).

SUBTLETIES AND FANCIES IN GAUGE THEORY NON TRIVIAL VACUUM [1]

A. R. LEVI
Department of Physics, Boston University
590 Commonwealth Avenue, Boston, MA 02215

Abstract

The one loop effective potential for a non-Abelian gauge configuration is analyzed using the background field method. The Savvidy result and the non-Abelian ansatz, the other alternative possible background that generates a constant color magnetic field configuration, are compared. This second possibility is very interesting because it avoids the possible coordinate singularity, $\mathrm{Det}B_i^a = 0$, and it is easy to implement in lattice simulations. We emphasize the interesting dependence of the potential by the gauge fixing parameter α, when the loop expansion is performed around a non trivial background configuration. Finally, we point out some crucial differences in analyzing the vacuum structure between non-Abelian gauge theories and the cases of scalar and Abelian gauge theories.

1. Introduction

Even after many years of intensive efforts, the vacuum structure of non-Abelian gauge theories still remains elusive to us. For a recent article where several aspects of the non-trivial vacuum for these theories have been analyzed see [1], and for a comprehensive revue see [2].

As a prelude to truly non-perturbative evaluation of the effective I.R. structure of non-Abelian theories in terms of lattice, the conventional approach is to analyze the effective action using the so-called background field method.

For scalar and Abelian gauge theories, we can assert that we understood very well their I.R. structures. In these cases, the vacuum is not ill-defined and this allows the construction of the complete physical space of the theory by acting with creator operators on the vacuum state. This procedure it is not ambiguous and it is formally well understood.

However, the situation of the non-Abelian case is very different, and very little has to do with the Abelian and scalar cases. For non-Abelian theories, there is no simple way to decompose the field into non-interacting "field oscillators"; they are non-linear and coupled together. To be more precise, there are two different regions: for large Euclidean momenta the effective coupling constant $g^2(p)$ is relatively small, and the perturbative picture of the plane gluonic waves is reasonable. Many perturbative results can be achieved in this region. For example, the reason why deep-inelastic scattering gives results which are so clear is not only because in high momenta transfer

[1]This work is supported in part by funds provided by the U. S. Department of Energy (D.O.E.) under contracts #DE-FG02-91ER40676

the gluonic gauge coupling is small, but also because the vacuum is so far away from the measured states with high momenta that the experimental data are independent of the vacuum structure.

The real problem is only with the momenta of the order of 1 GeV or smaller, where the infrared structure of the field theory shows up. For small momenta, the coupling is strong and it is not reasonable to assume that the non-Abelian gauge fields are distributed in space-time as plane waves. This suggests a classification of the vacuum models. The first model assume that the non-perturbative field is concentrated in some space-time region, which is the instanton-type vacuum. If it is concentrated in space, but on a line along time direction, is called a soliton-type vacuum. If it is concentrated on a two-dimensional surface, it is called a string-type vacuum. However, there is also the possibility that there is a homogeneous-type vacuum that will be referred to as the "non-trivial vacuum" which we will discuss here.

The SU(2) case should not be fundamentally different from other non-Abelian gauge theory. In fact, up to now, there is no reason why the QCD vacuum should not be in the same universality class of the SU(2) case. On the other hand, SU(2) allows explicit calculations that are still inaccessible for the SU(3) case.

Beside lattice computations, the only available method to investigate the properties of Yang-Mills theories vacuum is computation of the effective potential in the presence of a background gauge [3]. This manifest gauge invariant scheme is based on the observation that the loop expansion corresponds to an expansion in the parameter $\hbar$ which multiplies the entire action. Hence, a shift of the fields, or a redefinition of the division of the Lagrangian into free and interacting parts can be performed at any finite order of the loop expansion without violating the manifest gauge invariance.

Analogous conclusions can be obtained by studying the beta function. This is not surprising because the beta function is dependent on the ultraviolet part of the effective potential.

Further note that this procedure is straightforward only if the effective potential does not contain operators which are irrelevant at the ultraviolet fixed point but become relevant in the infrared. In this case, the β-function by construction, depends only on the ultraviolet relevant coupling constants and knowledge of it is not sufficient to investigate the infrared properties of the effective potential.

In doing the one-loop calculation, we encountered several technical points, which were known in the literature but not emphasized enough. We specially want to stress the gauge fixing α parameter dependence. Therefore, we decided to briefly go through the standard background field method in gauge theories with those technical points in mind.

2. Effective Potential Formalism

Before we launch an extensive numerical simulation, it is still worthwhile to calculate the one-loop effective potential for the non-Abelian background field Eq.(21). Even though it is not totally trustworthy, the loop expansion can still provide indicative information. Furthermore, to examine whether the qualitative feature of the one-loop effective potential strongly depends on the choice of the background field is also interesting. In particular, we would like to find out whether the coordinate singularity has any connection with the existence of the imaginary part in the effective potential. We will prove that the imaginary part cannot be eliminated by avoiding the coordinate singularity of $\mathrm{Det}B_i^a = 0$.

We will now outline the basic steps in the background field method in gauge theories [3]. There are two reasons to go through the well-known method. The first is to establish our own notations. The second is to emphasize two technical aspects; the question of whether it is necessary to require the background field to satisfy the classical equation of motion and the gauge choice in the evaluation of the functional integral. When possible we follow the convention of Abbott [3].

The generating functional in the background field method (in Euclidean space) is defined throughout

$$\tilde{Z}[J,A] = \int [dQ] \det\left[\frac{\delta G^a}{\delta\omega^b}\right] \exp\left\{-\int d^dx[\mathcal{L}(A+Q) + J^a_\mu Q^a_\mu]\right\} \prod_{x,a} \delta[G^a], \qquad (1)$$

where $G^a \equiv \partial_\mu Q^a_\mu + g f^{abc} A^b_\mu Q^c_\mu = D^{ab}_\mu(A) Q^b_\mu$ is the background field gauge condition and $\det\left[\delta G^a/\delta\omega^b\right]$ is the corresponding Jacobian. It is clear from the above definition the background field A^a_μ is fixed and should be regarded as an external parameter in the process of doing the Q–integral. Note also that we do not exponentiate the gauge constraint $\prod_{x,a}\delta[G^a]$, in contrast with what was usually done. The necessity of enforcing the gauge constraint with explicit delta-functions in Eq.(1) will be discussed in detail.

It is easy to verify that $\tilde{Z}[J,A]$ is invariant under the infinitesimal gauge transformations

$$\delta A^a_\mu = -f^{abc}\omega^b A^c_\mu + \frac{1}{g}\partial_\mu\omega^a\,; \qquad (2)$$

$$\delta J^a_\mu = -f^{abc}\omega^b J^c_\mu\,. \qquad (3)$$

As a consequence of this invariance, the effective action in the background field method, defined as the Legendre transform of $\tilde{Z}$:

$$\tilde{\Gamma}[\tilde{Q},A] = -\ln\tilde{Z}[J,A] + \int d^dx\, J^a_\mu \tilde{Q}^a_\mu\,, \qquad (4)$$

with $\tilde{Q}^a_\mu = \delta\ln\tilde{Z}/\delta J^a_\mu$, is invariant under

$$\delta A^a_\mu = -f^{abc}\omega^b A^c_\mu + \frac{1}{g}\partial_\mu\omega^a\,; \qquad (5)$$

$$\delta\tilde{Q}^a_\mu = -f^{abc}\omega^b\tilde{Q}^c_\mu\,. \qquad (6)$$

As shown by Abbott [3], when $\tilde{Q}^a_\mu = 0$, $\tilde{\Gamma}[0,A]$, it coincides with the usual effective action. Since Eq.() is a pure gauge transformation when $\tilde{Q}$ vanishes, $\tilde{\Gamma}[0,A]$ must be a gauge invariant functional of A^a_μ.

A standard way to evaluate $\tilde{Z}[J,A]$ explicitly is to make a loop expansion. For the purpose of convenience, we exponentiate the Jacobian $\det[\delta G^a/\delta\omega^b]$ and gauge constraints $\prod_{x,a}\delta[G^a]$ by introducing the Faddeev-Popov ghost fields (θ^a and $\bar{\theta}^a$) and a real scalar auxiliary field σ^a, respectively,

$$\begin{aligned}\tilde{Z}[J,A] &= \int [dQ][d\theta][d\bar{\theta}][d\sigma]\exp\Big\{-\int d^dx[\mathcal{L}(A+Q)+\bar{\theta}^a D_\mu^{ac}(A)D_\mu^{cb}(A)\theta^b + \\ &\quad +2i\sigma^a D_\mu^{ab}(A)Q_\mu^b + J_\mu^a Q_\mu^a]\Big\}. \end{aligned}\tag{7}$$

If we only want to calculate $\tilde{Z}[J,A]$ to one-loop order, then it is equivalent to evaluating Eq.(7) in the steepest descent approximation,

$$\begin{aligned}\tilde{Z}[J,A] &\approx \tilde{Z}_1[J,A]\equiv\int[dQ][d\theta][d\bar{\theta}][d\sigma]\exp\Big\{-\int d^dx[\mathcal{L}(A)+\mathcal{L}^{(1)}(A,Q)+ \\ &\quad \mathcal{L}^{(2)}(A,Q)+\bar{\theta}^a D_\mu^{ac}(A)D_\mu^{cb}(A)\theta^b+2i\sigma^a D_\mu^{ab}(A)Q_\mu^b+J_\mu^a Q_\mu^a]\Big\}, \end{aligned}\tag{8}$$

where

$$\mathcal{L}^{(1)}(A,Q) = -Q_\mu^a D_\mu^{ab}(A)F_{\mu\nu}^b(A)\,; \tag{9}$$

$$\mathcal{L}^{(2)}(A,Q) = Q_\mu^a\Big[-\frac{1}{2}\left(D_\rho(A)D_\rho(A)\right)^{ab}\delta_{\mu\nu}+igF_{\mu\nu}^{ab}(A)\Big]Q_\nu^b \equiv Q_\mu^a M_{\mu\nu}^{ab}(A)Q_\nu^b\,. \tag{10}$$

In arriving at the expression of $\mathcal{L}^{(2)}(A,Q)$, the gauge constraint condition $D_\mu^{ab}Q_\mu^b=0$ has been used.

We can now do the Q–integral in $\tilde{Z}_1[A,J]$ and make the Legendre transform as in Eq.(4). By setting $\tilde{Q}_\mu^a=0$ the final expression for the effective action to one-loop order can be written as $\Gamma[A]=-\ln\tilde{Z}_1[A]$, with

$$\begin{aligned}\tilde{Z}_1[A] &= e^{-\int d^dx\mathcal{L}(A)}\cdot\det[-D(A)D(A)]\cdot \\ &\quad \int[dQ][d\sigma]\exp\Big\{-\int d^dx[Q_\mu^a M_{\mu\nu}^{ab}(A)Q_\nu^b+2i\sigma^a D_\mu^{ab}(A)Q_\mu^b]\Big\} \\ &= e^{-\int d^dx\mathcal{L}(A)}\cdot\det[-D(A)D(A)]\cdot\Big\{\det M_{\mu\nu}^{ab}\cdot\det N^{ab}\Big\}^{-1/2}, \end{aligned}\tag{11}$$

with $N^{ab}=-D_\mu^{ac}(M^{-1})_{\mu\nu}^{ce}D_\nu^{eb}$. The first factor is the classical contribution. The second and third factors are the one-loop quantum corrections due to the ghost field (θ) and fluctuation field (Q), respectively. Notice that the linear term $\mathcal{L}^{(1)}(A,Q)$ drops out automatically in the calculation process. To remove the linear term there is no need to require that the background field satisfy the classical equation of motion $D_\mu^{ab}(A)F_{\mu\nu}^b(A)=0$. This independence of the linear term remains true to all orders in loop expansion, because the effective potential is a sum of all one-particle irreducible diagrams with A fields on the external lines and Q fields on internal lines (when $\tilde{Q}_\mu^a=0$). In fact, the linear term is always compensated by the source J_μ^a. Remember that the physical limit in the background field method is $\tilde{Q}_\mu^a=0$, not $J_\mu^a=0$. On the other hand, it should be pointed out that to require the background field to satisfy the equation of motion, in order to have a good chance to represent the true minimum of the action, is an entirely different matter.

Since the background field is a constant in space-time it is convenient to work in momentum-space. Using the formula $\det M = \exp(\mathrm{Tr}\ln M)$ combined with Eq.(11), we have the expression for the effective potential to one-loop order

$$V_{\rm eff}(h) = \mathcal{V}_{classical} - \int \frac{d^d p}{(2\pi)^d} \ln G(h;p) + \frac{1}{2}\int \frac{d^d p}{(2\pi)^d}\Big\{\ln M(h;p) + \ln N(h;p)\Big\}. \quad (12)$$

An eventual non-positive definiteness of $M(h;p)$ and $N(h;p)$ for small p^2 indicates that the background field is only a saddle point in configuration space and therefore is not stable.

Eq. (12) must be regularize to eliminate divergent constants. The standard method is to use the regularization of Salam and Strathdee [4], which is a variation of Schwinger's proper time method [5]. Peculiar tricks have to be implemented depending in how many dimensions the calculation is done. For example, because the three dimensional Yang-Mills theory is super renormalizable, the only divergence we encounter is an overall additive constant. The regularization procedure includes three steps. First, an integral representation for logarithmic function is used. For real E, positive or negative,

$$\ln(E - i\delta) = \frac{1}{\epsilon} - \frac{i^\epsilon}{\epsilon\Gamma(\epsilon)}\int dt\, t^{\epsilon-1}\, e^{-it(E-i\delta)}, \quad (13)$$

in the limit of $\epsilon \to 0^+$. The $i\delta$ in Eq.(13) with $\delta \to 0^+$ is to ensure the convergence of the integral. When E is explicitly complex, Eq.(13) can be easily generalized. Then the momentum integration can be done using

$$\int \frac{d^3 p}{(2\pi)^3} e^{-itp^2} = \frac{1}{8\pi^{3/2}}(it)^{-3/2}. \quad (14)$$

The remaining t–integral can be converted into a Gamma function through the contour integral technique. Finally, the limit of $\epsilon \to 0$ is taken.

At this point we would like to comment on the implementation of the gauge condition $\prod_{x,a}\delta[G^a]$ in Eq.(1). The standard approach is to exponentiate this factor,

$$\tilde{Z}[J,A;\alpha] = \int [dQ]\,\det\Big[\frac{\delta G^a}{\delta\omega^b}\Big]\exp\Big\{-\int d^d x[\mathcal{L}(A+Q) - \frac{1}{2\alpha}(G^a)^2 + J^a_\mu Q^a_\mu]\Big\}, \quad (15)$$

where α is the gauge fixing parameter. In order to write Eq.(1) into the form of Eq.(15), one has to generalize the gauge condition as $\prod_{x,a}\delta[G^a - f^a]$, with f^a being arbitrary, and then to integrate out f^a with the weighting factor $\exp[-(f^a)^2/2\alpha]$. If $\tilde{Z}[J,A]$ with the generalized gauge condition were independent of f^a, as in the case of zero background field $A^a_\mu = 0$, $\tilde{Z}[J,A;\alpha]$ would be independent of the gauge parameter α and therefore the exact equivalence between Eq.(1) and Eq.(15). However, in the presence of non-vanishing background field A_μ^a, $\tilde{Z}[J,A]$ in general would depend on f^a, as can be easily seen in Eq.(7) by adding a term $2i\sigma^a f^a$ to the exponential. Physically, this dependence on f^a is natural, because there would exist correlations between f^a and certain combinations of the background field, such as $(D_\mu^{ab}(A)f^b)^2$ when A_μ^a is non-vanishing. If indeed $\tilde{Z}[J,A]$ depended on f^a, then the integral

$$\tilde{Z}[J,A;\alpha] \equiv \int [df]\, \tilde{Z}[J,A]\Big|_{G^a=f^a} \cdot \exp\Big\{-\frac{1}{2\alpha}(f^a)^2\Big\} \tag{16}$$

would have to depend on α in general, which in turn implies that Eq.(15) could not be equivalent to Eq.(1) at all times.

This gauge dependence of the effective action calculation has long been recognized in the literature [6]. It was argued by Vilkovisky that the correct choice of the gauge parameter is to take $\alpha \to 0$ limit, or to choose the Landau background field gauge. Since $\exp[-x^2/2\alpha] \propto \delta(x)$ in the limit of $\alpha = 0$, Eq.(15) is formally equivalent to Eq.(1) in the Landau background field gauge. It should be emphasized that Eq.(15) would still define a gauge invariant effective action with respect to the background field A_μ^a. It is only the functional form of the effective action that depends on the gauge parameter when $\alpha \neq 0$. Since the β–function is gauge (or α) independent in the dimensional regularization and minimal subtraction scheme [7], even though the finite part is explicitly α dependent, this has generated much confusion. Finally, note that in the context of finite temperature QCD, this gauge dependence has been emphasized by Hansson and Zahed [8].

3. Savvidy vs non-Abelian

Because the effective potential is the truncation of the effective action at zero momenta, we are seeking solutions that give a constant magnetic field configurations. There are only two possibilities called, respectively, the Savvidy (or Abelian), and the non-Abelian ansatz. It has been proved [9] that there are no other ways to generate a constant magnetic field.

We will first analyze the Savvidy's case. As in Savvidy's and Matinyan's pioneer work [10], where they calculated up to one-loop order the effective potential for an Abelian background gauge field:

$$A_\mu^a(x) = \frac{1}{2} H \delta^{a3}(x_1\delta_{\mu 2} - x_2\delta_{\mu 1})\,, \tag{17}$$

which generates a constant color magnetic field $B_i^a = H\delta_{i3}\delta_{a3}$ in the SU(2) Yang-Mills theory. It was then found, remarkably, that the vacuum with this background field is energetically favored over the perturbative vacuum ($H = 0$). Unfortunately, a more careful analysis [11] soon revealed that there exists an imaginary part in the effective potential, indicating that the background field Eq.(17) is not a minimum but rather a saddle point in the configuration space in the context of the loop expansion. The subsequent work by the Copenhagen group [12], still within the framework of loop expansion, tried to remedy the physically appealing picture of Savvidy's vacuum by introducing inhomogeneity, leading to the so-called Copenhagen vacuum.

For the Savvidy ansatz, it is easy to solve and find the eigenvalues of the second derivative of the action, because is exactly the same operator of the Landau levels problem, which has similar symmetry. In fact, the corresponding M matrix is simply the Landau diamagnetic Hamiltonian. Therefore, their eigenvalues are:

$$\lambda(k) = k^2 + (2n+1)\, gH + 2gH\; S_z \tag{18}$$

where $k = k_0^2 + k_z^2$, $n = 0, 1, 2, ...$ and $S_z = \pm 1$. As expected from the symmetry of the problem, the modes in the x, y directions are quantized. However, the behavior of the modes along the z axis are labeled by an arbitrary continuous parameter, k_z. Moreover, note the difference from the Landau level where the quantum number S_z takes value $S_z = \pm 1/2$, which is the electron polarization; whereas, in this case, we have the polarization of the photon that give $S_z = \pm 1$, allowing the possibility of negative eigenvalue when $(S_z = -1, n = 0)$

Using the technique discussed in the previous section, it is possible to evaluate the effective potential. For the reader interested in an accurate discussion of the eigenfunction and of their multiplicity see [13]. For the Savvidy case, in (3+1) space-time dimensions the effective potential up to one loop contributions is given by

$$V(H) = \frac{1}{2} H^2 + \frac{11}{48\pi^2}\, g^2 H^2 \Big(\ln \frac{gH}{\mu^2} - \frac{1}{2} \Big) - i\, \frac{1}{8\pi}\, g^2 H^2 + \ldots . \tag{19}$$

It is also interesting to look at the 3 dimensional case. Because the theory in 3 dimensions is superrenormalizable, we should better control the full procedure. Moreover, Lattice simulations can be obviously performed easily, both because of the volume, and because the scaling window is very large due to the polynomial dependence of the renormalizable quantity throughout the β function, in contrast with the well known exponential dependence of the 4 dimensions. Therefore, several authors recently analyzed the 3 dimensional case both analytically and with extensive lattice simulations [15]. In 3 space-time dimensions, for the Savvidy ansatz, we have:

$$V(H) = \frac{1}{2} H^2 - \frac{1}{2\pi^2} \left[1 - \frac{\sqrt{2}-1}{4\pi} \zeta\Big(\frac{3}{2}\Big) \right] (gH)^{3/2} + i\, \frac{1}{12\pi}\, (gH)^{3/2} + \ldots , \tag{20}$$

where ζ is the Riemann's Zeta-function.

The non-Abelian ansatz give a similar result in the structure, but different results in the coefficients. Moreover, the physical picture is not the same of Landau levels, but is the same as a system with momenta coupled with spins. The Non-Abelian ansatz is:

$$A_0^a(x) = 0, \quad \text{and} \quad A_i^a(x) = h\delta_i^a\,, \tag{21}$$

where h is a constant in space-time. This choice was first considered in a work of Ambjørn, Nielsen and Olesen [16]. In respect to the Savvidy, the non-Abelian has the crucial vantage of being naturally implementable in lattice simulations. Moreover, recently, in an unrelated but very interesting work by Johnson and his collaborators [17], the Schrödinger functional approach in terms of magnetic field strength is applied to the SU(2) Yang-Mills gauge theory. In the explicit Schrödinger functional Hamiltonian they find that there is a factor $1/\mathrm{Det}B_i^a$ in the kinetic energy term, similar to the $1/r$ factor in the quantum mechanics Schrödinger equation in polar coordinate. This factor may signal a potential coordinate singularity for the color magnetic field in the configuration space, similar to that of the wavefunctions in quantum mechanics which have to satisfy certain boundary conditions at $r = 0$. Since the Savvidy ansatz yields $\mathrm{Det}B_i^a = 0$, due to its intrinsic Abelian nature, it is desirable to seek an alternative background field, which avoids this potential coordinate singularity. The corresponding color magnetic field for the non-Abelian ansatz is simply given by $B_i^a = gh^2\delta_i^a$ and obviously has $\mathrm{Det}B_i^a \neq 0$ when $h \neq 0$. It is easy to recognize that the constant magnetic field is generated by Eq.(21) through the commutator terms in the field strength, rather than the derivative terms as in the Savvidy case. Therefore, due to the non-trivial correlation between its directions in space and color space, the background field Eq.(21) evades the possible coordinate singularity.

In addition, since the non-Abelian background is space-time independent, it is much easier to put it on a lattice than the non-constant background field Eq.(17). For example, periodic boundary conditions are automatic.

For the non-Abelian case the second derivative of the action in respect to the gauge fields give the operator

$$M_{\mu\nu}^{ab} = \frac{1}{2}\Big[(p^2 + 2g^2h^2)\delta^{ab}\delta_{\mu\nu} + 2gh\,(\sigma_{ab})^c p_c \delta_{\mu\nu} - 2g^2h^2\,(\sigma^{ab})_c(\sigma_{c\mu})_\nu\Big], \tag{22}$$

which is a 9×9 matrix for a given momentum p. This matrix can be interpreted as the Hamiltonian of a relativistic spin-1 ($\vec{\sigma}_s$) and color spin-1 ($\vec{\sigma}_c$) boson, with the first term being the free particle part, the second the $\vec{p}\cdot\vec{\sigma}_c$ and the third $\vec{\sigma}_s\cdot\vec{\sigma}_c$. The last two terms can be negative for low momentum, depending on the relative orientations of $\vec{p}$, $\vec{\sigma}_s$ and $\vec{\sigma}_c$.

For the non-Abelian ansatz, in (3+1) dimensions the effective potential is

$$V(H) = \frac{1}{2}H^2 + \frac{21}{48\pi^2}g^2H^2\Big(\ln\frac{gH}{\mu^2} - \frac{1}{2}\Big) + i\,\frac{3}{50\pi}\,g^2H^2 + \ldots, \tag{23}$$

and in 3 dimension (see [1, 15, 18]), we have:

$$V(H) = \frac{1}{2}H^2 - \frac{2-\sqrt{2}}{3\pi^2}\,(gH)^{3/2} - i\,\frac{5}{6\pi}\,(gH)^{3/2} + \ldots. \tag{24}$$

4. Instability

When for some field configuration there is an imaginary part in the effective action this configuration is called unstable. In this case Unfortunately this is what happens for non-Abelian gauge theories where the expansion is around a saddle point in the functional space. The second derivative operator of the action has some negative eigenvalues and, when loop-expanded, it shows a complex effective potential. Therefore, the validity of the loop expansion in these cases has been questioned by Maiani et al [13]. These authors strongly argued that the calculation of the effective potential in the presence of a background field is actually non-perturbative. When the background is not realizing a true minimum there is no possibility for the existence of any reliable perturbative expansion. So far the only known non-perturbative technique in field theory that can be made systematic is the lattice approach. More recently several works with the lattice approach have appeared in the literature, both in four dimensions [14] and three dimensions [15]. Due to various technical reasons, mainly related to the fact that the Savvidy ansatz in Eq.(17) is non-uniform, all these lattice works have yet to yield a conclusive result.

Because we expand around a field configuration ϕ_c which is a saddle point in the functional space, the second derivative operator of the action will have at least one negative eigenvalue. We use this operator to choose a basis in the vicinity of ϕ_c that has as components the eigenvectors of this operator. Let ϕ^s and ϕ^u designate the stable and the unstable modes; these correspond, respectively, to the positive and negative eigenvalues. Hence, a configuration ϕ in the vicinity of ϕ_c can be written as:

$$\phi \;=\; \phi_c \;+\; \sum_i \, c_i^s \; \phi_i^s \;+\; \sum_i \, c_j^u \; \phi_j^u \tag{25}$$

where the index i runs over the set of stable modes $\{s\}$, and j runs over the set of unstable modes $\{u\}$.

We can hope that after splitting the stable from the unstable modes, we will be able to perform a perturbative expansion in the stable sector and handle non-perturbatively the unstable one. Due to the orthogonality of the stable and unstable modes, the measure is simply defined now as $(d\phi) = (d\phi^s)\,(d\phi^u)$, and we can write $W[J]$ as:

$$\begin{aligned} e^{\frac{i}{\hbar}W[J]} \qquad &= \int (d\phi)\; e^{\frac{i}{\hbar}\{S[\phi]+(J,\phi)\}} \;= \\ = \int (d\phi^s)(d\phi^u)\; &e^{\frac{i}{\hbar}\{S[\phi_c+\phi^u+\phi^s]+(J,\phi_c+\phi^u+\phi^s)\}} \;= \\ = \int (d\phi^u)\; &e^{\frac{i}{\hbar}\{S_{eff}[\phi_c+\phi^u]+(J,\phi_c+\phi^u)\}}\,, \end{aligned} \tag{26}$$

where we have supposed that we were able to handle the integration over the stable modes sector.

As carefully discussed in reference [13], this is not a correct way to proceed. It is true that we were formally able to write the connected Green's functions generating functional $W[J]$. However, the Legendre transformation is now different. In fact, now we are in presence of many solutions, and we must take the Legendre transform, choosing the J that minimizes $W[J] - (J,\phi)$ at fixed $\phi = \phi_c$. Therefore, the effective potential is defined as:

$$\Gamma[\phi_c] \;=\; \min_{\{J\}} \Big[W[J] - (J,\phi) \Big] \tag{27}$$

This definition reduces to the usual expression for stable configurations, but the choices of J may drive the system into a non-perturbative regime for unstable ϕ_c.

In the classical limit $\hbar \to 0$, the integral is still dominated by the saddle point configuration ϕ_0. Now, however, the expression above is no longer valid since there are no restoring forces for the unstable mode. What remains in equation is an effective action

$$S_{eff}[\phi_c + \phi^u] = S[\phi_c + \phi^u] + O(\hbar) \tag{28}$$

which comprises the effects of small fluctuations around $\phi_c + \phi^u$. This is not what we were doing in the previous section when we were took into account the small fluctuations around ϕ_0 looking for

$$S_{eff} = S[\phi_c + \phi^u + \phi^s] + O(\hbar) \tag{29}$$

Thus, we can conclude, following [13], that the problems are that, one, ϕ_0 will definitely not dominate $W[J]$ and, two, the small fluctuations around ϕ_0 and $\phi_c + \phi^u$ have nothing to do with each other.

A consequence of this is that the Legendre transform, evaluated at ϕ_c, deviates from the minimum by a dangerous term which is $O(\phi^u)$. Hence, we no longer have the possibility of using $\hbar$ as a parameter to expand the effective potential at ϕ_0. The non-perturbative effects become dominant and, as a result, non-perturbative techniques, such as the lattice regularization, must be used to find what replaces the one-loop analysis which, by definition, was obtained by the help of the perturbation expansion.

If this argument is correct, then the Savvidy solution, which was obtained perturbatively, no longer provides a good picture. Only a non-perturbative approach is reliable.

On the other hand, if the arguments of references [12] are correct, the Savvidy solution may be a good approximation, in the sense that even if quantum fluctuations alter the vacuum from the original chromomagnetic configuration, the transition might not be too drastic. If the imaginary part is negligible, in first approximation, the vacuum can be thought as microscopic domains of constant chromomagnetic fields. This is analogous to the case of ferromagnetic domains of iron below the Curie temperature. To be more precise, we cannot prove the existence of domains without studying the balance of the energy contributions of the walls and of the domain bulks. Nevertheless, from the fact that in the effective potential a domain scale of constant chromomagnetic field exists we might expect the existence of such domains. Hence, Monte Carlo simulations will be able to detect gH_{min}. However, if the fluctuations do occur very drastically, then the configurations will fluctuate so violently that it will be very difficult to collect evidence of a minimum.

A hybrid scenario might also occur. For intense fields, we might have a non-trivial vacuum, and for weak fields, the trivial vacuum, dominates. This hybrid scenario is the most difficult to investigate, because it is not clear quantitatively when the strong field regime is reached. Moreover, for lattice simulations, this is the region in which the samples are most difficult to collect. At the present stage of lattice simulations

[14, 15] we can not exclude consistence with this hybrid scenario, and it will be extremely important to to clarify this possibility with further analysis.

Note, however, that the existence of the imaginary part is also related to the asymptotic freedom of the pure gauge theory. In fact, in the absence of the imaginary part, the ultraviolet limit of the Savvidy solution implies that the beta function will be the same as the one of a scalar particle of mass $m^2 = 2gH$. It is just the imaginary part which prevents us after regulation from rotating the integration contour in the complex plane and thereby spoiling the asymptotic freedom.

5. Comparison with Coleman and Weinberg Potentials

In the discussion of the radiatively induced symmetry breaking, [19], the question of the trustworthiness of the non-trivial vacua is also addressed. In particular, Coleman and Weinberg found a non-trivial minimum in the one-loop effective potential of the $\lambda\phi^4$ theory at

$$\lambda \log \frac{\phi_c^2}{M^2} = -\frac{32}{3}\pi^2 + O(\lambda) \tag{30}$$

but they noted immediately that the effective potential is not calculable at this minimum in the loop expansion. In fact, higher loop contributions bring higher powers of $\lambda \log \frac{\phi_c^2}{M^2}$ and then, independently of how small the coupling constant is, the new minimum will lie outside the range of validity of the one-loop approximation.

In the same article, the case of the scalar electrodynamics theory was also addressed. In this case, the effective potential is found to be

$$V = \frac{3\, e^4}{64\pi^2}\, \phi_c^4 \log\left(\frac{\phi_c^2}{<\phi>^2} - \frac{1}{2} \right) \tag{31}$$

In this case the non-trivial vacuum must be trusted, the loop expansion can be trusted at the minimum for the appropriate choice of the coupling constants because the higher order terms in λ can be balanced by the contributions of the order $e^4 \log \frac{\phi_c}{M}$.

However, there is a fundamental difference between these cases and the non-Abelian gauge theories. For the above cases, the effective potential can be evaluated directly because there is no instability, while the opposite is true of the non-Abelian gauge theory. Consider for example the $\lambda\phi^4$ theory in the broken phase ($\mu^2 < 0$). A possible source of instability comes from the imaginary part of the contribution of the effective potential, given by:

$$\int dk \, \log(k^2 + V''(\phi_c)\,) = \int dk \, \log(k^2 + \mu^2 + \frac{\lambda}{2}\phi_c^2\,) \tag{32}$$

when the logarithm is integrated for small k. However, because $\phi_c = \phi_c[J]$ it is always possible to find a set of values of J sufficiently large for which there is a region where $\frac{\lambda}{2}\phi_c^2 > \mu^2$ and then it is possible to perform the integration without the imaginary part. Obviously, the existence of this region is not enough: we must reach the limit $J \to 0$ while remaining in the stable region. This is possible here by reaching the minimum via $\phi_c[J_1] > \phi_c[J_2 < J_1] > \phi_c[J = 0]$, while remaining away from the unstable region, that is $\phi < \frac{1}{2}\phi_c[J = 0]$. The lesson of this simple observation is that it is the quartic part, ϕ^4, of the effective potential which stabilizes the vacuum.

In non-Abelian gauge theories the situation is different. The same analysis might be performed with the vocabulary:

$$- \mu^2 \longleftrightarrow 2gH \tag{33}$$

$$\frac{\lambda}{2}\phi_c^2 \longleftrightarrow 6gA_c^2 \tag{34}$$

The first identification comes by inspection of the ultraviolet part of the effective potential, or equivalently, by neglecting the imaginary part and comparing the beta function with the one of the scalar theory. The second correspondence comes directly from the quartic part of the action.

For the non-Abelian theory the equivalent contribution of (32) is:

$$\log(\ k^2 \ - \ gH\ ) \tag{35}$$

and not the "safe"

$$\log(\ k^2 \ - \ gH \ + \ 6gA_c^2\ ) \tag{36}$$

because the last term is not gauge invariant. Hence, there is no region with stable minimum, as opposed to the scalar field theory. So unstable fluctuations must always be present.

This also teaches us that variational methods will not work for the vacuum properties of non-Abelian gauge theories. The variational method is very useful and powerful in the case of stable configurations. However, in my opinion, to use the variational method to analyze situations where unstable configurations are dominant is extremely hazardous.

The next natural question is if higher order contributions stabilize the instability. Unfortunately, we must perform a partial resummation of the loop expansion to clarify this point for non-Abelian models. At the present there is no evidence for a window for J where the argument of the logarithm could be made positive.

Consider now the case of SU(2) coupled to a fermion. In this case, there is a contribution to the one-loop effective potential from the graph with four gluonic legs connected by a fermion loop. The contribution of this diagram is analogous to the one in the Abelian case (see [20]), and for low momentum transfer, will give an extra term in the effective potential equal to $\frac{8}{135}\frac{(qe)^4}{m^4}H^4$, where m is the mass, and (qe) is the charge of the fermion. The difference in the coefficient of H^4 with respect to the Abelian case is due to the fact that for SU(2) there is an extra factor f^{bcd} at each vertex. Hence, for a sufficiently small, but non zero m, there is a region of J where the instability is not present. The issue in this case is if it is possible to reach $J = 0$ remaining in a stable region. There is a physical argument that might suggest that this is possible: adding gauge bosons at the same space-time point will strongly polarize the vacuum and eventually generate a chromomagnetic domain, but in the presence of fermions, the vacuum will soon create fermion anti-fermion pairs at the expense of the energy of the gluons. Depending on which of these two phenomena occurs first the vacuum structure will be different. It will be interesting to explore this situation in more detail. Unfortunately the realistic lattice simulations to investigate the vacuum structure with quarks are unfeasible by the present techniques.

6. Conclusions

To summarize, we have analyzed the effective potential for a non-Abelian background field in SU(2) Yang-Mills theory. The result is found to be qualitatively similar to the Savvidy ansatz, both in real part, which indicates a spontaneous generation of the color magnetic field, and imaginary part, which signals the instability of the background field as vacuum configuration under the loop expansion. Given the qualitative similarity between the three dimensional and four dimensional effective potential in the Savvidy ansatz, we conclude that the effective potential is insensitive to the coordinate singularity, $\mathrm{Det}B_i^a = 0$, if it indeed exists.

Rather, we realize that the presence of the imaginary part in non-Abelian theories is due to the incompatibility in the action between gauge invariance and stabilizing terms.

Technical questions related to the linear term and the gauge choice were illustrated, and in particular, we show the dependence of the one-loop effective potential of the gauge fixing parameter α when the loop expansion is performed around a non trivial background.

It is important to recognize that the effective potential for these backgrounds would be a well-defined problem if the functional were evaluated non-perturbatively. The appearance of the imaginary part in the effective potential is only caused by the loop expansion. In other words, when we were doing steepest descent approximation, we were expanding at a saddle point. In addition, even if the expansion point were a true minimum, the stationary solution of the effective potential calculated up to a finite loop order could not be trusted quantitatively, due to the fact that at the stationary point of the effective potential the higher order terms become as important as lower order terms and hence the loop expansion breaks down. The final answer has to be settled by a non-perturbative means, such as the lattice simulation as mentioned in the Introduction.

In fact, a lattice generalization of the background field method is rather straightforward. Let us consider

$$Z_L[J_\mu(x), B_\mu(x)] \equiv \int dU_\mu(x) \exp\Big\{-S[U_\mu(x)B_\mu(x)] + \mathrm{Tr}J_\mu(x)f[U_\mu(x)]\Big\}, \qquad (37)$$

where $U_\mu(x)$ is the standard link variable, $B_\mu(x)$ is the background link variable, S is the usual Wilson lattice action, $J_\mu(x)$ is a matrix valued external current, and $f[U]$ is an arbitrary function satisfying $gf[U]g^\dagger = f[gUg^\dagger]$ for any unitary matrix g. Using the property of the invariance under an unitary transformation for the Haar measure, one can easily verify the following:

$$Z_L[\tilde{J}_\mu(x), \tilde{U}_\mu(x)]\Big|_{\tilde{U}_\mu(x)=g(x)U_\mu(x)g^\dagger(x+\mu),\ \tilde{J}_\mu(x)=g(x)J_\mu(x)g^\dagger(x)} = Z_L[J_\mu(x), U_\mu(x)]\,, \qquad (38)$$

a lattice version of Eq.(2,3). A Legendre transform of $Z_L[J_\mu, U_\mu]$ would lead to a gauge invariant effective action, provided that the induced gauge field is constrained to have zero expectation value by adjusting $J_\mu(x)$, just as in the continuum case. The remarkable thing here is that we do not need to fix the gauge on a lattice and therefore the resulted effective action is unique for a given choice of the functional form of f.

As mentioned above the non-Abelian background field ansatz can be conveniently realized on a lattice. The constant nature of the ansatz avoids problems with the boundary condition and non-uniformness of the lattice constant effect due to the linear rising ansatz of Savvidy. The hope is that intensive lattice simulations with the non-Abelian background will clarify the picture. Since the non-Abelian background only involves one parameter h it may not be difficult to find a way to adjust the external current J_μ to ensure a vanishing of the expectation value for the induced quantum field.

Acknowledgements

This work was possible thanks to the collaboration of Suzhou Huang, Ken Johnson and Janos Polonyi to whom I am very grateful. It is my pleasure to thank Kerson Huang, H. B. Nielsen, H. Trottier, P. van Baal, for many fruitful discussions.

References

[1] S. Huang and A. R. Levi Phys. Rev. **D 49** (1994) 6849.

[2] W. Greiner, B. Müller and J. Rafelski, *Quantum Electrodynamics of Strong Fields*, (Spring-Verlag, New York, 1985); E.V. Shuryac, *The QCD Vacuum*, (World Scientific, Singapore 1988).

[3] B. S. DeWitt, Phys. Rev. **162** (1967) 1195; J. Honerkamp, Nucl. Phys. **B36** (1971) 130; G. ‘t Hooft, Nucl. Phys. **B62** (1973) 444; L. F. Abbott, Nucl. Phys. **B185** (1981) 189; D. G. Boulware, Phys. Rev. **D23** (1981) 389.

[4] A. Salam and J. Strathdee, Nucl. Phys. **B90** (1975) 203.

[5] J. Schwinger, Phys. Rev. **82** (1951) 664.

[6] G. A. Vilkovisky, Nucl. Phys. **B234** (1984) 125; A. O. Barvinsky and G. A. Vilkovisky, Phys. Rep. **119** (1985) 1.

[7] D. J. Gross, in *Methods in Field Theory*, (eds. R. Balian and J. Zinn-Justin, North-Holland, 1976).

[8] T. H. Hansson and I. Zahed, Nucl. Phys. **B292** (1987) 725; M. E. Carrington, T. H. Hansson, H. Yamigishi and I. Zahed, Ann. Phys. **190** (1989) 373.

[9] L. S. Brown and W. I. Weisberger, Nucl. Phys. **B157** (1979) 285.

[10] G. K. Savvidy, Phys. Lett. **71B**, (1977) 133; G. Matinyan and G. K. Savvidy, Nucl. Phys. **B134** (1978) 539; H. Pagels and E. Tomboulis, Nucl. Phys. **B143** (1978) 485.

[11] N. K. Nielsen and P. Olesen, Nucl. Phys. **B144** (1978) 376.

[12] H.B. Nielsen and M. Ninomiya, Nucl. Phys. **B156** (1979) 1; J. Ambjørn and P. Olesen, Nucl. Phys. **B170** (1980) 60 and 265; J. Ambjørn , B. Felsager and P. Olesen, Nucl. Phys. **B175** (1980) 349.

[13] L. Maiani, G. Martinelli, G. C. Rossi and M. Testa, Nucl. Phys. **B273** (1986) 275.

[14] J. Ambjørn *et al.*, Phys. Lett. **225B** (1989) 153, and **245B** (1990) 575; P. Cea and L. Cosmai, Phys. Rev. **D43** (1991) 620; Phys. Lett. **264B** (1991) 415; A. R. Levi and J. Polonyi, MIT report # 2161.

[15] H. D. Trottier, Phys. Rev. **D44** (1991) 464; H. D. Trottier and R. M. Woloshyn, Phys. Rev. Lett. **70** (1993) 2053; P. Cea and L. Cosmai, Phys. Rev. **D48** (1993) 3364.

[16] J. Ambjørn, N. K. Nielsen and P. Olesen, Nucl. Phys. **B152** (1979) 75.

[17] K. Johnson, in *QCD — 20 Years Later, Aachen, 1992*, (World Scientific, Singapore 1993); D. Z. Freedman, P. E. Haagensen, K. Johnson and J. I. Latorre, MIT preprint, CTP#2238.

[18] R. Parthasarathy, M. Singer and K. S. Viswanathan, Can. J. Phys. **61** (1983) 1442.

[19] S. Coleman and E. Weinberg, Phys. Rev. **D7** (1973) 1988.

[20] L.D. Landau, *Quantum Field Theory*, (translated from Russian, Pergamon Press, Oxford 1974).

FIXED POINTS, SADDLE POINTS AND UNIVERSALITY

JANOS POLONYI
Department of Physics,Louis Pasteur University,Strasbourg, France

ABSTRACT

It is pointed out that the universality might seriously be violated by models with several fixed points.

1. Multiple fixed points

The dependence of the fundamental equations and their parameters on the observational length scale is described by the renormalization group equations. In the vicinities of the fixed points the linearized version of the equations allow us to classify the relevant parameters in a systematic manner. This classification leads to renormalizable field theories as the largest class of models to consider.

Unfortunately the situation is more involved when the description of different interactions is sought in some unified theory. Consider for the sake of definiteness a Grand Unified Theory, GUT, defined at the energy range of 10^{15}GeV. The interaction vertices of the lower energy effective theories, such as the electro-weak theory below 100GeV, QCD around few GeV, nucleon-meson vertices of of nuclear physics and even the phenomenological coupling constants of condensed matter physics in the range of MeV and eV, respectively, are composite operators in terms of the elementary constituents of the GUT. The true coupling constant space where the renormalized trajectory generated by the Wilson-Kadanoff blocking procedure, [1], is given includes all of these parameters. The renormalized trajectory starts at the cut-off, $\Lambda > O(10^{15}GeV)$, at the GUT ultraviolet fixed point and arrives at the vicinity of the ultraviolet fixed point of the Standard Model for $\Lambda = O(10^3GeV)$. In fact, the theory exhibits asymptotic scaling in this energy range because the super heavy particles of the GUT scale have already decoupled and the high energy scaling of the Standard Model is manifest. When the energy is further decresed, at $\Lambda = O(1GeV)$, the heavy intermediate bosons and the Higgs particles decouple leaving behind the non-renormalizable Fermi contact interaction. The QCD evolution equations dominate the scale dependence in this range. Thus the renormalized trajectory is in the vicinity of the QCD fixed point at this energies.

This pattern repeats itself as we move towards lower energies. The higher energy (effective) theories generate effective vertices. The renormalizable vertices govern the renormalized trajectory at the infrared side of the lower energy fixed point. At the other, ultraviolet side of the low energy fixed point the relevant coupling constants are small and the irrelevant ones are dominant. The non-renormalizable, i.e. irrelevant vertices arising at the level of the lower energy effective theory are suppressed according to the decoupling theorem [2]. These vertices are nevertheless crucial for they are the seeds of the "new physics" as seen from the low energy effective theory. In fact, these are just the coupling constants which grow as we move towards high energy and deflect the renormalized trajectory from the ultraviolet fixed point of the low energy effective theory. In our case it is the Fermi type contact interaction which prevents the physics of few GeV energies to "fall" into the ultraviolet fixed point of QCD as the energy is raised. These fixed points which are in the vicinity of the renormalized trajectory generate cross-over phenomena when their characteristic scale is reached. It is worthwhile noting that there are additional fixed points as well which do not belong to either asymptotical regime of the effective theories. These correspond to critical phenomena occurring under special circumstances, such as finite temperature or densities. Finally there is an infrared fixed point, too.

The universality is the statement that the physics is insensitive to the initial, UV values of the irrelevant coupling constants. This is certainly true at the vicinity of the fixed point, more precisely in the infrared end of the renormalized trajectory within the range of the linearizability. But in the presence of several fixed points it is no longer clear that the irrelevant operators of a fixed point remain irrelevant at other fixed point. Thus we are left only with the "islands" of universality in the vicinities of different fixed points. The manner the sets of relevant coupling constants of different fixed points match is not at all described by universality.

There are two different ways to establish the connection between the coupling constants of the different fixed points. These will be briefly described below.

2. Scaling at the infrared fixed point

The straightforward procedure is to simply follow the evolution of the renormalized trajectory. If there are no further fixed points as we move towards lower energies then the irrelevance of the coupling constants at the last fixed point remain valid down to zero energy. But if there is a chain of fixed point in the vicinity of the renormalized trajectory as we seen before then an vertex which was irrelevant at high energy may become relevant at a lower energy fixed point. In this case the physics at the lower energy fixed point may in principle depend on the initial value of this coupling constant. Thus one has to introduce manifestly non-renormalizable coupling constants at the high energy fixed point in order to parametrize the physics at low energies.

In order to understand such a multi-fixed point situation better one should first consider a theory with two fixed points, one at the ultraviolet and the other at the infrared. The infrared fixed point is trivial, i.e. the its relevant operators are Gaussian in the presence of a mass gap. In fact, when the block size grows beyond of the Compton wavelength of the particle the fluctuations are suppressed and the evolution of the coupling constants slows down. In order to study the importance of a non-renormalizable coupling constant we need non-Gaussian relevant operator. This might be found in theories with massless excitations.

It has been argued in [3] by the help of the functional renormalization group equation [4] that the infrared fixed point of the linear O(N) model in the symmetry broken phase has nontrivial relevant operators which contain non-renormalizable vertices. The infrared divergences of the Goldstone modes pile up and make the beta function negative and divergent for the odd vertices of the massive particle. How many new important parameters are generated by this strong coupling vacuum of the Goldstone modes ? Unfortunately the linearization and the diagonalization of the renormalization group equation has not yet been done and the actual form of the scaling operators is not known. But it seems plausible to assume that the effect the new coupling constants generate is mainly in the deep infrared limit. This is because if they start with an O(1) value in the ultraviolet they are first suppressed as we leave the vicinity of the ultraviolet fixed point and are amplified only in the asymtotical infrared scaling regime. The non-perturbative infrared phenomenon which surely takes place in the vacuum is the spontaneous breakdown of the symmetry and its parameter is the magnitude of the condensate, $| < \phi_a > |$. Thus it is reasonable to assume that the only impact these new coupling constants have is that they influence the magnitude of the condensate. This influence is "hidden" for the observations made at intermediate energy where these coupling contants are extremely small. The value of the condensate naturally influences the dynamics of higher energies indirectly since it determines the mass gap in the massive sector. Thus the complete set of parameters of the theory should consists of the usual renormalizable coupling constants given at the ultraviolet regime and the values of the nontrivial condensates which are manifest in the infrared. By the help of these two sets of parameters we can parametrize the physics of the O(N) model.

3. Mini-instantons

Another less direct mechanisms which may couple the short and the long distance regime is operating in theories with non-homogenous saddle points. As an example consider a classically scale invariant theory with instantons [5]. The changing the instanton scale leaves the renormalized action invariant. The actual saddle point expansion is performed in the bare cut-off theory so it is the bare rather than the renormalized lagrangian what should be used. The bare regulated action does have a scale parameter, the cut-off itself. Thus it is not at all obvious that the action is indeed independent of the instanton scale. By adding appropriate irrelevant operators to the lagrangian, such as higher derivative terms, the action of the saddle point as the function of the instanton size may develop a minimum. The size of the saddle point at this minimum is in the order of magnitude of the only scale parameter, the cut-off. If this minimum is small enough then these "topological defects" [6] may form a gas with divergent density.

Qualitatively similar phenomenon, namely the population of the vacuum by small localized states has already been conjectured in strong coupling massless QED. There the e^-e^+ bound states with the size of the cut-off condense and change the long distance physics by the dynamical breaking of the chiral symmetry [7]. What is interesting in the mini-instanton gas is that it corresponds to time dependent coherent states close to the cut-off without relying on strong coupling constant. In fact, the mini-instantons are present in asymptotically free theories such as the O(3) model or QCD. The important question is whether they can modify the scaling at low energy like the singular e^-e^+ pairs of chiral QED.

Though a single mini-instanton can cause no harm to the renormalized trajectory at a finite i.e. cut-off independent energy, the gas of these saddle points might well modify the evolution of the renormalized coupling constants at finite energies. This may happen because the change of the cut-off generates two qualitatively new effects in the framework of the dilute instanton gas approximation. A tree-level cut-off dependence arises from the saddle-point action. The one-loop dependence comes from the small fluctuation determinant. When the instantons are well inside of the infrared regime compared to the cut-off then this dependence agrees with the usual renormalized trajectory. But the mini-instantons remain in the vicinity of the cut-off for a large part of the renormalized trajectory and they continue to modify the trajectory.

The irrelevant terms of the action come from the next fixed point in the ultraviolet direction as explained above. The higher energy theory cuts off the low energy effective model and provides the irrelevant operators which appear as regulator in the low energy theory. The dimension of the irrelevant coupling constants is given by the cut-off, i.e. the scale of the higher energy fixed point. The irrelevant operators represent the W and Z exchanges in QCD. Thus the mini-instantons saturate the path integral by the fluctuations of this higher energy fixed point. If their influence remain finite at low energy they represent a non-perturbative bridge between neighboring fixed points.

The ratio of the partition function in the one instanton and the flat sector, Z_1/Z_0, was computed in the two dimensional sigma model in the presence of higher order derivatives in [8]. If the higher order derivative terms are really irrelevant then this infrared quantity should be insensitive to the choice of the new coupling constants at the cut-off. This was not the case, Z_1/Z_0 displayed a clear dependence on the higher order terms in the action. Let us impose the renormalization condition that Z_1/Z_0 is cut-off independent. Then by assuming that the new coupling constants remain cut-off independent along the renormalized trajectory one can obtain the beta function for the usual coupling constant of the sigma model. The resulting beta function is negative and asymptotic freedom is preserved but its first coefficient which is given by the tree level mini-instanton action is not universal. For sufficiently large values of the higher derivative operators the mini-instanton action may even become negative. In this case the vacuum is a densely packed instanton-anti instanton gas, similarly an anti-ferromagnetic system. Saddle point expansion is unreliable and the asymptotic freedom is not guaranteed any more in this phase.

The phenomenon found in the sigma model is likely to be present in QCD, too. In the framework of the saddle point expansion one expects three new relevant parameters. These are the size of the min-instanton, its action and the second derivative of the action with respect to the instanton size. These parameters are determined at the cut-off by the dynamics of the higher fixed point. Their appearance in the beta function at finite energies represent the coupling between different fixed points.

Naturally one has to be careful and the final conclusion should be drawn only after computing Green functions in the infrared regime on the mini-instanton background. This is a formidable task in an asymptotically free model.

4. Conclusions

The nontrivial interplay of different fixed point of the renormalization group was briefly discussed in this talk from the point of view of the unified models. There might be another motivation as well to consider this problem in theories with nontrivial infrared dynamics. The taking into account such effect may provide a clue to follow the renormalized trajectory to regions where the usual perturbation expansion fails. Consider for example QCD [9]. The gauge coupling constant blows up according to the one-loop renormalization group equation at the cross-over separating the ultraviolet and the infrared scaling regime. The Wilson-Kadanoff renormalization group procedure suggests the inclusion of all coupling constant which might be relevant for the evolution into the action. This means the inclusion of the hadronic effective vertices together with the gauge coupling constant. This would be double counting in the hamiltonian formalism where we would use hadrons and quark-gluon states simultaneously. But this is not the case with the path integral and we should be able to pass the cross-over regime with this extended QCD without seeing diverging coupling constants. The onset of the hadronic world as we enter into the infrared scaling regime would be achieved by the hadronic vertices and the gauge coupling constant could remain finite.

ACKNOWLEDGEMENTS

It is pleasure to thank Vincenzo Branchina and Sen-Ben Liao for their collaboration which led to the results mentioned above.

REFERENCES

[1] K. Wilson and J. Kogut, *Phys. Rep.* **12C** (1975) 75; K. Wilson, *Rev. Mod. Phys.* **47** (1975) 773.

[2] T. Appelquist and Carazzone, *Phys. Rev* **D11** (1975) 2856.

[3] S. B. Liao, J. Polonyi, "Renormalization Group and Universality", hep-th 9403111, submitted to *Ann. Phys.*

[4] F.J. Wegner and A. Houghton, *Phys. Rev* **A8** (1972) 401.

[5] G. 'tHooft, *Phys. Rev.* **D 14** (1976) 3432.

[6] M. Luscher, *Nucl.Phys.* **B200** (1982) 61.

[7] P.I. Fomin, V.P. Gusynin, V.A. Miranski and YU.A. Sytenko, *Riv. Nuovo Cimento*, **6**, No. 5 (1983); C.N. Leung, S.T. Love , William A. Bardeen, *Nucl. Phys.* **B 323** (1989) 493 ; C.N. Leung, S.T. Love , William A. Bardeen, *Nucl. Phys.* **B 273** (1986) 649.

[8] V. Branchina, J. Polonyi, "Instantons in cut-off theories", hep-th 9401079 to appear in *Nucl. Phys.*

[9] J. Polonyi, in the Proceedings of the Workshop on "QCD Vacuum Structure and Its Applications", H.M. Fried and B. Müller eds., World Scientific, 1993, p. 3.

Elimination of Color in 1+1-dimensional QCD

Michael Engelhardt
Department of Condensed Matter Physics
Weizmann Institute of Science
Rehovot 76100, Israel

Bernd Schreiber
Center for Theoretical Physics
Massachusetts Institute of Technology
Cambridge, MA 02139, U.S.A.

Abstract

We discuss QCD in one space and one time dimension on a finite circle. Boundary conditions force the retention of quantum mechanical (zero-mode) gauge degrees of freedom in a Hamiltonian formulation. We explicitely consider systems of static quarks and find that the dynamics of the zero modes lead to an elimination of fermionic non-singlet states from the spectrum in the continuum limit.

1 Introduction

Quantum chromodynamics is well understood in the short distance domain due to asymptotic freedom. Not nearly as much success has been achieved, however, in the low energy region. Here, progress has mainly been achieved by the use of models such as for example the nonrelativistic quark model [1] or the M.I.T. bag model [2]. However, a rigorous justification of these models on the basis of QCD has so far not been possible.

In order to obtain a better understanding of 3+1-dimensional QCD, model field theories in one space and one time dimension have been examined, e.g. 1+1-dimensional QCD (QCD_{1+1}) [3]-[5]. We will show in the following that already QCD_{1+1} provides a mechanism for the elimination of states which carry fermionic color charge while at the same time being overall color singlets, and therefore exhibits a group-theoretical pattern in the observed hadron spectrum reminiscent of the one assumed in the nonrelativistic quark model and observed experimentally.

Despite much progress having been achieved in understanding QCD_{1+1}, this theory is up to now not solved completely. The reason for this lies mainly in the fact that the continuum formulation of QCD_{1+1} is plagued with infrared ambiguities; by solving the Poisson equation in the (usually adopted) axial gauge one obtains time-dependent background fields, the dynamics of which are completely unclear. The best way to treat the theory rigorously is by considering it in a box of finite volume and by specifying appropriate boundary conditions. It is this promise of a more rigorous and complete treatment of 1+1-dimensional gauge field theories which has recently sparked renewed interest in the compactified versions of these models. The Hamiltonian of QCD_{1+1} on a finite circle was derived in [6]. There it was found that the technically convenient axial gauge cannot be used because it is not compatible with the chosen boundary conditions for the fermionic degrees of freedom; one can only choose the Coulomb gauge.

2 The Hamiltonian

The starting point for our treatment is the Hamiltonian of QCD_{1+1} with an $SU(N_C)$ color group on a circle S^1 of length L with periodic boundary conditions [6]. In the Coulomb gauge $\partial_1 A_1 = 0$ it takes the form

$$\begin{aligned} H &= H^c_{kin} + \frac{2\pi}{L}\sum_{p,i}(p+1/2+c_i)(a_i^\dagger(p)a_i(p) - b_i^\dagger(p)b_i(p)) \\ &\quad + m\sum_{p,i}(a_i^\dagger(p)b_i(p) + b_i^\dagger(p)a_i(p)) + \frac{g^2L}{4\pi^2}\sum_{n,i,j}{}' \frac{j_{ij}(n)j_{ji}(-n)}{(n-c_j+c_i)^2}. \end{aligned} \quad (1)$$

The operators a_i and b_i are the annihilation operators for right-handed and left-handed quarks of color i, respectively, $j_{ij}(n)$ are the color components of the fermion current in momentum space and the prime on the last summation in (1) indicates that the sum over n excludes $n = 0$ for $i = j$. This Hamiltonian contains one novel feature: On the finite interval, $N_C - 1$ quantum mechanical gauge degrees of freedom must be retained; these are the zero-momentum, color-space diagonal modes of the A_1 gauge field which we have rescaled as $c_i = gL/2\pi \cdot A_{1ii}$. They appear in the denominator of the Coulomb interaction due to the non-abelian nature of the gauge group.

The Hamiltonian (1) must be supplemented by a color neutrality constraint upon the space of physical states

$$j_{ii}(0)|\text{phys}\rangle = 0 \quad (2)$$

(cf. the summation restriction in (1)). This condition is compatible with dynamics, since $j_{ii}(0)$ is conserved.

Note that fully taking account of the Euler-Lagrange equation for A_0 including the above constraint already implies physical states to be overall color singlets. This can be seen as follows: In QCD, the overall color charge not only consists of a fermionic, but also of a gluonic part,

$$j^0_{total} = j^0 - i[A_1, F_{01}] = -\frac{1}{g}\partial_1 F_{01}. \tag{3}$$

Integration of this equation leads to the value zero for the total color charge because of the periodic boundary condition on the A_0 field. However, it must be stressed that the formalism still allows for states of arbitrary fermionic color charge; in $SU(2)$ language, $j_{ii}(0) = 0$ only implies the z-component of the fermionic charge to be zero.

3 $SU(2)$ – Born-Oppenheimer approximation

In the case of $SU(2)$ color, there is only one gauge degree of freedom, namely $c := c_1 - c_2$. We examine the limit of very heavy quarks in the low-energy regime, such that the fermionic vacuum is trivial and the positions of the quarks can be treated as slow variables in the spirit of a Born-Oppenheimer approximation.

For the dynamics of the fast degrees of freedom, only the kinetic energy of the gauge zero mode and the Coulomb term are relevant; the former is represented in c-space by

$$H^c_{kin} = -\frac{g^2 L}{8\pi^2}\frac{1}{\sin^2 \pi c}\frac{d}{dc}\sin^2 \pi c\frac{d}{dc}, \tag{4}$$

where the Jacobian stems from the metric on the space of $SU(2)$ gauge field matrices. For the Coulomb term, we derive a position space representation in order to be able to project onto static fermion states:

$$\begin{aligned} H_{coul} &= \frac{g^2}{2}\sum_{i,j}\int_0^L dx \int_0^L dy\, j_{ij}(x) j_{ji}(y) \frac{L}{2\pi^2}\sum_n{}' \frac{\exp(2\pi i n(x-y)/L)}{(n - c_j + c_i)^2} \\ &= \frac{g^2}{2}\sum_{i,j}\int_0^L dx \int_0^L dy\, j_{ij}(x) j_{ji}(y) G_{ji}(x-y). \end{aligned} \tag{5}$$

The Coulomb propagator in position space can be evaluated with the result

$$G_{12}(x-y) = G^*_{21}(x-y) = e^{i2\pi c(x-y)/L}\left[-|x-y| + \frac{L}{2\sin^2 \pi c} - i(x-y)\cot \pi c\right] \tag{6}$$

$$G_{ii}(x-y) = -|x-y| + (x-y)^2/L. \tag{7}$$

Before we discuss the spectrum in the presence of quarks we calculate the eigenvalues and the corresponding eigenfunctions in the static limit when no quarks are present (pure Yang-Mills theory). In the fermionic sector one has the perturbative vacuum $|0>_f$ and the eigenstates are of the form $\psi_j(c)|0>_f$ with

$$H^c_{kin}\psi_j(c) = E_j \psi_j(c). \tag{8}$$

After the substitution $u_j(c) = \sin(\pi c)\psi_j(c)$ one obtains

$$-\frac{g^2L}{8\pi^2}\frac{d^2u_j}{dc^2} = (E_j + \frac{g^2L}{8})u_j. \tag{9}$$

Solving this equation gives

$$E_j = \frac{g^2L}{8}(j^2-1) \qquad j = 1, 2, \ldots \tag{10}$$

with $u_j(c) = \sin(j\pi c)$. Obviously, except for the vacuum state of lowest energy ($E_1 = 0$) all other excited eigenstates have no reasonable energy behavior in the continuum limit and are frozen out. We now treat the case of two static quarks in a relatively explicit way in order to exhibit the mechanisms at work. We need to evaluate the Hamiltonian on the subspace of two particles. As a basis in the fermionic sector we use

$$|\tilde{S}\rangle = \frac{1}{\sqrt{2}}(\psi_1^\dagger(x)e^{i\pi c(y-x)/L}\psi_2^\dagger(y) - \psi_2^\dagger(x)e^{-i\pi c(y-x)/L}\psi_1^\dagger(y))|0\rangle \tag{11}$$

$$|\tilde{T}\rangle = \frac{1}{\sqrt{2}}(\psi_1^\dagger(x)e^{i\pi c(y-x)/L}\psi_2^\dagger(y) + \psi_2^\dagger(x)e^{-i\pi c(y-x)/L}\psi_1^\dagger(y))|0\rangle. \tag{12}$$

In the limit $L \to \infty$, $|\tilde{S}\rangle$ and $|\tilde{T}\rangle$ correspond to the usual fermionic singlet and triplet combinations, i.e. color spin coupled to zero or one, respectively, with z-component zero due to the neutrality condition (2). In this basis, the Hamiltonian reads

$$H = -\frac{g^2L}{8\pi^2}\frac{1}{\sin^2\pi c}\frac{d}{dc}\sin^2\pi c\frac{d}{dc} - \frac{g^2(x-y)^2}{8L} + \frac{g^2}{8}\begin{pmatrix} 3|x-y| & 2i(x-y)\cot\pi c \\ -2i(x-y)\cot\pi c & \frac{2L}{\sin^2\pi c} - |x-y| \end{pmatrix} \tag{13}$$

It acts on linear combinations of the form $\psi_S(c)|\tilde{S}\rangle + \psi_T(c)|\tilde{T}\rangle$.

It is easy to check that $|\tilde{S}\rangle$ is the ground state of the Hamiltonian

$$H\ |\tilde{S}\rangle = \frac{3}{8}g^2|x-y|\ |\tilde{S}\rangle. \tag{14}$$

One obtains the expected linear potential between the quarks. More importantly, however, note that only for the singlet combination, the term proportional to L cancels in the potential matrix. Thus, states containing non-singlet admixtures are excited by energies of the order of L and are therefore frozen out in the continuum limit. In fact, in this limit, (14) is the only physically accessible state since all excited gauge wave functions $\psi(c)$ are also frozen out due to the kinetic energy rising proportionally to L.

This can be verified by explicit construction of the spectrum, which can be accomplished as follows: First note that for $x = y$, $|\tilde{S}\rangle$ and $|\tilde{T}\rangle$ decouple. The resulting equations for $\psi_S(c)$ and $\psi_T(c)$ yield the eigenvalues

$$E_j^S = \frac{g^2L}{8}(j^2-1) \quad \text{and} \quad E_j^T = \frac{g^2L}{8}((j+1)^2-1) \quad j = 1, 2, 3, \ldots \tag{15}$$

respectively, where the triplet equation can be reduced to a hypergeometric equation through the substitutions $\psi_T(c) = u(\sin^2 \pi c) \sin \pi c$ and $\psi_T(c) = u(\sin^2 \pi c) \sin \pi c \cos \pi c$ [7]. Note that $E_j^T = E_{j+1}^S$. Thus, for $x = y$, all excited states occur in doublets consisting of a singlet and a triplet configuration. If one now allows $x \neq y$, one notices that only states of the same doublet (i.e. the same L-behavior) can mix and one easily obtains the eigenvalues for $x \neq y$ by diagonalising the resulting 2×2 matrix. The full spectrum is given by

$$E_j = \frac{g^2}{8}|x-y|\,j\,(j+2) + \frac{g^2}{8}(L - |x-y|)\,(j-1)\,(j+1) \quad (16)$$

$$\text{and} \quad E_j' = \frac{g^2}{8}|x-y|\,(j-1)\,(j+1) + \frac{g^2}{8}(L - |x-y|)\,j\,(j+2). \quad (17)$$

It must be emphasized that this feature of a dynamical elimination of colored states is lost in the usual continuum formulations of QCD_{1+1}. This is due to the fact that only the linear potential is retained in the Coulomb propagator.

4 Conclusions and Outlook

In this work, we analysed the role of the dynamical gluonic zero modes within the framework of QCD_{1+1} on a finite circle. We have demonstrated in a specific example that the dynamics of these modes leads to an elimination of fermionic color non-singlet states from the spectrum in the continuum limit although these states are at the same time overall color-singlets.

The mechanism found in this work leading to the elimination of a certain sector of the Hilbert space seems interesting in a wider context due to the fact that in $3+1$ dimensions, the Coulomb interaction takes a form very similar to the one discussed here [8]. Therefore, it may to a certain extent be possible to translate the ideas presented here to the three-dimensional case.

Acknowledgements

We would like to acknowledge fruitful discussions with M.Burkardt, F.Lenz, S.Levit, R.Naus, D.Stoll, and M.Thies.

References

[1] N.Isgur and G.Karl, *Phys. Rev.* **D18** (1978) 4187.

[2] A.Chodos, R.L.Jaffe, K.Johnson, C.B.Thorn and V.F.Weisskopf, *Phys. Rev.* **D9** (1974) 3471.

[3] G.'t Hooft, *Nucl. Phys.* **B75** (1974) 461.

[4] I.Bars and M.B.Green, *Phys. Rev.* **D17** (1978) 537.

[5] M.B.Einhorn, *Phys. Rev.* **D14** (1976) 3451.

[6] F.Lenz, M.Thies, K.Yazaki and S.Levit, *Ann. Phys.* **208** (1991) 1.

[7] E.Kamke, *Differentialgleichungen I* (Teubner, Stuttgart 1983), p. 408.

[8] F.Lenz, H.W.L.Naus and M.Thies, *Ann. Phys.* **233** (1994) 317.

GAUGE INVARIANT REGULATOR AND RENORMALIZATION GROUP *

Sen-Ben Liao

Department of Physics
Duke University
Durham, North Carolina 27708 U.S.A.

ABSTRACT

We derive a manifestly gauge invariant renormalized action for Yang-Mills theory using the Schwinger proper-time formalism. In our scheme the regularization is achieved by the smooth suppression of the large eigenvalues in the quantum fluctuation determinants. With appropriate choice of the suppression factor this method incorporates the features of a sharp momentum cut-off and renders the calculation finite throughout. The similarity of this cut-off to the Pauli-Villars regulator in the ultraviolet regime is explicitly demonstrated. In addition, our approach makes it more feasible to implement Wilson's renormalization group method to Yang-Mills theory in a gauge symmetry preserving manner.

I. INTRODUCTION

The derivation of effective potential often involves functional determinants of differential operators, in particular, when approximation scheme such as loop expansion results in a Gaussian functional integration. However, since divergences are often encountered in evaluating functional determinants, one must further implement a regularization scheme which not only renders the theory finite, but also preserves all the symmetries of the original theory. For example, when studying gauge theories such as QCD, a sharp momentum cut-off regulator would not be appropriate since it explicitly violates gauge invariance. Therefore, one resorts to other gauge invariant methods such as dimensional regularization, ζ function regularization, Pauli-Villars procedure [1], or proper-time method [2].

Despite the shortcoming of its non-gauge invariant nature, the sharp momentum cut-off has proven to be a useful regulator, notably for its common usage in pure scalar theory. In addition to being simple and more physical, one readily derives the renormalization group (RG) equations which govern the behavior of the theory at different scales. The renown RG formalism pioneered by Wilson [3] is based on the use of this regulator.

* This work is supported in part by funds provided by the U. S. Department of Energy (D.O.E.) under contract #DE-FG05-90ER40592.

One way to achieve this separation is by the help of blocking transformation [4], in which one defines from the original field $\phi(x)$ an averaged field ϕ_{av} for each given block of size k^{-4} via a suitable choice of a smearing function $\rho_k(x)$:

$$\phi_{av}(x) = \phi_k(x) = \int_y \rho_k(x-y)\phi(y), \qquad \int_x = \int d^4x. \tag{1.1}$$

The desired scale is naturally set at $p = k$ when one uses a sharp momentum cut-off $\rho_k(p) = \Theta(k-p)$. From this particularly simple choice of regulator, one can derive the blocked potential, $U_k(\phi)$, an effective potential parameterized by the averaged field at the scale k. Subsequently, the RG flow behavior of the theory can be examine from the blocked potential which establishes a smooth connection between the bare classical potential $V_\Lambda(\phi)$ at $k = \Lambda$ and the Coleman-Weinberg effective potential $U^{\text{eff}}(\phi)$ at $k = 0$ [5].

With the advantages illustrated by the sharp momentum regulator, one then inquires how it is possible to implement an equivalent gauge invariant prescription which encompasses the features of blocking transformation. We shall demonstrate below how blocking is achieved for gauge theories without spoiling gauge invariance.

II. PROPER-TIME BLOCKED POTENTIAL

In the original formalism developed by Schwinger [2], the functional determinant is written as:

$$\mathcal{N} = \frac{\det\mathcal{H}}{\det\mathcal{H}_0} = \exp\left\{-\text{Tr}\int_0^\infty \frac{dt}{t}\left(e^{-\mathcal{H}t} - e^{-\mathcal{H}_0 t}\right)\right\}, \tag{2.1}$$

where counterterms are carefully defined to cancel the UV divergences which appear in the limit $t \to 0$. We introduce here a slight modification based on the construction of a smooth smearing function $\rho_k(t, \Lambda)$ in the proper-time integration variable t. While Λ acts as the usual UV regulator, k is the scale arising from the 'obscured' blocking. Moreover, (2.1) can be made finite with appropriate boundary conditions for $\rho_k(t, \Lambda)$.

To deduce its form, one first notes that in the usual computation of effective potential U^{eff} at $k = 0$, infinities can be easily eliminated if the smearing function takes on the values $\rho_{k=0}(t = 0, \Lambda) = 0$ and $\rho_{k=0}(t \to \infty, \Lambda) = 1$. On the other hand, since the blocked potential U_k has been shown to interpolate between U^{eff} and V_Λ, i.e., $U_{k=0} = U^{\text{eff}}$ and $U_{k=\Lambda} = V_\Lambda$. Thus, we must impose an additional constraint: $\rho_{k=\Lambda}(t, \Lambda) = 0$.

Assuming the smearing function to be monotonic with rapid convergence near $t = 0$, we see that in this UV limit where the kinetic term dominates in the eigenvalue spectra of $\mathcal{H}$, there exists a simple relation between Λ and the cut-off for the eigenvalues, $1/\epsilon$, namely: $\frac{1}{\epsilon} = \Lambda^2$. However, in the IR regime where the potential term gives an important contribution to the eigenvalues, the connection between these cut-offs is obscured. The task of searching for a special form of $\rho_k(t,\Lambda)$ which *retains* a relation between the two cut-offs can be accomplished by simply considering the scalar $\lambda\phi^4$ theory where the one loop correction to the blocked potential can be written as:

$$U_k^{(1)}(\phi) = \frac{1}{2\Omega}\ln\mathcal{N}_k(\phi) = -\frac{1}{32\pi^2}\int_0^\infty \frac{dt}{t^3}\rho_k(t,\Lambda)\Big(e^{-V_R''(\phi)t} - e^{-V_R''(0)t}\Big). \quad (2.2)$$

On the other hand, a direct momentum integration using sharp cut-off gives:

$$U_k^{(1)}(\phi) = \frac{1}{2}\int_p^{'} \ln\Big(\frac{p^2+V_R''(\phi)}{p^2+V_R''(0)}\Big) = \frac{1}{64\pi^2}\Bigg\{\Big(V_R''(\phi)-V_R''(0)\Big)\Big(\Lambda^2-k^2\Big)$$
$$+\Lambda^4\ln\Big(\frac{\Lambda^2+V_R''(\phi)}{\Lambda^2+V_R''(0)}\Big) - k^4\ln\Big(\frac{k^2+V_R''(\phi)}{k^2+V_R''(0)}\Big) - V_R''(\phi)^2\ln\Big(\frac{\Lambda^2+V_R''(\phi)}{k^2+V_R''(\phi)}\Big) + \cdots\Bigg\} \quad (2.3)$$

Imposing the above boundary conditions and demanding that both methods yield the same RG flow equation, we are lead to

$$\rho_k(t,\Lambda) = \Big[1-(1+\Lambda^2 t)e^{-\Lambda^2 t}\Big] - \Big[1-(1+k^2 t)e^{-k^2 t}\Big] = \rho(\Lambda^2 t) - \rho(k^2 t). \quad (2.4)$$

The insertion of the proper-time smearing function $\rho_k(t,\Lambda)$ has the similar physical origin as that of the Pauli-Villars regularization scheme, both of which are gauge invariant constructs in nature. Such an analogy can be seen from the manner the propagator is modified:

$$\frac{1}{\mathcal{H}} = \int_0^\infty dt\ e^{-\mathcal{H}t} \longrightarrow \int_0^\infty dt\ \rho(\Lambda^2 t)e^{-\mathcal{H}t} = \frac{1}{\mathcal{H}}\Big(1-\frac{\mathcal{H}}{\Lambda^2+\mathcal{H}}\Big)^2. \quad (2.5)$$

Other gauge invariant appraoches are considered in [6].

III. YANG-MILLS THEORY

Applying the above formalism to the pure Yang-Mills theory,

$$\mathcal{L}_\Lambda = \frac{1}{4}G_{\mu\nu}^a G_{\mu\nu}^a, \quad (3.1)$$

where

$$G_{\mu\nu}^a = \partial_\mu A_\nu^a - \partial_\nu A_\mu^a + g f^{abc} A_\mu^b A_\nu^c, \quad (3.2)$$

we are able to integrate out the irrelevant short-distance modes in studying its low-energy behavior. Writing

$$A_\mu^a(p) = \begin{cases} \bar{A}_\mu^a(p), & 0 \le p \le k \\ \xi_\mu^a(p), & k < p < \Lambda, \end{cases} \tag{3.3}$$

with $F_{\mu\nu}^a = G_{\mu\nu}^a(\bar{A})$ being the background field strength, and incorporating the covariant background gauge-fixing and ghost contributions respectively as [7]:

$$\mathcal{L}_{\rm GF} = -\frac{1}{2\alpha}(D_\mu A_\mu^a)^2, \tag{3.4}$$

and

$$\mathcal{L}_{\rm FPG} = \chi^\dagger D^2(\bar{A})\chi, \tag{3.5}$$

the blocked action becomes:

$$\tilde{S}_k(\bar{A}) = \int_x \frac{1}{4}F_{\mu\nu}^a(\bar{A})F_{\mu\nu}^a(\bar{A}) + \frac{1}{2}\mathrm{Tr}'\ln\Big(\frac{\mathcal{K}_{\bar{A}}}{\mathcal{K}_0}\Big) - \mathrm{Tr}'\ln\Big(\frac{\mathcal{O}_{\bar{A}}}{\mathcal{O}_0}\Big), \tag{3.6}$$

where the gauge field kernel and the ghost kernel are, respectively,

$$\mathcal{K}_{\mu\nu}^{ab} = \Big\{-\Big[D^2 g_{\mu\nu} - (1-\frac{1}{\alpha})D_\mu D_\nu\Big]^{ab} + 2gf^{abc}F_{\mu\nu}^c(\bar{A})\Big\}\delta^4(x-y), \tag{3.7}$$

and

$$\mathcal{O}^{ab} = -D^2(\bar{A})^{ab}\delta^4(x-y). \tag{3.8}$$

While Tr$'$ denotes trace sum over the restricted space-time, Lorentz indices, as well as the color indices, tr implies an operation over the latter two only.

Since the background field $\bar{A}$ is spatially dependent, the determinants can only be approximated. We shall adopt the perturbation formalism developed in [2], which is based on the expansion in the proper-time variable t. Writing $\mathcal{H} = \mathcal{H}_0 + \mathcal{H}_I$, where $\mathcal{H}_I$ contains all the interaction terms, we have:

$$\begin{aligned}\mathrm{Tr}\Big(e^{-\mathcal{H}t}\Big) = \mathrm{Tr}\Big\{ & e^{-\mathcal{H}_0 t} + (-t)e^{-\mathcal{H}_0 t}\mathcal{H}_I + \frac{(-t)^2}{2}\int_0^1 du_1 e^{-(1-u_1)\mathcal{H}_0 t}\mathcal{H}_I e^{-u_1\mathcal{H}_0 t}\mathcal{H}_I \\ & + \frac{(-t)^3}{3}\int_0^1 du_1 u_1 \int_0^1 du_2 e^{-(1-u_1)\mathcal{H}_0 t}\mathcal{H}_I e^{-u_1(1-u_2)\mathcal{H}_0 t}\mathcal{H}_I e^{-u_1 u_2\mathcal{H}_0 t}\mathcal{H}_I + \cdots\Big\}\end{aligned} \tag{3.9}$$

Concentrating only on the terms which are quadratic in $\bar{A}$ for coupling constant renormalization, the Yang-Mills blocked action can written as, in Feynman gauge whith $\alpha = 1$:

$$\tilde{S}_k = \frac{1}{4}\int_p \mathcal{Z}_k^{-1} F_{\mu\nu}^a(\bar{A})F_{\mu\nu}^a(\bar{A}), \tag{3.10}$$

where $\tilde{\Lambda} = \frac{\Lambda}{p}$, $\kappa = \frac{k}{p}$,

$$\mathcal{Z}_k^{-1} = 1 - \frac{g^2 C_2(G)}{48\pi^2}\Big[11\ \ln\Big(\frac{\tilde{\Lambda}^2}{\kappa^2}\Big) + \frac{31}{3} + 4\kappa^2 + (8\kappa^4 + 22\kappa^2 + 11)f(\kappa)\Big], \qquad (3.11)$$

and

$$f(\kappa) = \frac{1}{\sqrt{4\kappa^2+1}}\ \ln\Big(\frac{\sqrt{4\kappa^2+1}-1}{\sqrt{4\kappa^2+1}+1}\Big). \qquad (3.12)$$

In addition, the abelian approximation is used:

$$\frac{1}{2}F^c_{\mu\nu}(p)F^c_{\mu\nu}(-p) = (p^2 g_{\mu\nu} - p_\mu p_\nu)\bar{A}^c_\mu(p)\bar{A}_\nu(-p) + \cdots. \qquad (3.13)$$

Note that $f^{ac\ell}f^{\ell ma} = -\delta^{cm}C_2(G)$, where $C_2(G) = N$ for $SU(N)$ is a Casimir operator. We also see that the quadratic divergence based on naive power conuting completely disappears, as required by gauge invariance.

The above expression is not yet gauge invariant because the non-polynomial strcture in momentum variable p for $\mathcal{Z}_k$ implies appearance of non-local field strength coupling which manifestly violates gauge invariance for non-abelian system. However, in the large k limit, (3.11) takes on the following power-series expansion:

$$\mathcal{Z}_k^{-1} \approx 1 - \frac{g^2 C_2(G)}{48\pi^2}\Big[11\ \ln\big(\frac{\Lambda^2}{k^2}\big) + \frac{1}{3} - \frac{7}{6}\frac{p^2}{k^2} - \frac{1}{36}\big(\frac{p^2}{k^2}\big)^2\Big] + \cdots, \qquad (3.14)$$

which enables us to "upgrade" the polynomial terms in p to higher order gauge invariant operators with the substitution $p^2 \to \Pi^2$, where Π is the generalized momentum, i.e.,

$$\begin{cases} F^c_{\mu\nu}(p)\ p^2\ F^c_{\mu\nu}(-p) \longrightarrow F^c_{\mu\nu}(x)\ D^2\ F^c_{\mu\nu}(x) \\ F^c_{\mu\nu}(p)\ (p^2)^2\ F^c_{\mu\nu}(-p) \longrightarrow F^c_{\mu\nu}(x)\ (D^2)^2\ F^c_{\mu\nu}(x) \end{cases}. \qquad (3.15)$$

In this manner, the blocked action (3.10) is made gauge-invariant.

IV. RG EQUATION FOR YANG-MILLS THEORY

In considering the RG flow pattern for the Yang-Mills blocked action, we first note that via gauge invariance, the field renormalization $\mathcal{Z}_k$ found in (3.11) can be related to the coupling constant renormalization via $g^2 = \mathcal{Z}_k^{-1}g^2(\kappa)$, where $g^2(\kappa)$ is the running coupling constant, i.e.

$$\frac{1}{g^2(\kappa)} = \frac{1}{g^2} - \frac{C_2(G)}{48\pi^2}\Big[11\ \ln\Big(\frac{\tilde{\Lambda}^2}{\kappa^2}\Big) + \frac{31}{3} + 4\kappa^2 + (8\kappa^4 + 22\kappa^2 + 11)f(\kappa)\Big], \qquad (4.1)$$

where the scale $\kappa = \tilde{\Lambda}$ corresponds to the bare coupling g; and with $\kappa = 0$ one defines the renormalized coupling constant $g_R(\mu) = g(\kappa = 0, p = \mu)$ for some arbitrary subtraction scale μ^2. Similarly, the β function reads

$$\beta(g(\kappa)) = \frac{dg(\kappa)}{d\ln\kappa} = \frac{C_2(G)}{8\pi^2} g^3(\kappa) \frac{\kappa^2}{4\kappa^2+1} \Big\{ -3 + 4\kappa^2 + 2\kappa^2(5+4\kappa^2) f(\kappa) \Big\}, \tag{4.2}$$

which, being negative for all κ, exhibits the well-known property of asymptotic freedom. In the limit $\kappa \to \tilde{\Lambda}$, it is reduced to the standard result. The RG flow equation can readily be derived from (3.10):

$$k\partial_k \tilde{S}_k(F^2) = \frac{1}{4}\int_p k \frac{\partial \mathcal{Z}_k^{-1}}{\partial k} F^a_{\mu\nu}(\bar{A}) F^a_{\mu\nu}(\bar{A}), \tag{4.3}$$

which along with (3.11) and (4.1), implies that the RG flow equation for the Yang-Mills blocked action (3.10) in the "independent-mode" approximation scheme is simply coupling constant flow equation governed by the β function (4.2). This result is to be expected since in the perturbative evaluation of the Yang-Mills blocked action in the last Section, we have kept only terms which are quadratic in the background field $\bar{A}$; and it is precisely these additional one-loop quadratic terms which renormalize the coupling strength.

To obtain a renormalization group flow equation which contains a richer structure, we note that variation of S_k simply corresponds to variation of $\rho_k(t, \Lambda)$, i.e.,

$$\begin{aligned} k\partial_k \tilde{S}_k(\bar{A}) = -\frac{1}{2}\,\mathrm{Tr}\int_0^\infty \frac{dt}{t^3} k \frac{\partial \rho_k(t,\Lambda)}{\partial k} \Bigg\{ & \Big[e^{-\mathcal{K}^{ab}_{\mu\nu}(\bar{A})t} - e^{-\mathcal{K}^{ab}_{\mu\nu}(0)t} \Big] \\ & - 2\Big[e^{-\mathcal{O}^{ab}(\bar{A})t} - e^{-\mathcal{O}^{ab}(0)t} \Big] \Bigg\}. \end{aligned} \tag{4.4}$$

Since an expansion in small t corresponds to exploring the high energy behavior of the theory, it is not surprising that one can regain from the blocked action $\tilde{S}_k$ the property of asymptotic freedom. However, if one is interested in the infrared behavior of Yang-Mills theory, it then becomes questionable whether or not expanding in t is appropriate. Such dilemma can be resolved by a direct integration of the matrix representation (4.4). This gives, in the thin-shell limit:

$$\begin{aligned} k\partial_k \tilde{S}_k = k^4\,\mathrm{Tr}\Bigg\{ & \Big[\hat{1}^{ab}_{\mu\nu} k^2 + \Big(\frac{\partial^2 \tilde{S}_k}{\partial A^a_\mu \partial A^b_\nu} \Big) \Big]^{-2} - \Big[\hat{1}^{ab}_{\mu\nu} k^2 + \Big(\frac{\partial^2 \tilde{S}_k}{\partial A^a_\mu \partial A^b_\nu} \Big)_{\bar{A}=0} \Big]^{-2} \\ & - 2\Big[\delta^{ab} k^2 + \mathcal{O}^{ab}(\bar{A}) \Big]^{-2} + 2\Big[\delta^{ab} k^2 + \mathcal{O}^{ab}(0) \Big]^{-2} \Bigg\}, \end{aligned} \tag{4.5}$$

where Tr implies trace sum over the whole space-time, Lorentz and the color indices. One sees that this RG flow equation which contains contributions to *all* orders in t is quite different from (4.3).

V. SUMMARY

We have derived in this paper a Yang-Mills blocked action $\tilde{S}_k(F^2)$ using the proper-time perturbation technique. The complication of UV divergences is completely eschewed by inserting a proper-time smearing function (2.4) which, as an explicit construct, also plays the role of momentum regulator. The blocked potential computed in this manner agrees precisely with that obtained via a sharp momentum cut-off, up to $O(V_R''/\Lambda^2)$. However, unlike the sharp cut-off, the advantage of this proper-time smearing approach is that, when computed in the covariant background formalism, it yields a form of $\tilde{S}_k(F^2)$ which is manifestly gauge invariant. The equivalence between $\rho_k(t,\Lambda)$ and the Pauli-Villars regulator was explicitly demonstrated.

The conventional RG flow equation for the Yang-Mills blocked action derived in the independent-mode approximation was shown to correspond to the flow equation for the gauge coupling constant. The higher-order gauge invariant quantities such as $(F^2)^2$ and $f^{abc}F^a_{\mu\nu}F^b_{\nu\lambda}F^c_{\lambda\mu}$ which arise as consequence of integrating out short-distance modes are taken into consideration by the non-linear improved flow equation (4.5).

ACKNOWLEDGEMENTS

I would like to thank Professors H. Fried and Berndt Müller for organizing the workshop, and Professor Janos Polonyi for stimulating discussions.

REFERENCES

[1] W. Pauli and F. Villars, *Rev. Mod. Phys.* **21** (1949) 434.

[2] J. Schwinger, *Phys. Rev.* **82** (1951) 664.

[3] K. Wilson and J. Kogut, *Phys. Rep.* **12C** (1975) 75.

[4] S.-B. Liao and J. Polonyi, *Ann. Phys.* **222** (1993) 122.

[5] S. Coleman and E. Weinberg, *Phys. Rev.* **D7** (1973) 1888.

[6] D. G. C. McKeon and T. N. Sherry, *Phys. Rev* **D35** (1987) 3854; *Phys. Rev. Lett.* **59** 532; A. Rebhan, *Phys. Rev.* **D39** (1989) 3101.

[7] L. F. Abbot, *Nucl. Phys.* **B185** (1981) 189.

ADIABATIC GAUGE THEORY

JOHN P. RALSTON
Department of Physics and Astronomy
and
Kansas Institute for Theoretical and Computational Science
University of Kansas, Lawrence, KS 66045

ABSTRACT

I apply the Born-Oppenheimer approximation to a gauge theory and show how to reconcile it with gauge invariance. Wave functionals used in the adiabatic approximation necessarily break gauge invariance, but this symmetry can be restored after exploiting a novel local symmetry related to transformations of the Berry phase. I then give sufficient conditions for gauge invariance of the adiabatic approximation for the Hamiltonian. As an example I construct a 3+1 dimensional Abelian theory with unusual propagation characteristics, corresponding to one gauge invariant "massive" and one "massless" mode.

1. Gauge Invariance and the Born-Oppenheimer Approximation

A useful strategy for studying complicated quantum mechanical systems is the adiabatic approximation in which "fast" degrees of freedom are treated separately for "slow" ones. Here I study how the Born-Oppenheimer approximation[1] would be realized in a gauge field theory. Unlike path integral approaches to the generating functional, the approximation is a scheme to understand features of selected quantum states. Its validity has to be examined on a case by case basis. Thus, the nature of this procedure is to produce "model states" which may be good for a particular application.

The local symmetries of a gauge theory make the adiabatic approximation problematic. But we know in our hearts that the "infrared" is meanginful even if a bit arbitrarily defined. Closely related is the problem of ultraviolet regularization, which has an intrinsic arbitrariness. The usual "uv cutoff" approach to regularization consists of excluding selected degrees of freedom from calculations. A more physical approach[2] is to integrate out modes above a separation scale Λ, resulting in a "renormalized" Hamiltonian H_Λ which depends on the remaining modes and Λ in a proper way. Then the wave functionals depend on the cutoff but physical observables should not; the time independent Schroedinger equation $H_{\Lambda,s}\Phi_{\Lambda,s} = E_s\Phi_{\Lambda,s}$ with the energy E_s independent of Λ. I suppose that such a separation is made to break the variables into classes of "fast" and "slow". More complicated separations are of course possible.

My main concern here is not whether the approximation (as developed so far) is a good one for some favorite state. My goal, instead, is to answer a conceptual question. How can Born-Oppenheimer be reconciled at all with gauge invariance? The answer is interesting and in my opinion, beautiful.

I work in the functional Schroedinger picture. Denote the fast degrees of freedom by q and their momenta by p; the slow degrees of freedom will be Q with conjugate momenta P. The full wave functional Ψ_s is the overlap at fixed time of the abstract state $/E, s>$ onto field eigenstates:$\Psi_s(q;Q) \doteq < q; Q/E, s >$. Here s are parameters labeling the state with energy eigenvalue E.

Let $\zeta_s(q;Q)$ be the wave functional for the fast modes q solved as if Q were time-independent classical parameters, a background field. The Born-Oppenheimer[1,3] ansatz for the full wave functional is $\Psi_s(q;Q) = \zeta_s(q;Q)\Phi_s(Q)$. The effective Hamiltonian operator H_Q for the slow variables Q is given by taking the expectation value of the full Hamiltonian $H(q,p;Q,P)$ in the fast variable state ζ_s and integrating over the fast variables: $H_Q(Q,P) = (\zeta_s(q;Q)/H(q,p;Q,P)/\zeta_s(q;Q))$. Here the $(\ldots)$ bracket means the functional integral $\int d[q]$ at fixed time. The Hamiltonian functional of the slow variables is then $H_Q(\Phi_s(Q), \Phi_s^*(Q)) = \{\Phi_s(Q)/H_Q(Q,P_Q)/\Phi_s(Q)\}$ where the bracket $\{\ldots\}$ means the remaining fixed-time functional integrals, namely $\int d[Q]$.

2. Exploiting a Separate Symmetry

Already one is apparently in trouble in a gauge theory. Physically we know that there are slow modes ($Q's$) of the gauge fields, denoted by $\mathbf{A}$, with canonical momenta Π_A. These coordinates and their Hamiltonian H_A are intrinsically gauge dependent. Set $A^0 = 0$ to remove some ambiguity. Let us observe that the definition of "slow modes" is going to be convention-dependent, due to space-dependent gauge transformations. Then the renormalized wave functional $\Phi_{\Lambda s}$ of the slow modes is not supposed to be gauge invariant. Yet, in $A^0 = 0$ gauge the full wave functional, hopefully well-enough described by the ansatz, must be gauge invariant according to Gauss' law. How is this to be arranged?

The answer I propose hinges on the Berry phase[5,6]. Recall that the wave functional of any designated slow modes develops a phase, generated via the adiabatic connection $\mathbf{\Gamma_s}(\mathbf{A}) = (\zeta_s(q;\mathbf{A})/i\delta/\delta\mathbf{A}/\zeta_s(q;\mathbf{A}))$. The effects of the adiabatic connection reappear in the slow sector when one takes derivatives using $\Pi_A = -i\delta/\delta\mathbf{A}$; the rule is $H_A(\Pi_A, \mathbf{A}) \rightarrow H_A(\Pi_A - \mathbf{\Gamma_s}(\mathbf{A}), \mathbf{A})$. The slow modes thus get "gauged", in this case, super-gauged. This is the generalization of the result from quantum chemistry of Mead and Truhlar[3], whose work predates Berry's. In addition to the shift of the momenta, H_A also generally contains invariant new functionals of the slow coordinates,[7] denoted $W(\mathbf{A})$, corresponding to "molecular potentials".

Consider now the question of maintaining gauge invariance; for this use an Abelian gauge theory as an illustration. The basic Hamiltonian of the photon sector becomes $H = 1/2 \int d^3x[(-i\delta/\delta\mathbf{A} - \mathbf{\Gamma_s}(\mathbf{A}))^2 + \mathbf{B}^2(x) + \mathbf{j}\cdot\mathbf{A} + W(\mathbf{A})]$, where $\mathbf{B}(x) = \nabla\times\mathbf{A}$ is the magnetic field and $\mathbf{j}$ is the electromagnetic current. This looks bad because $\mathbf{\Gamma_s}(\mathbf{A})$ is not gauge invariant and also depends on the arbitrary separation $\Psi = \zeta\Phi$. However, through its very ambiguity, we have gained an enormous symmetry. Consider certain transformations $V(\Theta)$, which can be called adiabatic gauge transformations of the generalized "first and second" kind:

$$\begin{aligned}\Phi_s(A) &\rightarrow V(\Theta)\Phi_s(A) = exp(i\Theta(A))\Phi_s(A);\\ \Gamma_s(A) &\rightarrow \Gamma_s(A) + \delta\Theta(A)/\delta A.\end{aligned} \tag{1}$$

where $\Theta(\mathbf{A})$ is some functional of $\mathbf{A}$. Under these transformations the gauge-covariant functional derivative $-i\delta/\delta\mathbf{A} - \mathbf{\Gamma_s}(\mathbf{A})$ is invariant. This symmetry is a trivial property of the Born-Oppenheimer ansatz, $\Phi_s\zeta \rightarrow \Phi_s \exp(i\Theta(\mathbf{A}) \exp(-i\Theta(\mathbf{A}))\zeta_s$, before the integration over the fast variables. Naturally, the same symmetry has emerged after integrating.

Note the following observations: first, the full exact wave function is strictly gauge invariant, but the separate factors Φ_s and ζ_s cannot be so. Second, the result of the approximation after the fast variables have been eliminated has a gigantic local symmetry (1), which was never a part of the original theory. Can the system be invariant under the joint action of both transformations? If so, the effects of the real gauge transformations must be compensated by the adiabatic gauge transformations.

Let the standard gauge transformation be represented by an operator $U(\theta); U(\theta)$ $AU^{-1}(\Theta) = \mathbf{A} + \nabla\theta$. Suppose there is a subgroup of the transformations made jointly by the $U(\theta)$ and $V(\Theta)$ operators under which Φ_s is invariant. Then ζ_s and Γ_s must also be invariant. That is,

$$\mathbf{\Gamma_s}(\mathbf{A}) \rightarrow \mathbf{\Gamma_s}(\mathbf{A} + \nabla\theta) + \delta\Theta(\mathbf{A})/\delta\mathbf{A} = \mathbf{\Gamma_s}(\mathbf{A}).$$

It follows that the adiabatic connection's invariant curvature $\beta^{ij} = \partial\Gamma^i_s(\mathbf{A})/\partial A^j - \partial\Gamma^j_s(\mathbf{A})/\partial A^i$ satisfies $\beta_{ij}(\mathbf{A}) = \beta_{ij}(\mathbf{A}+\nabla\theta)$: *a sufficient condition is that the invariant curvature β_{ij} be gauge invariant.* A second condition comes from the definition of $\mathbf{\Gamma_s}(\mathbf{A})$, assuming that is space-translationally invariant. The space divergence $\nabla \cdot \mathbf{\Gamma_s}(\mathbf{A}) = (\zeta_s(q;\mathbf{A})/\nabla \cdot i\delta/\delta\mathbf{A}/\zeta_s(q;\mathbf{A}))$, which is the expectation value of $\nabla \cdot \mathbf{E}$, where $\mathbf{E}$ is the electric field. Then, $\nabla \cdot \mathbf{\Gamma_s}(\mathbf{A})$ *is also gauge invariant.*[†] Furthermore, the two conditions are compatible: if $\mathbf{\Gamma_s}$ is divergence-free under the space derivative, then adding a functional gradient $\delta\Theta(\mathbf{A})/\delta\mathbf{A}$ preserves it as divergence free. Generally, we have sufficient conditions for maintaining gauge invariance of the approximation if we require invariance under $aU + bV$, where depending on topology, a and b may be magic numbers.

3. Applications

I have investigated several applications. One is an Abelian theory, given by the Hamiltonian above with $W(\mathbf{A}) \rightarrow 0$ and $\mathbf{\Gamma_s}(\mathbf{A}) = m\mathbf{s} \times \mathbf{A}$. This corresponds to a "constant adiabatic magnetic field", which by power counting is likely to end up dominating the infrared, whatever $\mathbf{\Gamma_s}(\mathbf{A})$ is. Here $\mathbf{s}$ is some constant T- and P- odd

[†]In a non–Abelian theory $\nabla \cdot \Gamma$ can be gauge covariant.

vector characterizing the state, and m is a scale with dimensions of mass/hc. The resulting Hamiltonian is reasonably unique in depending on one external parameter, self-consistently modifying the "slow" (infrared) modes, while not unduely disturbing the ultraviolet. As anticipated, the Hamiltonian is not strictly gauge invariant, but the classical equations of motion are nicely gauge invariant:

$$\nabla\times\mathbf{B} = -m\mathbf{s}\times\mathbf{E}+\partial\mathbf{E}/\partial t+\mathbf{j}; \qquad \nabla\cdot\mathbf{E} = m\mathbf{s}\cdot\mathbf{B}+\rho; \qquad \nabla\times\mathbf{E} = -\partial\mathbf{B}/\partial t; \qquad \nabla\cdot\mathbf{B} = 0.$$

Here ρ is the charge density, added to the consistent Gauss's law constraint equation. This solvable theory has screened electric fields; the somewhat complicated gauge invariant dispersion relation reveals one "massive" mode (with a gap) and a "massless" (gapless) one. It is interesting to solve the classical electrostatistics problem of a point charge density $\rho = Q\delta^3(\vec{x})$. The new Gauss' law must be inserted into the new Ampere's law. After considerable algebra I find

$$\begin{aligned}
\mathbf{E}(\mathbf{x}) &= -Q\nabla \int \frac{d^3k}{(2\pi)^3} \frac{\mathbf{k}^2 e^{ik\cdot x}}{(\mathbf{k}^2)^2 + m^2(\mathbf{k}^2 - (\mathbf{k}\cdot\mathbf{s})^2)}\,; \\
\mathbf{B}(\mathbf{x}) &= mQ(\mathbf{s}\nabla^2 - \mathbf{s}\cdot\nabla\nabla) \int \frac{d^3k}{(2\pi)^3} \frac{e^{ik\cdot x}}{(\mathbf{k}^2)^2 + m^2(\mathbf{k}^2 - (\mathbf{k}\cdot\mathbf{s})^2)}\,.
\end{aligned}$$

One sees that the short distance behavior of the electric field is Coulomb–like, confirming the power counting quoted earlier. At long distances the fields fall off more quickly than the usual Coulomb case, and are complicated functions of direction. The electric flux escaping from the system vanishes: $\oint \mathbf{E}\cdot d\mathbf{S} = 0$, on a sphere at infinity. This occurs because Gauss' law in the new theory couples the longitudinal and transverse modes, allowing electric flux to disappear into the magnetic field, which tends to screen. This is more like a non–Abelian theory than an Abelian one.

The procedure is heuristic; once a connection is guessed it may be better to examine the equivalent Lagrangian for directly checking symmetry. The Lagrangian for the non-Abelian generalization of the above is $L = -Tr[F^{\mu\nu}F_{\mu\nu}] + ms\cdot k$, where $s^2 < 0$ and k^μ is the usual quantity defined so that $\partial_\mu k_\mu = \varepsilon^{\mu\nu\alpha\beta}Tr[F_{\mu\nu}F_{\alpha\beta}]$. This looks very Chern–Simons like, and in fact, one can view the result as a promotion of such theories to a higher dimension. Again, the states are not strictly gauge invariant, so it would not serve for a fundamental theory, but it is a fine candidate for invariance under the joint symmetries of the adiabatic approximation.

The perturbative expansion of the non–Abelian theory is conventional if one uses improved infrared self-screened propagators of the solvable Abelian theory. The theory is asymptotically free by power counting: the ultraviolet is not affected by operators $\mathbf{\Gamma_s}(\mathbf{A})$ which involve only one power of $\mathbf{A}$ and no derivatives. Very interestingly, the perturbation theory reveals an infrared Cerenkov instability, so that non-color singlet modes cannot propagate.[8] This occurs because the gapless mode is concave–upward. A "gluon" propagating with such characteristics has a higher phase velocity than other gluons with lower frequency, and decays spontaneously. The result is that color non–singlets cannot propagate. This, and the coupling of gluon topological current to the background vector **s**, suggests that the phenomenology of gluons in the proton state might possibly be modeled by the 'constant adiabatic magnetic field' ansatz, with the parameter s identified as the proton *spin*. This and other speculations are discusssed elsewhere.[8]

4. Acknowledgements

This work was supported in part under Department of Energy Grant Number DE FGO2 85 ER 40214 and by the *Kansas Institute for Theoretical and Computational Science.*

5. References

1. A. Messiah, *Quantum Mechanics*, (North Holland, Amsterdam 1962) Vol. 2.
2. See, e.g., K. Wilson, *Phys. Rev.* **D 2**, 1438 (1970).
3. C. A. Mead and D. G. Truhlar, *J. Chem. Phys.* **70**, 2284 (1979); I. J. R. Aitchison, *Physica Scripta* **T23**, 12,(1988)
4. For early work in the functional Schroedinger representation see J. Schwinger, *Nuovo Cimento* **30**, 278 (1963); H. Loos, *Phys. Rev.* **188**, 2342 (1969). For a more recent exposition, see H. Cheng and E.- R. Tsai, *Chinese J. Phys.* **25**, 95 (1987).
5. M. V. Berry, *Proc. R. Soc. London Ser.* **A292**, 45 (1984).
6. Y. S. Wu and A. Zee, *Nuc. Phys.* **B 258**, 157 (1985); A. Niemi and G. Semenoff, *Phys. Rev. Lett.* **55**, 927 (1985).
7. B. Zygelman, *Phys. Lett.* **A 125**, 476 (1987); J. P. Ralston, *Phys. Rev.* **A40**, 4872 (1989).
8. J. P. Ralston, in preparation.

ENTROPY AND VACUUM-INDUCED QUANTUM DECOHERENCE [1]

Hans-Thomas Elze

CERN-Theory, CH-1211 Geneva 23, Switzerland [2]

ABSTRACT

Or: "How to generate an ensemble in a single event?"– Following recent work on entropy in strong interactions, I explain the concept of environment-induced quantum decoherence in elementary quantum mechanics. The classically chaotic inverted oscillator becomes partially decoherent already in the environment of a single other oscillator performing only vacuum fluctuations. One finds exponential entropy growth in the subsystem with a Lyapunov exponent, which approaches the classical one for weak coupling.

Recently the long-standing "entropy puzzle" of high-multiplicity events in strong interactions at ultra-relativistic energies has been analysed from a new point of view [1]. This is related to the concepts of an *open quantum system* and *environment-induced quantum decoherence.* The problem dates back to Fermi and Landau and is intimately connected to understanding the rapid thermalization of high energy density ($\gg$ 1 GeV/fm^3) matter [2]. Why do thermal models work so well in reproducing global features of hadronic multiparticle final states? Why do they work at all?

Or, Why does high-energy scattering of pure initial states lend itself to a statistical description characterized by a large apparent *entropy* from a mixed-state density matrix describing intermediate stages in a space-time picture of parton evolution? Effectively, *unitary time evolution* of the observable part of the system breaks down in the transition from a quantum mechanically pure initial state to a highly impure (more or less thermal) high-multiplicity final state. This was discussed in detail in Refs. [1], where more references concerning formal aspects of this work can be found. Based on analogies with studies of the quantum measurement process [3] and motivated by related problems in quantum cosmology [4], I argued that **environment-induced quantum decoherence solves the entropy puzzle** of strong interactions.

A complex pure-state quantum system can show a quasi-classical behaviour, i.e. an impure density (sub)matrix together with decoherence of the associated pointer states in an observable subsystem [1, 3, 4]. I will demonstrate in the following that the decoherence process is uniquely correlated with entropy production. Considering strong interactions, in particular, there is a natural *Momentum Space Mode Separation* due to confinement. Thus, almost constant QCD field configurations form an *unobservable environment*, which interacts with the *observable subsystem* composed of

---

[1] Work supported by the Heisenberg Programme (Deutsche Forschungsgemeinschaft).

[2] E-mail address: ELZE@CERNVM.CERN.CH

partons. The environment modes are unobservable, since they can neither hadronize nor initiate hard scattering among themselves, whereas the partons are observable in the sense of parton-hadron duality or deep-inelastic scattering; equivalently, low-energy coloured vacuum fluctuations cannot propagate into asymptotic states.

Previously, I studied the induced quantum decoherence and entropy production in a non-relativistic single-particle model resembling an electron coupled to the quantized electromagnetic field, however, with a deliberately enhanced oscillator spectral density in the infrared. The remarkable result being that in the *short-time strong-coupling limit* the model parton behaves like a *classical particle* [1]: Gaussian parton wave packets experience *friction* and *localization*, i.e. no quantum mechanical spreading, and their coherent *superpositions decohere.*

Two coupled scalar fields representing partons and their non-perturbative environment, respectively, seem somewhat more realistic. In the functional Schrödinger picture I derived a Cornwall-Jackiw-Tomboulis (CJT) type effective action and the equations of motion for renormalizable interactions [1]. Thus, analysis of the entropy puzzle in strong interactions leads to study an observable field (open subsystem) interacting with a dynamically hidden one (unobservable environment), i.e. *quantum field Brownian motion.*

Summarizing, partons feel an unobservable (gluonic) environment, which manifests its strong non-perturbative interactions on a short time scale ($\ll 1$ fm/c) through decoherence of suitable partonic pointer states [3], their quasi-classical behaviour, and entropy production. The emergence of *structure functions* from initial-state wave functions can and will be further studied in this approach. – Instead of representing the formalism and more technical results from Refs. [1], I want to demonstrate here in simple quantum mechanical examples the basic Why and How of the solution to the entropy puzzle.

Consider a system that can be described in terms of two normalized discrete basisstates, $|1\rangle$ and $|2\rangle$. Forming a *pure state*, $|\psi\rangle \equiv a_1|1\rangle + a_2|2\rangle$, by a coherent superposition with amplitudes $a_1 \equiv p^{1/2}$ and $a_2 \equiv (1-p)^{1/2}$, the corresponding density matrix, $\hat{\rho} \equiv |\psi\rangle\langle\psi|$, is

$$\rho_{ij} = \begin{pmatrix} a_1^2 & a_1a_2 \\ a_1a_2 & a_2^2 \end{pmatrix} \longrightarrow \rho_{ij}^D = \begin{pmatrix} 1 & 0 \\ 0 & 0 \end{pmatrix} , \tag{1}$$

where $\hat{\rho}^D$ is obtained by diagonalization. Note the off-diagonal *interference terms* in ρ_{ij}. Furthermore, observe that $\hat{\rho}^D$ has only one non-vanishing eigenvalue. Introducing the **von Neumann** or **statistical entropy**,

$$S[\hat{\rho}] \equiv -\operatorname{Tr}\hat{\rho}\ln\hat{\rho} \ , \tag{2}$$

we find $S[\hat{\rho}] = S[\hat{\rho}^D] = -[1\ln 1 + 0\ln 0] = 0$, i.e. *no entropy in a pure state.* Secondly, forming a *mixed state* (ensemble) such that the system is in state $|1\rangle$ with probability

[3] In general, these are not single-particle states but rather coherent (Gaussian) wave functionals, as constructed in the second of Refs. [1].

p and in state $|2\rangle$ with probability $1-p$, the density matrix becomes $\hat{\rho}' = |1\rangle p\langle 1| + |2\rangle(1-p)\langle 2|$, i.e. a decoherent superposition. Hence, we obtain

$$\rho'_{ij} = \begin{pmatrix} p & 0 \\ 0 & 1-p \end{pmatrix} . \tag{3}$$

The density matrix (3) shows *no interference terms* and is diagonal per se.[4] Then, $S[\hat{\rho}'] \equiv S(p) = -[p \ln p + (1-p)\ln(1-p)] \neq 0$; generally, $0 \leq S(p) \leq S(1/2) = \ln 2$. Total ignorance ($p = 1/2$) corresponds to $\ln 2$ units of entropy in a two-state system.

One concludes that *entropy production can only occur if the interference terms of the density matrix representing a more or less pure state of the observed system decay dynamically.*[5] In a *closed system* evolving unitarily in time, however, there is no way to transform, for example, $\hat{\rho}$ into $\hat{\rho}'$, see Eqs. (1), (3). Only the interaction of the system with an environment can have such an effect [1, 3, 4], i.e. *environment-induced quantum decoherence.*

Next, consider a non-relativistic particle moving in a one-dimensional double-well potential presenting the observable subsystem. Let it be coupled translationally invariant to an environment consisting of a single oscillator. The classical action for unit masses and after a suitable rescaling is

$$S = \int dt \left\{ \tfrac{1}{2}\dot{x}^2 + \tfrac{1}{2}\dot{y}^2 - \tfrac{1}{2}\omega^2(y-x)^2 + \tfrac{1}{2}x^2 - \tfrac{1}{4!}\lambda x^4 \right\} , \tag{4}$$

in terms of dimensionless quantities and two coupling constants, $\omega^2, \lambda \geq 0$. For $\omega = 0$ the minima of the doublewell lie at $x_\pm = \pm(\lambda/3!)^{-1/2}$, at a depth of $-3/2\lambda$ (the local maximum is zero at $x = 0$). Presently, I want to study the case that the excitation energy of the x-particle (X) is smaller than the level spacing ω of the environment oscillator (Y), which is assumed to be in its ground state. Starting with a given initial state of X, I will calculate the time evolution of the corresponding density matrix $\hat{\rho}_X$ under the influence of the *vacuum fluctuations* of Y. Excited states of Y contribute only *virtually* here; they cannot become real due to energy conservation.

For illustration, I choose the metastable initial state of X, when classically the particle "rests on top of the hill" ($x = 0$). Quantum mechanically this can be represented by a minimum uncertainty Gaussian wave packet,

$$\psi(x, t=0) = \pi^{-1/4} w_0^{-1/2} \mathrm{e}^{-\frac{1}{2}x^2/w_0^2} , \tag{5}$$

with $w_0 \ll \lambda^{-1/2}$. Also, assume ω to be sufficiently larger than $3/2\lambda$.

First of all, let the system evolve **classically**. Nothing will happen. However, any infinitesimal perturbation of the fine-tuned initial conditions causes X to move "down the hill", left or right, dragging Y along. There is local chaos in the sense of extreme

[4]Note that $\mathrm{Tr}\,\hat{\rho} = \mathrm{Tr}\,\hat{\rho}' = 1$: the system is in *some* state with total probability 1.

[5]The argument does not depend on particular physical characteristics of the system; it holds for the two-state system as well as for an interacting quantum field.

sensitivity to the initial conditions at $x, y \approx 0$ leading to a *loss of predictability*. Then, the classical entropy is calculated after "coarse graining", i.e. by constructing a local probability density $f(t)$ in the phase space of X related to the ensemble average over initial conditions, $S_{c.g.}(t) \equiv \int dxdp\ f(t) \ln f(t) + C$. A chaotic loss of predictability from strongly diverging trajectories in phase space causes $S_{c.g.}$ to increase: as time passes, more and more cells of the coarse graining contribute to an effective growth of the volume occupied by the ensemble. – In conclusion, *in a classical system, be it chaotic or not (with or without coarse graining), entropy can only be produced IF there is a physically relevant ensemble of initial conditions.* Thus, one cannot explain altogether classically entropy production or thermalization in a single high-multiplicity event in strong interactions.

Secondly, let the system evolve **quantum mechanically**. To begin with, let there be *no* coupling to the environment ($\omega = 0$). Even with the fine-tuned initial condition, Eq. (5), the amplitude ψ to find X at a particular space-time point begins to flow "down the hill" symmetrically due to the quantum spreading of the wave packet. For a free unit mass particle $w_0 \rightarrow w(t) = (w_0^2 + w_0^{-2}t^2)^{1/2}$; here one expects an accelerated spreading "downhill", cf. Eq. (11) below. The system remains in a *pure quantum state.* The density matrix is $\hat{\rho}_X(t) = |\psi(t)\rangle\langle\psi(t)|$, $|\psi(t)\rangle = \exp[-i\hat{\mathrm{H}}_0 t]|\psi(0)\rangle$, where $\hat{\mathrm{H}}_0$ is the Hamiltonian of X from Eq. (4) with $\omega = 0$. Therefore, $S[\hat{\rho}_X(t)] = -1\ln 1 = 0$, cf. Eq. (2). Quantum mechanically one knows everything there is to know about a *closed system* (X), given any pure initial state and its Hamiltonian, which consistently yields $S[\hat{\rho}_X] = 0$. Even with an ensemble of initial states, i.e. an impure density matrix $\hat{\rho}_X(0)$, there is *no entropy production*, since $S[\hat{\rho}_X(t)] = S[\exp(-i\hat{\mathrm{H}}_0 t)\hat{\rho}_X(0)\exp(+i\hat{\mathrm{H}}_0 t)] = S[\hat{\rho}_X(0)]$ stays constant.[6]

The situation changes completely if the observable subsystem (X) evolves quantum mechanically coupled to the *vacuum fluctuations and virtual excitations of the environment* (Y). According to the above assumptions, the initial density matrix of the total system is:

$$\hat{\rho}(t=0) \ \equiv \ \hat{\rho}_X(0) \otimes \hat{\rho}_Y(0) \ , \tag{6}$$

with matrix elements $\rho_X(x, x'; 0) = \pi^{-1/2} w_0^{-1} \exp[-\frac{1}{2}(x^2 + x'^2)/w_0^2]$ and $\rho_Y(y, y'; 0) = (\omega/\pi)^{1/2} \exp[-\frac{1}{2}\omega(y^2 + y'^2)]$. The time evolution of the density matrix of the subsystem, $\hat{\rho}_X(t) = \mathrm{Tr}_Y \hat{\rho}(t)$, can be calculated with the Feynman-Vernon influence functional technique making use of general results obtained in the first of Refs. [1].

It should be remarked that the *final state* of the environment may contain virtual excitations of Y.[7] The relevance for a high-multiplicity reaction is the following: Even though the QCD vacuum "far away" conforms to the usual one before and after, the secondary hadrons all require a dressing of their valence quarks by localized virtual

---

[6]The unitary (time evolution) transformation does not change the eigenvalues of $\hat{\rho}_X$. Thus, the statistical entropy, Eq. (2), cannot possibly show a sign of classical chaos in a closed system.

[7]As a corollary to the *Schmidt decomposition* [1] it is easy to prove that starting with an overall pure state of the complex system, cf. Eq. (6), IF the final state of the environment is a pure state, THEN the observable subsystem ends up in a pure state too (without entropy production).

excitations of the vacuum or environment, which makes an essential difference as compared to the initial state.

Presently, the resulting density matrix $\hat{\rho}_X(t)$ is (cf. also the first of Refs. [1]):

$$\begin{aligned}\rho_X(z_-, z_+, t) &= \pi^{-1/2} w^{-1}(t)\, e^{-[z_+ - v(t)t]^2/w^2(t)} \\ &\times e^{-z_-^2\{C + \frac{1}{4}w_0^2 c^2 - d^{-2}[B + \frac{1}{2}w_0^2 bc]^2/w^2(t)\}} \\ &\times e^{iz_-\{(az_+ - 2d^{-1}[B + \frac{1}{2}w_0^2 bc][z_+ - v(t)t]/w^2(t)\}} \quad , \qquad (7)\end{aligned}$$

with the effective velocity $v(t) = 0$ for the zero-momentum initial wave packet, the effective width $w(t) \equiv 2\xi|d|^{-1}$, $\xi \equiv (A + \frac{1}{4}w_0^{-2} + \frac{1}{4}w_0^2 b^2)^{1/2}$, and with rather complicated time-dependent coefficients A, B, C, a, b, c, d, to be discussed elsewhere; the coordinates in Eq. (7) are $z_- \equiv x - x'$ and $z_+ \equiv \frac{1}{2}(x + x')$ in terms of ordinary one-dimensional ones. Since we are particularly interested in the decoherence process and entropy production, we consider only the simplest *off-diagonal density matrix elements* here, $\rho_X(x, x' = -x, t) = \rho_X(z_- = 2x, z_+ = 0, t)$. They can be directly related to the **linear entropy** produced in the observable subsystem (X):

$$\begin{aligned}S^{lin} &\equiv \mathrm{Tr}\,[\hat{\rho}_X - \hat{\rho}_X^2] = 1 - \int_{-\infty}^{\infty} dz_- \int_{-\infty}^{\infty} dz_+\, \rho_X(z_-, z_+, t)\, \rho_X(-z_-, z_+, t) \\ &= 1 - \tfrac{1}{2} c_1^{-1/2} w^{-1} \quad , \qquad (8)\end{aligned}$$

with $c_1 \equiv C + \frac{1}{4}w_0^2 c^2 - (B + \frac{1}{2}w_0^2 bc)^2/(dw)^2$ and independently of the initial wave packet momentum ($p = 0$ at present). Thus, inserting (8) into (7), one obtains:

$$\rho_X(x, -x, t) = \pi^{-1/2} w^{-1}(t) \exp\left\{-x^2 w^{-2}(t)[1 - S^{lin}(t)]^{-2}\right\} \quad , \qquad (9)$$

$$\int_{-\infty}^{\infty} dx\, \rho_X(x, -x, t) = 1 - S^{lin}(t) \geq e^{-S(t)} \quad . \qquad (10)$$

The inequality results from the fact that the linear entropy provides a *lower bound* for the relevant statistical entropy, cf. Eq. (2), as shown in [1].

At this point the attentive reader might wonder what happened to the non-linear interaction $\propto \lambda x^4$ of the double-well potential, see Eq. (4). Of course, it cannot be treated exactly. I employed a mean-field-type approximation, replacing $\frac{1}{4!}\lambda x^4$ by $\frac{1}{2}\lambda\langle x^2\rangle x^2 \equiv \frac{1}{2}\Lambda^2(t)x^2$. As long as one studies only the initial time-evolution over short periods, as compared to the time a classical particle would need to "roll down the hill", one may even set $\Lambda \approx 0$. For the following qualitative considerations, Λ plays the role of an adiabatically changing parameter.

To begin with, it can be checked explicitly that there is no entropy production for a vanishing coupling to the environment, $S^{lin}_{\omega=0}(t) = 0$, cf. Eq. (8). Next, calculating the effective width in the *long-time limit* [8], one finds:

$$w(t) \;=\; \left(w_0^2 + w_0^{-2} f_-^{-2} + \omega^3 f_-^{-2}[f_-^2 + \omega^2]^{-1}\right)^{1/2} \frac{f_-^2 + \omega^2}{f_-^2 + f_+^2} \;\exp\, t_- \quad , \tag{11}$$

with $t_- \equiv f_- t$, $f_\pm \equiv [\pm\frac{1}{2}\omega_+^2 + (\frac{1}{4}\omega_+^4 - \omega^2\omega_-^2)^{1/2}]^{1/2}$, and $\omega_\pm^2 \equiv \pm\omega^2 + \Lambda^2 - 1$. Assuming a sufficiently small coupling, $\omega^2 < 1$, note that $\omega_\pm^2$ is *negative* as long as $\Lambda^2(t) < 1 - \omega^2$. Thus, the *width grows exponentially* with an *effective Lyapunov exponent* f_-. For vanishing coupling to the environment, it reduces to the classical Lyapunov exponent of the inverted oscillator, $f_-^{\omega=0} = (1 - \Lambda^2)^{1/2}$, while the width becomes $w_{\omega=0}(t) = (w_0^2 + [w_0 f_-^{\omega=0}]^{-2})^{1/2} \exp(f_-^{\omega=0} t)$. – It is remarkable how the Lyapunov exponent reflects the dynamics: as the wave packet spreads "downhill", $\Lambda^2(t) \propto \langle x^2 \rangle$ increases until f_- reaches zero (becoming purely imaginary afterwards), when $\omega_- = 0$. At this point the behaviour becomes regular in the sense of being governed by harmonic motions close to the minima of the double-well potential with a correspondingly milder time-dependence of the width (cf. the model studied in the first of Refs. [1]).

The second dynamical time scale f_+^{-1} always stays real. It is relevant for certain non-Markovian effects generated by the interaction with the environment ($f_+^{\omega=0} = 0$). These become clearly visible in the entropy evaluated in the same limit as Eq. (11). Using $c_1 = C + \mathrm{O}(w_0^2)$, one obtains in leading order:

$$S^{lin}(t) \;=\; 1 - w^{-1}(t)\, \frac{f_-^2 + \omega^2}{\omega^{3/2} f_-} \;\times \tag{12}$$

$$\left[[(\frac{f_-[f_+^2 - \omega^2]}{f_+[f_-^2 + f_+^2} + \frac{f_+[f_-^2 + \omega^2]}{f_-[f_-^2 + f_+^2}) \sin t_+ + \frac{2\omega^2 \cos t_+}{f_-^2 + f_+^2}]^2 + \omega^2 [\frac{\sin t_+}{f_+} - \frac{\cos t_+}{f_-}]^2 \right]^{-1/2} ,$$

with $t_+ \equiv f_+ t$. Thus, the linear entropy approaches exponentially its saturation value 1 on the time scale set by the Lyapunov exponent, see Eq. (11), and the von Neumann entropy grows exponentially according to eq. (10), at least as fast. Note that the periodic function multiplying $w^{-1}(t)$ in Eq. (12) stays qualitatively the same even until the effective Lyapunov exponent becomes imaginary, when the stabilizing effect of the x^4-term in the potential is felt.

To summarize, in the above specified long-time limit and for the chosen initial conditions, Eq. (6), one obtains the observable subsystem (X) density matrix,

$$\rho_X(z_-, z_+, t) \;=\; \pi^{-1/2} w^{-1}(t)\, \mathrm{e}^{-\{z_+^2 + \frac{1}{4} z_-^2 [1 - S^{lin}(t)]^{-2}\} w^{-2}(t) + i z_- z_+ f_-} \; . \tag{13}$$

Even though this density matrix describes the exponential entropy production and its diagonal matrix elements with $z_- = 0$ grow rapidly, apart from the overall normalization factor, the (simplest) off-diagonal matrix elements ($z_+ = 0$) do *not* really

[8]Here, hyperbolic functions dominate over trigonometric ones in the time-dependent coefficients in Eq. (7) for times such that a classical particle would still be "rolling down the hill" of the potential; this restriction is presently assumed for simplicity.

decay here as in usual models of quantum decoherence [1, 3]. This is no surprise in view of the "poor environment" of one "frozen" degree of freedom considered at present, which is unable to accommodate all the phase information contained in the off-diagonal density matrix elements of the subsystem.

In conclusion, a strong observable entropy production in a quantum system, which shows a chaotic behaviour in the classical limit with exponentially growing modes, requires only a minimal decohering effect due to an environment of vacuum fluctuations coupled to it from a higher energy scale.

In particular, complementary to previous studies [3, 5], one observes here that the environment does not necessarily have to be at any finite temperature for this effect of *partial decoherence* to work. Furthermore, it should be realized that from the point of view of calculating the entropy, cf. Eq. (2), the roles of "subsystem" and "environment" can be interchanged [1] and it is a matter of practicability to decide which part of the total Hilbert space is integrated out to find the entropy of the *physically observed subsystem.* The environment-induced quantum decoherence and its relevance for entropy production have presently been illustrated by an elementary example, which, however, points out to interesting consequences for the quantum evolution of classically chaotic non-linear field theories [6].

I thank P. Carruthers, M. Danos, N. E. Mavromatos, and J. Rafelski for stimulating criticism and helpful discussions. I particularly thank H. Fried and B. Müller for discussions as well as for the invitation to this workshop.

References

[1] H.-Th. Elze, preprint CERN-TH.7131/93 (hep-ph/9404215), submitted to Nucl. Phys. B; CERN-TH.7297/94 (hep-th/9406085), submitted to Phys. Rev. Lett.

[2] E. Fermi, Progr. Theor. Phys. 5 (1950) 570; Phys. Rev. 81 (1951) 683;
L. D. Landau, Izv. Akad. Nauk SSSR, Ser. fiz., 17 (1953) 51;
E. Stenlund et al., eds., Proc. "Quark Matter '93", Nucl. Phys. A566 (1994).

[3] W. H. Zurek, Phys. Today 44, No. 10 (1991) 36;
R. Omnès, Rev. Mod. Phys. 64 (1992) 339;
H. D. Zeh, Phys. Lett. A172 (1993) 189.

[4] M. Gell-Mann and J. B. Hartle, Phys. Rev. D47 (1993) 3345;
J. Ellis, N. E. Mavromatos and D. V. Nanopoulos, Phys. Lett. B293 (1992) 37; preprint CERN-TH.7195/94; and references therein.

[5] W. H. Zurek and J. P. Paz, Phys. Rev. Lett. 72 (1994) 2508.

[6] B. Müller, contribution to these Proceedings.

Thermal Infrared Singularities

HARD THERMAL LOOPS AND COLLECTIVE EXCITATIONS

Jean-Paul BLAIZOT* and Edmond IANCU
Service de Physique Théorique†, CE-Saclay
91191 Gif-sur-Yvette, France

ABSTRACT

Hard thermal loops play a central role in the theory of long wavelength excitations of a quark-gluon plasma. Our recent derivation of their generating functional from the Dyson-Schwinger equations sheds light on their semi-classical nature and their simple gauge structure.

*CNRS
†Laboratoire de la Direction des Sciences de la Matière du Commissariat à l'Energie Atomique

1. Introduction

"Hard thermal loops" (HTL) are the dominant contributions at high temperature to n-point amplitudes with soft external momenta ($P \sim gT$, with T the temperature and g the gauge coupling, assumed to be small). Their role in perturbative calculations, and their remarkable properties, where first established via a systematic diagrammatic analysis.[1–3] Their gauge invariance has been used by Taylor and Wong to construct a generating functional.[4] Later, this generating functional has been rewritten in a manifestly gauge invariant effective action by Braaten and Pisarski[5] and also by Frenkel and Taylor.[6] It was also realized that this effective action could be understood as the eikonal of a Chern-Simons gauge theory.[7]

In a series of recent papers, we have shown that the hard thermal loops are intimately related to the long wavelength, collective, excitations of the hot gauge plasmas, much alike the non trivial dielectric properties of the electromagnetic plamas are related to the collective oscillations of their charge constituents. We have developed a theory of the collective excitations based on a perturbative analysis of the QCD Dyson-Schwinger equations.[8] It is obtained by consistently preserving the leading terms in an expansion in powers of the gauge coupling g. In doing so, it is important to take into account that g controls three types of approximations: the weak coupling approximation ($g \ll 1$), the long wavelength approximation ($\lambda \sim 1/gT \gg 1/T$), and the small amplitude approximation (the gauge field strength tensor is limited by $F \lesssim gT^2$, or, equivalently, the gauge potentials satisfy $A \lesssim T$). We present here a summary of this approach.

Mean field equation

The equation of motion for A_μ^a takes the form

$$[\,D^\nu,\, F_{\nu\mu}(x)\,]^a \;=\; j_\mu^a(x), \tag{1.1}$$

where $D_\mu = \partial_\mu + igA_\mu(x)$, ($A_\mu \equiv A_\mu^a t^a$), $F_{\mu\nu} = [D_\mu, D_\nu]/(ig)$ is the field strength tensor, and $j_\mu(x)$ is the induced current, which transforms as a color vector in the adjoint representation.

Structure of the induced current

It is convenient to separate in the induced current the contributions of the fermions from those of the transverse gluons, by writing $j_\mu^a = j_{\mathrm{f}\,\mu}^{\,a} + j_{\mathrm{b}\,\mu}^{\,a}$. For the fermions (with N_f flavors) we have

$$j_{\mathrm{f}\,\mu}^{\,a}(x) \;=\; gN_f \int \frac{d^3k}{(2\pi)^3}\, v_\mu \Big(\delta n_+^a(\mathbf{k},x) - \delta n_-^a(\mathbf{k},x)\Big), \tag{1.2}$$

where $v^\mu \equiv (1,\, \mathbf{v})$ and $\mathbf{v} \equiv \mathbf{k}/k$ is the velocity of the hard particle ($k \equiv |\mathbf{k}|$, and $|\mathbf{v}| = 1$). The quantities $\delta n_\pm^a(\mathbf{k},x)$ are the components of a color vector in the adjoint representation. They may be combined into a density matrix, $\delta n_\pm \equiv \delta n_\pm^a\, t^a$, ultimately related to the Wigner transform of the fermion propagator. Similarly, the current carried by the transverse gluons can be written in terms of a density matrix $\delta N(\mathbf{k},x) \equiv \delta N^a(\mathbf{k},x)\, T^a$:

$$j^{\,a}_{\mathrm{b}\,\mu}(x) = 2g \int \frac{d^3k}{(2\pi)^3}\, v_\mu \, \mathrm{Tr}\, T^a \, \delta N(\mathbf{k},x) = 2gN \int \frac{d^3k}{(2\pi)^3}\, v_\mu \, \delta N^a(\mathbf{k},x). \tag{1.3}$$

Kinetic equations for $\delta n_\pm$ and δN

The color density matrices $\delta n_\pm$ and δN are determined by the following kinetic equations[8]

$$[v\cdot D_x,\, \delta n_\pm(\mathbf{k},x)] = \mp\, g\, \mathbf{v}\cdot\mathbf{E}(x)\, \frac{dn_k}{dk}, \tag{1.4}$$

$$[v\cdot D_x,\, \delta N(\mathbf{k},x)] = -\, g\, \mathbf{v}\cdot\mathbf{E}(x)\, \frac{dN_k}{dk}, \tag{1.5}$$

where $E^i_a \equiv F^{i0}_a$ is the chromoelectric field. Furthermore, $N_k \equiv 1/(\exp(\beta k)-1)$ and $n_k \equiv 1/(\exp(\beta k)+1)$ are, respectively, the equilibrium boson and fermion occupation factors. Note that, in deriving these equations, we have used the isotropy (in $\mathbf{k}$ space) of the equilibrium distribution functions. In the abelian case, eq.(1.4) reduces to the well known linearized Vlasov equation.

For the subsequent analysis of these equations, it is convenient to express $\delta n_\pm$ and δN in terms of new functions, $W^\mu(x;v) \equiv W^\mu_a(x;v) t^a$, solutions to

$$[v\cdot D_x,\, W^\mu(x;v)] \;=\; F^{\mu\nu}(x)\, v_\nu. \tag{1.6}$$

The quantities $W^\mu_a(x;v)$ are the components of a color vector in the adjoint representation, and satisfy

$$v_\mu \, W^\mu_a(x;v) \;=\; 0, \tag{1.7}$$

so that $W^0_a = v^i W^i_a$. From the equations above, it is easily seen that

$$\delta n^a_\pm(\mathbf{k},x) = \mp\, g W^0_a(x;v)\, \frac{dn}{dk}, \qquad \delta N^a(\mathbf{k},x) = -g W^0_a(x;v)\, \frac{dN}{dk}. \tag{1.8}$$

Solving the kinetic equations

The equation (1.6) for W^μ_a is easily solved once boundary conditions are given. We assume here retarded conditions, such that the fields A_μ vanish as $x_0 \to -\infty$. The corresponding solution of eq. (1.6) reads

$$W^\mu(x;v) \;=\; \int_0^\infty d\tau\, U(x, x-v\tau)\, F^{\mu\nu}(x-v\tau)\, v_\nu \, U(x-v\tau, x), \tag{1.9}$$

where $U(x,y)$ is the parallel transporter along the straight line γ joining x and y,

$$U(x,y) = P\exp\{-ig\int_\gamma dz^\mu A_\mu(z)\}, \tag{1.10}$$

with P denoting the path-ordering operator. Thus,

$$U(x, x - v\tau) = P \exp\left\{-ig \int_0^\tau ds\, v \cdot A(x - v(\tau - s))\right\}. \tag{1.11}$$

Induced current in closed form

Once the solution of the kinetic equation is known, one can easily calculate the induced current in closed form. By inserting eqs. (1.8) in the expressions (1.2) and (1.3), and performing the integration over $k = |\mathbf{k}|$, we obtain

$$j_a^\mu(x) \;=\; 3\,\omega_p^2 \int \frac{d\Omega}{4\pi}\, v^\mu\, W_a^0(x; v). \tag{1.12}$$

Here ω_p is the *plasma frequency*, $\omega_p^2 \equiv (2N + N_{\rm f})g^2T^2/18$. The integral $\int d\Omega$ runs over all the directions of the unit vector $\mathbf{v}$. According to eqs. (1.9) and (1.12), the retarded current reads

$$j^\mu(x) \;=\; 3\,\omega_p^2 \int \frac{d\Omega}{4\pi}\, v^\mu \int_0^\infty d\tau\, U(x, x - v\tau)\, \mathbf{v} \cdot \mathbf{E}(x - v\tau)\, U(x - v\tau, x). \tag{1.13}$$

This expression can be viewed as a generating functional of the hard thermal loops with retarded boundary conditions.[8] The current (1.12) is covariantly conserved,

$$\left[D_\mu, j^\mu(x)\right] \;=\; 0, \tag{1.14}$$

as necessary for the consistency of the mean field equation (1.1). The "abelian-like" Ward identities relating the HTL's, which were first identified in the diagrammatic approach,[2,3] are a direct consequence of eq. (1.14).

Physical interpretation of W_μ

Consider first a QED plasma, for which eq. (1.6) reduces to

$$(v \cdot \partial_x)\, W^\mu(x; v) \;=\; F^{\mu\nu}(x)\, v_\nu. \tag{1.15}$$

This is very much similar to the equation of motion for a particle in an electromagnetic field, viz.

$$\frac{dk^\mu}{dt} \;=\; e\, F^{\mu\nu}(x)\, v_\nu, \tag{1.16}$$

where k_μ is the *kinetic* momentum of the particle. To make the analogy closer, we remark that the line derivative in the l.h.s. of eq. (1.15), i.e. $v \cdot \partial_x = \partial_t + \mathbf{v} \cdot \boldsymbol{\nabla}$, may be interpreted as the total time derivative along the trajectory $\mathbf{x}(t) = \mathbf{x}_0 + \mathbf{v}t$ of a fictitious particle. That is, eq. (1.15) may be rewritten as

$$\frac{d}{dt}\, W^\mu(t, \mathbf{x}(t); v) \;=\; F^{\mu\nu}(t, \mathbf{x}(t))\, v_\nu, \tag{1.17}$$

showing that $e\,dW^\mu(x;v)$ is the kinetic 4-momentum acquired during the time dt by a charged particle due to its interaction with the electromagnetic field. Note, however, that in eq.(1.15) the velocity $\mathbf{v}$ is a constant, while in the Lorentz equation (1.16), k^μ and $\mathbf{v}$ are related quantities ($k^\mu = k^0 v^\mu$). This is in line with our assumption that the typical momenta of the thermal particles are "hard", $k \sim T$, while those of the background fields are "soft", $P \sim gT \ll k$. Thus the hard particles follow straight line trajectories at the speed of the light, and these trajectories are not altered, to leading order, by the interaction with the background gauge field.

In the non-abelian case, the fluctuations $\delta n_\pm$ and δN are matrices in color space, that is, $W^\mu = W_a^\mu t^a$. The color vector of components W_a^μ precesses in the background gauge field. This precession is induced by the covariant derivative in eq. (1.6). Viewing this precession as an additional source of time-dependence for the color vector W_a^μ, one can write

$$\frac{d}{dt}\,W_a^\mu(t,\mathbf{x}(t);v) \;=\; \Big[(\partial_t+\mathbf{v}\cdot\boldsymbol{\nabla})\delta_{ac} \;-\; g f_{abc}\,(v\cdot A_b)\Big]\,W_c^\mu, \tag{1.18}$$

so that eq. (1.6) may be given a form similar to eq. (1.17).

A simple gauge structure

The equation of motion for a particle in an electromagnetic field can also be written in terms of the *canonical* momentum $p^\mu = k^\mu + eA^\mu$:

$$\frac{dp^\mu}{dt} \;=\; e\,v^\nu\partial^\mu A_\nu. \tag{1.19}$$

The same possibility exists for our kinetic equation. This follows by noticing that

$$F^{\mu\nu}\,v_\nu \;=\; \partial^\mu(v\cdot A) \;-\; \Big[v\cdot D, A^\mu\Big]. \tag{1.20}$$

Thus, if one sets

$$a^\mu(x;v) \equiv A^\mu(x) \,+\, W^\mu(x;v), \tag{1.21}$$

one gets

$$[v\cdot D,\, a^\mu] \;=\; \partial^\mu(v\cdot A), \tag{1.22}$$

showing in particular that a^μ is a functional of $v\cdot A$ only. In the abelian case, eq. (1.22) can be rewritten in a form analogous to eq. (1.19) for p^μ:

$$\frac{d}{dt}\,a^\mu(t,\mathbf{x}(\mathbf{t});v) = (\partial_t+\mathbf{v}\cdot\boldsymbol{\nabla})\,a^\mu = \partial^\mu(v\cdot A), \tag{1.23}$$

showing that ea^μ may be indeed understood as the change in the canonical momentum of a fictitious particle following the trajectory $\mathbf{x}(t) = \mathbf{x}_0 + \mathbf{v}t$.

Under a gauge transformation induced by the operator $h(x) = \exp(-ig\theta^a(x)t^a)$, $W^\mu = W_a^\mu t^a$ transforms as $W^\mu(x) \to h(x)\,W^\mu(x)\,h^{-1}(x)$, while $A_\mu = A_\mu^a t^a \to$

$h\,A_\mu\,h^{-1} - (i/g)\,h\partial_\mu h^{-1}$. Thus $a^\mu(x;v)$ transforms as a gauge potential. Let $\mathcal{D}_\mu \equiv \partial_\mu + ig\,a_\mu$ be the covariant derivative constructed from a_μ, and $f_{\mu\nu}(x;v) \equiv [\mathcal{D}_\mu, \mathcal{D}_\nu]/(ig)$ the corresponding field strength tensor. Then it may be shown that[11]

$$f_{\mu\nu} = \partial_\mu a_\nu \,-\, \partial_\nu a_\mu \,+\, ig\,[a_\mu, a_\nu] \,=\, 0, \tag{1.24}$$

so that the field a^μ is a pure gauge. This result follows essentially from the assumption that the trajectories of the plasma particles are not deviated by the interaction with the gauge fields.[11]

Gauge invariance of the generating functioanl

In the calculation of the induced current, only the component $\mu = 0$ of eq. (1.22) is needed. Explicitly,

$$(v\cdot\partial)a_0 \,-\, \partial_0(v\cdot A) \,+\, ig\,[v\cdot A, a_0] = 0. \tag{1.25}$$

This is, within minor notational changes, the equation which has been obtained by Taylor and Wong as the condition of gauge invariance of the effective action.[4]

In order to appreciate the nature of the constraint imposed on the effetive action by gauge invariance, let us consider an infinitesimal gauge transformation

$$A_\mu \to A_\mu + \delta A_\mu, \qquad \delta A_\mu = [D_\mu, \theta]. \tag{1.26}$$

Under such a transformation, the effective action transforms into $S \to S + \delta S$, with

$$\delta S = \int d^4x\, j^\mu_a\, \delta A_\mu = -\int d^4x\, \theta^a \left[D_\mu, j^\mu\right]_a. \tag{1.27}$$

The second equality follows after an integration by parts (the surface term has been assumed to vanish). Clearly, the conservation law for j^μ, eq. (1.14), ensures the gauge invariance of S. Now,

$$\left[D_\mu, j^\mu\right] = 3\omega_p^2 \int \frac{\mathrm{d}\Omega}{4\pi}\{\partial_+\, a_0 \,-\, \partial_0\, A_+ \,+\, ig\,[A_+,\, a_0]\}, \tag{1.28}$$

where $A_+ \equiv v\cdot A$ and $\partial_+ \equiv v\cdot\partial$. (To obtain this result, use eq.(1.15) for j_μ with $W_0 = a_0 - A_0$ and verify that $[v\cdot D, A_0]$ gives the same result as $\partial_0 A_+$ after the angular integration.) The terms in braces in eq.(1.28) are the left hand side of eq.(1.25). Thus, only the *angular average* of the equation of motion is actually constrained by the gauge invariance. The full equation can only be recovered by calling for other properties of the HTL's.[4]

Relation to Chern-Simons theory

The induced current has also been obtained by Jackiw and Nair in Ref.[9] from the effective action S of Taylor and Wong. To make contact with their work, we replace eq. (1.25) with an equivalent equation for the new function $a_- \equiv 2a_0 - A_+$:

$$\partial_+\, a_- \,-\, \partial_-\, A_+ \,+\, ig\,[A_+,\, a_-] \,=\, 0, \tag{1.29}$$

where we have set $\partial_- \equiv v' \cdot \partial$, with $v'^\mu = (1, -\mathbf{v})$. Then, the current (1.12) becomes

$$j^\mu(x) = 3\,\omega_p^2 \int \frac{d\Omega}{4\pi}\, v^\mu\,(a_0 - A_0) = \frac{3}{2}\,\omega_p^2 \int \frac{d\Omega}{4\pi}\, v^\mu\,(a_- - A_-). \tag{1.30}$$

Eqs. (1.29) and (1.30) coincide with the expression of j^μ proposed in Ref..[9]

It can be easily verified using eqs. (1.7) and (1.21) that a_- just defined is also equal to $v' \cdot a$. This is not surprising, since eq. (1.29) also follows from eq. (1.22) for a^μ after multiplication with v'_μ. Let us further define $a_+ \equiv v \cdot a$. Because of eqs. (1.7) and (1.21), we have $a_+ = A_+$, so that eq. (1.29) may be finally written as

$$\partial_+\, a_- - \partial_-\, a_+ + ig\,[a_+, a_-] = 0. \tag{1.31}$$

This coincides with the projection of the zero-curvature condition (1.24) onto the hyperplane defined by the 4-vectors v^μ and v'^μ; that is, eq. (1.31) is the same as $v_\mu\, f^{\mu\nu} v'_\nu = 0$. The equation (1.31) may also be viewed as the equation of motion for a 2+1 dimensional Chern-Simons theory, in a specific gauge. This relation has been explored in Refs..[7,9]

References

1. R.D. Pisarski, Phys. Rev. Lett.**63** (1989) 1129.
2. E. Braaten and R.D. Pisarski, Nucl. Phys. **B337** (1990) 569; *ibid.* **B339** (1990) 310; Phys. Rev. Lett.**64** (1990) 1338.
3. J. Frenkel and J.C. Taylor, Nucl. Phys. **B334** (1990) 199.
4. J.C.Taylor and S.M.H.Wong, Nucl. Phys. **B346** (1990) 115.
5. E. Braaten and R. Pisarski, Phys. Rev. **D45** (1992) 1827.
6. J. Frenkel and J.C. Taylor, Nucl. Phys. **B374** (1992) 156.
7. R. Efraty and V.P. Nair, Phys. Rev. Lett.**68** (1992) 2891; Phys. Rev. **D47** (1993) 5601.
8. J.P. Blaizot and E. Iancu, Nucl. Phys. **B390** (1993) 589; Phys. Rev. Lett.**70** (1993) 3376; Nucl. Phys. **B417** (1994) 608; Nucl. Phys. **B421** (1994) 565.
9. R. Jackiw and V.P. Nair, Phys. Rev. **D48** (1993) 4991.
10. R. Jackiw, Q. Liu and C. Lucchesi, Phys. Rev. **D49** (1994) 6787.
11. J.P. Blaizot and E. Iancu, "Gauge structure and semi-classical aspects of hard thermal loops", Saclay preprint T94/041.

TANGUY ALTHERR

1963-1994

Tanguy Altherr was killed in a rock-climbing accident, which occurred last July in Brittany, the region of France where he was born and which he loved. The news of his death came as terrible shock to his friends and colleagues. In the name of the participants to this Conference, I would like to express our deepest sympathy to his wife, Hélène, who is expecting a baby next November, and to his parents and his relatives.

Tanguy was a bright young physicist, who leaves, in spite of his short career, a remarkable amount of scientific work. He had broad scientific interests in various domains of particle physics and field theory, but his work deals mainly with finite temperature and/or density field theory. He started his PhD thesis in Annecy, under the supervision of Patrick Aurenche; and together with T. Becherrawy, they were able to obtain a generalization of the Kinoshita-Lee-Nauenberg theorem in a particular, but important case: the production of lepton pairs from a quark-gluon plasma [1]. They showed, to lowest non trivial order in perturbation theory, the cancellation of all collinear and infrared singularities in the production rate, and they were even able to compute explicitly the finite terms.

Among Tanguy's other contribution to this field, I would like to mention:

- his study of the infrared behaviour of ϕ^4-theory at finite temperature [2];

- his calculation of axion emission from a QED plasma with subsequent applications to astrophysics [3];

- his contribution to the difficult problem of the fermion damping rate [4];

- his remark that in a QCD plasma out of chemical equilibrium, an important source of quark-antiquark pairs would be the decay of the gluonic plasmon [5].

Although Tanguy made important contributions to the formal aspects of finite temperature field theory (experts will find in his PhD thesis many of useful remarks), he always wanted to stay close to physical phenomena, and he chose astrophysics as a domain where finite-temperature field theory could find relevant applications. He quickly became a recognized expert in this subject, and quite naturally I asked him to give a review talk at this Conference. Unfortunately we shall never see the written version of the brilliant talk he gave in Paris. From his transparencies, I'll attempt to summarize in a few lines the contents of his lecture "Thermal field theory and its applications to astrophysics"$\cdots$ Michel Le Bellac

THERMAL FIELD THEORY AND ITS APPLICATION TO ASTROPHYSICS

TANGUY ALTHERR
Theroy Division, CERN
CH-1211, Geneva 23, Switzerland
(Submitted by Michel Le Bellac)

Many systems in astrophysics reach temperatures and/or densities such that thermal field theory becomes a necessary tool. Among those systems, one may quote: the core of red giants and of white dwarfs, which corresponds to an almost degenerate QED plasma ($\mu \gg T$), and the core of supernovae where a hot QED plasma is relevant. In this latter case, one may even think that a quark-gluon plasma is present in the center of the star.

After a review of star evolution and of typical orders of magnitude, Tanguy developed the aspects of thermal field theory particularly relevant to astrophysics. In the case of degenerate plasmas, the resummation program of Braaten and Pisarski relies on the existence of two scales: the "hard scale" μ and the "soft scale" $e\mu$, where e is the electron charge. Resummation of the photon propagator is necessary for soft photon momenta; however, due to Pauli-blocking, there is no need to resum the electron lines. It is important that there exists very good approximate analytical expressions for the photon propagator in all situations [6]. These expressions lead to simple formulae for the dispersion laws of the transverse and longitudinal photon excitations, often called plasmons. For the transverse plasmons one finds the dispersion law $\omega_T = f(k)$ in the following implicit form

$$\omega_T^2 = k^2 + \omega_P^2 \frac{3\omega_T^2}{2v_*^2 k^2}\Big(1 - \frac{\omega_T^2 - v_*^2 k^2}{\omega_T^2}\,\frac{\omega_T}{2v_* k}\ln\frac{\omega_T + v_* k}{\omega_T - v_* k}\Big) \tag{1}$$

while for longitudinal plasmons one has

$$\omega_L^2 = \omega_P^2 \frac{3\omega_L^2}{v_*^2 k^2}\Big(\frac{\omega_L}{2v_* k}\ln\frac{\omega_L + v_* k}{\omega_L - v_* k} - 1\Big) \tag{2}$$

where v_* is a typical velocity: for example v_* is the Fermi velocity $v_F = p_F/E_F$ in the case of degenerate plasmas, $v_* = 1$ for ultrarelativistic plasmas.The plasmon frequency ω_P is given in all cases of interest by

$$\omega_P^2 = \frac{4\alpha}{\pi} \int_0^\infty \mathrm{d}p \, \frac{p^2}{E}\Big(1 - \frac{1}{3}v^2\Big)[n_F(E) + \overline{n}_F(E)] \tag{3}$$

Equation (1) holds for $0 \le k < \infty$ while in equation (2), k is limited by a value k_{max} given by the point where the dispersion relation crosses the light cone.

Then Tanguy turned to two important applications of the formalism. First he discussed particle production (e.g. axion production through photon exchange). He showed that in the ultrarelativistic regime, transverse exchange is dominant, while longitudinal exchange dominates in the non relativistic regime. In this latter case, the dispersion relation crosses the light cone (Langmuir waves) and this phenomenon can be important in axion emission from stars on the horizontal branch of the Herzprung-Russell diagram.

The second topic treated by Tanguy was that of the cooling of stars from neutrino emission. The most important process is plasmon decay into neutrino pairs through a virtual $e^+ - e^-$ pair. This process can be interpreted by attributing to the neutrino an effective charge e_ν (the precise meaning of this statement is explained in a paper by Tanguy and P.Salati [7])

$$e_\nu = \frac{2\sqrt{2}}{e} G_F C_V \omega_\beta^2 \tag{4}$$

where G_F is the Fermi constant and $C_V = 1/2 + 2\sin^2\theta_W$; ω_β^2 is determined by the properties of the plasma. For example in the case of a degenerate electron gas

$$\omega_\beta^2 = \frac{e^2}{2\pi^2} p_F E_F \Big(1 - \frac{1 - v_F^2}{2v_F} \ln\frac{1 + v_F}{1 - v_F}\Big) \tag{5}$$

The energy loss of the star R_T arising from the decay of transverse plasmons can be written in terms of e_ν^2 and of the residue $Z_T(\omega_T(k), k)$ at the pole

$$R_T = \frac{e_\nu^2}{24\pi^3} \int_0^\infty \frac{k^2 \mathrm{d}k}{\omega_T(k)} Z_T(\omega_T(k), k)(\omega_T^2(k) - k^2) n_B(\omega_T(k)) \tag{6}$$

where $\omega_T(k)$ is the dispersion law (1) of transverse excitations and n_B is the Bose-Einstein factor.A similar equation holds for longitudinal plasmons.

In the case of white dwarfs, taking into account the correct dispersion laws (1) and (2) allows to bring the theoretical predictions in very good agreement with the data. One also predicts interesting effects in the case of red giants, in relation with the so-called helium flash.

7. References

1. T. Altherr, P. Aurenche and T. Becherrawy, Nucl. Phys. **B315** (1989) 436.
2. T. Altherr, Phys. Lett. **238** (1990) 360.
3. T. Altherr, Zeit. Phys. **C47** (1991) 706 and Ann. Phys. (NY) **207** (1991) 374.
4. T. Altherr, E. Petitgirad and T. del Rio Gaztelurrutia, Phys. Rev. **D47** (1993) 703.
5. T. Altherr and D. Seibert, Phys. Lett **313** (1993) 149 and CERN preprint CERN-TH 7038/93.
6. T. Altherr, E. Petitgirad and T. del Rio Gaztelurrutia, Astroparticle Phys. **1** (1993) 289; E. Braaten and D. Segel, Phys. Rev. **D48** (1993) 1478.
7. T. Altherr and K. Kainulainen, Phys. Lett. **B262** (1991) 79; T. Altherr and P. Salati, ENS LAPP preprint (1993).

THE INFRARED SENSITIVITY OF NONABELIAN DEBYE SCREENING

A. K. REBHAN
DESY, Gruppe Theorie,
Notkestr. 85, D-22603 Hamburg, Germany

In memory of Tanguy Altherr

ABSTRACT

It is shown that a perturbative treatment of nonabelian Debye screening at high temperature suffers from infrared problems already at the next-to-leading order, which is given by ring-resummed one-loop diagrams. Superficial infrared power counting would let one expect a sensitivity to the magnetic mass scale only at higher orders, but the form of Debye screening depends on the analytic structure of the correlation functions, which is strongly sensitive to the existence of screening of static magnetic fields.

1. Introduction

High-temperature gauge theories[1] become increasingly complicated as the infrared regime is approached. At the momentum scale gT, where g is the coupling and T the temperature, the spectrum of the theory begins to deviate considerably from the one of the free theory. Gauge bosons as well as fermions acquire new, collective degrees of freedom described by the gauge-invariant (nonlocal) effective action generated by the so-called "hard thermal loops" (HTL)[2,3], the high-temperature limit of all one-loop Feynman diagrams proportional to T^2. The resulting leading-order dispersion laws of the quasi-particle excitations can be understood in classical terms[4,5], but already at next-to-leading order the dispersion laws receive contributions from all orders of the conventional perturbation series. In order to restore perturbation theory it is necessary to resum all HTL corrections[3], and in this way some corrections to the HTL dispersion laws have been obtained[6,7,8]. However, in nonabelian theories, the restitution of perturbation theory is not complete. Static chromomagnetic fields are not screened at the HTL level, and the self-interactions of these lead to a breakdown of perturbation theory at a certain loop order. In particular, the magnetic screening mass, which is expected to arise at the order g^2T, is not calculable in perturbation theory[9,10]; HTL resummation does not help in this respect.

On the other hand, HTL resummation should allow one to calculate corrections to the classical chromoelectric (Debye) screening mass m_0, which is of the order gT, to wit,

$$m_0^2 = \frac{e^2T^2}{3} \tag{1}$$

with $e^2 = (N + N_f/2)g^2$ for color group SU(N) and N_f flavors. Upon HTL resummation, perturbation theory is organised in powers of g rather than g^2, and the relative order g correction is determined by resummed one-loop diagrams. According to superficial infrared power-counting[1], problems coming from the perturbatively vanishing magnetic mass would be expected to set in at two-loop order.

In this note, I shall demonstrate, however, that already the next-to-leading order term of the resummed perturbation theory is crucially sensitive to the magnetic mass scale.

2. The electrostatic gluon propagator at resummed one-loop order

In momentum space, the 'electrostatic' gluon propagator is given by

$$D_L(k) = \frac{1}{k^2 + \Pi_{00}(k_0 = 0, k)} \tag{2}$$

and its Fourier transform $\Phi(r)$ determines the chromoelectric field induced by a single external (conserved) source J through

$$\langle E^i(\mathbf{x})\rangle = \frac{\partial}{\partial x_i}\Phi(|\mathbf{x}|), \tag{3}$$

where we have suppressed all color indices. Despite the nonabelian nature, this is an exact result of linear response theory[11], since with only one direction in color space introduced by the single source J, the gauge potentials all point in the same direction, which eliminates the nonabelian commutator terms trivially.

At high temperature, the leading contribution to $\Pi_{00}(k_0 = 0, k)$ equals the constant m_0^2 of Eq. (1), and $\Phi(r) \propto e^{-m_0 r}/r$. However, beyond leading order Π_{00} is a gauge fixing dependent quanitity. At one-loop resummed order one obtains[12,13] in covariant gauges with gauge parameter α

$$\Pi_{00}(0,k) = m_0^2 + \frac{g\sqrt{3N'}m_0^2}{2\pi}\left[\frac{m_0^2 - k^2}{m_0 k}\arctan\frac{k}{m_0} + \frac{\alpha - 2}{2}\right] + O(g^2), \tag{4}$$

where $N' = N/(1 + \frac{N_f}{2N})$. Computing the Fourier transform of $D_L(k)$ with the thus corrected gluon self-energy, one finds a surprising behavior[14]: $\Phi(r)$ has the form of a Yukawa potential multiplied by an oscillating function which approaches a negative value asymptotically. Its details are gauge parameter dependent except for the screening mass which characterises the exponential decay for very large distances, which is still given by m_0.

The reason behind this peculiar result is displayed in Fig. 1, where the analytic structure of $D_L^{-1}(k)$ is rendered for $t \equiv g\sqrt{3N'}/8 = 0.25$ and gauge parameter $\alpha = 1$.

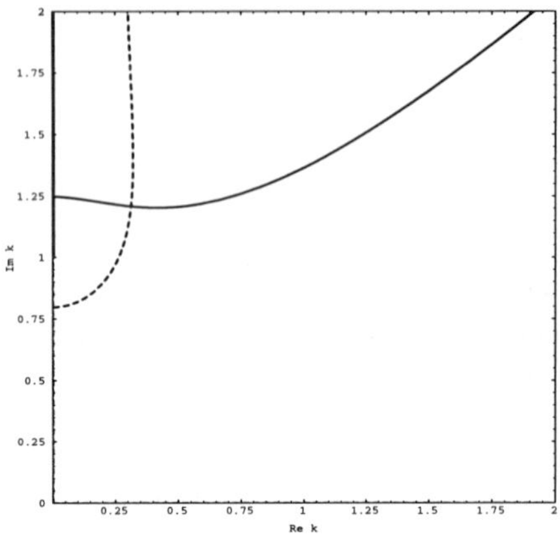

Figure 1: The analytic structure of $D_L^{-1} \equiv k^2 + \Pi_{00}(0,k)$ in units where $1 = m_0$ and with coupling $t \equiv g\sqrt{3N'}/8 = 0.25$. The full line corresponds to $\Re e D_L^{-1} = 0$ and the dashed one to $\Im m D_L^{-1} = 0$. The latter extends to all of the real and imaginary axes except for the thick part of the imaginary axis which marks the location of the branch cut. The intersection of the full and the dashed line corresponds to a pole of D_L and is located at $k/m_0 \approx 1.208 + 0.313i$.

Shown is the first quadrant of the complex k plane; the others are given by reflection on the real and imaginary axes. At leading order, $D_L(k)$ was determined by a simple pole at $k/m_0 = i$, but including the next-to-leading order, D_L^{-1} no longer has a simple zero there, but a logarithmic branch singularity followed by a branch cut from i to ∞. The original zero of D_L^{-1}, however, still exists: it has moved to the right (and also left) of the imaginary axis. With the above parameters, the corresponding poles of D_L contribute to Φ a term proportional to $\cos(0.313x)e^{-1.208x}/x$, where $x \equiv rm_0$. There is, however, also the contribution from the cut, which adds a term $-f(x)e^{-x}/x$ with a strictly positive function f, so that asymptotically, for very large x, the behavior is that of a repulsive Yukawa potential with screening mass m_0.

However, the singularity of the next-to-leading correction to $\Pi_{00}(0,k)$ at $k = \pm i m_0$ indicates a breakdown of perturbation theory in the vicinity of these points in momentum space, for there the correction term becomes larger than the supposed leading one. In Ref. [13] I have pointed out that this singularity might be cut off by a nonvanishing magnetic mass. Indeed, introducing a small magnetic mass changes $\Pi_{00}(0,k)$ significantly around $k = \pm i m_0$. Denoting the magnetic mass by m_m and the full Debye mass by m, Π_{00} becomes

$$\Pi_{00}(0,k) = m_0^2 + \frac{g\sqrt{3N'}m_0}{2\pi}\Big[-\frac{1}{2}m - \frac{1}{2}m_m \qquad (5)$$

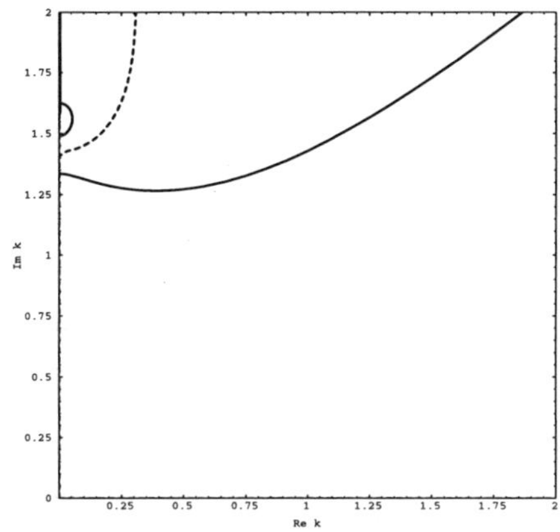

Figure 2: As in Fig. 1, but with a finite magnetic mass $m_m = 0.25m$.

$$+\frac{1}{k}(m^2 - \frac{1}{2}m_m^2 - k^2)\arctan\frac{k}{m+m_m}$$
$$+\left(k^2+m^2\right)\Big\{\frac{k^2+m^2}{m_m^2 k}\Big(\arctan\frac{k}{m+\sqrt{\alpha}m_m}$$
$$-\arctan\frac{k}{m+m_m}\Big)+(\sqrt{\alpha}-1)\frac{1}{m_m}\Big\}\Big],$$

where m_m has been introduced in a gauge-invariant manner by mimicking the Higgs mechanism[15]. Since the branch singularity now has moved to $k = \pm i(m + m_m)$, one can use a self-consistent definition of the Debye mass through

$$m^2 = \Pi_{00}(0,k)\Big|_{k^2=-m^2}, \tag{6}$$

which restitutes a pole at purely imaginary $k = \pm im$. From (6) one finds that the mass m thus defined is gauge parameter independent, and is given by $m^2 = m_0^2(1+\delta)$ with[13]

$$\delta \equiv \frac{\delta m^2}{m_0^2} = \frac{\sqrt{3N'}g}{2\pi}\left(\ln\frac{m}{m_m} + \ln 2 - \frac{1}{2}\right) + O(g^2). \tag{7}$$

The full analytic structure of $D_L^{-1}(k)$ following from (6) is given in Fig. 2 for the same parameters as in Fig. 1, but now with nonvanishing $m_m/m = 0.25$. The lines $\Re e D_L = 0$ and $\Im m D_L = 0$ no longer intersect at complex values of k, but only on the imaginary axis. There is also a gauge dependent zero very close to the branch singularity, but the dominant singularity in D_L is the one at $k = im$. Thus $\Phi(r)$ no longer changes sign but decays exponentially for large x with gauge independent screening mass m.

I would like to thank H. Fried, B. Müller, and M. LeBellac for organizing this stimulating workshop, and R. Baier for continuous discussions. This research is supported in part by the EEC Programme "Human Capital and Mobility", Network "Physics at High Energy Colliders", contract CHRX-CT93-0357 (DG 12 COMA).

References

1. J. I. Kapusta, *Finite Temperature Field Theory* (Cambridge University Press, Cambridge, 1989).
2. J. Frenkel and J. C. Taylor, *Nucl. Phys.* **B334** (1990) 199.
3. E. Braaten and R. D. Pisarski, *Phys. Rev. Lett.* **64** (1990) 1338; *Nucl. Phys.* **B337** (1990) 569.
4. V. P. Silin, *Zh. Eksp. Teor. Fiz.* **38** (1960) 1577 [*Sov. Phys.* **11** (1960) 1136]; J.-P. Blaizot and E. Iancu, *Nucl. Phys.* **B417** (1994) 608.
5. G. Barton, *Ann. Phys. (N.Y.)* **200** (1990) 271; J. Frenkel and J. C. Taylor, *Nucl. Phys.* **B374** (1992) 156.
6. E. Braaten and R. D. Pisarski, *Phys. Rev.* **D42** (1990) 2156; *ibid.* **D46** (1992) 1829; R. Kobes, G. Kunstatter and K. Mak, *Phys. Rev.* **D45** (1992) 4632.
7. H. Schulz, *Nucl. Phys.* **B413** (1994) 353.
8. U. Kraemmer, A. K. Rebhan and H. Schulz, DESY-report 94-034 (1994), to appear in Ann. Phys. (N. Y.).
9. A. D. Linde, *Phys. Lett.* **B96** (1980) 289.
10. D. Gross, R. Pisarski and L. Yaffe, *Rev. Mod. Phys.* **53** (1981) 43.
11. H. A. Weldon, *Phys. Rev.* **D26** (1982) 1394.
12. S. Nadkarni, *Phys. Rev.* **D33** (1986) 3738.
13. A. K. Rebhan, *Phys. Rev.* **D48** (1993) R3967.
14. R. Baier and O. K. Kalashnikov, *Phys. Lett.* **B328** (1994) 450.
15. A. K. Rebhan, to appear in the proceedings of the NATO Advanced Research Workshop "Electroweak Physics and the Early Universe", March 1994, Sintra, Portugal.

EXACT SUM RULES AT FINITE TEMPERATURE AND CHEMICAL POTENTIAL AND THEIR APPLICATION TO QCD *

SUZHOU HUANG

Department of Physics, FM-15, University of Washington, Seattle, Washington 98195, U.S.A.

and

MARCELLO LISSIA

Istituto Nazionale di Fisica Nucleare, via Negri 18, Cagliari, I-09127, ITALY

ABSTRACT

Within the framework of the operator product expansion (OPE) and the renormalization group equation (RGE), we show that the temperature and chemical potential dependence of the zeroth moment of a spectral function (SF) in an asymptotically free theory is completely determined by the one-loop structure of the theory. This exact result constrains the qualitative shape of SF's, and implies striking phenomenological effects near phase transitions.

Our present understanding of QCD at finite temperature (T) and baryon density (or chemical potential μ) is mainly limited in the Euclidean realm, due to the lack of non-perturbative and systematic calculating tools directly in the Minkowski space. The typical methods, with QCD Lagrangian as the starting point, are the OPE and lattice simulation. Because of these two formulations are intrinsically Euclidean, only static quantities are conveniently studied. In order to gain dynamical informations, which are more accessible experimentally, the analytic structure implemented through dispersion relations often have to be invoked within the theory of linear response.

The real-time linear response to an external source coupled to a renormalized current $J(x)$ is given by the retarded correlator:

$$K(x;T,\mu) \equiv \theta(x_0)\langle[J(x),J(0)]\rangle_{T,\mu}\,, \tag{1}$$

where the average is on the grand canonical ensemble specified by (T,μ). Disregarding possible subtraction terms, the following dispersion relation for the frequency dependence of the retarded correlator can be written:

$$\tilde{K}(\omega,\mathbf{k};T,\mu) = \int_0^\infty du^2\,\frac{\rho(u,\mathbf{k};T,\mu)}{u^2-(\omega+i\epsilon)^2}\,. \tag{2}$$

*Talk presented by S. Huang.

For convenience, we discuss only the uniform limit ($\mathbf{k}=0$). Upon analytic continuation, $\omega \to iQ$, the dispersion relation in principle connects the correlator in Euclidean region to the SF, which embodies all the real-time information. However, given only approximate knowledge of $\tilde{K}(iQ;T,\mu)$, either via the OPE or lattice calculations, it is extremely difficult technically to "invert" the dispersion relation. Only when we have enough understanding of the SF involved this "inverting" can be made practical. In fact, the success of the so-called QCD sum rule approach and analyzing lattice data under many situations rests, to some extend, fortuitously on the successful parameterization of SF's.

On the other hand, the lack of adequate understanding of SF's at finite (T,μ) severely limits our chances to extract physical information using these approaches. It is likely that physical results at finite (T,μ) can be strongly biased, if one naively assumes the same parameterizations of the SF's that work at zero (T,μ). The purpose of this work is to derive exact sum rules that constrain the variation of SF's with (T,μ). The derivation, based on the OPE and the RGE, has close analogies to the analysis of deep inelastic lepton scattering experiments. In addition, we apply these sum rules to the chiral phase transition, and demonstrate that SF's in some channels are drastically modified compared both to their zero (T,μ) and perturbative shape.

In an asymptotically free theory, the OPE yields, *e.g.* in the $\overline{\text{MS}}$ scheme, the large-Q^2 asymptotic expansion

$$\tilde{K}(iQ;T,\mu,\kappa) \sim \tilde{K}_0(iQ,\kappa) + \sum_n C_n(Q^2, g^2(\kappa), \kappa)\langle [O_n]_\kappa \rangle_{T,\mu}\,, \tag{3}$$

where $g^2(\kappa)$, $[O_n]_\kappa$'s and C_n's are, respectively, the coupling constant, the renormalized composite operators and their corresponding Wilson coefficients at the subtraction mass scale κ. It is important to notice that the information of the ensemble average is encoded in the matrix elements of the composite operators, while the Wilson coefficients and $\tilde{K}_0$ are independent of T and μ. Although the matrix elements $\langle [O_n]_\kappa \rangle_{T,\mu}$ cannot be determined perturbatively, the Q^2-dependence of Wilson coefficients C_n is dictated by the renormalization group equation and given by

$$C_n(Q^2, g(\kappa), \kappa) = \frac{c_n(g^2(Q))}{Q^{d_n}} \left[\frac{g^2(Q)}{g^2(\kappa)}\right]^{(2\gamma_J - \gamma_n)/2b} \left\{1 + \mathcal{O}(g^2(Q))\right\}, \tag{4}$$

where d_n is the canonical dimension of the operator O_n minus the dimension of $\tilde{K}$ and $c_n(g^2(Q))$ is calculable perturbatively. The pure numbers γ_i ($i = J, n$) and b are related to the anomalous dimensions of J, O_n and to the β-function as follows

$$\Gamma_i = -\gamma_i g^2 + \mathcal{O}(g^4)\,, \text{ and } \beta = -bg^4 + \mathcal{O}(g^6)\,. \tag{5}$$

To study the dependence of $\tilde{K}$ on (T,μ) we only need to consider the difference $\Delta\tilde{K}(iQ) \equiv \tilde{K}(iQ;T,\mu) - \tilde{K}(iQ;T',\mu')$ and

$$\Delta\tilde{K}(iQ) = \int_0^\infty du^2 \frac{\Delta\rho(u)}{u^2+Q^2}\,, \tag{6}$$

where $\Delta\rho(u) \equiv \rho(u;T,\mu) - \rho(u;T',\mu')$. This subtraction is crucial to remove $\tilde{K}_0(iQ,\kappa)$, which contains all the terms not suppressed by a power of $1/Q^2$, and also to make $\Delta\tilde{K}(iQ)$ independent of the renormalization point κ. Finite masses give corrections of order $m^2(Q)/Q^2$, with $m^2(Q)$ that runs logarithmically and hence can be ignored, if we are only interested in the lowest moment of the subtracted SF.

At this point we have expressed the left-hand side of Eq. (6) as an asymptotic expansion of the form:

$$\Delta\tilde{K}(iQ) \sim \sum_{n,\nu=0}^{\infty} \frac{c_n^{(\nu)}(\kappa)\Delta\langle[O_n]_\kappa\rangle}{Q^{d_n}}[g^2(Q)]^{\nu+\eta_n}\,, \tag{7}$$

where $\Delta\langle[O_n]_\kappa\rangle$ denotes the difference between the expectation values of $[O_n]_\kappa$ in the ensembles specified by (T,μ) and (T',μ') and the exponent η_n and the Q^2-independent coefficients $c_n^{(\nu)}(\kappa)$ are known perturbatively.

We proceed by making an analogous asymptotic expansion of $\Delta\rho(u)$:

$$\Delta\rho(u) \sim \sum_{n=0}^{\infty} \frac{[g^2(u)]^{\xi_n}}{u^{2(n+1)}} \sum_{\nu=0}^{\infty} a_n^{(\nu)}[g^2(u)]^\nu\,. \tag{8}$$

For notational clarity, we have ignored exponentially suppressed terms and the fact that there can be more than one η_n and ξ_n for each n. We then obtain the sum rules by imposing that the asymptotic expansion of the right-hand side of Eq. (6), which we get by inserting Eq. (8) in the dispersion integral, matches the left-hand side obtained by the OPE. In a long paper[1] we shall present those technical details that make the procedure sketched above rigorous. At the moment we only need to consider the leading terms in the expansions of, respectively, the left-hand side and the right-hand side of Eq. (6):

$$[g^2(Q)]^{\eta_n}\frac{\Delta\langle[O_n]_\kappa\rangle}{Q^{d_n}}[c_n^{(0)}(\kappa) + c_n^{(1)}(\kappa)g^2(Q)]\,, \tag{9}$$

$$\frac{1}{Q^2}\left[\overline{\Delta\rho} + \frac{a_0^{(0)}}{(1-\xi_n)}[g^2(Q)]^{\xi_n-1}\right]. \tag{10}$$

In Eq. (10), if $\xi_n > 1$, $\overline{\Delta\rho}$ can be shown to be equal to the zeroth moment of the subtracted SF; note that the zeroth moment of a function whose asymptotic expansion is Eq. (8) is infinite, if $\xi_n \leq 1$, since $\int_A^\infty dx\, x^{-1}(\log x)^{-\xi_n} = \infty$.

For the sake of concreteness, let us examine the consequences of matching Eq. (9) and Eq. (10) in the case we are concerned with, *i.e.* $n = 1$ and $d_1 = 2$. If the OPE calculation produces $\eta_n = 0$, then ξ_n must be an integer greater than one, the zeroth moment exists and is given by

$$\int_0^\infty du^2\, \Delta\rho(u) = c_1^{(0)}(\kappa)\Delta\langle[O_1]_\kappa\rangle\,. \tag{11}$$

If $\eta_n > 0$, then $\xi_n = 1 + \eta_n$, the zeroth moment is again finite and equal to zero:

$$\int_0^\infty du^2\, \Delta\rho(u) = 0\,. \tag{12}$$

Our main results, Eq. (11) and Eq. (12), can be expressed in physical terms as follows. The zeroth moment of a SF for a current J whose OPE expansion yields $\eta_n > 0$ is independent of T and μ, while the same moment for a current with $\eta_n = 0$ changes with T and μ proportionally to the corresponding change(s) of the condensate(s) of leading operator(s). When $\xi_n < 1$ or $\xi_n = 1$ in Eq. (8) terms such as $[g^2(Q)]^{-1}$ or $\ln[g^2(Q)]$ in Eq. (10) would be produced. Hence the appearance of the inverse power of the $g^2(Q)$ or $\ln[g^2(Q)]$ in the OPE series is a indication that the zeroth moment of the subtracted SF is infinite.

At this point several more general comments are appropriate: 1) It is essential to take into account the QCD logarithmic corrections, for the logarithmic corrections not only dictate whether $\overline{\Delta\rho}$ satisfy Eq.(11) or Eq.(12), but also control the very existence of $\overline{\Delta\rho}$. 2) The derivation of sum rules for higher moments of the SF requires the complete cancelation of all the lower dimensional operator terms, not just the leading $g^2(Q)$ terms; in particular, we also need current quark mass corrections to the Wilson coefficients. An appropriate subtraction is the prerequisite for the convergence of higher moments. 3) The (T,μ)-dependent part of the leading condensate appearing in Eq. (11) does not suffer from the infrared renormalon ambiguity, because only the perturbative term $\tilde{K}_0$ can generate contributions to the leading condensate that are dependent on the prescription used to regularize these renormalons. But $\tilde{K}_0$ is independent of T and μ and any prescription dependence cancels out when we make the subtraction in Eq. (6). On the contrary, unless we generalize Eq. (6) and make other subtractions, sum rules that involve non-leading condensates are, in principle, ambiguous. 4) We have explicitly verified the correctness of our results in a soluble model,[1] the Gross-Neveu model in the large-N limit, where we obtain exactly all the relevant quantities, such as SF's at arbitrary (T,μ), Wilson coefficients, β- and Γ-functions in pseudoscalar and vector channels.

Now let us specialize to QCD and consider four correlation functions: two involving the non-conserved scalar $J_S = \bar{\psi}\psi$ and pseudoscalar $J_P = \bar{\psi}\gamma_5\psi$ currents ($\gamma_{J_S} = \gamma_{J_P} = 1/4\pi^2$) and two involving the conserved vector $J_V = \bar{\psi}\gamma_\mu\psi$ and axial-vector $J_A = \bar{\psi}\gamma_\mu\gamma_5\psi$ currents ($\gamma_V = \gamma_A = 0$). Since the leading operators (dimension four) have non-positive anomalous dimensions, the two non-conserved currents have $\eta_n \geq (2\gamma_J - \gamma_n)/2b > 0$ and Eq. (12) applies, *i.e.* the zeroth moments of their SF's are independent of T and μ. On the other hand, the two conserved currents have $\eta_n = 0$ and a generalization of Eq. (11) applies.[1] Since there are three dimension-four operators with zero anomalous dimension, the sum rules for the vector and axial-vector currents are

$$\int_0^\infty du^2\, \Delta\rho(u) = a\Delta\langle[m\bar{\psi}\psi]\rangle + \frac{\Delta\langle[\alpha_s G^2]\rangle}{2\pi} + 8\Delta\langle[\theta_{00}]\rangle\,, \tag{13}$$

where $\theta_{\mu\nu}$ is the traceless stress tensor and $a = 6$ for vector and $a = -10$ for axial channels respectively. This exact sum rule should not be contaminated explicitly by instantons, although the value of the condensates certainly have instanton contributions. The reason is that the instanton singularities in the Borel-plane are located on the positive axis starting at $8\pi^2$, and, therefore, contribute to correlation functions only with higher order terms of $1/Q^2$ in the OPE series.

Finally, let us discuss some of the phenomenological consequences of these exact sum rules. In the pseudoscalar channel, $\overline{\Delta\rho} = 0$ implies that, in the broken-chiral-symmetry phase, the change of the pion pole induced by T or μ is exactly compensated by a corresponding change of the continuum part of the SF. Next let us consider the scalar correlation function at $Q^2 = 0$, the chiral susceptibility,

$$\chi(T,\mu) \equiv \int d^4x\, \theta(x_0)\langle[J_S(x), J_S(0)]\rangle_{T,\mu} = \int_0^\infty du^2\, \frac{\rho(u;T,\mu)}{u^2}\,, \tag{14}$$

which diverges when (T,μ) approaches the phase boundary, provided the chiral restoration is a continuous transition. The divergence of the chiral susceptibility near phase transition only can be produced in Eq.(14) by singularities very close to the origin, when the exact sum rule $\overline{\Delta\rho} = 0$ is simultaneously taken into account. Thus the spectral function would have to have a vanishing threshold (since there is no massless poles in the chirally symmetric phase) and develop a strong peak right above the threshold, when (T,μ) is very close to the phase boundary. Because in the chirally symmetric phase the pseudoscalar and scalar channels are degenerate, the same would also happen to the pseudoscalar SF. This strong peak in the pseudoscalar and scalar SF's, which is intimately connected with the critical phenomenon of diverging susceptibility and the correlation length near the phase transition, can be interpreted as some kind of quasi-particle, thus confirming the qualitative picture, originally proposed in the context of the Nambu-Jona-Lasinio model,[2] of the appearance of soft modes near the chiral phase transition. Similar situation, though less drastic, can be argued to occur in vector and axial channels, based on the facts that $\Delta\langle[\theta_{00}]\rangle$, $\Delta\langle[m\bar{\psi}\psi]\rangle$ and $\Delta\langle[\alpha_s G^2]\rangle$ behave smoothly across the critical line and the rapid increase of the so-called baryon number susceptibility at the chiral restoration point in lattice simulations.[3] If the chiral restoration turns out not to be a second order phase transition, but rather a cross-over or weak first order transition (finite but large correlation length), as the lattice data seem to indicate,[4] we expect the same qualitative features, though less pronounced.

In summary, we used OPE and RGE to derive exact sum rules at finite (T,μ) in asymptotically free theories. These exact sum rules strongly constrain the qualitative shape of SF's, especially near phase transition region. We urge whoever parameterizes a SF, *e.g.* in the QCD sum rule calculations or to interpret lattice simulations, to incorporate these exact constraints. In the future, we plan to generalize these results to baryonic currents and analyze their phenomenological consequences in greater detail.

We thank Profs. Fried and Müller for organizing a very enjoyable workshop. This work is supported in part by the U.S. Department of Energy.

References

1. S. Huang and M. Lissia, hep-ph/9404275, and a manuscript in preparation.
2. T. Hatsuda and T. Kinihiro, Phys. Rev. Lett. **55**, (1985) 158.
3. S. Gottlieb et al., Phys. Rev. **D38**, (1988) 2888.
4. F. Brown et al., Phys. Rev. Lett. **65**, (1990) 2491.

Soft Pomeron Physics

HARD DIFFRACTION AND RAPIDITY GAPS

MICHAEL G. ALBROW
FERMILAB, P.O.Box 500, Wilson Road
Batavia, IL 60510, USA

Abstract

I describe the evolution of experiments at hadron colliders on (a) high mass diffraction (b) double pomeron exchange, from the ISR through the $Sp\bar{p}S$ to the Tevatron. I emphasize an experimental approach to the question: "What is the pomeron?".

1. Introduction

For over 30 years we have associated an exchanged entity with diffractive scattering, called the pomeron after I.Y.Pomeranchuk. Despite the fact that diffractive processes are a large fraction of all high energy hadron-hadron collisions, and directly relate to the total cross section, we still do not have an acceptable understanding of what the pomeron is. Experiments on inelastic pomeron interactions, both with protons and with other pomerons, should shed light on this question and perhaps teach us how to extend the domain of calculability of QCD. We may find that the pomeron is intimately related to the very low-x structure function of the proton which is now being probed by experiments at HERA and elsewhere. Quite new and perhaps unexpected phenomena may manifest themselves for example in pomeron-pomeron collisions (if that is permissible language), including changes of the vacuum state. That is admittedly very speculative, but would clearly be very fundamental. Compared with the major effort currently spent in detailed experimental tests of perturbative QCD, where we know how to calculate theoretical predictions, pitifully little effort is spent in exploring beyond its frontiers.

2. Prehistory: Up to the ISR

The angular distributions of elastically scattered hadrons in the few GeV/c region typically show a 0 deg peak followed by a dip and secondary maximum (a ring in 2-D) very much like the diffraction of light through a hole, hence the terminology. The phenomenology of elastic scattering and other reactions such as

$$\pi^- + p \rightarrow \pi^0 + n$$

was developed using the exchange in the t-channel of "Reggeons", virtual (negative M^2, usually denoted t) hadronic states whose (complex) angular momentum J changes linearly with t. The graph of J vs t is called a Regge trajectory, and when this is extrapolated to the positive M^2 region passes through the J,M^2 values of known mesons (or baryons). The Regge trajectory in the negative t region can be measured from the s-dependence (s is the square of the c.m. collision energy) at each t-value. Those associated with mesons have slopes around 1 GeV^{-2} and intercepts $J(0)$ less than about 0.5. This means that the cross section for scattering by meson exchange must decrease with s. In fact the cross section rises, due to the exchange of a "high, flat" trajectory with intercept $J(0)$ just above 1.0 and slope around 0.2 GeV^{-2} which had no known mesons associated with it. (I shall refer later to a new candidate state.) This trajectory is called the pomeron, and it is a long-standing goal to elucidate its nature. For high $\sqrt{s}$ or equivalently for exchanges over large intervals of rapidity y the lower lying meson exchanges will be negligible and pomeron exchange dominates.

It is important not only in elastic scattering but in inelastic processes whenever there is a large rapidity gap, i.e. an interval of rapidity containing no hadrons; by "large" we mean typically 3 or 4 units, depending on how pure you want the pomeron to be. The leading meson trajectory falls off rapidly relative to the pomeron as Δy increases. Prior to the ISR, which turned on in 1971, it was known that one could "diffractively excite" a hadron into a state of mass around 2-3 GeV; this is called single diffractive excitation. One of the two incident hadrons emerges from the collision "quasi-elastically" scattered, having lost just a few percent of its energy in exciting the other (e.g. $p \to p\pi^+\pi^-$) and emerging with small momentum transfer t like in elastic scattering. There is necessarily, for kinematic reasons, a large rapidity gap in these events. It could be shown that the exchanged trajectory was the same as that in high energy elastic scattering, namely a pomeron.

3. ISR : High mass diffraction

The ISR, the first hadron collider, gave us a very dramatic increase in collision energy $\sqrt{s}$ and equivalently in the rapidity span

$$\Delta y = 2 \times ln \left(\frac{\sqrt{s}}{m_p} \right) .$$

When the proton spectrum was measured at small angles to the incident beam, it was found that apart from the elastic δ-function peak at $x = p/p_{beam} = 1.0$ there was a large high-x peak from inelastic collisions that extends down to about $x = 0.95$. When plotted as the Lorentz invariant cross-section at fixed small values of the 4-momentum tranfer squared, t, it was found to scale (be independent of s). The peak is due to the diffractive excitation of the other proton to masses well defined through the formula

$$M^2 = s(1 - x).$$

You see that the masses M to which a proton can be diffractively excited grow proportional to the center-of-mass energy $\sqrt{s}$, and at the top of the ISR energy range ($\sqrt{s} = 63$ GeV) reach about 15 GeV. If this continues up to the LHC ($\sqrt{s} = 14$ TeV) up to 3000 GeV can be "injected" into a proton by diffraction, i.e. by a pomeron. These numbers are simply given by the approximate rule of thumb that pomeron exchange dominates for scattered protons with $x \geq 0.95$; below that meson exchanges are more important. The identity of the "high mass diffraction" exchange and the pomeron of elastic scattering could be shown also by the shape of the peak in x (or M) which goes approximately like $\frac{1}{M^2}$ but with a correction depending on the difference of the pomeron trajectory from 1.0. This matched the "elastic pomeron" well. Perhaps we can learn about the nature of this object from its collisions, either with a proton (as in these diffractive excitation experiments) or with itself?

4. ISR: double pomeron exchange

The first observation of double pomeron exchange was at the CERN ISR. Both incident protons are quasi-elastically scattered with small p_T and Feynman x, $x_F > 0.95$, and a low mass hadronic system is created in the central region, as if from a pomeron-pomeron collision. This central system necessarily has very constrained quantum numbers: B = Q = S = 0 and $I^G J^{PC} = 0^+ even^{++}$. Its mass can be calculated from the x_F of the fast protons through

$$M^2 = s(1 - x_1)(1 - x_2).$$

For the highest ISR energy of $\sqrt{s} = 63$ GeV this corresponds to $M \leq 3$ GeV. The quantum number filter, mass region and possibility that the pomeron might be dominated by gluons makes this process an excellent hunting ground for "glueballs" as well as more standard mesons. Experiments showed that the process exists, with reasonable cross sections, and that the dominating channel which is $\pi\pi$ contains interesting structures, see Fig. 1 [1].

The $\pi^+\pi^-$ mass spectrum shows no ρ as it should not if pomeron-pomeron is dominating; at lower energies similar experiments show a ρ signal from (e.g.) pomeron-ρ exchanges. The next striking feature of the data is a sudden drop in the cross section by an order of magnitude around 950 - 1000 MeV which is due to the $f_0(975)$ state (or states). This is a perfect example of a narrow resonance appearing not as a bump but as a sharp dip in a mass distribution! Although the nature of the $f_0(975)$ is not known, it would be a mistake to say that the possibly glue-rich nature of the pomeron argues for a gluonic f_0. Final state interactions are strong and cause some "censorship" of the pomeron. However, absence of structure (dips) below the $f_0(975)$ allows one to rule out a lighter scaler glueball unless it were unreasonably narrow (e.g. 5 MeV). There are further structures in the higher mass region of the $\pi\pi$ spectrum, around 2 GeV, which have not been explained. The four-π mass spectrum appears to have an enhancement in the region of 2 GeV which was not clear enough to permit any special claims, but which could be the same object (Mass = 1.93 GeV, Width = 0.3 GeV, spin 2) found in similar central production experiments [2] [3] with lower energy but much higher statistics. As Landshoff and Donnachie have mentioned at this meeting this mass is compatible with the intersection of the pomeron trajectory and the line $J = 2$ so this might be a meson associated with the pomeron, presumably a glueball. Will there be a $J = 4$ state near 3.3 GeV (assuming the pomeron trajectory to be linear)?

Before leaving the subject of low mass double pomeron exchange I would just like to point out that the Axial Field Spectrometer experiment at the ISR showed that one could replace the protons with α particles and find the same central $\pi^+\pi^-$ spectrum. The fact that pions can be created while the α-particles fly past, not even excited, demonstrates the coherence of the phenomenon, that it is probably independent of the nature of the colliding hadrons and perhaps only to do with the vacuum. I think *VACUUM EXCITATION* is an appropriate term to describe this (although now the word *pomeron* is becoming more acceptable again we may have less need for alternatives!).

5. What is the pomeron?

I will take a purely experimental approach to the above question. This is something like: "If you don't know what it is, hit it" and see what comes out. Consider single diffractive excitation at the Tevatron energy $\sqrt{s} = 1800$ GeV. Requiring a quasi-elastic p with $x_F \geq 0.95$ allows excitation masses up to 400 GeV, and we view this, perhaps naively, as a pomeron-proton collision at that $\sqrt{\hat{s}}$. This (sub-)energy is certainly enough for hard parton phenomena such as high p_T jet production, high mass Drell-Yan pairs and heavy flavor production in hadron-hadron collisions, and we can use them to map out the parton content, or structure functions, of the colliding hadrons. The argument was applied to pomeron-proton collisions by Ingelman and Schlein [4] in 1985, and tested at the CERN $p\bar{p}$ Collider by Experiment UA8 which coupled the central detectors of UA2 to some very forward spectrometers. They indeed found [5] events with both a high-x ($x_F \geq 0.9$) antiproton and a pair of central jets, not very high p_T by collider standards but above 8 GeV/c. Knowing the structure function of the proton, measurements of the central jets give a parton distribution for the exchanged object, assumed to be a pomeron. The data seem to show that this is a hard distribution with high-x partons and even a *"superhard"* component where essentially all the momentum of the pomeron participates in the hard scatter: pomeron + q/g $\Rightarrow$ jet + jet. That would clearly be an important observation if confirmed but I do not find the UA8 data convincing on this point; there is no visible δ-function component, and any would be washed out by resolution effects, jet hadronization and so on. The UA8 group have done a good job in pioneering this study, but the energy even of the CERN Collider was marginal. It is important to separate protons scattering diffractively, through pomeron exchange, from a possibly large background of scatters by reggeized meson exchange. The best way to show that the "UA8-effect" has something to do with the pomeron is to repeat the experiment at higher energies. At the Fermilab Tevatron a mass of 200 GeV is associated with x_F of 0.988 compared with 0.90 at the S$p\bar{p}$S and the ratio of pomeron:reggeon should be much larger (roughly 5:1 compared with 1:2, according to fits by CDF [6]).

6. Double pomeron exchange, CERN Collider

Two experiments at the CERN Collider looked at double pomeron exchange but

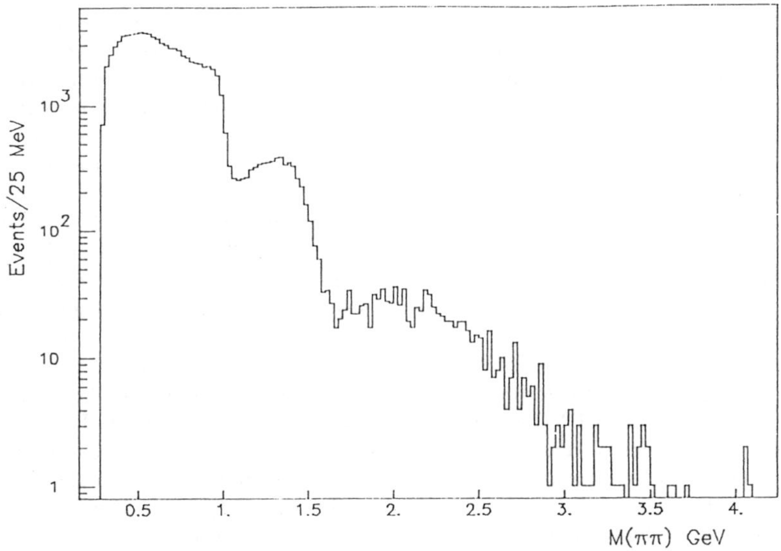

Figure 1: Mass spectrum of $\pi^+\pi^-$ produced in pomeron-pomeron interactions at the CERN ISR.

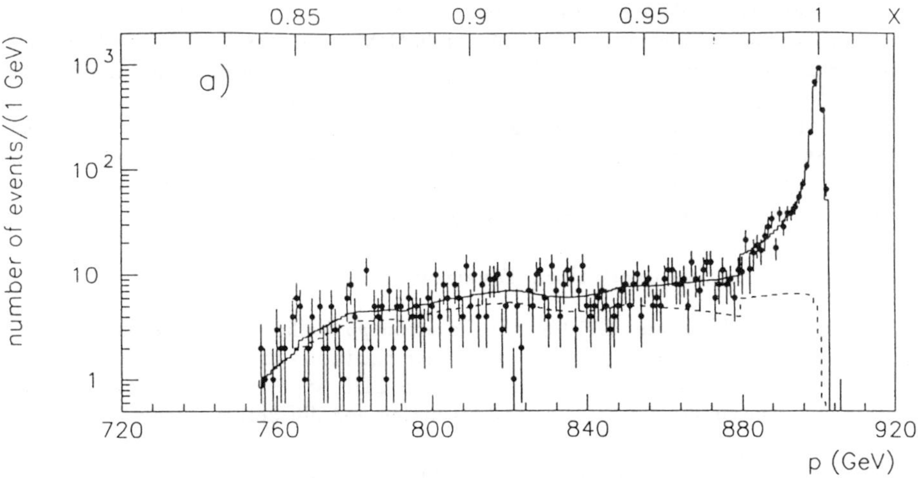

Figure 2: Longitudinal momentum spectrum of forward protons at Tevatron (CDF). The dashed line is the estimated non-diffractive background.

as a very peripheral activity. Experiment UA8, just discussed, selected events with two non-colinear (both "up") forward protons with $p_T \approx 1$ GeV/c. The central hadronic system has a mass distribution peaking at 3 GeV and there are no events above 10 GeV. This mass is determined by the calorimeter and the mass resolution is not sufficient to allow resonance and glueball studies, but it will be interesting to have information on the general structure of this central system. I understand these studies are underway.

These UA8 studies are complemented by an experiment done in UA1 [7] which selected higher mass (10 - 70 GeV) events with a much smaller cross-section (about 0.3 μb compared to 30 - 150 μb) still consistent with DPE topology. There were no forward (anti-)proton detectors but events were selected with *two* rapidity gaps over the ranges $-6 \leq \eta \leq -3$ and $3 \leq \eta \leq 6$ together with some minimum transverse energy in the central region. This is admittedly not very clean: on the one hand inefficiencies of tracking and calorimetry in the gap regions could give artificial gaps, on the other hand a genuine DPE event can veto itself if hadrons are emitted forwards (one would rather select on true rapidity y than pseudorapidity η). Also the gap should be a region with no *primary* hadrons but the experiment selects on the decay products, e.g. the γ from η decay. Nevertheless the use of (pseudo-)rapidity gaps to select pomeron exchange is now becoming widespread. The UA1 mass spectrum peaks around 20 GeV (lower masses are suppressed by the trigger) and extends up to about 100 GeV (the highest masses surely include more background). I would like to mention a couple of observations made on the central hadronic system which, if not the result of above-mentioned biasses, I find very interesting. The first is that the mean charged multiplicity rises with mass, M, with a much steeper slope than in e^+e^- or $p\bar{p}$ collisions plotted at $\sqrt{s} = M$. At 60 GeV a pomeron-pomeron collision has on average nearly twice the charged multiplicity of a $p\bar{p}$ collision. The multiplicity distribution is also odd in that, unlike most interactions, it does not obey KNO scaling. Making the standard KNO plot of $< n > P(n)$ vs $n/ < n >$ the distribution shrinks and becomes more Poisson-like as the mass increases. Perhaps something new and interesting is happening here, and the next step should again be to improve and repeat the experiment at the higher energy Tevatron.

7. Single diffractive excitation, Tevatron

Experiments CDF and E710 at the Tevatron $p\bar{p}$ Collider have measured [6] [8] the mass or x_F dependence of single diffractive excitation at $\sqrt{s}$ both 540 GeV and 1800 GeV. Fig. 2 shows a spectrum from the CDF experiment showing the data with a fit to the sum of a diffractive and estimated non-diffractive (shown separately) component.

Such fits support the conclusion that pomeron exchange only dominates for x_F greater than about 0.95. The integrated total SDE cross section at 1800 GeV is about 9 mb (both CDF and E710 agree on this), not having increased much with energy since the ISR ($\sqrt{s} = 63$ GeV) where it was about 7 mb.

It is unfortunate that when CDF took the above data, with a very forward spec-

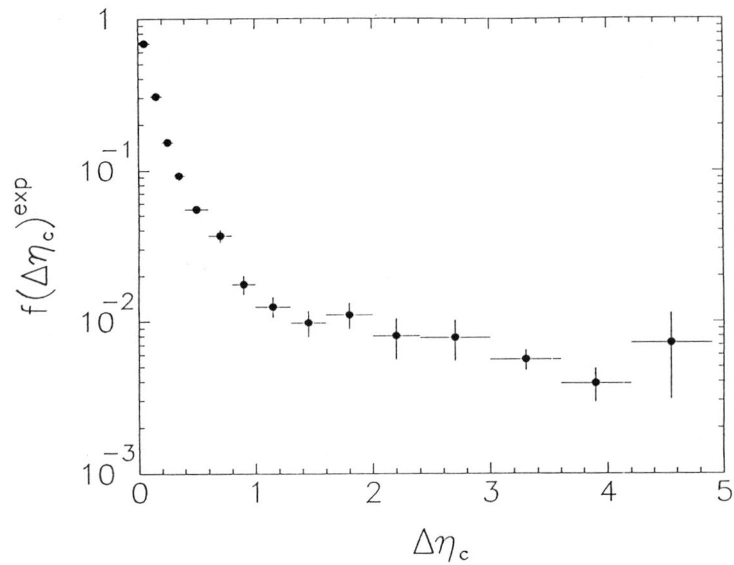

Figure 3: Probability f(0) of no clusters in a rapidity interval of size $\Delta\eta_c$ between two high E_T jets (D0).

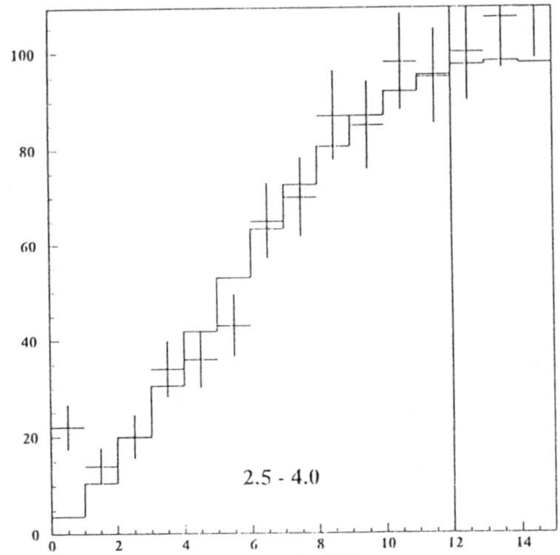

Figure 4: Multiplicity distribution of charged tracks with p_T above 400 MeV/c in rapidity intervals 2.5 - 4.0 units wide between high E_T jets (CDF). Crosses = data, histogram = fit.

E_T from 56 GeV to 85 GeV. This CDF study does not use calorimeter clusters or low p_T tracks, but assumes the tracks observed are a representative sample of all hadrons in the interval. They claim a rapidity gap signal at the above level, a little below the D0 upper limit, saying that it is "consistent with the exchange of color-singlet di-gluons". These events look like classic double diffractive dissociation except that the momentum transfer between the left- and right-moving systems is not $t \approx 1\ GeV^2$ but $t \approx 5000\ GeV^2$! A high E_T jet in one system is balanced by one in the other. What carries this huge momentum transfer? The cross section is much too high for γ, W or Z exchange, and some form of di-gluon seems a possibility. More information about these gap events would be valuable and should be forthcoming over the next year or two.

Another rapidity gap study by CDF uses W production as a probe of the pomeron. In a model [12] in which the pomeron has a structure function, either $q\bar{q}$ or gluon dominated, one can estimate the fraction of all W events that are diffractive. If the pomeron is predominantly $q\bar{q}$ as much as 17% of all W should be diffractively produced at the Tevatron, having a rapidity gap between the W and a very forward particle or particles. One should also find twice as many W^+ as W^- in the proton fragmentation, and the inverse for antiproton fragmentation. The preliminary results [13] show a ratio of diffractive to non-diffractive W production of $(0.56 \pm 1.0)\%$, ruling out a hard-q-dominated pomeron. We need to understand the compatibility between these new results and the results on dijets from UA8, but the authors claim [13] that "our result disagrees with the interpretation that the UA8 dijets are due to a hard-$q\bar{q}$ component of the pomeron".

9. Central vacuum excitation, Tevatron

Among the many possible future experiments in this field, one that I find particularly exciting is to study central vacuum excitation, or double pomeron exchange, at the Tevatron and eventually at the LHC. Central masses at the Tevatron should extend to about 100 GeV, and at LHC to about 700 GeV. At these center of mass (pomeron-pomeron) energies we should certainly be able to use hard phenomena (jets, Drell-Yan, W,Z and heavy flavors) to probe the pomeron and measure its structure function (if that is a valid concept). In CDF we are planning to do some preliminary studies by triggering on events with two forward rapidity gaps together with central activity. Later we may be able to add detectors to measure the high-x (anti-)protons. Perhaps the pomerons will sometimes behave like single "hard" gluons, producing a pair of central high-p_T jets with little additional activity. Gluon dominated pomerons will produce little Drell-Yan, W or Z but plenty of heavy flavors (including $t\bar{t}$ at LHC). Perhaps there will be events with a large number of very soft pions. This is purely speculation, but it could be that events where the pomeron "excites the vacuum" are a good place to look for exotic vacuum phenomena like a disoriented chiral condensate. This might show up as regions of $\eta\phi$ with very abnormal charged:neutral ratios, as seen in the Centauro cosmic ray events.

Conclusions

It is an exciting challenge to try to understand the pomeron at a fundamental level, such as QCD. Theory and experiment need guidance from each other. Experimentally the new possibilities of studying hard diffraction at Colliders, especially single diffractive excitation and double pomeron exchange, allow us to investigate the relation between the pomeron and quarks and gluons. There is much to be done both at the Tevatron and HERA, and it is a field full of potential for new phenomena.

I would like to thank the organizers for inviting me to this conference, especially Peter Landshoff who also made helpful comments on this written version of my talk.

References

[1] AFS Collaboration, T. Akesson et al.,Nucl. Phys. **B264** (1986) p 154.

[2] T.A.Armstrong et al., Phys. Lett **22B**, (1989) p.536

[3] WA91 Collaboration, S. Abatzis et al., CERN/PPE 94-28 (1994).

[4] G.Ingelman and P.Schlein, Phys. Lett. **152B** (1985) p.256

[5] UA8 Collaboration, R.Bonino et al., Phys.Lett. **B 211** (1988) p.239; A.Brandt et al., Phys.Lett **B 297** (1992) p.417.

[6] CDF Collaboration, F. Abe et al., Measurement of $p\bar{p}$ Single Diffractive Dissociation at $\sqrt{s} = 546$ and 1800 GeV, Phys. Rev.D (to be published).

[7] UA1 Collaboration, D. Joyce et al., Phys.Rev.D **48**, (1993) p.1943

[8] E710, N.A.Amos et al., Phys. Lett. **B 301**, (1993) p.313

[9] J.D.Bjorken, Int.J.Mod.Phys. **A7**, (1992) p.4179

[10] D0 Collaboration, S. Abachi et al., Phys.Rev.Lett **72** (1994) p.2332

[11] CDF Collaboration, F. Abe et al., Observation of Rapidity Gaps in $p\bar{p}$ Collisions at 1.8 TeV, Phys.Rev.Lett (submitted).

[12] P.Bruni and G. Ingelman, Phys. Lett **B311** (1993) p.317

[13] S.Bagdasarov and K.Goulianos, Search for Diffractive $W^{\pm}$ Production at the Tevatron, CDFNOTE 2580 (1994)

Phenomenology of the Soft Pomeron

A. Donnachie
Department of Physics and Astronomy, University of Manchester
Manchester M13 9PL, UK

August 11, 1994

Abstract

The standard phenomenolgy of the soft pomeron is reviewed. The evidence for the soft pomeron having a well-defined parton content is outlined. The role of the pomeron in deep inelastic scattering at small-x is discussed, and it is suggested that the standard phenomenology is incompatible with the HERA data. It is shown how minijet production can arise naturally as part of the soft pomeron, and hence that minijets do not contribute separately to total cross sections.

1 Introduction

The soft pomeron initially was a response to the experimental observation of rising hadronic total cross sections. It is an exchange, governed by the rules of Regge theory [1], and can be represented by the diagram of Fig.1.

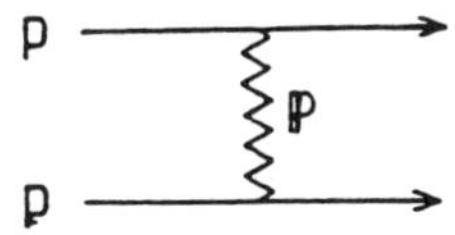

Figure 1. Pomeron exchange

The pomeron interacts with valence quarks, with an effective interaction

$$\beta F_1(t)\bar{u}\gamma_\mu u \tag{1}$$

where β is the strength of the coupling of the pomeron to a quark, with $\beta^2 \sim 3.5$ GeV^{-2}, and $F_1(t)$ is the elastic (Dirac) form factor of the proton [2] [3]. Given the structure of Eq.1, it can be seen that the pomeron behaves like a C = +1 photon.

The amplitude for high energy proton-proton scattering at a centre-of-mass energy $\sqrt{s}$ and momentum transfer t is then of the form

$$T(s,t) \sim \beta^2 F_1(t)^2 \xi_\alpha(t) s^{\alpha(t)} \tag{2}$$

where $\xi_\alpha(t)$ is a definite phase calculated from the Regge trajectory $\alpha(t)$. The total cross section can be calculated from the elastic amplitude by the optical theorem

$$\sigma_T = \frac{1}{s} ImT(s,t=0) \sim s^{\alpha(0)-1} \tag{3}$$

Experimentally it is found that although total cross sections initially decrease with increasing energy they then start to rise, and it is this rise which required the introduction of the soft pomeron, with

$$\alpha_P(t) = 1 + \epsilon + \alpha' t \tag{4}$$

and with $\epsilon > 0$.

An immediate extension is to diffractive dissociation in hadronic collisions, depicted in Fig.2, in which one of the incoming protons emerges with only a very small change of momentum. A small fraction

$$\xi = 1 - x_F \tag{5}$$

with ξ typically less than 5% or so, is carried off by the pomeron which collides with the other proton i.e. the process can be considered as a pomeron-proton interaction.

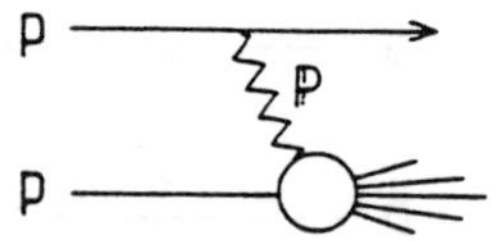

Figure 2. Diffractive dissociation

The mass M of the pomeron-proton subsystem is given by $M^2 = \xi s$, and the cross section is of the form

$$\frac{d^2\sigma}{dt d\xi} = F_{P/p}(t,\xi)\sigma^{Pp} \tag{6}$$

where σ^{Pp} is the total pomeron-proton cross section, t is the momentum transfer from the initial to the final fast proton and $F_{P/p}(t,\xi)$ is the "pomeron flux factor" [4]:

$$F_{P/p}(t,\xi) = \frac{9\beta^2}{4\pi^2}F_1^2(t)\xi^{1-2\alpha(t)} \tag{7}$$

Of course both incoming protons can "radiate" a pomeron, as in Fig.3, which leads directly to pomeron-pomeron interactions.

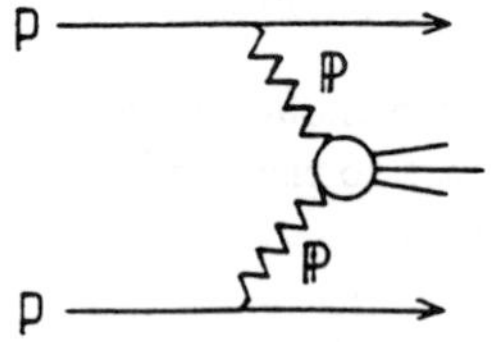

Figure 3. Double pomeron exchange

In the pomeron-proton interaction of Fig.2, it is possible for a hard process to occur as in Fig.4, giving rise to the phenomenon of diffractive hard scattering. This is now exploring the parton content, both gluon and quark, of the pomeron and is naturally described in terms of gluon and quark structure functions of the pomeron, $G_{g/P}(t,x)$ and $G_{q/P}(t,x)$.

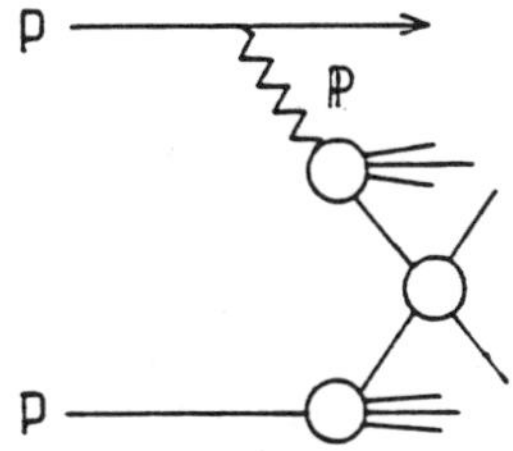

Figure 4. Diffractive hard scattering

The pomeron also appears in deep inelastic scattering at small x, and indeed dominates at sufficiently small x. Consider νW_2 at moderate Q^2 so that perturbative evolution has not yet set in and the parton model applies: Fig.5. From elementary kinematics, when a parton of momentum k is pulled out of a proton of momentum p with $k = xp + ...$ the squared invariant mass of the system it leaves behind is

$$s_0 \sim -\frac{k^2 + k_T^2}{x} \tag{8}$$

and so s_0 is large when x is small. The lower bubble in Fig.5 is thus an elastic strong-interaction amplitude and so is governed by Regge theory: it is just a sum of terms $s_0^{\alpha(0)}$. If this is inserted into the calculation of νW_2 from Fig.5 then a sum of terms $(1/x)^{\alpha(0)-1}$ is obtained [5].

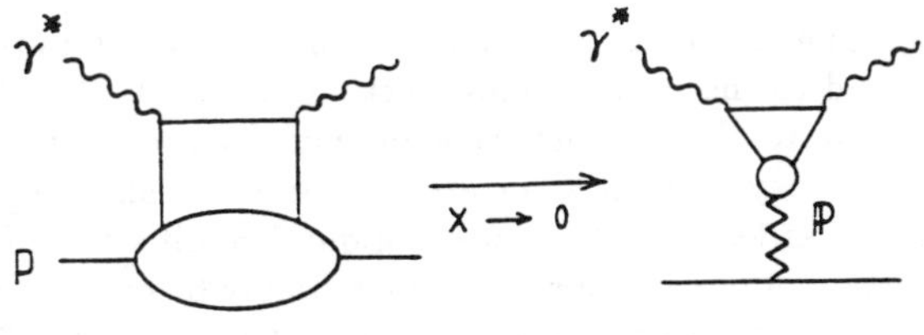

Figure 5. The pomeron in deep inelastic scattering

Finally the quark structure function of the pomeron can be obtained in diffractive deep inelastic scattering, Fig.6, which is expected to contribute 5–10% of the events at small x. Explicitly the quark structure function in these events is given by [4]

$$x\frac{d^2q(x)}{dtd\xi} = G_{q/P}(t,x')F_{P/p}(t,\xi) \tag{9}$$

with $x' = x/\xi$.

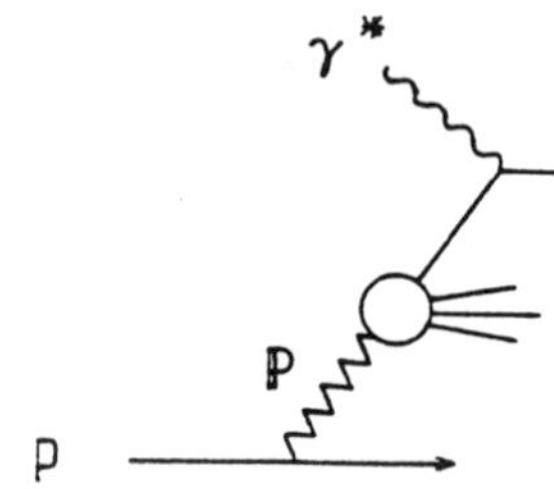

Figure 6. The quark structure function of the pomeron

2 Total and elastic cross sections

Figure 7 shows the Regge trajectory associated with the ρ, ω, f, a families from which we see that $\alpha_R(0) \sim \frac{1}{2}$, so from Eq.3 the contribution to σ_T has a behaviour close to $s^{-\frac{1}{2}}$ i.e. it gives a decreasing cross section. It was this that prompted the creation of the pomeron.
The precise value of $\alpha_R(0)$ from Fig.7 is 0.4525, and fitting total cross sections with

$$\sigma_T = Xs^{\alpha_R(0)-1} + Ys^{\epsilon} \tag{10}$$

yields $\epsilon = 0.0808$ [6].
The fits to the total pp and $p\bar{p}$ cross sections are shown in Fig.8, to the total $\pi^{\pm}p$ cross sections in Fig.9 and to the total γp cross section in Fig.10. The HERA data [7] [8] appeared subsequent to the fit, and were predicted correctly by the model. The pomeron/photon analogy and the associated assumption that the pomeron interacts with the valence quarks leads to the concept of quark additivity [9] [10], which relates the total cross section of one process to another simply by quark counting. From the fits to the total pp and πp cross sections we see that the expected 3:2 ratio for pp to πp total cross sections holds to better than 7% for the pomeron contribution.

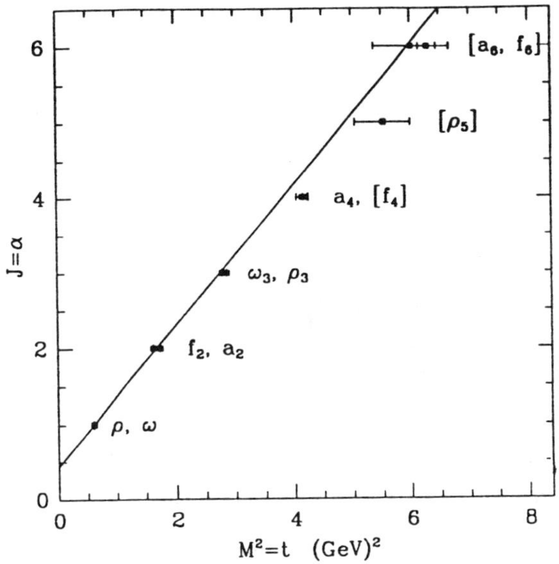

Figure 7. The ρ, ω, f, a trajectory

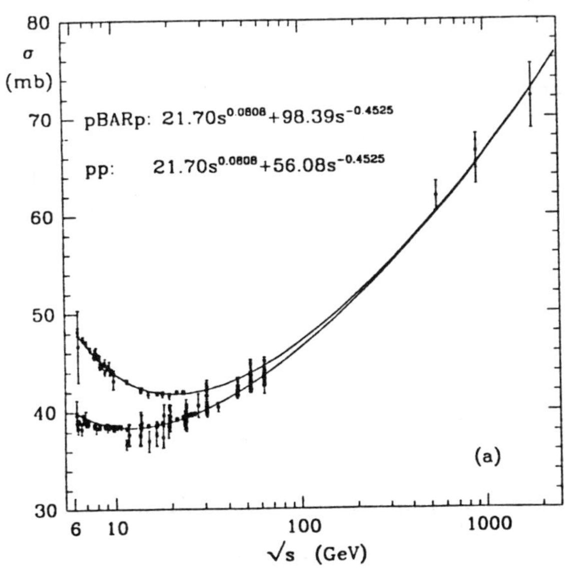

Figure 8. Fits to the pp and $p\bar{p}$ total cross sections

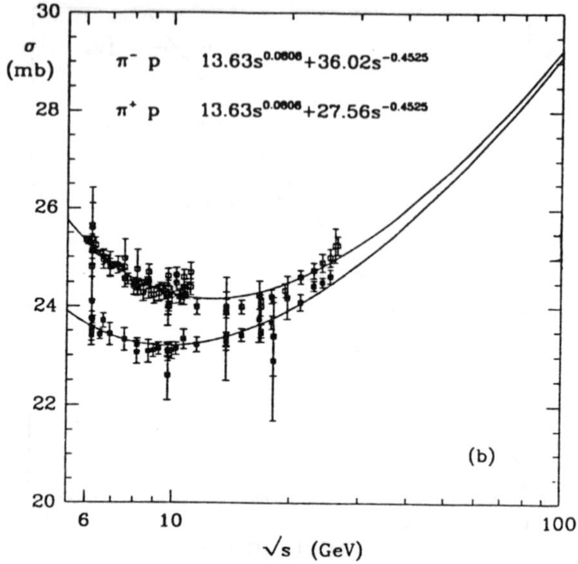

Figure 9. Fits to the πp total cross sections

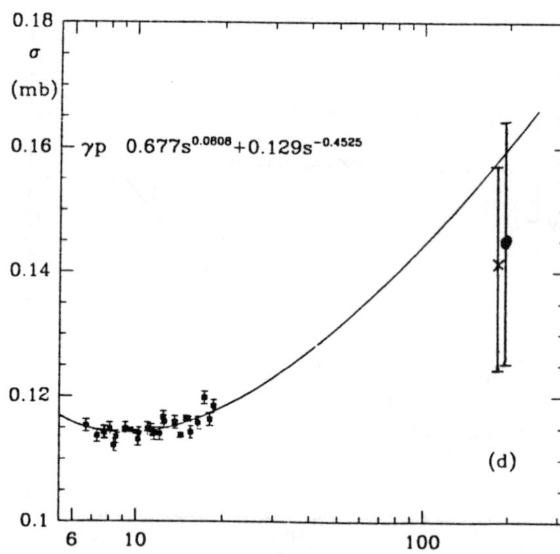

Figure 10. Fit to the γp total cross section

However it has to be said that this fit is not unique, and equally valid fits can be obtained with different parameters based on a different theoretical input [11]. The data at present energies do not provide sufficient information between pomeron models with very different high energy behaviour.

It should be noted that ϵ represents an effective power. It is not sufficient to consider the exchange of a single pomeron: multiple exchanges should be taken into account. Their magnitude is *apriori* incalculable, but phenomenologically their contribution to the total cross section appears to be small [3]. This "weak absorption" conclusion is compatible with the use of $\alpha_R(0)$ in Eq.10: if absorption were strong, then this would be significantly modified.

Figure 8 shows the published measurement of the E710 experiment at the Tevatron [12]. The CDF experiment has recently reported a much larger value: $\sigma = 80.6 \pm 2.3$ mb. The model presented here cannot be readily reconciled with this result, so if the conflict between the two measurements is resolved in favour of CDF, then it possibly indicates the onset of new physics.

The form of Eq.4 may be compared with elastic scattering data. In 1974 [13] the ISR pp data yielded the value of $\alpha' = 0.25$. Subsequent detailed analysis [3] of the ISR data led to remarkably successful predictions for $p\bar{p}$ elastic scattering at the CERN collider and at the Tevatron, which are shown in Fig.11. This is a real triumph for Regge theory: between ISR and Tevatron energies the change in the forward slope is more than 3 GeV^{-2} and is a direct confirmation that the value of α' is correct. Further, the phase of the forward amplitude is well predicted, in good agreement with the final value of ρ obtained by the UA4/2 collaboration at the CERN collider [14].

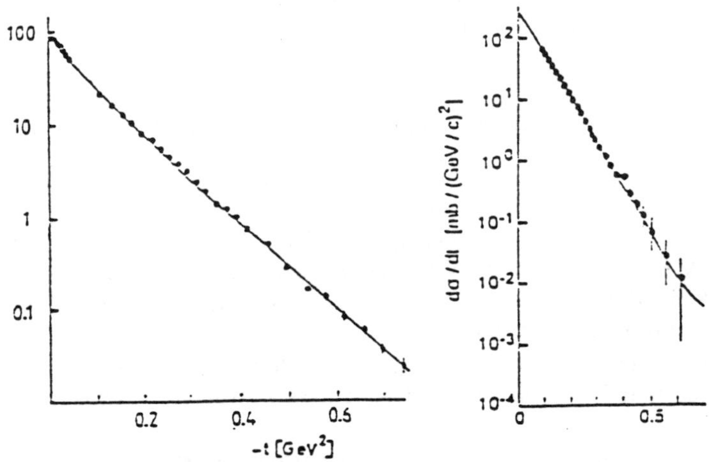

Figure 11. ISR data for pp and Tevatron data for $p\bar{p}$ elastic scattering with 1984 curves from [2]

3 Pomeron structure

The process of diffractive dissociation in hadron-hadron collisions is defined by Eqs.5 and 6. Because there is a very fast proton involved in the process, the kinematics lead to a rapidity gap in the final state. However the presence of a rapidity gap alone is not sufficient to guarantee that pomeron-exchange is involved, and it could be ρ, ω, f, a, or even a pion. In order to exclude the possibility of such contamination it is necessary to confirm the presence of the factor $\xi^{1-2\alpha(t)}$ with $\alpha(t)$ the pomeron trajectory of Eq.4, rather than with the ρ, ω, f, a trajectory of Fig.7 or the pion trajectory. To minimise the contamination ξ should be as small as possible.
The standard pomeron phenomenology requires that the pomeron has to couple to single quarks. The first direct experimental evidence for this came from the R608 experiment [15] at the CERN ISR in reactions of the type

$$p + p \rightarrow p_f + X \tag{11}$$

where p_f is a fast final state proton, implying pomeron exchange. Striking linear structures were found in the X systems $\Lambda^0\phi^0K^+$ and $\Lambda^0\bar{\Lambda}^0p$ which could only be explained by the pomeron interacting with a single valence quark in the beam proton. The first of these examples is illustrated schematically in Fig.12.

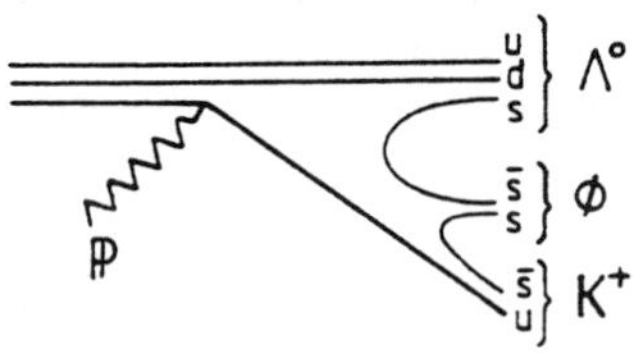

Figure 12. Diagram for diffractive forward $\Lambda^0\phi K^+$ production

Evidence for global linear structure was found in the UA8 experiment [16] at the CERN $p\bar{p}$ collider, in reactions analogous to Eq.11 but with M_X in the mass-range 50–200 GeV rather than $\sim$ 4 GeV for the ISR experiment. The energy-flow quantity $dE/d(cos\theta)$ in the X c.m. was found to possess sharp "polar peaks" looking like pomeron and proton "beam jets".
These results imply that the pomeron can "hit hard" and has a parton content. If the pomeron is indeed hadron-like, does it have a structure function and can it be measured?
Ingelman and Schlein [17] suggested that the partonic structure of the pomeron could be established in diffractive dissociation by studying the pomeron-proton system for evidence of hard scattering. This was verified by the UA8 experiment at the CERN

collider, which found [18] events containing jets with $p_T > 8GeV/c$. To calculate the cross section expected for such events, it is necessary to know the pomeron structure functions. The coupling of the pomeron to quarks is well-known, and the quark structure function of the pomeron is readily calculable [4]. The relevant diagram for the UA8 process is that of Fig.4, where one of the quarks to which the pomeron couples participates in the hard scattering that produces the high-p_T jets and the other is a longitudinal spectator.

The quark structure function of the pomeron, for each light quark or antiquark, is found to be

$$xq_P(x) \sim 0.075\pi x(1-x) \tag{12}$$

Note that x is the Bjorken fractional momentum variable relative to the momentum of the pomeron. Summing over all quarks yields x(1-x) with a multiplying coefficient a little more than 1.

Assuming that their data are dominated by quark jets, UA8 find a coefficient $2.76 \pm 0.48 \pm 1.44$, although the data can be equally well fitted assuming only a gluon structure function of the same form but with a coefficient $1.14 \pm 0.18 \pm 0.60$. The data cannot determine whether the pomeron structure function is quark or gluon (or a combination of both) but the data do rule out the possibility that the structure functions saturate a momentum sum rule. This is not particularly surprising, as the pomeron is not a state but only something that is exchanged so there is no reason to have saturation. As interesting feature is that in a significant fraction of events all, or nearly all, the energy of the pomeron-proton collision is taken by the pair of high-p_T jets. A natural explanation is that these events are generated by a process of "gluon-pomeron fusion" with no longitudinal spectator jet coupled to the pomeron.

4 The pomeron in deep inelastic scattering

We have seen that at sufficiently small x νW_2 should be given by a sum of terms $(1/x)^{\alpha(0)-1}$ [5]. NMC data [20] at moderate Q^2 and not-too-small x show that νW_2 contains such Regge terms at small x: a slowly varying term from soft pomeron exchange and close to $\sqrt{s}$ from f, a exchange.

Another important feature of νW_2 is that, at $Q^2 = 0$, it vanishes linearly with Q^2. Specifically

$$\sigma_{Tot}^{\gamma p} = \frac{4\pi^2\alpha}{Q^2}\nu W_2(Q^2 = 0) \tag{13}$$

The simplest fit to the small-x data which incorporates these features is analogous to Eq.10:

$$\nu W_2 = Xx^{-0.0808}[\frac{Q^2}{Q^2+a}]^{1.0808} + Yx^{0.4525}[\frac{Q^2}{Q^2+b}]^{0.5475} \tag{14}$$

Figure 13 shows the result of such a fit [19] to the small-x NMC data [20] up to $Q^2 = 10GeV^2$. It is only a two-parameter fit, as for each choice of X and Y the parameters a and b are determined so as to reproduce, via Eq.13, the γp fit shown in Fig.10.

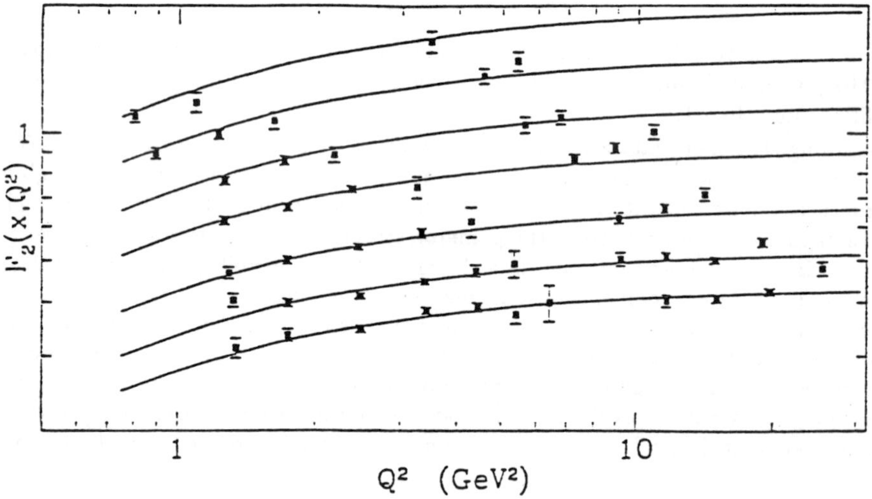

Figure 13. NMC data [20] for νW_2. with fit of the form (14). The successive values of x from top to bottom are 0.008. 0.0125.0 .025. 0.035. 0.05.0.07

If one sets $Q^2 = 8.5GeV^2$ and extrapolates the fit down to x = 2×10^{-4}, the fit to the NMC data gives $\nu W_2 \sim 0.6$. This is much smaller than the measured value of about 1.4 ± 0.5 reported by H1 [21]. To reproduce this requires that $\nu W_2 \sim x^{-0.3}$. This is incompatible with the behaviour expected from the phenomenological pomeron exchange with weak rescattering as the power of x is a function of the exchanged object (if it is a Regge pole) and not of Q^2. One way out of this dilemma is to suggest that the weak rescattering assumption is incorrect. In a strong absorption model, it is possible to choose $\epsilon \sim 0.3 - 0.5$ for the bare pomeron and reduce this to the effective $\epsilon \sim 0.1$ by multiple rescattering, although for this to be convincing would require a successful reworking of all soft-pomeron phenomenology. In deep inelastic scattering, as Q^2 increases the interaction time decreases and so does the absorption. A much more exciting possibility is that the discrepancy may be attributed to the presence of the Lipatov pomeron [22]. In order to decide whether this is true, it is necessary to learn how the two pomerons live together. One possibility is that, as Q^2 increases, the soft pomeron goes over smoothly to the Lipatov pomeron, that is $x^{-0.08}$ changes to something like $x^{-0.5}$. Alternatively, the two terms could be added together.

Whatever the nature of the pomeron in deep inelastic scattering, important information about its structure function will come from HERA, since γ^*-pomeron collisions probe directly the quark structure function of the latter. With the structure function of Eq.8, some 10% of the total events that make up νW_2 at small x should have [4] a very fast proton. They would appear as "single jet" events, the deep-inelastic analogue of the UA8 "beam-jet" events. The analogue of events corresponding to Fig.4b should also be present i.e. fast-proton events with two high-p_T jets and nothing else. Real photons can also be used to look for these, and the cross section should be quite large [23]: $\sigma^{\gamma p} \sim 1nb.$ for $p_T > 5GeV/c$.

5 Minijets

We have seen that Regge theory successfully predicted the γp total cross section measure at HERA. There were other predictions that were very much larger [24] and it is interesting to discuss why they were wrong. They were based on perturbative calculations of inclusive minijet production, and these should reproduce the inclusive cross section $d\sigma/dq_T$ reasonably successfully down to $q_T^{min} \sim 1GeV$, with a K-factor somewhere between $\frac{1}{2}$ and 2. Now

$$\int_{q_T^{min}} dq_T \frac{d\sigma^{pair}}{dq_T} = \bar{n}\rho(q_T^{min})\sigma_T \tag{15}$$

where $\rho(q_T^{min})$ is the fraction of events containing minijets and $\bar{n}$ is the average number of minijet pairs in these events. The inclusive cross section is calculated from the familiar diagram of Fig.14.

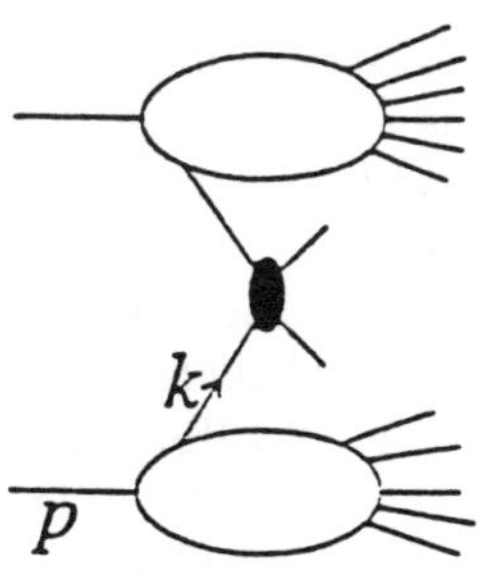

Figure 14. Diagram for single inclusive minijet pair production

The naive expectation is that exactly two minijets are produced i.e. that $\bar{n} = 1$. However evaluating the integral in Eq.15 and comparing with the HERA measurements leads to the erroneous conclusion that $\rho(q_T^{min}) > 1$, which is wrong by definition. It follows that $\bar{n} > 1$, which is not really a surprise, as it has long been known that it is correct to use Fig.14 to calculate the inclusive cross section $d\sigma/dq_T$ but that this cannot be expected to reproduce correctly other features of the event structure. One way in which it can be greater than 1 is from multiple parton-parton scatterings [25] involving two or more partons from each of the initial particles. While this will almost certainly be very important in high-energy nucleus-nucleus collisions, it is doubtful whether it can be the main mechanism in pp or γp collisions.

Rather there is another mechanism [26] which is intrinsically non-perturbative. In order to calculate $d\sigma/dq_T$ from Fig.14, the two structure functions are required down to fractional-momentum values x of order $x \sim q_T/\sqrt{s}$, which is very small when $\sqrt{s}$ is large. The argument that led to Eq.8 and hence to s_0 being large at small x is equally applicable here. That is the upper and lower clusters of residual fragments of the initial hadrons in Fig.14 each have large invariant mass and so are very likely to contain minijets.

At small x there is no k_T ordering, so that perturbative and non-perturbative effects are inevitably intermingled. This can be understood from a ladder model for the total cross sections in which there is an integral over the transverse momentum round each loop in the ladder. If there is a contribution from a "hot" loop somewhere in the middle of the ladder, i.e. from a loop with large transverse momentum, then cutting the ladder down the middle and absorbing the rungs above and below the hot loop reproduces exactly the form of Fig.14.

However it may well be that another loop of the diagram is hot, or even several other loops, corresponding to several pairs of minijets. This leads to proposing Fig.15 as the dominant mechanism for producing minijets [27].

A calculation based on this mechanism is in good agreement with the UA1 minijet data [28] and predicts a potentially high rate of production for minijets. The result is naturally strongly dependent on the value of q_T^{min} used in the calculation: at Tevatron energies the inclusive cross section for the production of a minijet pair is about a factor of 10 more than the total cross section for $q_T^{min} = 1$ GeV, but is only about 50% for $q_T^{min} = 3$ GeV. The model is equally valid if one of the hadrons is replaced by a real photon. This is because, in minijet production, as in the total cross section, the photon behaves like a hadron.

The same argument can be applied to νW_2 at small x since, as we have already indicated, the energy of the lower bubble in Fig.5 is large and pomeron exchange dominates the small-x behaviour of νW_2. Minijets are predicted to be produced copiously [27] by this mechanism in deep inelastic scattering at small x, although the answer again depends on the choice of q_T^{min}: for $q_T^{min} = 3$ GeV some 10% of deep inelastic events are predicted to contain minijets.

Figure 15. Diagrams for single- and double-inclusive minijet pair production, with alternating soft pomeron exchange and hard scattering

6 Conclusions

Pomeron physics is once again a very active and exciting area of study. The soft pomeron is non-perturbative in origin and is almost certainly a consequence of non-perturbative gluon exchange [29]. However a simple model of the exchange of two non-perturbative gluons is compatible with much of pomeron phenomenology, which indicates the difficulty of establishing the true nature of the pomeron. Just as the reggeon exchange is associated with physical meson states, it would be nice to be able to associate the pomeron with a glueball state. If $\alpha_R(t)$ remains linear for positive t, then there should be a 2^{++} glueball at 1900 MeV. A candidate has been reported [30] at precisely this mass. It has the appropriate $I(J^{PC}) = 0(2^{++})$ but it is not clear whether it is a single broad state ($\Gamma = 370 \pm 70 MeV$) or possibly two separate states.

A significant development in recent years has been the recognition that the pomeron has partonic structure with a hard structure function, although whether this is due primarily to quarks or gluons, or to some combination of the two, is as yet unknown. As the soft pomeron can couple to a hard system, minijets can be seen to be an integral part of the pomeron, which is itself the controlling mechanism for the total cross section. It is thus incorrect to ascribe the rise in the photon-hadron or hadron-hadron total cross section to the onset of minijet production as additional to the pomeron.

The pragmatic development of soft pomeron phenomenology has been remarkably successful, and there are only two pieces of data with which it is in serious conflict.

One is the latest total cross section measurement at the Tevatron, but there is a genuine experimental inconsistency to be resolved before any comment on the soft pomeron phenomenology can be made. The second is the measurement of νW_2 at small x at HERA. The soft pomeron phenomenology as currently developed cannot incorporate these data, and there has to be a genuinely new effect present. Whether it is the Lipatov pomeron is still open to question, and clearly much more data will be required on all aspects of the hard interaction of the pomeron to resolve this.

References

[1] P.D.B.Collins: Introduction to Regge Theory, Cambridge University Press (1977)

[2] A.Donnachie and P.V.Landshoff: Nucl.Phys.B231(1984)189

[3] A.Donnachie and P.V.Landshoff: Nucl.Phys.B267(1986)690

[4] A.Donnachie and P.V.Landshoff: Nucl.Phys.B303(1988)634

[5] P.V.Landshoff, J.C.Polkinghorne and R.D.Short: Nucl.Phys.B28(1970)210

[6] A.Donnachie and P.V.Landshoff: Phys.Lett.B296(1992)227

[7] ZEUS Collaboration. M.Derrick et al: Phys.Lett.B293(1992)465

[8] H1 Collaboration, T.Ahmed et al.: Phys.Lett.B299(1993)374

[9] E.M.Levin and L.L.Frankfurt: JETP Letters(1965)65

[10] H.J.Lipkin and F.Scheck: Phys.Rev.Lett.16(1966)71

[11] H.J.Lipkin: Weizmann Institute preprint WIS-94/25/May-PH

[12] E710 Collaboration, N.Amos et al: Phys.Lett.B243(1990)158

[13] G.A.Jaroskiewicz and P.V.Landshoff: Phys.Rev.D10(1974)170

[14] UA4/2 Collaboration, C.Augier et al: Phys.Lett.B316(1993)448

[15] A.M.Smith et al: Phys.Lett.B163(1985)267; B167(1988)248; T.Henkes et al: Phys.Lett.B283(1992)155

[16] UA8 Collaboration: A.Brandt et al: Proc. Joint Int. Lepton-Photon Symp. and Europhysics Conf. on High Energy Physics, Geneva 1991 (ed. S.Hegarty, K.Potter, E.Quercigh), p.741

[17] G.Ingelman and P.Schlein: Phys.Lett.B152(1985)256

[18] UA8 Collaboration, A.Brandt et al; Phys.Lett.B297(1992)417

[19] A.Donnachie and P.V.Landshoff: Z.Phys.C61(1994)139

[20] NMC Collaboration, P.Amaudruz et al: Phys.Lett.B295(1992)159

[21] H1 Collaboration, T.Ahmed et al: preprint DESY 93-117

[22] E.A.Kunaev, L.V.Lipatov and V.Fadin: Soviet Physics JETP 45(1977)199

[23] A.Donnachie and P.V.Landshoff: Phys.Lett.B285(1992)172

[24] M.Drees and F.Halzen: Phys.Rev.Lett.61(1988)275; R.Gandhi and I.Sarceric: Phys.Rev.D44(1991)R10

[25] J.R.Forshaw and J.K.Storrow: Phys.Rev.D46(1992)4955; R.S.Fletcher, T.Gaisser and F.Halzen: Phys.Lett.B298(1993)442

[26] M.Jacob and P.V.Landshoff: Mod.Phys.Lett.A1(1986)657

[27] A.Donnachie and P.V.Landshoff: Phys.Lett. (to be published)

[28] UA1 collaboration, C.Albajar et al: Nucl.Phys.B309(1988)405

[29] P.V.Landshoff and V.Nachtmann: Z.Phys.C35(1987)405: F.Halzen, G.Krein and A.Natale: Phys.Rev.D47(1993)295; J.R.Cudell and B.V.Nguyen: McGill preprint 93-25

[30] T.Armstrong et al: Phys.Lett.B228(1989)536; S.Abatzis et al: CERN/PPE 94-28

THE THEORY OF THE SOFT POMERON

D.A. ROSS
Department of Physics
University of Southampton
Southampton S09 5NH,
U.K.

ABSTRACT

I consider some attempts which have been made to describe soft pomeron behaviour within the context of QCD.

1. Introduction

I want to address (although I shall not be able to answer) the question of what the soft pomeron "is" in terms of QCD, which is now almost universally accepted at the correct theory of the strong interactions and must therefore give us all observed strong interaction behaviour. Firstly, however I would like to take a step back and consider the question of whether or not a soft pomeron exists at all since there are "pomero-sceptics", who would argue that one can satisfactorily fit all diffractive data without recourse to a pomeron. For example we look at the data for the proton-proton and proton-antiproton total cross-sections (Figure 1.).

The fit on the left is given by the Donnachie-Landshoff formula[1]

$$\begin{aligned} \sigma^{pp}(s) &= 21.7s^{0.08} + 56.1s^{-.45} \\ \sigma^{p\bar{p}}(s) &= 21.7s^{0.08} + 98.4s^{-.45} \end{aligned} \tag{1}$$

This is the soft pomeron with intercept just above zero and it is important that the coefficient of the leading term is the same for the $p-p$ and $p-\bar{p}$ cross sections, indicating that the pomeron couples equally to proton and antiproton. The parameters of the daughter (subleading) trajectory were fit to data. The fit on the right is the fit due to Levin and Ryskin[2]

$$\sigma^{pp}(s) = 21.4 + 46.8s^{-.25} - 16.3s^{-4.3} + 0.24\ln^2 s \tag{2}$$

This has more parameters, but fits the data almost as well † without introducing any terms which grow with s faster than $\ln^2 s$, thereby not violating the unitarity bound[3].

† There is one data point from the CDF collaboration which is somewhat higher than the other data, and which is missed by both of these fits

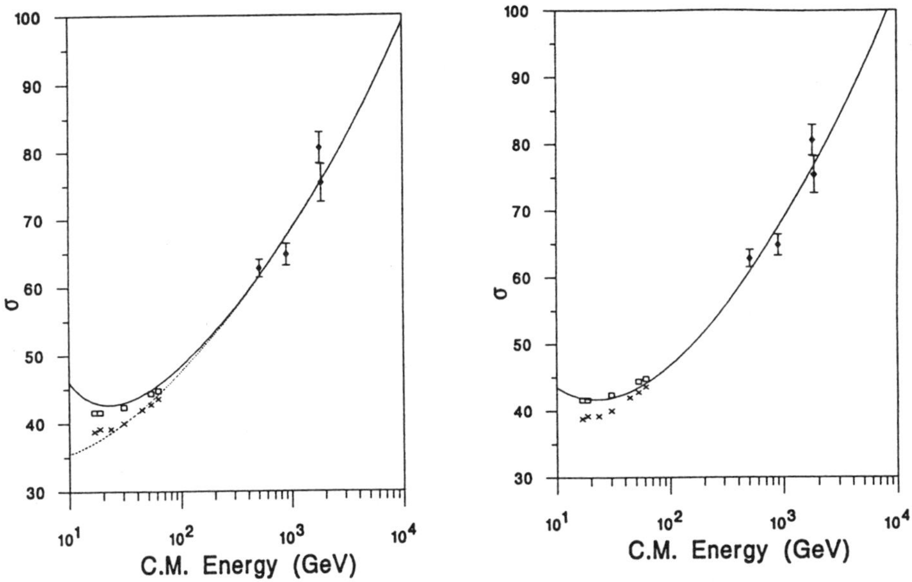

Figure 1: Two different fits for $p-p$ and $\bar{p}-p$ scattering

It is argued in ref.2 that this $\ln^2 s$ can be obtained from the "hard" pomeron of BFKL[4], but that this behaviour is only valid for impact parameter, b greater than a certain critical value.

The amplitude as a function of s and b is written as

$$a(s,b) \;=\; s^{12\ln 2\,\alpha/\pi} e^{-b^2/(c\ln s)} \tag{3}$$

The first factor does indeed come from the leading singularity of the BFKL pomeron, but the dependence on b is not related to the BFKL pomeron at all. It is obtained form a diffusion model and it leads to a linear trajectory for the pomeron. Although data for the differential cross-sections are consistent with linear trajectories, I am going to suggest that it is highly unlikely that QCD will give rise to such linear trajectories. In fact the dependence on the behaviour of the BFKL pomeron with impact parameter (or equivalently with momentum transfer) is likely to be very soft and for small impact parameter is determined by the running of the strong coupling constant with impact parameter[5]. The next step in the argument is that the above amplitude cannot be valid for sufficiently large s and small b since this would violate unitarity. The cutoff for b is thus chosen to be given by

$$b^2_{crit} = 12\ln 2\frac{\alpha}{\pi} c\, ln^2 s$$

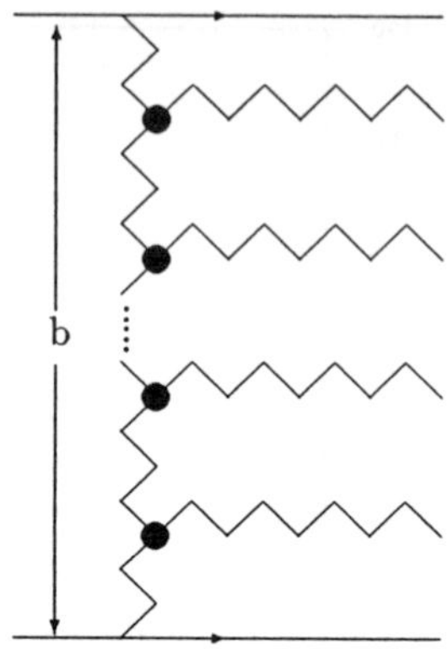

Figure 2: Scattering amplitude for impact parameter b

Below this impact parameter the proton behaves like a black disc of radius given by b_{crit} and therefore proportional to $\ln s$.

I have a few comments about this approach

- Unitarity is effectively imposed by hand by choosing to cut off the impact parameter above which the BFKL result is valid, precisely where unitarity starts to be violated - it is therefore hardly surprising that a result consistent with unitarity is obtained.
- All memory of the BFKL pomeron has effectively been lost - the result obtained would have been obtained by imposing unitarity on any initial amplitude, assuming the same dependence on the impact parameter. so this does not tell us anything about QCD and pomerons
- The impact parameter dependence is that which gives rise to linear trajectories, but is not in any way derived from QCD.
- The "black disc" picture, which is supposed to dominate at large s is one in which several partons are interacting at the same time, and this would violate the quark counting rule which tells us that the pomeron coupling is proportional to the number of quarks it couples to and leads to the relation

$$\frac{\sigma^{\pi p}}{\sigma^{pp}} \underset{s \to \infty}{\longrightarrow} \frac{2}{3}$$

This is experimentally verified as can be seen from Figure 3. Notice that the

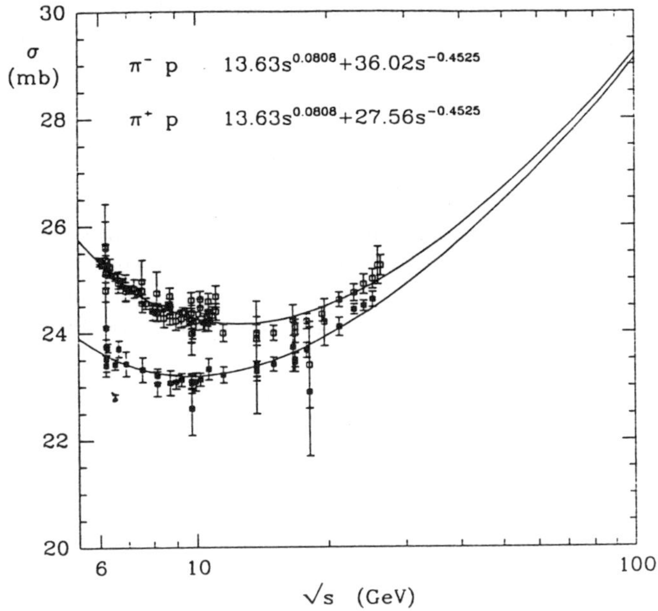

Figure 3: $\pi - p$ Scattering

coefficient of the leading term is almost 2/3 times the coefficient of the leading term in Equation(1).

I therefore conclude that although eventually (for sufficiently high s) screening effects must become important in order to render the cross sections consistent with unitarity, the total cross sections and other diffractive processes at currently accessible energies contain a substantial contribution from the soft pomeron.

At this stage I should point out that there are two opposing schools of thought. On the one hand there are those who like to draw a clear distinction between what is called the "hard" pomeron, that is the solution to the BFKL equation derived purely from perturbative QCD, and the "soft" pomeron which is responsible for hadronic diffractive processes. The argument here is that the "hard" pomeron, although it may be relevant in some kinematic regime in deep inelastic scattering is almost entirely screened out from diffractive hadronic processes and the pomeron which gives rise to the behaviour correctly predicted by the old Regge theory is an entirely different object. This is a particularly depressing approach from the point of view of a field theorist since it leaves us with absolutely no idea of what aspects of QCD could possibly lead to this "soft" pomeron behaviour. I prefer to take the view that there is no real distinction between the "hard" and "soft" pomeron, but that the "soft" pomeron could be obtained from a similar approach to that of BFKL (and others) but with the added ingredient of non-perturbative infrared behaviour of QCD about which we know so little, with the possible exception of calculations that have been performed using lattice techniques. An alternative approach has been taken by Levin and Tan[6] who develop a "heterotic" pomeron using a hybrid of ingredients from perturbative QCD and the ideas of parton diffusion, which can reproduce the behaviour of either the "hard" or "soft" pomerons depending on the kinematic regime.

The idea that the infrared behaviour of QCD is necessary for a complete description of the pomeron is not new. Indeed if we study the BFKL equation we notice that the effective two-gluon ladder which generates the hard pomeron requires integration over transverse momenta which go down to zero. The off-shellness of the vertical reggeized

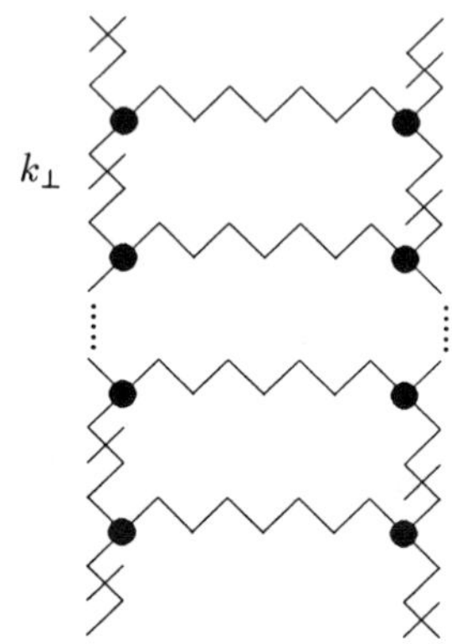

Figure 4: The BFKL pomeron drawn schematically in terms of a ladder in which the dashes indicate reggeized gluons and the blobs are non-local effective vertices

gluons in the ladder are of order $k_\perp^2$. Even though there are no infrared divergences, it would be surprising if we could describe diffractive process with very small momentum transfers without some knowledge of the (non-perturbative) infrared behaviour of the gluon propagators.

Indeed the BFKL equation in its pure form, does not even give us an isolated pole in the Mellin transform, which can be identified as a pomeron; the leading singularity is a cut. This problem was addressed by Lipatov[7], who pointed out that there were *two* effects not considered in the purely perturbative calculation (to leading order in $\ln s$) which can have the effect of changing the cut singularity into a pole (see Figure 5). One of these is the effect of the running coupling constant, which slows down the oscillation frequency of the solution to the BFKL equation as a function of $k_\perp^2$, eventually turning the oscillatory behaviour into an exponentially decaying behaviour. The other is some unknown infrared behaviour which is necessary to fix the phase of the oscillatory solution for some value of $k_\perp$. It can easily be shown[8] that

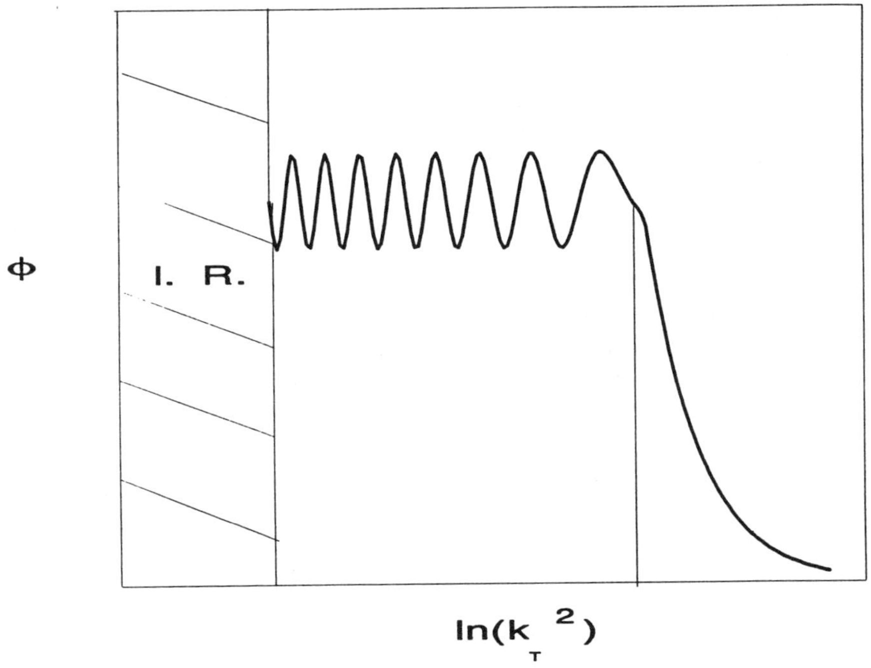

Figure 5: The solution to the BFKL equation for the two-gluon amplitude, $\phi(k_\perp^2)$ as a function of $\ln k_\perp^2$

if and only if we have both of these effects then there is only a discrete spectrum of allowed eigenvalues (leading to isolated poles in the Mellin transform of the scattering amplitude).

2. Soft Gluon Propagators from the Dyson Schwinger Equation

A complete non-perturbative treatment is clearly quite out of the question. In fact the picture of a pomeron as being associated with two (reggeized) gluon exchange subsumes a perturbative approach. We make the assumption therefore that the infrared end of QCD will have its largest effect in the soft gluon propagators in the pomeron and we investigate the possibility that we can get make some progress towards obtaining soft pomeron behaviour, by considering the soft gluon propagator.

One of the useful tools for doing this is the Dyson-Schwinger equation. This is an exact equation, derivable without recourse to perturbation theory, but unfortunately insoluble, since it is really an infinite tower of equations relating the n-point Green function with the (n+1)-point Green function. An attempt to get around this problem by making a reasonable ansatz for the triple gluon vertex was started in the early eighties by Baker, Ball, and Zachriasen[9]. They looked at the Dyson-Schwinger equation for the gluon propagator in axial gauge. This gauge has two advantages. Firstly the terms in the Dyson-Schwinger equation involving the four-point gluon vertex can be eliminated. Secondly since we are in a gauge which is free from Faddeev-Popov ghosts there is a Ward-Takahashi identity which relates the longitudinal part of the three-point vertex to the self energy functions. Now using a theorem due to Mandelstam[10], which tells us that for small momentum the propagator is dominated by the longitudinal part of the vertex one now has a consistent ansatz for the vertex function and the Dyson-Schwinger equation becomes an implicit equation for the self-energy, which can be solved, at least numerically.

If we write the gluon propagator in the axial gauge (axis n^μ) as

$$D_{\mu\nu}(q^2) \;=\; \frac{Z(q^2)}{q^2}\Delta_{\mu\nu}$$

where

$$\Delta_(\mu\nu) \;=\; -g_{\mu\nu} + \frac{q_\mu n_\nu}{n\cdot q} + \frac{q_\nu n_\mu}{n\cdot q} - \frac{n\cdot n q_\mu q_\nu}{(n\cdot q)^2}$$

Baker, Ball and Zachariasen found a solution, which for small q behaved like

$$Z(q^2) \;\propto\; \frac{1}{q^2}$$

This may well have something to do with confinement as it is the Fourier transform of a confining potential, but such a soft propagator is a disaster for describing diffractive processes which involve the exchange of soft propagators. In some work with Cudell[11] we exploited the fact that the Dyson-Schwinger equation obtained from making the ansatz for the vertex of ref.9, was a nonlinear integral equation and as such admits more than one solution (even with the same boundary conditions). Thus we conclude that there could be other solutions that had a more gentle behaviour as $q^2 \to 0$, and that these solutions would be the ones relevant for describing diffractive data. It was shown in ref. 9 that in the axial gauge the propagator had to have a singularity at $q^2 = 0$, but this singularity could be much softer than a pole - this softening being due to the infrared behaviour of QCD and would introduce a dimensionful scale μ.

With this in mind we parametrized the function $Z(q^2)$ as

$$Z(q^2)^{-1} \;=\; a_1\left(\frac{\mu^2}{q^2}\right)^{b_1} + a_2\left(\frac{\mu^2}{q^2}\right)^{b_2} + c\ln\left[d\left(\frac{\mu^2}{q^2}\right)+e\right]$$

and by inserting into the integral equation and minimizing with respect to the parameters we found a very good fit for the values $a1 = 0.88,\ b_1 = 0.78,\ a_2 = -0.95,\ b_2 = 0.14,\ c = 0.59,\ d = 2.1,\ e = 4.1$

Note that this solution gives rise to a propagator which behaves like $(q^2)^{-.22}$ as $q^2 \to 0$, so we still have a singularity, but not a pole. The parameter μ is a parameter that we fit from phenomenology, but we choose it to be the value at which $Z(\mu^2) = 1$ so that there is one constraint between the above parameters.

3. Applying the Solution

The first "test" of this soft propagator that we tried was to look at proton-proton elastic scattering. We look at the simplest and most naive contribution to this process, namely the exchange of two gluons, not because we expect accurate results this way, but to get some idea of the usefulness or otherwise of this soft propagator. If we were to use perturbative gluon propagators and consider on-shell quarks then the cross section has infrared divergences, when the momentum of either gluon vanishes.

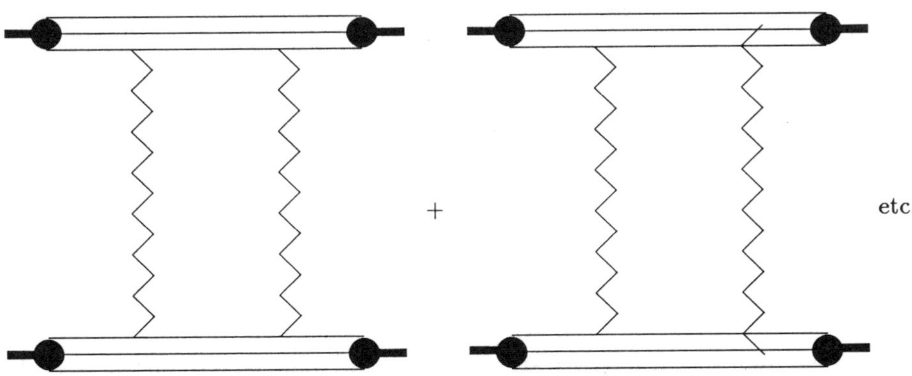

Figure 6: Two gluon exchange contribution to p p scattering

However if we take into account the wavefunction of the hadrons[12] thereby allowing the quarks to be off shell by an amount of the order of the primordial $k_\perp$ of the quarks inside the hadron, then since the hadrons are colour singlets the infrared divergence cancels between graphs in which the gluons end on different quarks lines (two of which are shown in Figure 6.). Reasonable estimates of the wavefunctions can be obtained by comparison with the electromagnetic form factor of the proton. However the differential cross section is not rendered infrared finite, by this mechanism and the slope at the origin still comes out infinite. Clearly by using a propagator, which has softer behaviour at the origin will solve this problem. We have applied our soft propagator to the graphs shown in Figure 6. and obtain results shown in Figure 7. The agreement with data turns out to be much better than one could reasonable have

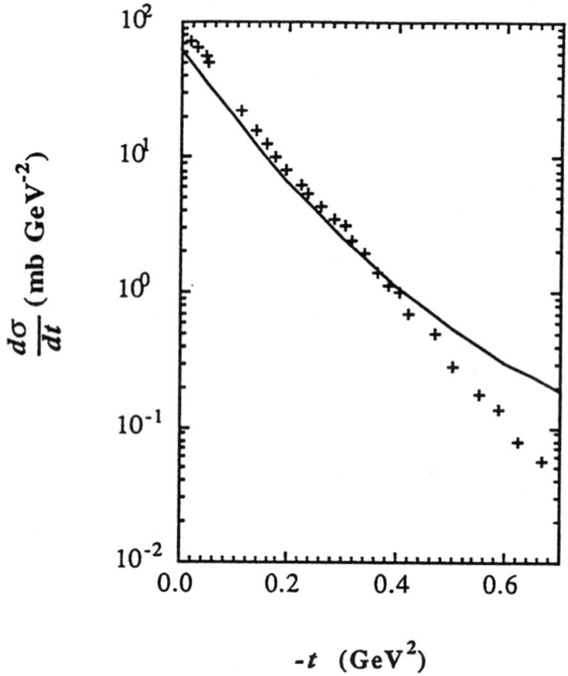

Figure 7: The differential cross section $d\sigma^{pp}/dt$ against t. The crosses indicate some of the data

expected from such a crude treatment, and should only be interpreted as evidence that the soft-propagators used are likely to give rise to sensible results.

The next step is to see what happens to the trajectories if this propagator, together with a running of the coupling constant is inserted into the BFKL equation[13]. This is of course not a complete treatment, but is based on the assumption that the largest sensitivity to infrared effects comes from the propagators of the reggeized gluons. The results are shown in Figure 8. We see as expected that for large t the trajectories obtained converge to those from the purely perturbative "hard" pomeron with its very gentle (logarithmic) dependence on t. This is expected since for sufficiently large t the infrared effects should become negligible. On the other hand near the forward direction the soft propagator effects result in a much steeper trajectory. If this is indeed the correct road to the matching of the "soft" and "hard" pomerons then a linear trajectory appears to be excluded. The intercept of the leading trajectory is suppressed a little from its rather large value of $12 \ln 2\, \alpha/\pi$, for the "hard" pomeron,

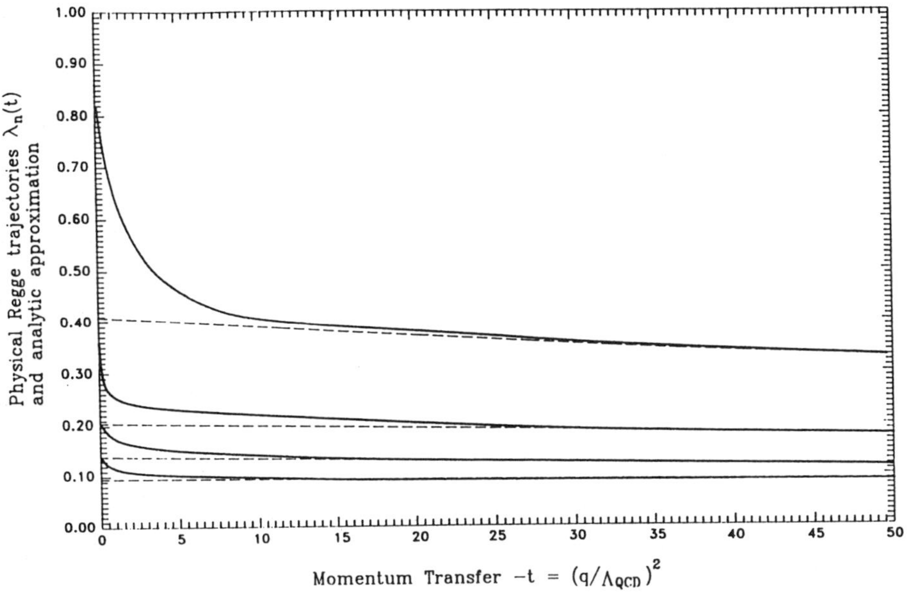

Figure 8: The solid lines show the Regge trajectories with the soft propagator inserted into the BFKL equation, and the dotted lines are the results of the purely perturbative treatment.

but nowhere nearly enough to match the phenomenologically fit value of 0.08. A similar analysis has been conducted for other estimates of the soft gluon propagtor[14], which have come either from lattice calculations or from attempts to solve the Dyson-Schwinger equation in other gauges. It turns out that the results are not particularly sensitive to the exact form of the non-perturbative propagator - all these propagators have the common feature that some mass scale has been introduced and the trajectories are of course sensitive to this mass scale, but somewhat less sensitive to the exact way they are introduced.

Finally I would like to mention a somewhat better attempt due to Nikolaev, Zakharov, and Zoller[15] to incorporate non-perturbative gluons into the BFKL equation. They use a Fock space expansion. We can consider for simplicity quark meson scattering. The first term in the Fock space expansion is just a quark-antiquark pair, which is treated as a colour dipole and the cross section for scattering off this dipole is $\sigma_0(r)$, where r is the separation of the quark and antiquark in impact parameter space. Next one considers the quark, antiquark, gluon state and obtains a cross section

$\sigma_3(r,\rho_1,\rho^2)$, where $\rho_1,\ \rho_2$ are impact parameter separations of the fermions from the gluon. This must be multiplied by a probability (square of a wavefunction) to find these partons inside the meson. When this is done it is found the at the contribution to the cross section from this Fock space state can be written schematically as

$$\frac{\partial \sigma_1(r)}{\partial \ln s} = \mathcal{K} \otimes \sigma_0(r)$$

It turns out that this generalises to relate the (n+1)-gluon state contribution to the n-gluon state contribution

$$\frac{\partial \sigma_{(n+1)}(r)}{\partial \ln s} = \mathcal{K} \otimes \sigma_n(r)$$

Now in the case of massless gluons the Kernel $\mathcal{K}$ is precisely the BFKL Kernel, but it can also be calculated for other propagators, in particular a propagator in which a gluon mass has been inserted - this may not be very realistic, but it is the simplest way of encoding some infrared effect, which softens the infrared behaviour of the gluon and probably give good estimates for other soft propagators, which have the same mass scale. From Figure 9. we can see that taking the largest credible value for the gluon mass of 1 GeV, the intercept has fallen to 0.4, which is in the right direction, but still nowhere near small enough.

4. Conclusions

- The infrared behaviour of QCD is certainly needed to describe pomeron behaviour. Non-perturbative contributions to the gluon propagators influence the solutions to the BFKL equation.
- The Dyson-Schwinger equation in the axial gauge (with a suitable ansatz for the triple-gluon vertex) admits solutions which give sensible predictions for the proton-proton elastic differential cross-section.
- Attempts to reproduce the pomeron by inserting non-perturbative propagators into the BFKL equation give an intercept which is still too high and non-linear trajectories.
- A Fock space expansion leads to a more systematic way of including non-perturbative propagators. Using a massive gluon model the intercept is further reduced, but still a long way off 0.08.

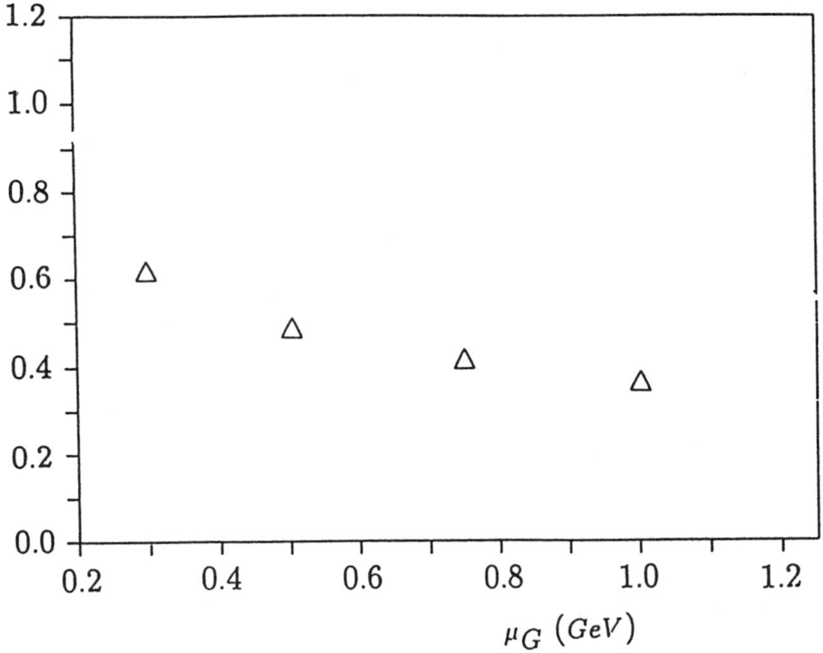

Figure 9: The intercept of the pomeron calculated in ref. 15 as a function of gluon mass (μ_g)

References.

1. A. Donnachie and P.V. Landshoff, *Phys. Letts.* **B296** (1992) 227

2. E.M. Levin and M.G. Ryskin, *Sov. Phys. Usp.* **32** (1989) 479

3. M. Froissart, *Phys. Rev.* **123** (1961) 1053
   A. Martin, *Phys. Rev* **129** (1963) 1432

4. H. Cheng and C.Y. Lo, *Phys. Rev.* **D13** (1976) 1131 and **D15** (1977) 2959
   E.A. Kurayev, L.N. Lipatov and V.S. Fadin *Sov. Phys. JETP* **45** (1977) 199
   Y.Y. Bailitski and L.N. Lipatov, *Sov. J. Nucl. phys.* **28** (1978) 822

5. R. Kirschner and L.N. Lipatov, *Z. Phys.* **C465** (1990) 477

6. E.M. Levin and C.I. Tan it FERMILAB CONF-**92-391-T** (1992)

7. L.N. Lipatov, *Sov. Phys. JETP* **63** (1986) 904

8. R.E. Hancock and D.A. Ross, *Nucl. Phys.* **B383** (1992) 575
   J.G. Daniell and D.A. Ross, *Phys. Letts* **B22** (1989) 166

9. M. Baker, J.S. Ball and F. Zachriasen, *Nucl. Phys.***B186** (1981) 531, 560

10. S.Mandelstam *Phys. Rev.* **D20** (1979) 3223

11. J.R. Cudell and D.A. Ross, *Nucl. Phys.* **B359** (1991) 247

12. J.F. Gunion and D. Soper, *Phys. Rev* **D15** (1977) 2617
    E.M. Levin and M.G. Ryshkin *Sov. J. Nucl. Phys.* **34** (1981) 619

13. R.E. Hancock and D.A. Ross, *Nucl. Phys.* **B394** (1993) 200

14. U. Häbel, R. Könning, H.G. Reusch, M. Stingl and S. Wigard, *Z. Phys.* **A336** (1990) 435
    J.E. Mandula, and M. Oglivie, *Phys. Letts.* **185** (1987) 127
    D. Zwanziger, *Nucl. Phys.* **B323** (1989) 513
    P. Marizoni, G. Martinelli, N. Stella, and M. Testa, *Phys. Letts* **B318** (1993) 511

15. N.N. Nikolaev and B.G. Zacharov, *Z. Phys.* **C49** (1991) 607 and **C53** (1992) 331
    N.N. Nikolaev, B.G. Zacharov, and V.R. Zoller, *JETP Lett.* **59** (1994) 6
    N.N. Nikolaev, B.G. Zacharov. and V.R. Zoller *preprint no.* **KFA-IKP-TH-1994-1** and **KFA-IKP-TH-1994-2** (1994)

DUAL PARTON MODEL FOR SOFT HADRONIC COLLISIONS

UDAY P. SUKHATME
Department of Physics, University of Illinois at Chicago,
845 W. Taylor Street,
Chicago, Illinois 60607-7059, U.S.A.

ABSTRACT

The Dual Parton Model (DPM) provides a complete, widely used, phenomenological description of all the dominant features of multiparticle production in soft hadronic collisions at high energies. The model incorporates most of the generally accepted theoretical ideas like duality, unitarity, Regge behavior and the parton structure of hadrons. DPM is motivated by a suitable large N expansion of QCD which gives a topological classification of scattering diagrams. The basic theoretical ideas underlying DPM and some of its many phenomenological successes are described.

1. Introduction

The Dual Parton Model (DPM) was first introduced in 1979 at Orsay[1]. Many physicists have since worked on the model, expanded its scope and tested its consequences for a wide variety of experimental data [2,3]. Like good French wine, DPM has aged well, and today, 15 years after its introduction, it continues to provide a good overall description of multiparticle production in soft high energy hadronic and nuclear collisions. Here, my aim is to give a simple, phenomenological review of the main features of DPM - a more detailed and comprehensive treatment can be found in ref. 3.

In a high energy hadronic collision, most events have a large number of produced particles with small transverse momenta with respect to the collision axis. In principle, it should be possible to compute these "soft" multiparticle events directly from the Lagrangian of quantum chromodynamics. However, for soft processes, the running coupling constant is clearly much too large for ordinary perturbation theory to be sensible, and consequently alternative non-perturbative procedures must be used. Such approaches have yielded many qualitative features of hadronic amplitudes, but one is far from making quantitative calculations. So, at the present time, the best one can do in describing soft hadronic physics is to construct models which incorporate all available theoretical ideas motivated both by non-perturbative studies of QCD as well as the general properties of the S-matrix. It is ironic that although so much is known about soft physics experimentally, it remains hardest to calculate theoretically. Nevertheless, the wide variety of good quality data put stringent constraints on phenomenological models.

A widely studied non-perturbative approach to QCD consists of taking various large-N limits, where N can be either N_c, the number of colors or N_f, the number of flavors[4]. This procedure immediately gives rise to a topological expansion, which implies that interactions represented by topologically complicated diagrams are suppressed by powers of $1/N$. The dual topological unitarization scheme[5] incorporates the constraints of unitarity and duality in a topological expansion. This scheme provides the basic ideas and motivation underlying DPM.

DPM gives a quantitative description of all the dominant features of soft multiparticle production, like inclusive single-particle rapidity distributions, rising rapidity plateaus, charged-particle multiplicity moments, violations of KNO scaling, charge distributions in rapidity, correlations between average transverse momentum and multiplicity, forward-backward hemisphere rapidity correlations, antiproton-proton annihilation, diffraction dissociation processes, two-particle inclusive reactions, production of heavy flavors, minijet production data, etc.[3]. Furthermore, the model can be used for both hadronic as well as nuclear beams and targets. After describing DPM, I will focus on a few of its numerous phenomenological successes.

2. Dual Parton Model

The Dual Parton Model provides a concrete recipe for associating an experimentally observable cross section with each diagram in a topological expansion. At high energies, these diagrams involve t-channel exchanges of Pomerons. In fact, each diagram has a one-to-one correspondence with diagrams used in Reggeon Field

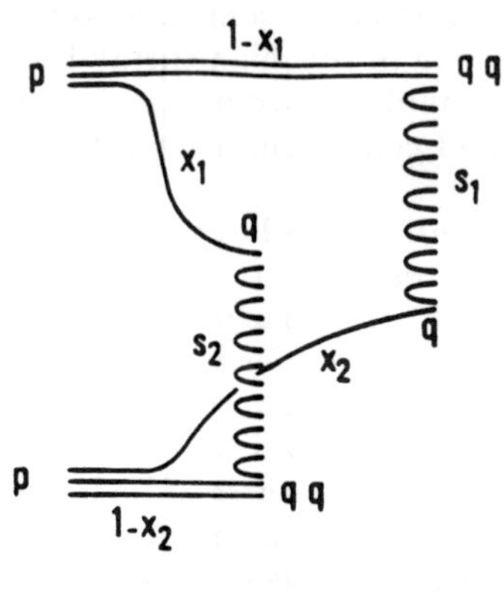

Fig. 1: Two chain (single Pomeron exchange) diagram for pp collisions.

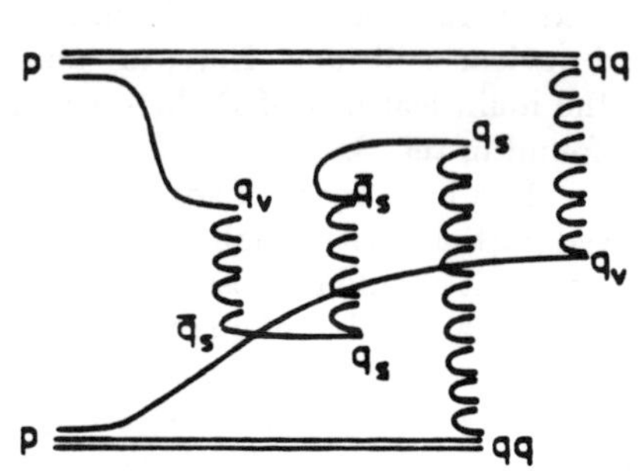

Fig. 2: Four chain (two Pomeron exchange) diagram for pp collisions.

Theory[6]. Multiparticle production results from making unitarity cuts. The simplest contribution to the elastic scattering amplitude is a single dual Pomeron, which has the topology of a cylinder. A unitarity cut gives two chains of hadrons (Fig. 1). These chains are stretched between valence quarks of the initial hadrons so as to form color singlets. Similarly, cutting a two Pomeron exchange diagram gives four chains, which end on valence and ocean quarks and antiquarks from the initial hadrons (Fig. 2).

To proceed further, we make use of the partonic structure of hadrons, and treat each contributing diagram as a two step process: (i) separation of color in the initial collision, and (ii) fragmentation of colored objects which results in the production of hadronic chains. Consequently, the two ingredients one needs are structure functions of partons in hadrons and fragmentation functions of partons into hadrons[1]. For definiteness, let us first consider the leading contribution to multiparticle production in high energy proton-proton (pp) collisions. The collision separates the valence quarks of each incident proton into two colored systems - a quark and a diquark. The fragmentation occurs in the form of two quark-diquark chains. Note that each chain is an overall color singlet which contains a quark fragmentation region and a diquark fragmentation region.

In order to make the Dual Parton Model quantitative, it is necessary to specify the probability that the interaction separates the protons into two quarks with momentum fractions x_1 and x_2 and two diquarks with the remaining momentum fractions $(1-x_1)$ and $(1-x_2)$. This probability, which we denote by $\rho(x_1, x_2)$, is taken to be proportional to the valence quark structure functions in a proton: $\rho(x_1, x_2) = \frac{1}{9}v(x_1)v(x_2)$. Structure functions are peaked near $x = 0, v(x) \propto x^{-1/2}$. This peaked behavior, which follows from Regge considerations, results in two "held-back" quarks near $x_1, x_2 = 0$.

Typically in pp collisions, the "effective" values[1] of the momentum fractions x_1 and x_2 are around 0.05 .The total energy $\sqrt{s}$ in the overall pp center of mass (CM) frame is shared between the two chains shown in Fig. 1. If $\sqrt{s_1}$ and $\sqrt{s_2}$ denote the energies of chain 1 and chain 2 in their respective CM frames, we have $s_1 = sx_2(1-x_1), s_2 = sx_1(1-x_2)$. The rapidity shifts Δ_1 and Δ_2 necessary to go from the overall pp CM frame to the CM of chain 1 and chain 2 are given by $\Delta_1 = \frac{1}{2}\ln[(1-x_1)/x_2]$ and $\Delta_2 = -\frac{1}{2}\ln[(1-x_2)/x_1]$ respectively, and they are essentially independent of energy.

The second ingredient which one needs are quark and diquark fragmentation functions. These can be taken either (i) from hard processes (universality of fragmentation hypothesis) or (ii) from iterative cascade models[7] e.g. Lund model[8].

The above two ingredients can be readily combined to obtain physically observable quantities. For example, the one Pomeron exchange contribution to the single particle inclusive cross section $pp \to h + X$ is given by the superposition of two chains (see Fig. 3).

$$\frac{dN}{dy}(s,y) = \int_0^1 \int_0^1 dx_1 dx_2 \frac{1}{9} v(x_1)v(x_2) \left\{ \frac{dN}{dy}\bigg|_1 (s_1, y-\Delta_1) + \frac{dN}{dy}\bigg|_2 (s_2, y-\Delta_2) \right\}. \quad (1)$$

The individual chain contributions $\left.\frac{dN}{dy}\right|_{1,2}$ come from quark and diquark fragmentation functions into hadron h. Corresponding formulae for other hadronic processes and quantities of physical interest can be easily obtained. Corrections to the leading single Pomeron (two chain) diagram come from multiple Pomeron exchanges (e.g. Fig. 2). The weights of these diagrams are fixed by unitarity requirements[6]. Note that the additional chains end on ocean quarks and antiquarks, and hence mainly contribute to multiparticle production in the central region of rapidity. It is clear that each chain must at least have a minimum threshold CM energy in order for physical hadrons to materialize from it. Therefore, on the average, up until the highest CERN-ISR energies ($\sqrt{s} \leq 63$ GeV), the two chain contribution dominates. However, at higher energies (like those at the CERN collider or the Fermilab Tevatron) there is sufficient energy available to share between four (or more) chains, and multiple Pomeron exchanges give an important contribution. Indeed, they are responsible for the continuous rise of the central plateau height with energy and violations of KNO scaling.

3. Phenomenology

In this section, we discuss three examples of how DPM can describe experimental measurements pertaining to soft hadronic collisions. A complete description of other applications can be found in Ref. 3. The applications chosen here are: (i) Inclusive and charge distributions; (ii) Particle ratios in pp and $\bar{p}p$ collisions; (iii) Multiplicity distributions and KNO scaling violations.

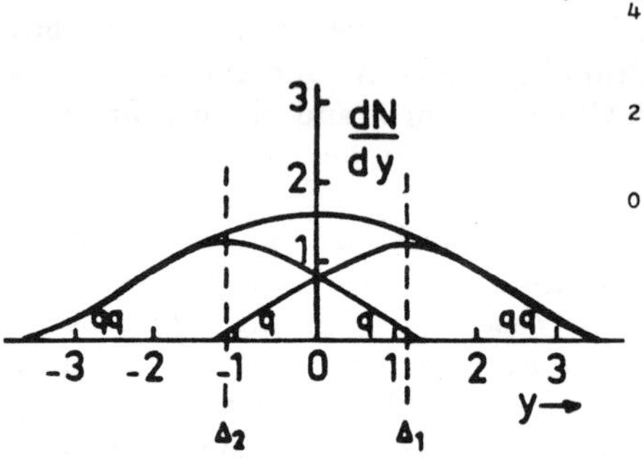

Fig. 3: Diagram illustrating how the contributions from the two quark-diquark chains of Fig. 1 add up to give the complete inclusive single particle distribution in pp collisions.

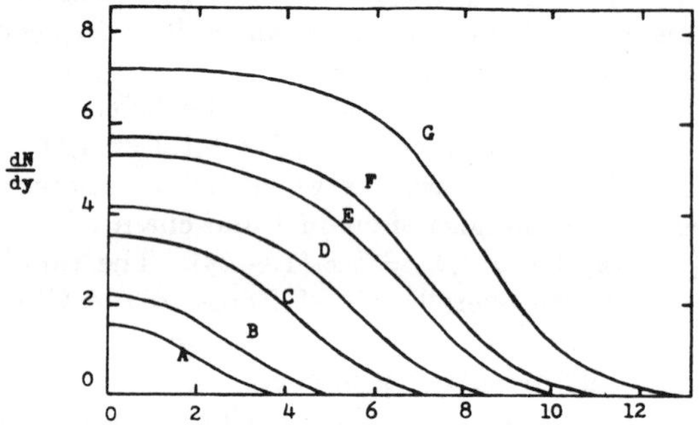

Fig. 4: DPM calculations for the single particle inclusive distributions in pp and $\bar{p}p$ collisions. Note the rise of the central plateau height with energy. Experimental data (not shown) are available for energies up to 2 TeV and are in excellent agreement with the above curves. The seven curves marked A to G correspond to energies $\sqrt{s}$ = 20 GeV, 53 GeV, 540 GeV, 2 TeV, 10 TeV, 20 TeV and 100 TeV.

3.1. Inclusive and Charge Distributions

The first successful application of DPM was to describe the shape and energy dependence of single particle inclusive distributions[3,4]. As the energy increases, the increasing overlap of the two chains in single Pomeron exchange diagram as well as multi-chain contributions produce a rising central plateau (see Fig. 3). This prediction of DPM was confirmed by CERN $\bar{p}p$ collider data (Fig. 4). Two particle distributions are also well-described[9].

The distribution of charge versus rapidity in a hadronic collision provides a sensitive test of any model since the charge distribution involves differences of fragmentation functions into positive and negative particles.

$$\frac{dQ}{dy} = \sum_h q_h \frac{dN_h}{dy}, \tag{2}$$

where q_h is the charge carried by hadron h. Furthermore, the shape and energy dependence of $dQ/dy(s,y)$ contains information about parton dynamics and the flow of quantum numbers, that is how the charge carried by the valence quarks in the incident colliding hadrons emerges in the multiparticle final state.

The charge distributions in a variety of hadronic collisions have been computed using DPM[10]. The agreement with available high energy data is very good. For example, a comparison with the charge distribution in π^+p collisions is shown in Fig. 5. Note that the data cannot be described by a fragmentation model with a single long chain without a "held back" valence quark. DPM also accounts for the forward net charge in π^-p and K^-p collisions as a function of energy (see Fig. 6).

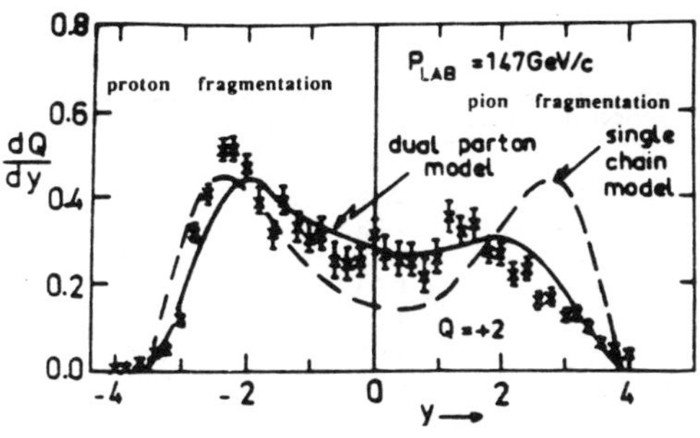

Fig. 5: The distribution of charge versus rapidity for π^+p collisions. The solid line comes from a two chain DPM calculation[10], whereas the dashed line follows from fragmentation via a single long chain.

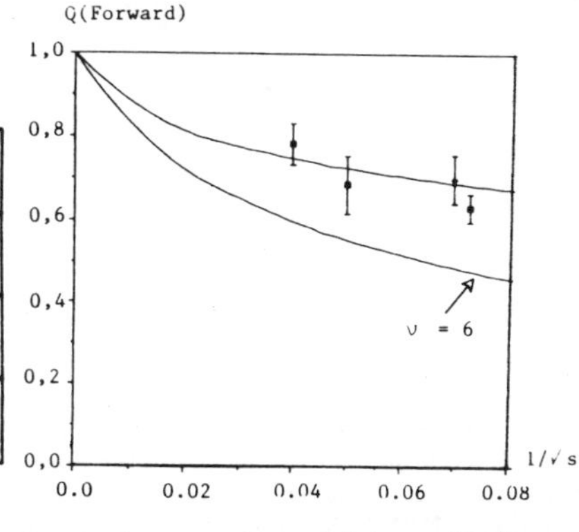

Fig. 6: A comparison of data on the forward net charge in π^-p and K^-p collisions as a function of energy with a calculation based on DPM[10]. The lower curve is the DPM prediction for π^- nucleus collisions with $\nu = 6$ interactions[11].

It is useful to analyze π^+p and π^-p collisions a little more. On the proton side, as a consequence of the peaking of momentum distributions near $x = 0$, there are two dominant configurations: $(uu)_F d_S$ with probability $\frac{1}{3}$ and $(ud)_F u_S$ with probability $\frac{2}{3}$, where F denotes fast and S denotes slow. The average charge of the slow valence quark is $Q_S^p = (Q_d + 2Q_u)/3 = \frac{1}{3}$ and that of the fast diquark is $Q_F^p = (2Q_d + 4Q_u)/3 = \frac{2}{3}$. Applying the same argument to the pion side, one finds that the average charge of the slow and fast constituents are equal, that is $Q_F^{\pi^+} = Q_S^{\pi^+} = \frac{1}{2}$ and $Q_F^{\pi^-} = Q_S^{\pi^-} = -\frac{1}{2}$. The asymmetry between the forward and backward directions seen in Fig. 5 follows from the fact that $Q_F^p > Q_F^{\pi^+}$. Similarly, in π^-p collisions, the experimental observation that the integrated charge in the π^- hemisphere is smaller than unity (in absolute value), is a consequence of the average charges of the slow quark systems which can cross from their own hemispheres into the opposite ones. This crossover probability decreases as s increases (see Fig. 6). The calculation can be readily generalised to hadron-nucleus collisions. The variation with energy of the charge in the forward hemisphere $|Q_F|$ is substantially amplified by multiple inelastic collisions in a nuclear target. The prediction[11] for $\nu = 6$ collisions is also shown in Fig. 6. This amplification can be increased even further by having several nuclear interactions, as is the case for extensive air showers (EAS) initiated by a light primary in the upper atmosphere. Using DPM and well understood Monte Carlo simulations for EAS, one can compute the muon and electron content in EAS as a function of altitude[12]. The DPM expectations are in very good agreement with data (see Fig. 7).

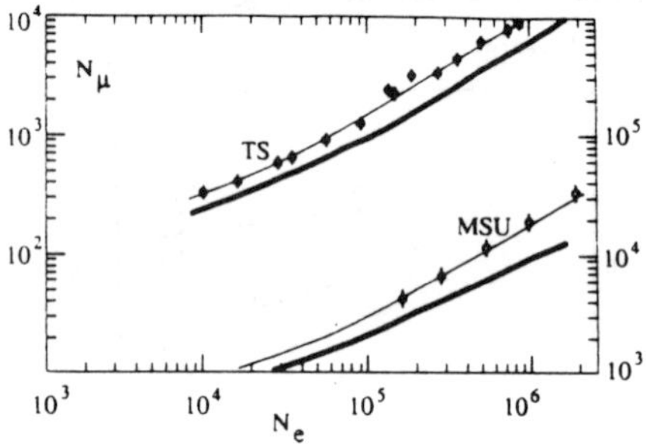

Fig. 7: Extensive Air Shower calculations[12] yielding the number of muons N_μ vs. the number of electrons N_e using two phenomenological models: DPM (thin lines) and the Lund Fritiof model (thick lines). The left axis is for the upper pair of curves and corresponds to high altitude data from Tian Shan Laboratory. The right axis is for the lower pair of curves and corresponds to Moscow State University sea level measurements.

Table 1: Comparison of the data[13] with DPM calculation[14] for the hadronic ratio R^h in $\bar{p}p$ and pp collisions.

Detected particle h	R^h (experiment)	R^h (DPM)
all (±)	1.00 ± 0.01	1.02
only positives	0.96 ± 0.01	0.98
only negatives	1.03 ± 0.01	1.06
$K^+ + K^-$	0.98 ± 0.04	1.06
K^+	0.90 ± 0.05	0.97
K^-	1.08 ± 0.06	1.17
$p + \bar{p}$	$1.05 \pm .04$	1.06
p	0.93 ± 0.04	0.92
$\bar{p}$	1.25 ± 0.07	1.26

3.2. Particle Ratios in pp and $\bar{p}p$ Collisions

The ratio of the inclusive distributions in pp and $\bar{p}p$ collisions for a variety of detected charged particles (pions, kaons, protons, antiprotons) has been measured by the Axial Field Spectrometer Collaboration[13] at the CERN-ISR at $\sqrt{s} = 53$ GeV in the rapidity range $|y| < 0.8$. The experimentally obtained values of the ratios

$$R^h = \frac{dN/dy\ (\bar{p}p \to h)}{dN/dy\ (pp \to h)} \tag{3}$$

for various detected particles h are shown in Table 1. At a qualitative level, these results are a straightforward consequence of the valence quark-diquark separation with the valence quark being held back near $x = 0$. Let us consider, for concreteness, the case in which the detected hadron h is either a π^+ or a π^-. In a pp collision, we have 4u and 2d valence quarks. The fragmentation of a u-quark (d-quark) into $\pi^+(\pi^-)$ is favored whereas d-quark (u-quark) fragmentation into $\pi^+(\pi^-)$ is unfavored. In a $\bar{p}p$ collision, we have three valence constituents (u, u, $\bar{\text{d}}$) with favored fragmentation into π^+ and three constituents ($\bar{\text{u}}$, $\bar{\text{u}}$, d) with unfavored fragmentation. Since the ratio of unfavored to favored fragmentation is smaller than unity, one can immediately see that $R^{\pi^+} < 1, R^{\pi^-} > 1$. Indeed, to do a naive, back of the envelope calculation, we denote the average ratio of unfavored (U) to favored (F) fragmentation by just a single parameter $\omega \equiv U/F$. Then,

$$R^{\pi^+} = \frac{3F + 3U}{4F + 2U} = \frac{3(1+\omega)}{2(2+\omega)} \quad . \tag{4}$$

Similarly, all other ratios R^h can be expressed in terms of ω. Taking ω to be about 0.8 at $\sqrt{s} = 53$ GeV gives good agreement with data. A more complete DPM calculation[14] gives essentially the same results (see Table 1). With increasing energy, the held back valence quarks will move away from the central region, and one expects all ratios to go to one. This phenomenological application provides strong support for held back valence quarks, which is one of the key ideas underlying DPM.

3.3. Multiplicity Distributions and KNO Scaling Violations

In this section, we shall examine multiplicity distributions and moments for proton-proton collisions using DPM. As discussed before, one needs inputs from various types of single chains (q-$\bar{\text{q}}$, q-qq,qq-$\bar{\text{q}}\bar{\text{q}}$). For this, we will use lognormal distributions which have recently been shown[15] to describe e^+e^- multiplicity data very accurately. This is accomplished with just two parameters over a wide energy interval from relatively low values all the way to LEP energies. The average charged particle multiplicity has a power law behavior $\bar{n}_{ch} \propto \sqrt{s}^{\ \alpha}$, characteristic of a branching process.

In DPM, the multiplicity moments for pp collisions are readily obtained. For now, let us focus on the dominant leading single Pomeron contribution. The probability of producing n hadrons from two q-qq chains with energies $\sqrt{s_1}$ and $\sqrt{s_2}$ is given by

$$P_n^{pp}(\sqrt{s}) = \int\int dx_1 dx_2 \rho(x_1, x_2) \sum_{l=0}^{n} P_l^{q-qq}(\sqrt{s_1}) P_{n-l}^{q-qq}(\sqrt{s_2}) \quad ; \quad \sum_n P_n^{pp}(\sqrt{s}) = 1 \ . \quad (5)$$

From this, the average values of n, n^2, etc. are given by $\bar{n^q} = \sum n^q P_n^{qq}$. In particular. the average multiplicity is

$$\bar{n}^{pp}(\sqrt{s}) = 2 \int\int dx_1 dx_2 \rho(x_1, x_2) \bar{n}^{q-qq}(\sqrt{s_1}) \ . \quad (6)$$

Clearly, one needs q-qq chain inputs. We obtain them by making the simplest model relating q-qq and q-$\bar{q}$ chains i.e. one assumes that the diquark goes into a baryon (with a flat momentum distribution) and a leftover antiquark. This gives[16]

$$P_n^{q-qq}(\sqrt{s}) = \int_0^1 dx P_{n-\frac{2}{3}}^{q-\bar{q}}(\sqrt{sx}) \quad , \quad \bar{n}^{q-qq}(\sqrt{s}) = \frac{2}{3} + \int_0^1 dx \ \bar{n}^{q-\bar{q}}(\sqrt{sx}) \ . \quad (7)$$

Thus, in a DPM approach, all moments and multiplicity distributions for pp collisions can be readily expressed in terms of the corresponding quantities for e^+e^- collisions.

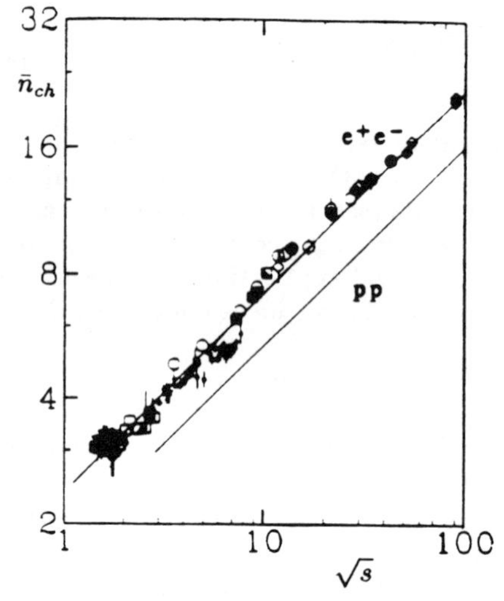

Fig. 8: The average charged particle multiplicity $\bar{n}_{ch}$ versus the CM energy $\sqrt{s}$ for e^+e^- and pp collisions. Only e^+e^- data are shown.

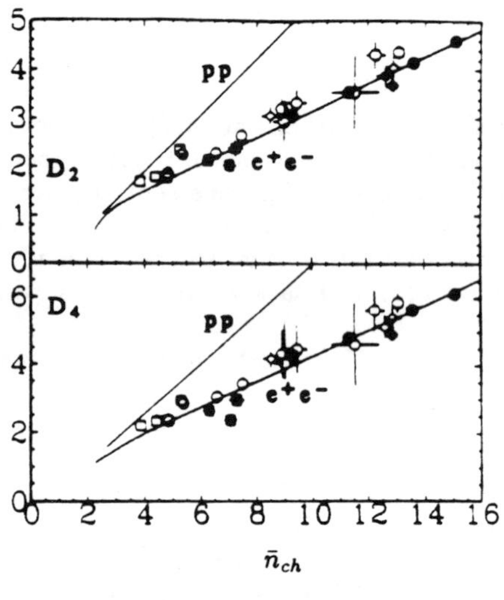

Fig. 9: The dispersions D_2 and D_4 versus the average charged particle multiplicity $\bar{n}_{ch}$ for e^+e^- and pp collisions. Only e^+e^- data are shown.

For the average charged particle multiplicity, one gets a universal $\sqrt{s}^{\,\alpha}$ behavior for all processes (see Fig. 8). The available pp multiplicity data fall rather precisely on the curve shown in Fig. 8. The universal power law behavior follows from the branching character of the fragmentation process into hadrons. Note that one understands why pp collisions are far less efficient than e^+e^- in producing charged hadrons. This happens because a considerable amount of the incident energy goes into overall motion of the centers of mass of the two chains, and is thus unavailable for producing hadrons. The dispersions D_2 and D_4 for pp collisions are also in good agreement with the DPM expectations shown in Fig. 9 .

For describing the rising tails of the multiplicity distributions with energy, one needs to take into account multi-Pomeron diagrams in DPM[17]. These diagrams become increasingly more important as energy increases, and have been shown to account for not only the rise of the central rapidity plateaus, but also the observed violations of KNO scaling[17]. It is plausible that the small experimentally observed bump in the KNO plot around $z = 2$ may be due to the two Pomeron exchange diagram.

4. Conclusions

The Dual Parton Model is a phenomenological model based on sound theoretical principles (duality, Regge behavior, unitarity, topological expansion). It is an attempt to provide a concrete realisation of the main features of large-N QCD. It is easy to apply to experimental situations. The underlying parton dynamics involves color separation and re-arrangement into color neutral hadronic chains. The only input ingredients are parton structure and fragmentation functions. Here, we have reviewed some of the many phenomenological successes of DPM. At the present time, the Dual Parton Model certainly provides the most successful and complete framework for quantitatively describing soft hadronic collisions.

5. Acknowledgments

It is a real pleasure to take this opportunity to thank my three co-originators of the Dual Parton Model: Alfons Capella, Chung-I Tan and J. Tran Thanh Van. This work was supported in part by the U. S. Department of Energy under grant DE-FGO2-84ER40173.

6. References

1. A. Capella, U. Sukhatme, C-I Tan and J. Tran Thanh Van, *Phys. Lett.* **81B** (1979) 68; A. Capella, U. Sukhatme and J. Tran Thanh Van, *Zeit. Phys.* **C3** (1980) 329.
2. H. Minakata, *Phys. Rev.* **D20** (1979) 1656; G. Cohen-Tannoudji et al., *Phys. Rev.* **D21** (1980) 2689; P. Aurenche and F. Bopp, *Phys. Lett.* **114B** (1982)

363; A. Kaidalov, *Phys. Lett.* **116B** (1982) 459; A. Kaidalov and K. A. Ter-Martirosyan, *Phys. Lett.* **117B** (1982) 247.

3. A. Capella, U. Sukhatme, C-I Tan and J. Tran Thanh Van, *Phys. Rep.* **236** (1994) 225.
4. G. 't Hooft, *Nucl. Phys.* **B72** (1974) 461; E. Witten, *Nucl. Phys.* **B160** (1979) 57; G. Veneziano, *Nucl. Phys.* **B74** (1974) 365 and **B117** (1976) 519; S. R. Das, *Rev. Mod. Phys.* **59** (1987) 235.
5. Chan Hong-Mo et al., *Nucl. Phys.* **B86** (1975) 470 and **B92** (1975) 13; G. F. Chew and C. Rosenzweig, *Nucl.Phys.* **B104** (1976) 290 and *Phys. Rep.* **41** (1978) 263.
6. V. N. Gribov, *Sov. Phys. JETP* **26** (1968) 414; V. A. Abramowski, V. N. Gribov and O. V. Kancheli, *Yad. Fiz.* **18** (1973) 595 and *Sov. J. Nucl. Phys.* **18** (1974) 308; M. Baker and K. A. Ter-Martirosyan, *Phys. Rep.* **28** (1976) 1.
7. A. Krzywicki and B. Petersson, *Phys. Rev.* **D6** (1972) 924; J. Finkelstein and R. D. Peccei, *Phys. Rev.* **D6** (1972) 2606; R. Feynman and R. Field, *Nucl. Phys.* **B136** (1978) 1; U. Sukhatme, *Phys. Lett.* **76B** (1978) 478.
8. B. Andersson, G. Gustafson, G. Ingelman and T. Sjöstrand, *Phys. Rep.* **97** (1983) 31.
9. A. Capella, U. Sukhatme and J. Tran Thanh Van, *Phys. Lett.* **119B** (1982) 220; C. Pajares and A. Varias, *Zeit. Phys.* **C19** (1983) 69.
10. A. Pagnamenta and U. Sukhatme, *Zeit. Phys.* **C14** (1982) 79; K. Fialkowski and A. Kotanski, *Zeit. Phys.* **C20** (1983) 1; C. Pajares and A. Varias, *Zeit. Phys.* **C19** (1983) 143.
11. A. Capella, C. Merino and J. Tran Thanh Van, *Zeit. Phys.* **C43** (1989) 663.
12. J. N. Capdevielle, *Jour. Phys.* **G15** (1989) 909; *ibid.* **G16** (1990) 769.
13. T. Akesson et al., *Phys. Lett.* **108B** (1982) 58.
14. A. Capella, U. Sukhatme and J. Tran Thanh Van, *Phys. Lett.* **125B** (1983) 330.
15. R. Szwed, G. Wrochna and A. Wroblewski, *Mod. Phys. Lett.* **A5** (1990) 1851 and **A6** (1991) 245.
16. U. Sukhatme, in *Proc. XXII Int. Symposium on Multiparticle Dynamics*, Santiago de Compostela, Spain (1992), ed. C. Pajares.
17. A. Capella and J. Tran Thanh Van, *Zeit. Phys.* **C23** (1984) 165; A. Capella, A. Staar and J. Tran Thanh Van, *Phys. Rev.* **D32** (1985) 2933; P. Aurenche, F. Bopp and J. Ranft, *Zeit. Phys.* **C23** (1984) 67 and **C26** (1984) 279.

HIGHER-ORDER LIPATOV KERNELS AND THE QCD POMERON

Alan R. White
High Energy Physics Division, Argonne National Laboratory, Argonne, IL 60439.

Abstract

Three closely related topics are covered. The derivation of $O(g^4)$ Lipatov kernels in pure glue QCD. The significance of quarks for the physical Pomeron in QCD. The possible inter-relation of Pomeron dynamics with Electroweak symmetry breaking.

1. Introduction

I will spend most of my time in this talk describing my recent derivation[1] of higher-order Lipatov kernels, for pure-glue QCD, directly from t- channel unitarity. However, I also want to put this analysis in the context of my general study[2] of the physical, or *soft*, Pomeron in full QCD with quarks. I shall outline why quarks play such an important part in the emergence of confinement in the low transverse momentum region. I will also describe how study of the full unitarity properties of the Pomeron leads naturally to the introduction of a new higher-color quark sector which can provide a very attractive picture of dynamical electroweak symmetry breaking.

2. Higher-order Lipatov Kernels

Currently there is much excitement about the small-x behavior of structure functions. In particular it seems that at HERA the "Lipatov Pomeron"[3] may have been seen i.e.

$$F_2(x,q^2) \sim x^{1-\alpha_0} \sim x^{-\frac{1}{2}} \tag{1}$$

The Lipatov equation was originally derived from extensive leading and next-to-leading log calculations in the Regge limit of (massive) Yang-Mills theories[4, 5] and is applied as an evolution equation for parton distributions at small-x i.e.

$$\frac{\partial}{\partial(ln1/x)}F(x,q^2) = \tilde{F}(x,q^2) + \frac{1}{(2\pi)^3}\int\frac{d^2k}{k^4}\,K(k,q)F(x,k^2) \tag{2}$$

$K(k,q)$ is given in terms of the $O(g^2)$ Lipatov kernel by

$$\frac{2}{3g^2}K(k,q) = \frac{2}{3g^2}K^{(2)}_{2,2}(k,-k,q,-q) \tag{3}$$

where

$$g^{-2}K^{(2)}_{2,2}(k_1,k_2,k_3,k_4) = \sum_{1<->2}\Bigg((2\pi)^3\, k_1^2J_1(k_1^2)k_2^2\,\left(k_3^2\delta^2(k_2-k_4) + k_4^2\delta^2(k_2-k_3)\right) - \frac{k_1^2k_4^2 + k_2^2k_3^2}{(k_1-k_3)^2} - (k_1+k_2)^2\Bigg) \tag{4}$$

To obtain non-leading corrections to the $O(g^2)$ kernel, it appears that (very complicated) very non-leading log **Regge limit calculations** are required.

Back in the days when Pomeron Reggeon Field Theory (RFT) was studied intensely, it was understood that to satisfy **multiparticle t-channel unitarity** in the Regge limit a theory must be describable in terms of **reggeon diagrams** that satisfy **reggeon unitarity**[6]. For an even signature Pomeron, reggeon unitarity is very simple - as we now briefly review.

Introducing RFT variables, $E = 1 - \ell$ and $k^2 = -t$, and writing $\alpha_{\mathbb{P}}(t) = 1 - \Delta(k^2)$, a "partial-wave amplitude" $a^+(\ell, t)$ will satisfy reggeon unitarity if we can write

$$a^+(\ell,t) \equiv F\left(E,k^2\right) = \sum_{n,m=1}^{\infty} F_{nm}(E,k^2), \tag{5}$$

with

$$\begin{aligned} F_{nm}\left(E,k^2\right) &= \frac{1}{(2\pi)^{3n+3m}} \int \left[\prod_{i,j} d^2k_i d^2k_j \delta^2\left(k - \sum_{i=1}^{n} k_i\right) \delta^2\left(k - \sum_{j=1}^{m} k'_j\right)\right] \\ & g_n\left(k_1,\dots k_n\right) g_m\left(k'_1,\dots k'_m\right) G_{nm}\left(E,k_1,\dots k_n,k'_1,\dots k'_m\right), \end{aligned} \tag{6}$$

where the G_{nm} are reggeon "Green's functions" satisfying a "unitarity equation" -

$$\begin{aligned} G_{nm}\left(E+i\epsilon,k\right) - G_{nm}\left(E-i\epsilon,k\right) &= \frac{1}{(2\pi)^{3M}} \sum_M (-1)^{M-1} \int \prod_s d^2k_s \\ \delta^2\left(k - \sum_{s=1}^{M} k_s\right) \delta\left(E - \sum_{s=1}^{M} \Delta(k_s^2)\right) & G_{nr}\left(E+i\epsilon,k\right) G_{rm}\left(E-i\epsilon,k\right). \end{aligned} \tag{7}$$

This equation is satisfied by writing Pomeron reggeon diagrams. For an n-Pomeron state we write the ("non-relativistic") propagator

$$\Gamma_n = \frac{1}{\left(E - \sum_{i=1}^{n} \Delta(k_i^2)\right)} \tag{8}$$

A minimal unitary set of diagrams is obtained by iterating this propagator with transverse momentum conserving $1 \to 2$ and $2 \to 1$ triple Pomeron interactions and then integrating over all transverse momenta. The arbitrariness arising from the possible existence of an infinite number of unknown higher-order Pomeron couplings is overcome by demonstrating that (when $\alpha_{\mathbb{P}}(0) = 1$) there is a fixed point where **only the triple coupling is relevant**. The result is the well-known Critical Pomeron scaling theory[7].

For an odd-signature Regge pole the corresponding reggeon diagrams are much more complicated - due to the presence of **vector particle** poles. It has been known for some time[8] that the Lipatov equation can be written as a **reggeon "Bethe-Salpeter equation"** with the $O(g^2)$ kernel as a **singular** reggeon interaction. (The singularity of the kernel is due to gluon poles). Nevertheless, the complete set of reggeon diagrams should determine (and be determined by) **all Regge limit logs**. Therefore, if we could construct such diagrams directly we could predict the results of arbitrary higher-order log calculations.

At first sight, such a construction looks impossible both because of **singular reggeon interactions due to gluon poles** and **undetermined parameters in higher-order interactions**. The first problem is resolved in general by the construction of reggeon loops via multiple discontinuities - this explicitly involves **no singular interactions**. The second problem is resolved for Yang-Mills theories by the **imposition of Ward identity constraints directly on reggeon amplitudes**. The result is a powerful technique for constructing higher-order reggeon diagrams, and therefore **higher-order Lipatov kernels**, without going through the very, very, complicated underlying Feynman diagram calculations.

We will illustrate the method by outlining a derivation of the Lipatov equation directly from reggeon diagrams, then present the $O(g^4)$ kernels. We first discuss general properties of the diagrams.

A general reggeon amplitude is gauge-invariant and (in a suitable multi-Regge limit) can be embedded in a multigluon S-Matrix element as illustrated in Fig. 1. If $k_\perp \to 0$ for one reggeon, then we obtain an amplitude A_ν for a gluon with zero **four-momentum** k_μ to couple to a physical multigluon state. This gluon amplitude satisfies the Ward identity[9]

$$k_\mu A_\mu = 0 \tag{9}$$

which, **provided there are no massless fermions in the theory**, implies A_ν vanishes at $k_\mu = 0$. Consequently, the reggeon amplitude must also vanish **when** $k_\perp = 0$.

To use multiple discontinuity theory we introduce an α' for gluons from the outset - we will show that $\alpha' \to 0$ gives perturbative results. The gauge group is manifest via group factors in reggeon vertices and gauge invariance is imposed by the Ward identity constraint we have just discussed. In addition to reggeon propagators, diagrams will now contain a particle pole or "signature factor" $[\alpha' k^2]^{-1}$ for each **uncut** reggeon line. The triple reggeon vertex contains a **"nonsense zero"** i.e.

$$\Gamma_{12} \sim g\,\sqrt{\alpha'}\,[E - \alpha' k_1^2 - \alpha' k_2^2] \underset{(E = \alpha' k^2)}{\sim} [k^2 - k_1^2 - k_2^2] \underset{(k_1^2 = k_2^2 = 0)}{\sim} k^2 \tag{10}$$

The simplest reggeon diagrams are the set of "self-energy" diagrams. These give gluon reggeization. In the cut diagrams - **all two reggeon propagators are cancelled by nonsense zeroes**, leaving one zero per loop to provide the reggeization. The result is a series which sums up to

$$[E-\alpha' k^2 - g^2\ k^2 J_1(k^2)]^{-1} \underset{\alpha'\to 0}{\longrightarrow} [E-g^2\ k^2 J_1(k^2)]^{-1} \tag{11}$$

where

$$J_1(k^2) \;=\; \frac{1}{(2\pi)^3}\int d^2q \frac{1}{q^2(k-q)^2} \tag{12}$$

which is the **the perturbative reggeization result.**

To construct multi-loop diagrams we have to proceed loop-by-loop, utilising multiple discontinuities. E.g. for the two-loop diagram shown in Fig. 2, we first construct the **triple discontinuity** of a three-reggeon vertex diagram. As illustrated in Fig. 3, **nonsense zeroes at the vertices cancel the reggeon propagators Γ_2 and Γ_3** (and produce an external zero factor). The cuts also remove an internal signature factor. **Only a transverse momentum integral remains** which is then used as a vertex in the one-loop bubble diagram to obtain the two-loop diagram.

The two-loop diagrams in the **color zero**, even-signature, channel are shown in Fig. 4. Now **only the three reggeon propagators are cancelled by nonsense zeroes.** With external couplings $\alpha' g^2$, the sum of such diagrams gives

$$\begin{aligned} &\frac{g^4}{(2\pi)^6}\int \frac{d^2k_1}{k_1^2}\frac{d^2k_2}{k_2^2}\delta^2(k-k_1-k_2)\int \frac{d^2k_3}{k_3^2}\frac{d^2k_4}{k_4^2}\delta^2(k-k_3-k_4) \\ &\qquad \frac{1}{(E-\alpha' k_1^2-\alpha' k_2^2)}\frac{1}{(E-\alpha' k_3^2-\alpha' k_4^2)}\ K_{2,2}^{(2)}(k_1,k_2,k_3,k_4) \end{aligned} \tag{13}$$

where $K_{2,2}^{(2)}(k_1,k_2,k_3,k_4)$ is a **sum of transverse momentum diagrams.** The relative weight of the diagrams is uniquely determined by demanding both **infra-red finiteness** of the (integral) kernel and the **Ward identity** vanishing when $k_i \to 0,\ i=1,..,4$. We thus obtain, **without calculating a single Feynman diagram,** the Lipatov kernel, to $O(g^2)$, given in (4) above. The limit $\alpha'\to 0$ of (13) gives directly the sixth-order perturbative result (if we identify g with the gauge coupling). Iteration of the construction procedure to obtain the full Lipatov equation is straightforward.

The $O(g^4)$ contribution to the (2-2) Lipatov kernel originates from three-loop reggeon diagrams of the form shown in Fig. 5 and the resulting kernel is a sum of the corresponding transverse momentum diagrams. **If the vanishing at $k_i\to 0,\ i= 1,..,4,$ is imposed together with infra-red finiteness, the relative weights of the distinct transverse momentum diagrams is uniquely determined** and the result is

$$g^{-4}K^{(4)}_{2,2}(k_1,k_2,k_3,k_4) \;=\; \sum_{1<->2}\Bigg(\frac{2}{3}(2\pi)^3k_1^2J_2(k_1^2)k_2^2\Big(k_3^2\delta^2(k_2-k_4)+\;k_4^2\delta^2(k_2-k_3)\Big)$$

$$-\;\left(\frac{k_1^2J_1(k_1^2)k_2^2k_3^2\;+\;k_1^2J_1(k_1^2)k_2^2k_4^2\;+\;k_1^2k_3^2J_1(k_3^2)k_4^2\;+\;k_1^2k_3^2k_4^2J_1(k_4^2)}{(k_1-k_3)^2}\right)$$

$$+\;J_1((k_1-k_3)^2)\Big(k_2^2k_3^2+k_1^2k_4^2\Big)\;+\;k_1^2k_2^2k_3^2k_4^2\;I(k_1,k_2,k_3,k_4)\Bigg) \tag{14}$$

with

$$I \;=\; \frac{1}{(2\pi)^3}\int d^2q\frac{1}{q^2(q+k_1)^2(q-k_3)^2(q+k_1-k_4)^2}\;,\quad J_2 \;=\; \frac{1}{(2\pi)^3}\int d^2q\frac{1}{(k-q)^2}J_1(q^2) \tag{15}$$

Similiarly a **new (2-4) kernel** is generated which is also **uniquely determined by the Ward identity and infra-red finiteness constraints**

$$K^{(4)}_{2,4}(k_1,..,k_6) = \sum_{1<->2}2\pi^3k_2^2\Bigg(\delta^2(k_2-k_6)K^{(4)}_{1,3}(k_1,k_3,k_4,k_5)+\delta^2(k_2-k_5)K^{(4)}_{1,3}(k_1,k_3,k_4,k_6)$$

$$+\;\delta^2(k_2-k_4)K^{(4)}_{1,3}(k_1,k_3,k_5,k_6)\;+\;\delta^2(k_2-k_3)K^{(4)}_{1,3}(k_1,k_4,k_5,k_6)\Bigg)\;-\;K^{(4)}_{2,4}(k_1,..,k_6)_c \tag{16}$$

where $K^{(4)}_{1,3}$ and $K^{(4)}_{2,4}(k_1,..,k_6)_c$ are given by

$$K^{(4)}_{1,3}(k,k_3,k_4,k_5) \;=\; \frac{1}{(2\pi)^3}\int\frac{d^2k_1}{k_1^2}\frac{d^2k_2}{k_2^2}\;k^2\delta^2(k-k_1-k_2)\;K^{(4)}_{2,3}(k_1,k_2,k_3,k_4,k_5) \tag{17}$$

with

$$g^{-4}K^{(4)}_{2,3}(k_1,..,k_5) \;=\; \sum_{1<->2}\Bigg((k_1+k_2)^2\;-\;\left(\frac{k_1^2(k_4+k_5)^2}{(k_1-k_3)^2}\;+\;\frac{k_1^2(k_3+k_5)^2}{(k_1-k_4)^2}\right.$$

$$\left.+\;\frac{k_1^2(k_3+k_4)^2}{(k_1-k_5)^2}\right)\;+\;\frac{1}{3}\left(\frac{k_1^2k_5^2}{(k_2-k_5)^2}\;+\;\frac{k_1^2k_4^2}{(k_2-k_4)^2}\;+\;\frac{k_1^2k_3^2}{(k_2-k_3)^2}\right)$$

$$+\;\frac{2}{3}\left(\frac{k_1^2k_2^2k_4^2}{(k_1-k_3)^2(k_2-k_5)^2}\;+\;\frac{k_1^2k_2^2k_5^2}{(k_1-k_4)^2(k_2-k_3)^2}\;+\;\frac{k_1^2k_2^2k_3^2}{(k_1-k_5)^2(k_2-k_4)^2}\right)\Bigg) \tag{18}$$

and

$$
\begin{aligned}
g^{-4}K^{(4)}_{2,4}(k_1,k_2,k_3,k_4,k_5,k_6)_c \; = \; & \sum_{1<->2} \Bigg((k_1+k_2)^2 - \Bigg(\frac{k_1^2(k_4+k_5+k_6)^2}{(k_1-k_3)^2} \\
& + \frac{k_1^2(k_3+k_5+k_6)^2}{(k_1-k_4)^2} + \frac{k_1^2(k_3+k_4+k_6)^2}{(k_1-k_5)^2} + \frac{k_1^2(k_3+k_4+k_5)^2}{(k_1-k_6)^2} \Bigg) - \frac{1}{4}\Bigg(\frac{k_1^2k_3^2}{(k_2-k_3)^2} \\
& + \frac{k_1^2k_4^2}{(k_2-k_4)^2} + \frac{k_1^2k_5^2}{(k_2-k_5)^2} + \frac{k_1^2k_6^2}{(k_2-k_6)^2} \Bigg) + \frac{1}{2}\Bigg(\frac{k_1^2(k_5+k_6)^2}{(k_2-k_5-k_6)^2} + \frac{k_1^2(k_5+k_4)^2}{(k_2-k_5-k_4)^2} \\
& + \frac{k_1^2(k_4+k_6)^2}{(k_2-k_4-k_6)^2} + \frac{k_1^2(k_3+k_6)^2}{(k_2-k_3-k_6)^2} + \frac{k_1^2(k_5+k_3)^2}{(k_2-k_5-k_3)^2} + \frac{k_1^2(k_3+k_4)^2}{(k_2-k_3-k_4)^2} \Bigg) \\
& + \frac{1}{2}\Bigg(\frac{k_1^2k_2^2(k_4+k_5)^2}{(k_1-k_3)^2(k_2-k_6)^2} + \frac{k_1^2k_2^2(k_3+k_5)^2}{(k_1-k_4)^2(k_2-k_6)^2} + \frac{k_1^2k_2^2(k_3+k_4)^2}{(k_1-k_5)^2(k_2-k_6)^2} \\
& + \frac{k_1^2k_2^2(k_3+k_6)^2}{(k_1-k_4)^2(k_2-k_5)^2} + \frac{k_1^2k_2^2(k_4+k_6)^2}{(k_1-k_3)^2(k_2-k_5)^2} + \frac{k_1^2k_2^2(k_5+k_6)^2}{(k_1-k_3)^2(k_2-k_4)^2} \Bigg) \\
& - \frac{1}{4}\Bigg(\frac{k_1^2k_2^2k_4^2}{(k_1-k_3)^2(k_2-k_5-k_6)^2} + \frac{k_1^2k_2^2k_5^2}{(k_1-k_3)^2(k_2-k_4-k_6)^2} + \frac{k_1^2k_2^2k_6^2}{(k_1-k_3)^2(k_2-k_4-k_5)^2} \\
& + \frac{k_1^2k_2^2k_3^2}{(k_1-k_4)^2(k_2-k_5-k_6)^2} + \frac{k_1^2k_2^2k_5^2}{(k_1-k_4)^2(k_2-k_3-k_6)^2} + \frac{k_1^2k_2^2k_6^2}{(k_1-k_4)^2(k_2-k_3-k_5)^2} \\
& + \frac{k_1^2k_2^2k_4^2}{(k_1-k_5)^2(k_2-k_3-k_6)^2} + \frac{k_1^2k_2^2k_3^2}{(k_1-k_5)^2(k_2-k_4-k_6)^2} + \frac{k_1^2k_2^2k_6^2}{(k_1-k_5)^2(k_2-k_4-k_3)^2} \\
& + \frac{k_1^2k_2^2k_4^2}{(k_1-k_6)^2(k_2-k_5-k_3)^2} + \frac{k_1^2k_2^2k_5^2}{(k_1-k_6)^2(k_2-k_4-k_3)^2} + \frac{k_1^2k_2^2k_3^2}{(k_1-k_6)^2(k_2-k_4-k_5)^2} \Bigg) \Bigg)
\end{aligned} \tag{19}
$$

This technique explicitly combines the defining Ward identities of the theory with unitarity in the Regge limit. Since this should be sufficient to define the theory it appears that **Yang-Mills reggeon theories can be constructed directly**. That is, the Feynman diagram expansion can be bypassed completely. It certainly looks **straightforward to construct even higher-order kernels**. There are, of course, many issues to be studied already with the $O(g^4)$ kernels, including the new Pomeron intercept, consequences for the anomalous dimensions of two and four gluon operators, consequences for deep-inelastic high-mass diffraction etc.[10].

3. Confinement and Quarks

The two defining properties we have used for pure glue reggeon theories are actually sufficient to show that such a theory **can not produce a confining Pomeron RFT**. While there is an **extensive exponentiation** of infra-red divergences in color non-zero channels, which removes all color non-zero amplitudes from the theory, **all color zero reggeon amplitudes are infra-red finite**. Also, the Ward identity constraint that reggeon amplitudes vanish when any $k_\perp \to 0$ guarantees that

phase-space integrals for multi-reggeon intermediate states are finite. Consequently **color zero** multigluon **reggeon states** necessarily survive and give singularities (but not divergences) in all amplitudes at zero transverse momentum. This clearly implies there is **no confinement.**

As we now discuss, this conclusion is avoided, **when quarks are present,** in a manner which does lead to confinement. To argue that the Ward Identity (9) requires that A_ν vanishes at $q = 0$ we simply differentiate

$$A_\nu + \frac{\partial A_\mu}{\partial q_\nu} q_\mu = 0 \;=>\; A_\nu \underset{(q_\mu \to 0)}{\longrightarrow} 0 \;\; unless \;\; \frac{\partial A_\mu}{\partial q_\nu} \to \infty \tag{20}$$

The Ward identity follows directly from gauge invariance and can not be violated, **but massless quark infra-red divergences** of A_ν - arising from the Regge limit - can prevent the conclusion that A_ν vanishes at $q = 0$. There are **quark reggeon diagrams** for which this is the case. In the color zero two gluon channel, quark loops of the form shown in Fig. 6 satisfy the Ward Identity by "$A_\mu \equiv A_-$" and "$q_\mu \equiv q_\perp$", but produce a vertex (for quark mass m)

$$V_c(q,k,k') \sim \int d^2p \; \frac{Tr\left[(-\not p + \not q + m)\not k(-\not p - \not q + m)\not k'\right]}{[(p+q)^2 + m^2]\,[(p-q)^2 + m^2]}$$

which satisfies

$$V_c \underset{\substack{m \sim 0 \\ q \sim 0}}{\longrightarrow} 8\pi \left[\frac{2(q \cdot k)(q \cdot k')}{q^2} - k \cdot k'\right] \underset{q \to \pm k, \pm k'}{\longrightarrow} 8\pi k \cdot k' \tag{21}$$

and so the **Ward identity constraint is not satisfied.** (This is consistent with (20) because $[\frac{\partial V_c}{\partial q_0}]_{q_0 \sim 0}$ $(\equiv [\frac{\partial V_c}{\partial m}]_{m \sim 0})$ $\sim \frac{1}{q_0}$ $\to \infty$ due to quark propagator infra-red divergences).

V_c (as a partial-wave amplitude) is distinct from the Lipatov kernel, but " V_c^2 " does contribute to this kernel. Consequently **a new infra-red divergence** is generated which produces the **exponentation to zero** of the two-gluon reggeon state. In fact, in SU(2) gauge theory all color zero reggeon states produce divergences which are similarly exponentiated[2], apart from a scaling divergence associated with the "anomalous Odderon" three gluon state. This divergence couples only via a **"triangle quark loop anomaly"** and so does not exponentiate. Instead it produces a **reggeon** ("winding number") **condensate** which gives **a confinement spectrum.** In **SU(3)** gauge theory, a reggeized gluon in the background of this condensate gives **a Regge pole Pomeron.**

4. Pomeron Dynamics and Electroweak Symmetry Breaking

A Regge pole Pomeron gives $\sigma_T \to 0$ **unless** $\alpha_{\mathbb{P}}(0) = 1$. Therefore, (confining) Pomeron RFT is, asymptotically, inconsistent with the validity of perturbative

QCD at large $k_\perp$ **unless the Pomeron is Critical !!** My detailed analysis[2] of QCD Pomeron RFT suggests that $\alpha_{\mathbb{P}}(0) = 1$ if, and only if, the number of flavors N_f is a maximum. This requires that $N_f = 16$ **or** $N_f = 6$ **for color triplet quarks and** $N_f = 2$ **for color sextet quarks.** It is a remarkable coincidence that two flavors of color sextet quarks can provide **a natural form of dynamical symmetry-breaking[11] for the electroweak interaction which meshes perfectly with the observed experimental features.**

Add a massless doublet (u_6, d_6) with the **usual quark quantum numbers** to the Standard Model (with no scalar Higgs sector). The axial $U(2) \otimes U(2)$ chiral symmetry breaks spontaneously, producing four pseudoscalar Goldstone bosons (π_6^+, π_6^-, π_6^0 and η_6) which couple "longitudinally" to the sextet axial currents i.e.

$$< 0|A_\mu^\tau|\pi_6^\tau(q) > \sim F_{\pi_6} q_\mu \quad , \quad < 0|a_\mu|\eta_6(q) > \sim F_{\eta_6} q_\mu \tag{22}$$

The $SU(2)$ gauge fields B_μ^τ couple via

$$\mathcal{L}_I = g W^{\tau\mu}\Big(V_\mu^\tau - A_\mu^\tau\Big) \tag{23}$$

As a result π_6^+, π_6^- and π_6^0 are "eaten" and provide **masses for the** W^+, W^- **and** Z^0, with

$$M_W \sim g\ F_{\pi_6} \tag{24}$$

where F_{π_6} **is a QCD scale**. The "Casimir Scaling" rule[11] gives

$$C_6\ \alpha_s(F_{\pi_6}^2)\ \sim\ C_3\ \alpha_s(F_\pi^2) \tag{25}$$

which is consistent with $F_{\pi_6} \sim 250$ GeV ! Consequently, **the electroweak scale is naturally explained as a second** QCD **scale.** The restriction to a sextet flavor doublet, necessarily gives

$$\rho = \ (M_W^2/M_Z^2 cos^2\theta_W)\ =\ 1 \tag{26}$$

as also required by experiment.

Finally we note that the sextet sector may also be deeply tied to the issue of Strong CP conservation and the origin of CP violation. The η_6 is a Goldstone boson associated with the $U(1)$ axial chiral symmetry and may be **the axion**. It is a conventional (Peccei-Quinn) axion[12] except that it can aquire an electroweak scale mass as a result of enhanced electroweak scale color instanton interactions[13].

Not only may the sextet sector be the explanation of **Strong** CP **conservation in the triplet sector** (via the η_6 axion), but it may also be responsible for CP **violation at the weak scale.** Because the sextet sector has no axion the QCD interactions at this scale will naturally be "Strong CP-violating". The normal triplet quark hadrons will contain a **small admixture of sextet quark states** which could **provide their** CP **violating interactions.**

5. Conclusion

The QCD Pomeron may be the key to many of the remaining puzzles of the Standard Model.

References

[1] A. R. White, Argonne preprint, ANL-HEP-PR-94-23, to be published in *Phys. Lett.* **B**.

[2] A. R. White, *Int. J. Mod. Phys.* **A8**, 4755 (1993).

[3] L. N. Lipatov, in *Perturbative QCD*, ed. A. .H. Mueller (World Scientific, 1989).

[4] E. A. Kuraev, L. N. Lipatov, V. S. Fadin, *Sov. Phys. JETP* **45**, 199 (1977) ; Ya. Ya. Balitsky and L. N. Lipatov, *Sov. J. Nucl. Phys.* **28**, 822 (1978).

[5] H. Cheng and C. Y. Lo, *Phys. Rev.* **D13**, 1131 (1976) and **D15**, 2959 (1977).

[6] A. R. White, *Int. J. Mod. Phys.* **A6**, 1859 (1990) and references therein.

[7] A. A. Migdal, A. M. Polyakov and K. A. Ter-Martirosyan, *Zh. Eksp. Teor. Fiz.* **67**, 84 (1974); H. D. I. Abarbanel and J. B. Bronzan, *Phys. Rev.* **D9**, 2397 (1974).

[8] J. B. Bronzan and R. L. Sugar, *Phys. Rev.* **D17**, 585 (1978); J. Bartels, *Nucl. Phys.* **B151**, 293 (1979), **B175**, 365 (1980)

[9] G. 't Hooft, *Nucl. Phys.* **B33**, 173 (1971).

[10] J. Bartels, **DESY** *preprint*, DESY 91-074 (1991) and *Zeitsch. Phys.* **C60**, 471 (1993), J. Bartels and M. Wusthoff, **DESY** *preprint*, DESY 94-016 (1994).

[11] W. J. Marciano, *Phys. Rev.* **D21**, 2425 (1980) ; E. Braaten, A. R. White and C. R. Willcox, *Int. J. Mod. Phys.* **A1** 693 (1986).

[12] R. D. Peccei and H. R. Quinn, *Phys. Rev. Lett.* **38**, 1440 (1977); S. Weinberg, *Phys. Rev. Lett.* **40**, 223 (1978); F. Wilzcek, *Phys. Rev. Lett.* **40**, 279 (1978).

[13] A. R. White, ANL-HEP-CP-93-56 (1993).

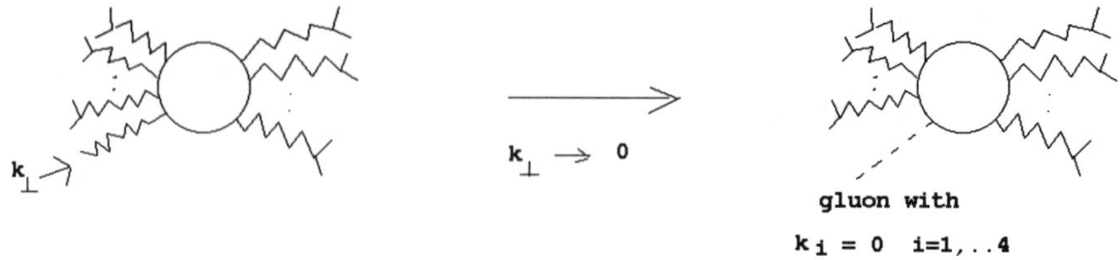

Fig. 1 A reggeon amplitude gives a gluon amplitude as $k_\perp \to 0$.

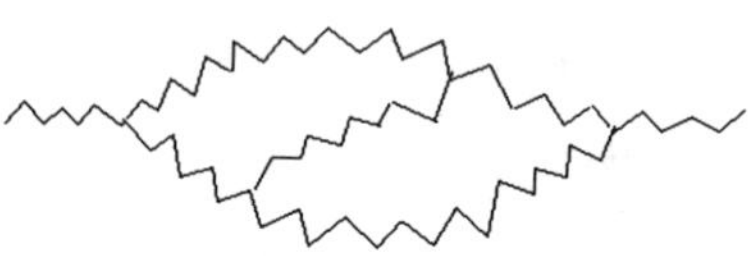

Fig. 2 A two-loop reggeon diagram.

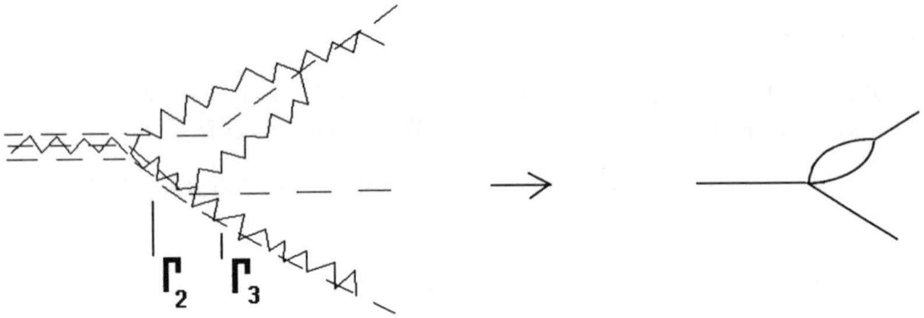

Fig. 3 The reduction of a three-reggeon vertex diagram to a transverse momentum diagram.

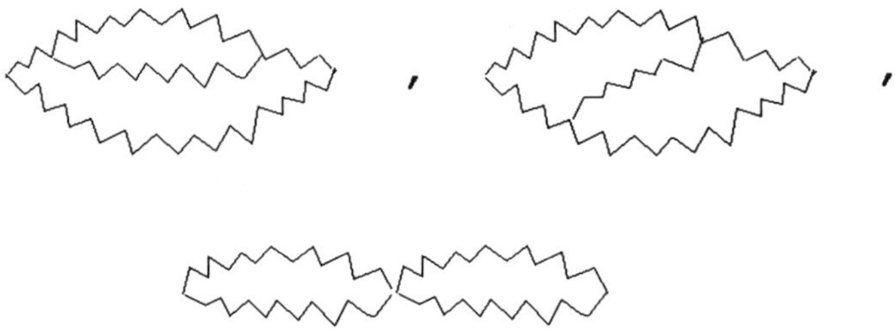

Fig. 4 Two-loop even-signature reggeon diagrams.

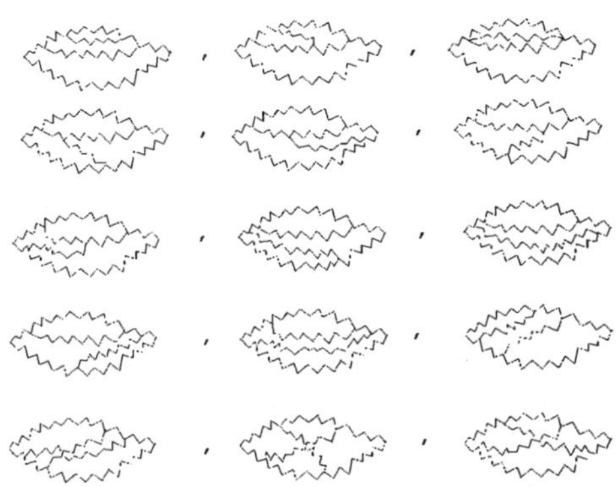

Fig. 5 Three loop diagrams giving the $O(g^4)$ kernel.

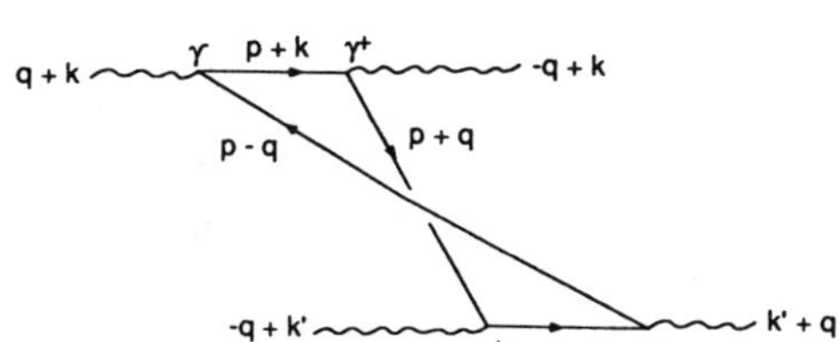

Fig. 6 A one loop quark reggeon diagram.

An angle representation of QCD

Dieter Stoll
Department of Physics, University of Tokyo
Bunkyo–ku, Tokyo 113, Japan

Abstract

For the sake of eliminating gauge variant degrees of freedom we discuss the way to introduce angular variables in the hamiltonian formulation of QCD. On the basis of an analysis of Gauss' law constraints a particular choice is made for the variable transformation from gauge fields to angular field variables. The resulting formulation is analogous to the one of Bars in terms of corner variables and it is closely related to the hamiltonian lattice QCD formulation. Therefore the corner or angle formulation may constitute an useful starting point for the investigation of the low energy properties of QCD in terms of gauge invariant degrees of freedom.

1 Introduction

One of the long standing problems in contemporary physics is understanding confinement of quarks and gluons from first principles. The difficulty in dealing with the infrared properties of QCD is on the one hand due to the non–linear gluonic interaction and on the other due to the constraints on the dynamics of the fundamental degrees of freedom which originate from the requirement of gauge invariance. In spite of the general belief that the non–linear interaction gives rise to confinement it has been conjectured recently that in fact the non–abelian constraints may be most important [1]. Aiming at an understanding of the low energy properties of QCD we should therefore try to develop approximations to the full QCD dynamics after the gauge variant degrees of freedom have been identified and isolated.

The aforementioned constraints are given by Gauss' law operators, which generate a compact group in each point in space, telling us that the gauge variant degrees of freedom are "angle" variables. Due to the fact that field theory deals with an infinite number of degrees of freedom, the choice of unphysical variables is to a large extent arbitrary. Therefore various decompositions into unphysical "angle" variables and remaining physical variables are possible to arrive at the desired separation of unphysical degrees of freedom [1,2,3]. Although successful in that respect the variables chosen in this way to parametrize the physical Hilbert space may be inadequate to account in a simple way for the dynamics relevant for the low energy properties of QCD. For further variable changes, on the other hand, the complexity of the so derived hamiltonians constitutes a basic obstacle.

Therefore our starting point is the assumption that not only the unphysical but all variables are "angle" variables and the hamiltonian should be expressed first in terms of these angular degrees of freedom before making a separation into gauge variant and gauge invariant ones. To find a definition of "angle" variables in terms of gauge or electric fields we concentrate on an analysis of Gauss' law operators. It will be shown that the form of these operators suggests the introduction of "angles" which are non–locally related to the gauge fields. By a variable transformation the originally quantized gauge fields and electric fields in the Hamiltonian can be replaced by "angle" and angular momentum operators respectively. The resulting formulation is analogous to the one in terms of corner variables obtained by Bars [4]. In contrast to similar approaches [1,2] the separation into gauge variant and gauge invariant degrees of freedom is not made from the outset in the "angle" or corner variable formulation. Therefore it may constitute an useful starting point for the search of approximations to the full QCD dynamics intended to understand its nonperturbative aspects.

2 The hamiltonian in terms of angle variables

We consider a Hamiltonian formulation of SU(N) gauge theories on a d–dimensional torus. Choosing the Weyl gauge $A_0 = 0$ we have the following Hamiltonian density

$$\mathcal{H} = \sum_i \bar{\psi}(x)\gamma_i\left(i\partial_i + gA_i\right)\psi(x) + m\bar{\psi}(x)\psi(x) + \frac{1}{2}\sum_i E_i^a(x)E_i^a(x) + \frac{1}{2}\sum_{ij}\mathrm{Tr}\left\{F_{ij}F^{ij}\right\} .$$

Imposing periodic boundary conditions for the gauge and anti–periodic ones for the fermion fields we quantize canonically ($E_i^a(x) = \partial_0 A_i^a(x)$)

$$\begin{aligned}\left[E_i^a(x), A_j^b(y)\right] &= -i\delta_{a,b}\delta_{ij}\delta^d(x-y); \quad a,b = 1,\ldots,N^2-1;\ i,j = 1,\ldots,d \\ \left\{\psi^\dagger_{k,\alpha}(x), \psi_{l,\beta}(y)\right\} &= \delta_{k,l}\delta_{\alpha,\beta}\delta^d(x-y); \quad k,l = 1,\ldots,N;\ \alpha,\beta = \text{Spinor indices}\end{aligned}$$

where it is understood that the δ–functions are periodic, as well. Since we have not fixed the gauge classically, Gauss' law operator is the quantum mechanical generator of the gauge symmetry. It commutes with the Hamiltonian and therefore physical, i.e. gauge invariant, eigenstates must be annihilated by the generators of the symmetry

$$G^a(x)|\,phys.> \; = \; 0 \tag{2.1}$$

$$G^a(x) \; = \; \sum_i \left[\partial_i E_i^a(x) + gf^{abc}A_i^b(x)E_i^c(x)\right] + g\psi^\dagger(x)\frac{\lambda^a}{2}\psi(x) \tag{2.2}$$

$$\left[G^a(x), G^b(y)\right] \; = \; igf^{abc}G^c(x)\delta^d(x-y) \; . \tag{2.3}$$

Since the generators obey the Lie algebra of the gauge group it is understood that out of $d{\cdot}(N^2-1)$ gauge degrees of freedom only a set of N^2-1 "angle" variables in each point in space is changed by gauge transformations . Consequently the constraints are satisfied if these "angles" have been identified and Gauss' law operator has been transformed such that it is the angular momentum operator only with respect to these unphysical "angles". Physical states then correspond to s–wave states which are annihilated by these angular momentum operators and the Hamiltonian after transformation will not contain the unphysical variables anymore.

Since we want to replace gauge and electric fields by "angles" and angular momenta in such a way that the constraints can easily be implemented, we study the form of Gauss' law operators first in 1+1 dimensions. In this case the contribution in eq.(2.2)

$$f^{abc}A^b(x)E^c(x) \tag{2.4}$$

acts locally as an angular momentum operator on $N(N-1)$ "angle" variables in either the gauge field or the elctric field representation. The missing $(N-1)$ "angle" variables could not be identified, if this was the complete Gauss' law operator already. Therefore we must conclude that in 1+1 dimensions the full number of (N^2-1) variables in each point in space can only be eliminated due to the presence of $\partial_x E(x)$ in the Gauss' law operators. This term not only distinguishes the gauge fields as source of the additional unphysical variables but also introduces a non-locality into the Gauss' law operators. Therefore it seems natural to assume that the "angle" variables which are unphysical are nonlocally related to the gauge field variables. Although this argument is rigorous only in 1+1 dimensions we assume it to be an useful hypothesis for introducing "angle" variables in any dimensions.

An expression for the gauge fields satisfying this requirement is[1]

$$\begin{aligned} A_i(x) &= \frac{i}{g}V_i(x)\partial_i V_i^\dagger(x) \quad \text{(no summation)} && (2.5)\\ V_i(x) &= P\exp\left[ig\int_0^{x_i} dz_i A_i(x_i^\perp, z_i)\right] && (2.6)\\ V_i(x) &= \exp\left[i\xi_i(x)\right], \quad \xi_i(x) = \xi_i^a(x)\frac{\lambda^a}{2}; \quad 0 < x_i \le L && (2.7) \end{aligned}$$

where $V_i(x)$ is a SU(N) matrix parametrized in terms of "angles" $\xi_i^a(x)$, P denotes path ordering and $x_i^\perp$ stands for all coordinates orthogonal to x_i. Since this definition together with the specific choice of paths in eq.(2.6) leads to an unique relation[2] between $\xi_i(x)$ and $A_i(x)$ a change of variables from $A_i(x), E_i(x)$ to $\xi_i(x)$ and the corresponding angular momenta $J_i(x)$ becomes feasible. Using eq.(2.5) we rewrite fermionic and magnetic part of the hamiltonian

$$\bar{\psi}(x)\left\{\gamma_i\left[i\partial_i + gA_i(x)\right] + m\right\}\psi(x) = \left[\bar{\psi}(x)U_i(x)\right]\left\{\gamma_i i\partial_i + m\right\}\left[U_i^\dagger(x)\psi(x)\right], \tag{2.8}$$

[1]Note that throughout the paper spatial indices are not summed over unless explicitly indicated.

[2]With the choice $0 < x_i \le L$ the "angles" are uniquely determined from gauge fields by eq.(2.6) if the derivative is taken from one side only $\partial_i f(x_i) = \lim_{\epsilon\to 0}\frac{1}{\epsilon}\left[f(x_i) - f(x_i-\epsilon)\right]$; $0 < x_i \le L$. In this way it is possible to work with periodic, although not continuous "angles" and angular momenta.

$$[D_i, D_j] = [i\partial_i + gA_i, i\partial_j + gA_j] = -U_i \left\{\partial_i \left[\left(U_i^\dagger U_j\right)\partial_j\left(U_j^\dagger U_i\right)\right]\right\} U_i^\dagger \ , \quad (2.9)$$

$$\Rightarrow \mathrm{Tr}\left\{F_{ij}F_{ij}\right\} = \frac{-1}{g^2}\mathrm{Tr}\left\{\partial_i \left[\left[\left(U_i^\dagger U_j\right)\partial_j\left(U_j^\dagger U_i\right)\right]\right]\left[\partial_i\left[\left(U_i^\dagger U_j\right)\partial_j\left(U_j^\dagger U_i\right)\right]\right]^\dagger\right\}.$$

In order to reformulate the electric part of the Hamiltonian which contains the conjugate momenta of the gauge fields, we introduce the angular momentum operators $J_k^c(z)$ [8]. These operators generate translations in the space of "angles" ξ_k as may be seen from the commutation relations

$$\begin{aligned} [J_i^a(x), V_j(y)] &= \delta_{i,j}\delta^d(x-y)V_j(y)\frac{\lambda^a}{2} \\ \left[J_i^a(x), J_j^b(y)\right] &= \delta_{i,j} i f^{abc} J_i^c(x)\delta^d(x-y)\ . \end{aligned} \quad (2.10)$$

We note that due to the periodicity of V_i, J_i the δ-functions in these expressions are periodic, as well. Introducing furthermore the orthogonal matrices N_i

$$\begin{aligned} N_i^{ac}(x) &= \mathrm{Tr}\left\{V_i^\dagger(x)\frac{\lambda^a}{2}V_i(x)\lambda^c\right\} \\ \left[J_i^b(z), N_j^{ac}(x)\right] &= i f^{bce} N_j^{ae}(x)\delta_{i,j}\delta^d(x-z) \end{aligned} \quad (2.11)$$

one can derive the following expression for the electric part of the hamiltonian density [8]

$$\begin{aligned} E_i^a(x) &= g\int d^dz\,\delta^{d-1}(z_i^\perp - x_i^\perp)\theta(z_i - x_i)\theta(x_i)N_i^{ac}(x)J_i^c(z) \\ \frac{1}{2}E_i^a(x)E_i^a(x) &= \frac{g^2}{2}\int_{x_i}^L dz_i J_i^b(x_i^\perp, z_i)\int_{x_i}^L dz_i' J_i^b(x_i^\perp, z_i')\ . \end{aligned} \quad (2.12)$$

Collecting all the results the hamiltonian density reads

$$\begin{aligned} \mathcal{H} &= \sum_i \left[\bar\psi(x)U_i(x)\right]\left[\gamma_i i\partial_i + m\right]\left[U_i^\dagger(x)\psi(x)\right] \\ &+ \frac{g^2}{2}\sum_i \int_{x_i}^L dz_i J_i^b(x_i^\perp, z_i)\int_{x_i}^L dz_i' J_i^b(x_i^\perp, z_i') \\ &+ \frac{1}{2g^2}\sum_{ij}\mathrm{Tr}\left\{\left[\partial_i\left[\left(U_i^\dagger U_j\right)\partial_j\left(U_j^\dagger U_i\right)\right]\right]\left[\partial_i\left[\left(U_i^\dagger U_j\right)\partial_j\left(U_j^\dagger U_i\right)\right]\right]^\dagger\right\} \end{aligned} \quad (2.13)$$

which is the "angle" representation we have been looking for. Note that the locality of the Hamiltonian has been lost although we have not been fixing the gauge yet. The electric part of the Hamiltonian is non–local and it shows already the linear "potential" $|z_i - z_i'|$ characteristic for both the axial gauge formulation and the strong coupling limit in lattice gauge theory. We observe also that the Hamiltonian in QED, corresponding to eq.(2.13), is obtained by dropping the summations over color indices which shows the similarity of abelian and non–abelian gauge theories in the "angle" formulation.

Finally we want to consider the form of Gauss' law operator in these variables. We find

$$\begin{aligned} g f^{abc} A_i^b(x) E_i^c(x) &= -g\left[\partial_i N_i^{ad}(x)\right]\int dz_i \theta(z_i - x_i)\theta(x_i) J_i^d(x_i^\perp, z_i) && (2.14)\\ \partial_i E_i^a(x) &= g\left[\partial_i N_i^{ad}(x)\right]\int dz_i \theta(z_i - x_i)\theta(x_i) J_i^d(x_i^\perp, z_i) \\ &\quad + g N_i^{ad}(x)\int dz_i J_i^d(x_i^\perp, z_i)\partial_i\left[\theta(z_i - x_i)\theta(x_i)\right] && (2.15) \end{aligned}$$

and taking the sum of these two contributions and the charge density operator we obtain for Gauss law the following expression

$$G^a(x) = -g\sum_i\left[N_i^{ad}(x)J_i^d(x) - \delta(x_i)\int dz_i J_i^a(x_i^\perp, z_i)\right] + g\psi^\dagger(x)\frac{\lambda^a}{2}\psi(x)\,. \qquad (2.16)$$

3 The connection with the lattice QCD hamiltonian

The task in this section is to show that the similarity between the variables in our formulation and those of the lattice hamiltonian approach to QCD can be made explicit. To this extend we replace the continuous spatial variables x_i by lattice points x_i^m and differential operators are replaced by finite differences. In order not to violate gauge invariance, we have to guarantee that the derivative operators only act on gauge invariant functions. This is the case for the fermionic part of the hamiltonian. Using the simplest prescription for the derivative operation we find

$$\partial_i\left[V_i^\dagger(x)\psi(x)\right] \to \frac{1}{a^{5/2}}\left[V_i^\dagger(x_i^\perp, x_i^m)\psi(x_i^\perp, x_i^m) - V_i^\dagger(x_i^\perp, x_i^m - a)\psi(x_i^\perp, x_i^m - a)\right]\,. \qquad (3.1)$$

Introducing the link variables in terms of "angles" $\xi_i^m(x_i^\perp)$

$$\begin{aligned} L_i(x_i^{m-1}, x_i^\perp) &= P\exp\left[ig\int_{x_i^m - a}^{x_i^m} dz_i A_i(x_i^\perp, z_i)\right] = \exp\left[i\xi_i^m(x_i^\perp)\right] && (3.2)\\ V_i(x_i^\perp, x_i^m) &= L_i(x_i^{m-1}, x_i^\perp)L_i(x_i^{m-2}, x_i^\perp)\ldots L_i(0, x_i^\perp) \end{aligned}$$

the fermionic hamiltonian can be cast into the familiar form

$$\begin{aligned} \mathcal{H}_{ferm} &= \frac{i}{2a^4}\sum_i\left[\bar\psi(x_i^\perp, x_i^{m-1})\gamma_i L_i^\dagger(x_i^{m-1}, x_i^\perp)\psi(x_i^\perp, x_i^m)\right.\\ &\quad\left. -\bar\psi(x_i^\perp, x_i^m)\gamma_i L_i(x_i^{m-1}, x_i^\perp)\psi(x_i^\perp, x_i^{m-1})\right] + \frac{m}{a^3}\bar\psi(x_i^\perp, x_i^m)\psi(x_i^\perp, x_i^m)\,. && (3.3) \end{aligned}$$

Defining angular momentum operators $\mathcal{J}_i^a(x_i^\perp, x_i^m)$ with respect to the link "angles" $\xi_i^m(x_i^\perp)$ obeying the following commutation relations

$$\begin{aligned} \left[\mathcal{J}_i^a(x_i^\perp, x_i^m), L_j(z_j^{n-1}, z_j^\perp)\right] &= \delta_{i,j}\delta_{n,m}\delta_{x_i^\perp, z_i^\perp} L_j(z_j^{n-1}, z_j^\perp)\frac{\lambda^a}{2}\\ \left[\mathcal{J}_i^a(x_i^\perp, x_i^m), \mathcal{J}_j^b(z_j^\perp, z_j^n)\right] &= \delta_{i,j}\delta_{x_i^\perp, z_i^\perp}\delta_{n,m} i f^{abc}\mathcal{J}_i^c(x_i^\perp, x_i^m) && (3.4) \end{aligned}$$

we arrive at the following expression for the integral over "angular momentum" operators appearing in eq.(2.13)

$$\int_{x_i}^{L} dz_i J_i^c(x_i^\perp, z_i) \to \frac{1}{a^2}\sum_{k=m}^{N} J_i^c(x_i^\perp, x_i^k) = \frac{1}{a^2} N_i^{ac}(x_i^\perp, x_i^{m-1})\mathcal{J}_i^c(x_i^\perp, x_i^m) \; . \tag{3.5}$$

The electric part of the hamiltonian eq.(2.13) then becomes

$$\mathcal{H}_{elec} = \frac{g^2}{2a^4}\sum_i \sum_{k=m}^{N}\sum_{l=m}^{N} J_i^a(x_i^\perp, x_i^k) J_i^a(x_i^\perp, x_i^l) = \frac{g^2}{2a^4}\sum_i \mathcal{J}_i^a(x_i^\perp, x_i^m)\mathcal{J}_i^a(x_i^\perp, x_i^m) \tag{3.6}$$

which is the usual lattice form of this contribution to the hamiltonian. In particular the non–locality has disappeared and the existence of the linear "potentials" $|z_i - z_i'|$ has become less obvious. The operator which remains to be discretized is the magnetic part of the hamiltonian. Introducing the following modification

$$V_i^\dagger(x_i, x_j)V_j(x_i, x_j) \to U_{ij}^\dagger(x_i, x_j) = V_j^\dagger(x_i = 0, x_j)V_i^\dagger(x_i, x_j)V_j(x_i, x_j)V_i(x_i, x_j = 0)$$

it can be rewritten such that the derivatives are acting on gauge invariant functions. Therefore we obtain

$$\begin{aligned} \partial_i \left[U_{ji}(x_i, x_j)\partial_j U_{ji}^\dagger(x_i, x_j)\right] \quad \to \quad & \frac{1}{a^2}\Big[U_{ji}(x_i - a, x_j)U_{ji}^\dagger(x_i - a, x_j - a) \\ & - U_{ji}(x_i, x_j)U_{ji}^\dagger(x_i, x_j - a)\Big] \; . \end{aligned} \tag{3.7}$$

Multiplying this result with its hermitian conjugate and using the link matrices L_i defined in eq.(3.2) the result for the magnetic part of the Hamiltonian eq.(2.13) becomes

$$\begin{aligned} \mathcal{H}_{magn} &= \frac{-1}{2g^2a^4}\sum_{i,j} \mathrm{Tr}\Big\{L_i^\dagger(x_i^{m-1}, x_j^n)L_j(x_i^m, x_j^{n-1})L_i(x_i^{m-1}, x_j^{n-1})L_j^\dagger(x_i^{m-1}, x_j^{n-1}) \\ &\quad + L_i^\dagger(x_i^{m-1}, x_j^{n-1})L_j^\dagger(x_i^m, x_j^{n-1})L_i(x_i^{m-1}, x_j^n)L_j(x_i^{m-1}, x_j^{n-1}) - 2\Big\} \\ &= \frac{1}{2g^2a^4}\sum_{ij} \mathrm{Tr}\Big\{2 - U_\square(x_i^{m-1}, x_j^{n-1}) - U_\square^\dagger(x_i^{m-1}, x_j^{n-1})\Big\} \end{aligned} \tag{3.8}$$

where $U_\square(x_i^{m-1}, x_j^{n-1})$ is seen to be the unitary matrix obtained from the path ordered integral along the boundary of a plaquette in the (i,j)–plane starting from the point (x_i^{m-1}, x_j^{n-1}). Thus collecting the results eq.(3.3,3.6,3.8) we find the standard Kogut–Susskind lattice hamiltonian density [5]

$$\begin{aligned} \mathcal{H}_{lattice} &= \frac{1}{a^4}\Bigg\{\frac{g^2}{2}\sum_i \mathcal{J}_i^a(x)\mathcal{J}_i^a(x) + \frac{1}{2g^2}\sum_{i,j}\mathrm{Tr}\Big\{2 - U_\square(x_i^{m-1}, x_j^{n-1}) - U_\square^\dagger(x_i^{m-1}, x_j^{n-1})\Big\} \\ &\quad + \frac{i}{2}\sum_i \Big[\bar\psi(x_i^\perp, x_i^{m-1})\gamma_i L_i^\dagger(x_i^{m-1}, x_i^\perp)\psi(x_i^\perp, x_i^m) \\ &\quad - \bar\psi(x_i^\perp, x_i^m)\gamma_i L_i(x_i^{m-1}, x_i^\perp)\psi(x_i^\perp, x_i^{m-1})\Big] + (ma)\bar\psi(x_i^\perp, x_i^m)\psi(x_i^\perp, x_i^m)\Bigg\} \; . \end{aligned} \tag{3.9}$$

To complete the comparison with the lattice formulation we determine also the functional integration measure in our formulation. For this calculation we have to evaluate the Jacobian of the transformation from the gauge fields A_i to the "angles" ξ_j for which we find

$$\mathcal{R}_{ij}^{ab}(x,y) = \frac{\delta A_i^a(x)}{\delta \xi_j^b(y)} = \frac{1}{g}\delta_{i,j}\delta^{d-1}(x_i^\perp - y_i^\perp) M_i^{bc}(y) N_i^{ac}(x) \partial_i \delta(x_i - y_i) \ . \tag{3.10}$$

The Jacobian is obtained from the determinant of $\mathcal{R}$ which can be written as

$$det(\mathcal{R}) = det(\mathcal{R}^{-1}) = \prod_i det\left[M_i(z)\right] \ . \tag{3.11}$$

The expression eq.(3.11) is just the volume of the gauge group and after discretization this result gives the usual integration measure in the lattice formulation (see e.g. [6]).

We note that in establishing the relation between the lattice and the angle formulation of QCD an expansion in powers of the lattice spacing a only occurs due to the necessity to replace differential operators by finite differences. As a consequence we naturally obtain the change from the flat integration measure in the space of gauge fields $A_i(x)$ to the group integration measure in the space of unitary matrices $V_i(x)$ or links $L_i(x_i^m, x_i^\perp)$. It also implies that continuum operators can be translated into lattice operators. In general this will not lead to simple expressions in terms of plaquettes (e.g. in the case of $\bar{\psi}\sigma_{\mu\nu}F^{\mu\nu}\psi$ or $\mathrm{Tr}\{\epsilon_{\mu\nu\rho\tau}F^{\mu\nu}F^{\rho\tau}\}$) unlike in the special combination ($\mathrm{Tr}\{F_{ij}(x)F_{ij}(x)\}$) appearing in the hamiltonian. With regard to the widely discussed differences between compact and non–compact formulations of lattice gauge theories we clearly observe that the existence of the strong coupling limit is related to the replacement of an unbounded by a bounded magnetic operator

$$U_{ji}(x_i,x_j)\partial_j U_{ji}^\dagger(x_i,x_j) \to \left\|\frac{1}{a}\left[1 - U_{ji}(x_i,x_j)U_{ji}^\dagger(x_i,x_j - a)\right]\right\| \leq \frac{2}{a} \ .$$

This suggests to use a non–compact continuum formulation, separate gauge variant degrees of freedom and introduce a lattice for the gauge invariant degrees of freedom as a procedure leading to a meaningful non–compact lattice formulation of non–abelian gauge theories.

4 Summary and conclusion

In summary we have arrived at a continuum formulation of non–abelian gauge theories entirely in terms of angular degrees of freedom. The objective for doing so was the wish to have a formulation that leaves us freedom in choosing appropriate unphysical "angle" variables. This may prove importantfor being able to develop useful approximations to understand the low energy properties of QCD. The resulting formulation can be shown to be equivalent in a finite volume to Bars corner variable formulation and is like Bars formulation [7] closely related to the lattice hamiltonian approach to

QCD [5] as we have shown explicitly. In addition to this it is a formulation which has the built–in freedom in selecting unphysical variables. Furthermore it has been shown for the axial gauge representation of QCD [3] already that this reformulation leads to great technical simplifications in the elimination of unphysical variables [9]. All these advantages together may render the "angle" formulation an useful starting point for investigating non–perturbative aspects of QCD in terms of gauge invariant degrees of freedom.

Acknowledgements

We thank Profs. K.Ohta, M.Thies and K.Yazaki for helpful discussions and acknowledge gratefully the financial support by the "Japan Society for the Promotion of Science".

References

[1] K.Johnson, L.Lellouch and J.Polonyi, Nucl. Phys. **B 367** (1991) 675
K.Johnson, Invited talk at the 1992 Aachen conference on "QCD, 20 Years Later"; eds.: P.M.Zerwas and H.A.Kastrup, World Scientific 1993

[2] J.Goldstone and R.Jackiw, Phys. Lett. **74 B** (1978) 81
V.Baluni and B.Grossmann, Phys. Lett **78 B** (1978) 226
A.G.Izergin, M.A.Semenov and L.D.Fadeev, Theor. Mat. Fiz. **38** (1979) 3
M.Creutz, I.J.Muzinich and T.N.Tudron, Phys. Rev. **D 19** (1979) 531
N.H.Christ and T.D.Lee, Phys.Rev. **D 22** (1980) 939
Yu.A.Simonov, Sov. J. Nucl. Phys. **41** (1985) 1311

[3] F.Lenz, H.W.L.Naus and M.Thies, The axial gauge representation of QCD, to appear in Ann. Phys.

[4] I.Bars, Phys. Rev. Lett. **40** (1978) 688
I.Bars, Nucl. Phys. **B 148** (1979) 445

[5] J.B.Kogut and L.Susskind, Phys. Rev. **D 11** (1975) 395

[6] M.Creutz, Quarks, gluons and lattices; Cambridge University press 1990

[7] I.Bars, Nucl. Phys. **B 149** (1979) 39

[8] D.Stoll, The hamiltonian formulation of QCD in terms of angle variables, preprint

[9] D.Stoll, Gauge fixing by unitary transformations in QCD, preprint

Non-Perturbative Solutions of Schwinger Equation *

Vladimir E. Rochev
Institute for High Energy Physics
Protvino, 142284, Moscow Region, Russia

Abstract

Some features of essentially non-perturbative expansions are discussed. A general solution for the Schwinger equation for the scalar field theory has been constructed.

The Schwinger equations for the generating functional are one of the ways of representing dynamical principle of quantum field theory (QFT). In the framework of perturbative QFT this formalism is equivalent to other well known methods, such as operator formalism, functional integral technique, etc. However, the other situation arises in the non-perturbative region. The construction of the essentially non-perturbative expansions, such as the strong coupling expansion (SCE), by Schwinger equation method requires an additional information into the iterative scheme. In present report the situation is analysed for the theory of scalar field with quadric interaction. An origin of the problem can be clarified by the consideration of a general solution of the Schwinger equation. A method for the construction of such solution is presented. The general solution contains an arbitrary functional and may be useful for construction of new non-perturbative approximation schemes.

Consider the complex scalar field $\phi(x)(x \in E_d)$ with the self-action $\frac{\lambda}{2}(\phi^*\phi)^2$. The Schwinger equation for the generating functional G has the form:

$$\delta(x-y)G + \int dx'\eta(xx')\frac{\delta G}{\delta\eta(yx')} = \lambda\frac{\delta^2 G}{\delta\eta(yx)\delta\eta(xx)}. \qquad (1)$$

Here $\eta(xy)$ is a bilocal source. The source switching off corresponds to $\eta = \eta_0 = m^2 - \partial^2$, where m^2 is the mass of the free field. The generating functional G is normalized by the condition $G[\eta_0] = 1$. The propagator is

$$\Delta(x-y) =< \phi(x)\phi^*(y) >= -\frac{\delta G}{\delta\eta(yx)}\,|_{\eta=\eta_0}\,,$$

*Supported in part by the Russian Foundation for Fundamental Investigation under grant No.94-02-03545-a the two-particle function is

$$\mathcal{F} =< \phi\phi\phi^*\phi^* >= \frac{\delta^2 G}{\delta\eta^2}|_{\eta=\eta_0} ,$$

and so on.

At $\lambda = 0$ Eq.(1) has the solution $G_0 = \det \frac{\eta_0}{\eta}$ which is the generating functional of the free field. The G_0 based iteration scheme reproduces perturbation series. The normalization condition $G[\eta_0] = 1$ suffices to define uniquely the generating functional in any order in λ because the corresponding equations are first order ones with respect to the functional derivatives. In the SCE, however, the situation is different. For the SCE

$$G = G^{(0)} + \frac{1}{\sqrt{\lambda}} G^{(1)} + \frac{1}{\lambda} G^{(2)} + ... \tag{2}$$

one has to solve second-order equations and one more boundary condition is required to be set. Problems of this kind are, in general, typical for non-perturbative solutions of Schwinger equations[1]. My claim is that Schwinger equations in the non-perturbative region possess more freedom in choosing solutions than an arbitrariness connected with boundary conditions.

Let us introduce an additional boundary condition, for example

$$\frac{\delta G}{\delta\eta}|_{\eta=\eta_0} = -\Delta \, . \tag{3}$$

Then we get the first two terms of SCE:

$$G^{(0)} = 1, \quad G^{(1)} = tr[(\eta_0 - \eta)\Delta^{(1)}]. \tag{4}$$

Here $\Delta = \frac{1}{\sqrt{\lambda}}\Delta^{(1)} + \frac{1}{\lambda}\Delta^{(2)} + ...$ The first nontrivial term of SCE is[2]

$$G^{(2)} = tr[(\eta_0 - \eta)\Delta^{(2)}] +$$

$$\frac{1}{2}\int dx_1 dx_2 dx_3 dx_4 (\eta - \eta_0)(x_1 x_2)(\eta - \eta_0)(x_3 x_4)\delta(x_1 - x_2 + x_3 - x_4) + \bar{G},$$

where $\bar{G}$ is a solution of homogeneous equations

$$\frac{\delta^2 \bar{G}}{\delta\eta(yx)\delta\eta(xx)} = \frac{\delta^2 \bar{G}}{\delta\eta(yx)\delta\eta(yy)} = 0.$$

The functional $\bar{G}$ describes new type of arbitrariness that cannot be fixed by the above boundary conditions.

An origin of the arbitrariness can be clarified by the consideration of a general solution of the Schwinger equation. Since the Schwinger equation (1) is functional-differential, one can impose the problem of integrating this equation. When speaking about the integration of the Schwinger equation we will imply, by analogy with integration of ordinary differential equations, the reduction of these equations to some simplest equations in the functional derivatives. Such non-perturbative integration may be useful in studying the problems going beyond the perturbation theory.

Let us consider more general model with the nonlocal interaction

$$S_{int} = \frac{1}{2}\int dxdy\phi^*(x)\phi(x)\lambda(xy)\phi^*(y)\phi(y).$$

The local limit corresponds to $\lambda(xy) = \lambda\delta(x-y)$. This model describes also the interaction $g\phi^*\phi\chi$ of the field ϕ with a real scalar field χ ("scalar Yukawa coupling"). In the last case $\lambda(xy) = g^2 D_c(x-y)$, where D_c is a free propagator of the field χ.

We shall use notations:

$$A * B(xy) \equiv \int dx' A(xx')B(x'y), \;\; AB(xy) \equiv A(xy)B(xy).$$

In these notations Schwinger equation for the model is

$$G + \eta * \frac{\delta G}{\delta\eta} = \lambda\frac{\delta}{\delta\eta} * \frac{\delta G}{\delta\eta}. \tag{5}$$

Let us introduce the functional W

$$W(xy) = R(xy)\cdot G - \frac{\delta G}{\delta\eta(yx)}, \tag{6}$$

where R is any exact solution of the equation

$$\lambda R * \frac{\overleftarrow{\delta}}{\delta\eta} + \lambda R * R = 1 + \eta * R. \tag{7}$$

At least one such solution exists there $R(xy) = \int dx'dy'\lambda^{-1}(yy')\eta(y'x')$.

Substituting into Eq.(5) we obtain the equation for W

$$\eta * W = \lambda W * \frac{\overleftarrow{\delta}}{\delta\eta} + \lambda R * W. \tag{8}$$

Eq.(8) is linear one of the 1-st order w.r.t. the functional derivatives over the source η , which means that substitution (6) does reduce the order of Eq.(5).

Make one more substitution

$$W(xy|\eta) = V(xy|\eta)\exp\{Z[\eta]\}, \tag{9}$$

where the functional Z is normalized by the condition $Z[\eta_0] = 0$. Then from Eq.(8) we obtain for Z

$$\frac{\delta Z}{\delta\eta} = (\lambda V)^{-1} * [\eta * V - \lambda R * V - F]. \tag{10}$$

Here $F \equiv \lambda V * \overleftarrow{\frac{\delta}{\delta\eta}}$.

Eqs.(6)-(10) describe the general solution of the Schwinger equation (5): for *arbitrary* functional V we can calculate Z using Eq.(10), and consequently the functional W. Then Eq.(6) allows to calculate all the derivatives of G in the point η_0, and therefore we obtain an exact solution of Schwinger equation (5). Note that the arbitrariness in the choice of the functional V cannot be removed by the set of usual boundary conditions (such as Eq.(3)).

Simplest solution corresponds to $\delta V/\delta\eta = 0$. At the time the functional $Z = \mathcal{O}(\lambda^{-1})$ is a quadratic polynomial over η.

It is doubtful that such model represents some physical interest. Notice, however, that this simple solution is consistent with the principal order of SCE (see Eqs.(2)-(4)) in the limit $\lambda \to \infty$.

The consistency condition with some given approximant of G can be used for the construction of a new non-perturbative approximation scheme based on the above general solution of Schwinger equation [3].

Let us $W = W_0 + W'$, where the functional W_0 is defined by some approximant (e.g. perturbative one or something else) of generating functional G in correspondence with Eq.(6). Then the functional W' satisfies the inhomogeneous equation

$$\lambda W' * \overleftarrow{\frac{\delta}{\delta\eta}} + \lambda R * W' - \eta * W' = f, \tag{11}$$

where $f \equiv \eta * W_0 - \lambda W_0 * \overleftarrow{\frac{\delta}{\delta\eta}} - \lambda R * W_0$.

Any solution of Eq.(11) defines the exact solution of the Schwinger equation (5). Such solution of Schwinger equation will contain an non-perturbative information which the initial functional W_0 cannot contain. On the other hand, the functional W_0 can be considered as a boundary condition for the solution constructed.

***

I am grateful to P.A. Saponov for useful discussions. I would like to thank Prof. H.M. Fried and the International Science Foundation for making my participation in the Workshop possible.

References

1. C.M. Bender, F. Cooper and L.M. Simmons, *Phys. Rev.* **D39** (1989) 2343.
2. V.E. Rochev, *J. Phys.* **A26** (1993) 1235.
3. V.E. Rochev and P.A. Saponov, in preparation.

DIMENSIONAL TRANSITION IN GAUGE THEORIES

A. FERRANDO, A. JARAMILLO and V. VENTO
Departament de Física Teòrica and I.F.I.C. Centre Mixt Universitat de València – C.S.I.C.
E-46100 Burjassot (València), Spain.

ABSTRACT

We use pure gauge QCD defined on a strip of width L as a model interpolating between 2D and 3D QCD. We analyze the properties of the model for small widths. In particular we calculate the glueball spectrum and the string tension. Their evolution with the width L is analyzed. A confinement mechanism based on the continuity of the dimensional transition is proposed. This mechanism is supported by finite temperature lattice results.

1. Introduction

The reason why 2D Quantum Chromodynamics shares so many features with its higher dimensional counterparts is still poorly known. Confinement[1], existence of massless pseudoscalars[2] and asymptotic freedom[3] are some of its most relevant properties. The dimensionality of space-time allows the unveiling of the non-trivial mechanism behind these in an exact manner. However the constraint induced by 2D space-time kinematics is so strong that fix the limits of applicability of the previous mechanisms in many cases. In this sense the drastic reduction in dimension, the only approximation when considering QCD_2 as a model for QCD , is too extreme. There is no good reason to believe *a priori* that the relation between the 2D version of QCD and the real theory is nothing but purely coincidental.

In here we develop a continuous procedure to lower the dimensionality of a gauge theory. We start by considering QCD in 2+1 dimension. One of the spatial dimensions is considered of finite length L (*the strip*) and we study how the three dimensional theory approaches the two dimensional one as the width of the strip is made to vanish. We pay special attention to the glueball spectrum and analyze the evolution with L of our findings.

2. The glueball spectrum at small widths

Our starting point is the pure gauge QCD_3 action. We compactify the transversal degrees of freedom by requiring periodic boundary conditions on the gauge field. Periodicity implies that the gauge field can be expanded in Fourier modes, exactly in the same manner as in finite temperature QCD. The only difference comes from the fact we compactify the transversal coordinate instead of the Euclidean time. From the mathematical point of view we deal with the same technical issue. Physically though both theories represent different problems and comparisons have to be done in a careful way.

The great advantage of the previous procedure is that it yields to an effective 2D theory mimicking 3D physics[4]. After the choice of the strip version of the so called static gauge $\partial_2 A_2 = 0$, the theory can be written in terms of the gauge field Fourier modes. They are 2D fields. The resulting action is still invariant under 2D *longitudinal* gauge transformations. These transformations are compatible with the "static" gauge condition just implemented. The theory on the strip contains besides the conventional QCD_2 (2D) gluons an infinite set of covariant (non-gauge) interacting fields $(\phi, \{V_\alpha^n, \alpha = 0{,}1;\ n \in Z,\ n \neq 0\})$ belonging to the adjoint representation of the color group.

The 2D gluon field is the zero mode of the *longitudinal* components of the gauge field $(A_\alpha,\ \alpha = 0{,}1)$. The ϕ field is a massless gauge-covariant degree of freedom. It is the remnant of the gauged-away *transversal* field A_2. This scalar field ϕ couples to the 2D gluons through the covariant derivative to preserve the 2D *longitudinal* gauge invariance of the strip action. The V_n fields of the non abelian strip theory self-couple with couplings which are not arbitrary, but determined by the 3D gauge invariance, lost after the implementation of the gauge fixing condition. The 3D gauge invariance also defines originally the coupling of the V_n fields with the other covariant non-gauge particle of the reduced action, the scalar field ϕ. All these couplings are not established by the 2D gauge symmetry because neither the V_n modes nor the ϕ field are 2D gauge particles (V_n are massive). The 2D gauge symmetry, however is instrumental in fixing the couplings of the 2D gluons with the V fields through covariant derivatives.

At a given width L and coupling constant g_3 we can characterize our theory by fixing the 2D gauge coupling constant $g_2 = g_3/L^{1/2}$ and the dimensionless width $\epsilon = g_3 L^{1/2}$. As far as the effective 2D theory is concerned, ϵ acts as a dimensionless coupling constant. The non-zero mode fields V_n become infinitely heavy ($m_n \sim 1/\epsilon$) in the $\epsilon \to 0$ limit and a good description of the system can be given in terms of the lighter degrees of freedom. That is in terms of the remnant *classically* massless degrees of freedom, the 2D gluon and the scalar field ϕ. The scalar field ϕ undergoes a mass renormalization process which provides it with mass. Unlike the 2D gluons, whose masslessness is protected by the 2D gauge invariance, there is no custodial symmetry for the scalar field. The integration of the heavy modes turns out to be non-trivial. They decouple eventually. However vacuum V_n fluctuations in the form of one-loop contributions provide the scalar field ϕ with a very heavy mass for small widths ϵ's. The Higgs field in high temperature QCD gets a mass through the same mechanism[5]. As a consequence the action describing dynamics in very narrow strips (small ϵ) is given by,

$$\mathcal{S}_\epsilon = \int d^2x \left\{ \frac{1}{2} f_{\alpha\beta}^2 + (D_\alpha \phi)^2 + \mu_\phi^2(\epsilon)\phi^2 \right\}, \quad \mu_\phi^2(\epsilon) = \frac{g_2^2 N}{\pi} + \frac{g_2^2 N}{2\pi} \ln\left(\frac{1}{\epsilon^2}\right) \qquad (1)$$

We can take advantage of the fact that the scalar field is very heavy to give a non-relativistic (NR) treatment to the action (1). Our interest lies in the calculation of (very massive $\sim \mu_\phi$) scalar bound states which do not show up in the $\epsilon \to 0$ limit (pure QCD_2). A systematic application of the $1/\mu_\phi$ expansion yields to a considerably simplification of the bound states formalism. The bound state mass equation is nothing but a one dimensional Schrödinger equation for a $\phi_{(1)}\phi_{(2)}$ system interacting through a color gauge field. All the contributions are suppressed by powers of the scalar mass with the exception of the linearly confining OGE (One Gluon Exchange) potential. There also exists a dressing of the ϕ mass by the gluon interaction giving rise to a renormalized scalar mass ($\mu_R^2(\epsilon) = [g_2^2 N/4\pi] \ln[1/\epsilon^2]$)[5]. The 1D Schrödinger equation for a linear potential has an exact solution in terms of the Airy function[6]. The glueball spectrum is given by the eigenvalues of the bound state equation in terms of the zeroes of the Airy function $-\varepsilon_n$ and the $\phi\phi$ string tension $\sigma_\phi = \frac{1}{2} g_2^2 C_N$ (C_N is a color invariant constant)

$$E_n(\epsilon) = 2\mu_R(\epsilon) + \varepsilon_n \left[\frac{\sigma_\phi^2}{\mu_R(\epsilon)} \right]^{1/3} \tag{2}$$

3. Confinement mechanism

The theory on a narrow strip is still confining. The Wilson loop between two static charges is blind to the presence of the scalar field. The external static charges do not couple to this degree of freedom. Thus the Wilson loop is ruled by a pure 2D gluonic action. The integration of the non-abelian scalar field in Eq. (1) influences the 2D gauge dynamics by dressing the kinetical gluon term. This is nothing but a gluon wave function renormalization. In 2D the latter can be likewise interpreted as a dressing of the 2D coupling constant (g_2 renormalization). The dressed gluon kinetical term ($\sim f_{\alpha\beta}^2$) rules the long distance behavior of the Wilson loop since is the lowest order gauge invariant gluon operator in the derivative expansion $-\Delta/g_2^2 \ll 1$. Confinement is thus linear because is identical to that of pure gauge QCD_2. The only difference comes from the appearance of a ϵ-dependent $q\bar{q}$ string tension arising from the renormalization of the 2D coupling constant,

$$\sigma_q(\epsilon) = \frac{6\, g_2^2\, c_N\, N\, \log(\epsilon^{-2})}{1 + 12\, N\, \log(\epsilon^{-2})} \tag{3}$$

The 2D gluonic long distance dominance in the Wilson loop is also present at any finite width[4]. The role of the transversal fields, V_n and ϕ, is to dress the *longitudinal* effective interaction driven by the 2D gluons. The string tension can be obtained from the 2D gluon effective action just by looking at the constant ($Z(\epsilon)$) dressing the gluon kinetic term ($f_{\alpha\beta}^2$). There is a simple relation between them, $\sigma_q(\epsilon) = (1/2) Z(\epsilon) c_N$.

In this context the string tension given by Eq. (3) is just the leading order approximation in a perturbative expansion in the dimensionless width ϵ. Exactly the same happens to the glueball spectrum previously presented Eq. (2). Both equations show a flow in the dimensionless width ϵ. This flow goes beyond the perturbative regime when $\epsilon \geq 1$.

The prevalence of 2D gluon dynamics at long distances on the strip ensures that the dominant contribution to the static $q\bar{q}$ energy is linear $\sigma_q(\epsilon)R$. However this is not a guarantee that confinement will survive in the infinite width limit. Not even at finite L. The evolution of the string tension flow above $\epsilon=1$ is not reachable from perturbation theory. The string tension can vanish at same value of L spoiling the linear confining behavior in favor of some other non-leading contribution.

Certainly the behavior of the string tension in QCD on the strip is highly non-trivial, as the low ϵ result Eq. (3) already exhibits . Even though we can outline the behavior of the ϵ flux beyond the perturbative regime. This will define a mechanism describing the dimensional transition between 2D and 3D gauge QCD.

The mechanism we propose here is the following:

1. The flow of the string tension with ϵ is continuous and different from zero. That is, one can extrapolate the perturbative result in Eq. (3) to the $\epsilon\geq 1$ regime in a continuous way providing a non-zero value for the QCD_3 string tension in the $\epsilon\rightarrow\infty$ limit.

2. There is a stabilization of the ϵ flow at a critical width L^*. The string tension and the glueball spectrum tend quickly to a constant , their QCD_3 value, above the critical point .

There is a strong evidence supporting the previous mechanism arising from finite temperature lattice calculations[7]. The way to connect the finite temperature formalism with that of the strip is through a Euclidean rotation (R). We can describe confinement by means of Wilson loops. The rotated standard finite T Wilson loop is the strip *transversal* loop, given in terms of the A_2 zero mode ϕ ($[A_2]_S = R\,[A_0]_{FT}R^{\top}$). The real timelike Wilson loop on the strip arises from the rotation of the FT *spatial* Wilson loop ($[A_0]_S = R\,[A_1]_{FT}R^{\top}$). According to this we can find a simple relation between timelike (σ_0) and spacelike (σ_α) string tensions in both formalisms, $\sigma_2^S(\epsilon)=\sigma_0^{FT}(T)$, $\sigma_0^S(\epsilon)=\sigma_1^{FT}(T)$. It is known in FT gauge theories that the spatial Wilson loop is *not* an order parameter for confinement. There is no reason for the spatial string tension to be different from zero even in the deconfined phase. Unlike $\sigma_0^{FT}(T)$, the real order parameter, no discontinuity in the first derivative of the spatial string tension is expected (no energy gap).Consistency requires the spatial and the time string tension to be the same at $T=0$. Thus at low temperatures, confinement phase, we expect $\sigma_1^{FT}(T)$ to be different from zero. At high temperatures we can resort to our calculation of $\sigma_0^S(\epsilon)$ on the strip for small ϵ and to the strip-FT equivalence to show that the spatial string tension *grows* at high temperatures.

This qualitative picture, continuity and non-zero value of the spatial string tension for all T, is completely confirmed by the lattice simulation. Interpreted now in terms of the real string tension on the strip, $\sigma_0^S(\epsilon)$, the lattice result gives a solid support to the point 1. of the dimensional transition mechanism.

Lattice results seems to corroborate also a second feature. The spatial string tension reveals a plateau below the phase transition point* The spatial string tension

*There also results of the same sort for glueball bound states.

is flat in this region and gets stabilized reaching its zero temperature value. This constant behavior for $T<T_c$ indicates the existence of a discontinuity in the high order derivatives of the spatial string tension (non-analyticity) at T_c. When interpreted on the strip frame, this behavior shows up as a stabilization of the ordinary string tension above a critical length $L^* \sim 1/T_c$. This gives an evidence for the point 2. of the dimensional transition mechanism. The critical length L^* has a nice interpretation in the strip framework. It can be understood mathematically as an IR fixed point of the scaling transformations of the strip width . The string tension holds constant after reaching the critical value L^*. Further scaling transformation do not modify its value. Physically the interpretation of L^* as the flux tube width arises in a natural way. If physics is frozen at L^*, it means that increasing the strip width, and consequently enlarging the gauge field configurations out of the critical strip of with L^*, has no effect on observables. Thus the only relevant configurations to the functional generator are confined in the critical strip. The rest can be neglected. Therefore even though we enlarge artificially our strip width, the theory knows that a *dynamically* generated width exists. For $L > L^*$ there exists a physical mass scale $1/L^*$. surviving even to the infinite width limit $L \to \infty$.

4. Acknowledgments

One of us (A.F.) thanks H.J. Fried and B. Müller by their kindness in the organization of the Workshop and their presence at 10 P.M. seminars.

5. References

1. G. 't Hooft, *Nucl. Phys.* **B72** (1974) 461; **B75** (1974) 461.
   M.B. Einhorn, *Phys. Rev.* **D14** (1976) 3451.
   A. Ferrando and V. Vento, *Phys. Rev.* **D49** (1994) 3044.
2. D. Amati, Kuang-Cho Chou, S. Yankielowicz, *Phys. Lett.* **110B** (1982) 309.
   W. Buchmüller, S.T. Love, R.D. Peccei, *Phys. Lett.* **108B** (1982) 426.
3. A. Ferrando and A. Jaramillo, *University of Valencia Preprint* FTUV/94-41, IFIC/94-36.
4. A. Ferrando, A. Jaramillo, V. Vento, work in progress.
5. E. D'Hoker, *Nucl. Phys.* **B201** (1982) 401.
6. A. Galindo, P. Pascual, "Mecánica Cuántica". Ed. Alhambra 1978.
   M. Abramovitz and I.A. Stegun (ed.), *Handbook of Mathematical Functions with Formulas, Graphs and Mathematical Tables.* Dover Pub. Inc. New York (1964).
7. M. Teper, *Phys. Lett.* **B311** (1993) 223 ;
   G.S. Bali, J. Fingberg, U.M. Heller, F. Karsch, K. Schilling, *Wuppertal Preprint* HLRZ-93-43, June 1993.
   M. Caselle, R. Fiore, F. Gliozzi, P. Guaita and S. Vinti, *Nucl. Phys.* **B422** (1994) 397.

Multiple solutions in Quantum Field Theory

Santiago García and G.S. Guralnik
Department of Physics, Brown University,
Providence RI 02912, USA

ABSTRACT

A quantum field theory can be defined by the path integral of an action. With the usual convention of measure and contour, the value of this integral constitutes one solution of the corresponding canonical theory. Spontaneous symmetry breaking, when allowed, can be described using a path integral formulation and evaluating the same integral over a different contour in the complex function space plane. This is often (incompletely) simulated by changing the sign of the bare mass in the action forming the integrand.

In addition to these two solutions, the possibility remains for additional solutions depending on the structure of the interaction. This becomes evident by examining the Schwinger-Dyson equations with external sources for the fields. If, for example, a pure quartic scalar interaction is considered, the partial differential equations are cubic in source derivatives. It should be expected that these have (at least) three solutions. We show these can be classified as follows: One solution is the usual *path integral* evaluated on the real field line. This solution is linked to the weak coupling perturbation series. The second and third (real) solutions are obtained by evaluating the path integral over appropriate complex contours. Both of these can be associated with non-vanishing odd Green's functions. One has an asymptotic expansion corresponding to the loop expansion while the other solution has an essential singularity for vanishing coupling. Both of these solutions exist regardless of the sign of the bare mass squared. In this note, we outline the *other* solutions to full Φ^4 using the ultralocal approximation to display some detailed results. Our conclusions about multiple solutions apply to an arbitrary number of space time dimensions and can be stated in differential or path integral form.

1. Introduction

While Monte Carlo methods have been very successful, there are still problems in simulating Quantum Field Theory on a lattice, especially for fermionic theories. Our group at Brown University (S. García, G. Guralnik, S. Hahn, J. Lawson, and K. Platt) is constructing an alternative approach to numerical QFT that is not based on the computation of a path integral. This method treats bosons and fermions in a relatively symmetric manner and, as a consequence, avoids many of the problems with fermionic systems associated with traditional numerical methods. Our first results have been published elsewhere.[1]

The starting point in our formalism is the functional differential equation satisfied by the generating functional $Z[J]$.

$$\left(\left. \frac{\delta S}{\delta \Phi(x)} \right|_{op} - J(x) \right) Z[J] = 0, \quad \Phi(x) \mapsto \frac{\delta}{\delta J(x)} \tag{1}$$

For an Euclidean Φ^4 theory with action

$$S = \int d^D x \left(\frac{1}{2} \partial^\mu \Phi(x) \partial_\mu \Phi(x) + \frac{1}{2} M^2 \Phi^2(x) + \frac{g}{4} \Phi^4(x) - \Phi(x) J(x) \right) \tag{2}$$

we have, in the continuum,

$$\left(-\Box + M^2 \right) \frac{\delta Z[J]}{\delta J(x)} + g \frac{\delta^3 Z[J]}{\delta J^3(x)} - J(x) Z[J] = 0. \tag{3}$$

Any solution of this equation (constrained by equations (10)) is a solution to the canonical quantum field theory which follows from equation (2). These equations can be rewritten using a (hypercubic, periodic) D-dimensional space time lattice

$$\left(-\Box + M^2 \right)_{x,y} \frac{\partial Z}{\partial J_y} + g \frac{\partial^3 Z}{\partial J_x^3} - J_x Z = 0 \tag{4}$$

Here, we have used the representation

$$\left(-\Box + M^2 \right)_{x,y} \equiv \left(2D + M^2 \right) \delta_{x,y} - \sum_{\hat{\mu}} \delta_{y,x+\hat{\mu}}. \tag{5}$$

The μ summation is taken over nearest neighbors. Even though there is one equation per lattice point, we need only consider all equations for just one point since all other points are equivalent for solutions invariant under the lattice symmetry group. Even with this observation the problem still is very complex since there remain as many variables as sites.

In this note, we study solutions to these equations. We use our knowledge of a simple model (the ultralocal model[2]) to separate space-time and source dependence and obtain useful analytic representations. We illustrate some of our general formalism with specific ultralocal results.

All lattice symmetric solutions to equations (4) can be written as

$$Z(J) = e^{\xi} \prod_x Z^{(ul)}(J_x) \qquad \xi \equiv \frac{1}{2} \sum_{x,y} \Box_{x,y} \frac{\partial^2}{\partial J_x \partial J_y} \tag{6}$$

The function $Z^{(ul)}(J_x)$, is the same for all lattice points, and satisfies the third order ordinary differential equation

$$\left(g \frac{d^3}{dJ^3} + M^2 \frac{d}{dJ} - J \right) Z(J) = 0. \tag{7}$$

$Z^{(ul)}$ is called the ultralocal generating function.

2. Basis for ultralocal Φ^4

It is clear, from equation (6), that knowledge of the solutions of the ordinary differential equation (7) provides useful information about the lattice symmetric solutions of the partial differential equations (4).

We can easily find a basis for the solution space of equation (7) by a Laplace integral representation. The complex contours are determined[6] by the condition $\mathrm{Re}(g\Phi^4) > 0$, where Φ is the integration variable and the coupling g is positive. We obtain

$$Z_1^* = \int_{-\infty}^{+\infty} d\Phi\, e^{-A^+ + J\Phi}$$
$$Z_2^* = \int_{-\infty}^{+\infty} d\Phi\, e^{-A^- + iJ\Phi} \tag{8}$$

and

$$Z_3^* = \int_0^{\infty} d\Phi\, e^{-A^+ - J\Phi} - \int_0^{\infty} d\Phi\, e^{-A^-} \sin(J\Phi) \tag{9}$$

where

$$A^{\pm} \equiv \frac{g}{4}\Phi^4 \pm \frac{1}{2}M^2\Phi^2$$

The basis function $Z_1^* = Z_1^*(M^2, g; J)$ is even in J, and clearly corresponds to the conventional path integral solution of Φ^4. $Z_2^*(-M^2, g; J)$ has the same moments (*Green's functions*) as $Z_1^*(M^2, g; J)$ except for minus sign factors.* Notice that only Z_3^* generates odd Green's functions, which suggests that is extension to the full D-dimensional Φ^4 theory connects with symmetry breaking solutions.

The general solution of (7) is then

$$Z^{(ul)} = aZ_1^* + bZ_2^* + cZ_3^* \tag{10}$$

where, in principle, a, b, c could still be arbitrary functions of g and M^2. However, the equations

$$\frac{\partial Z^{(ul)}}{\partial M^2} = -\frac{1}{2}\frac{\partial^2 Z^{(ul)}}{\partial J^2} \qquad \frac{\partial Z^{(ul)}}{\partial g} = -\frac{1}{4}\frac{\partial^4 Z^{(ul)}}{\partial J^4} \tag{11}$$

required by the Schwinger action principle, restrict a, b, c to be constants. This leads, using equation (6) and lattice symmetry, to the following tri-parametric general lattice symmetric solution of equations (4)

$$Z_{a,b,c}(J) = e^{\xi} \prod_x (aZ_1^*(J_x) + bZ_2^*(J_x) + cZ_3^*(J_x)). \tag{12}$$

The conclusion that the theory is only specified up to a three arbitrary numbers is confirmed by direct analysis of the set of coupled linear differential equations given by equation (4). Complete specification of the boundary conditions (this is usually implicit) fixes these numbers.

* $Z_2^*(-M^2, G; J)$ is called *characteristic function*, already discussed in R.P. Feynman and A.R. Hibbs, *Quantum Mechanics and Path Integrals*, (New York, N.Y., 1965).

3. Truncation, path integral and existence of the weak coupling expansion.

It has been noted by several authors[2,3] that truncation of the Schwinger-Dyson equations leads to a solution analytic in the coupling g for positive M^2. The analogue to the Schwinger-Dyson equations for the ultralocal model are the recursion relations obtained when an infinite series

$$Z_\infty(J) = \sum_{k=0}^{\infty} \frac{A_k}{k!} J^k \tag{13}$$

is substituted in equation (7)

$$g A_{k+3} + M^2 A_{k+1} - k A_{k-1} = 0 \qquad k = 0, 1, 2 \ldots \tag{14}$$

It is straightforward to check[6,3] that truncation of the above equations for $M^2 > 0$ forces all odd *Green's functions* A_{2k+1} to vanish, whereas the even Green's functions can be computed from a Stieltjes continued fraction.

$$R_k = \cfrac{1}{1 + \cfrac{(2k+1)\lambda}{1 + \cfrac{(2k+3)\lambda}{1 + \cfrac{(2k+5)\lambda}{1 + \ldots}}}} \tag{15}$$

Here we have defined

$$b_n = \left(\frac{1}{M^2}\right)^n A_{2n}, \quad \lambda = \frac{g}{M^4} \qquad\qquad R_n \equiv \frac{1}{2n+1} \frac{b_{n+1}}{b_n} \tag{16}$$

Note that $R_k \in [0, 1]$ and that the continued fraction converges for all values of λ in the complex plane cut at $[-\infty, 0]$. In terms of parabolic cylinder functions[5]

$$R_n = t \frac{U(n,t)}{U(n-1,t)} \qquad \lambda = \frac{1}{2t^2} \quad , \qquad n = 1, 2, 3, \ldots \tag{17}$$

Using integral representations for $U(n,t)$, we find:

$$A_{2n} = \frac{\int_{-\infty}^{+\infty} d\Phi\, \Phi^{2n}\, e^{-(\frac{1}{2} M^2 \Phi^2 + \frac{1}{4} g \Phi^4)}}{\int_{-\infty}^{+\infty} d\Phi\, e^{-(\frac{1}{2} M^2 \Phi^2 + \frac{1}{4} g \Phi^4)}} \tag{18}$$

These are the usual Green's functions for the ultralocal *path integral* solution Z_1^*. When these are extended to arbitrary dimension through equation (12), they form the usual path integral solution for quartic scalar quantum field theory.

4. The *other* solutions

Of the three independent solutions of equation (7), only one, Z_1^*, has a well defined *free field limit* $g \to 0^+$. The other two solutions diverge in this limit, and hence cannot be obtained by successive truncation of a polynomial series. The Source Galerkin method[6] is suitable to obtain approximations to these other solutions even for the case of arbitrary dimensional Φ^4 theory.

For the special case of the ultralocal model, the recursion relations (14) together with the canonical equations (11) suffice to uniquely characterize the solutions[2]. The most general expression for the expectation of the *field* Φ is

$$\left.\frac{\partial \log Z}{\partial J}\right|_{J=0} = \frac{A_1}{A_0} \equiv \langle\Phi\rangle = \frac{\alpha}{M}\frac{e^{\frac{1}{4}t^2}t^{1/2}}{U(0,t)+\rho V(0,t)}\,,, \quad t \equiv \frac{1}{\sqrt{2\lambda}} \tag{19}$$

Where U and V are parabolic cylinder functions. For the limit

$$\alpha = \frac{2}{\sqrt{\pi}}\rho \qquad\qquad \rho \to \infty, \tag{20}$$

we obtain the usual *double well* solution, with the leading order obtained when the field is initially at the minimum of the classical potential ($M^2 > 0$)

$$V_c(\Phi) = -\frac{1}{2}M^2\,\Phi^2 + \frac{g}{4}\Phi^4 \tag{21}$$

$$\langle\Phi\rangle = \frac{1}{M}\frac{2}{\sqrt{\pi}}\frac{e^{\frac{1}{4}t^2}t^{1/2}}{V(0,t)} \simeq \frac{1}{M}\frac{1}{\sqrt{\lambda}}(1-\frac{3}{4}\lambda-\frac{87}{32}\lambda^2-\frac{2889}{128}\lambda^3-\ldots) \qquad \lambda \simeq 0^+ \tag{22}$$

The asymptotic expansion of $\langle\Phi\rangle$ when $\lambda \to 0^+$, for the above parameter values, agrees order by order with the solutions obtained using the loop expansion. It is important to emphasize that this result is obtained for positive bare mass squared in the action. A similar solution can be obtained with the opposite choice of sign. This is not a contradiction to usual results but a reflection of a more careful and general analysis. This result can, of course, also be derived from integral representations, equations (8) and (9). We find

$$\langle\Phi\rangle = \frac{\int_0^\infty d\Phi\,\Phi\, e^{-\frac{1}{4}g\Phi^4}\left(e^{-\frac{1}{2}M^2\Phi^2} + e^{+\frac{1}{2}M^2\Phi^2}\right)}{\int_0^\infty d\Phi\, e^{-\frac{1}{4}g\Phi^4+\frac{1}{2}M^2\Phi^2}} \tag{23}$$

This corresponds to the expectation value obtained from $Z_3^*(J)/Z_2^*(J=0)$. Note that $\langle\Phi\rangle$ has contributions from $+M^2$ **and**$-M^2$. This is in contrast with what a usual numerical simulation of *symmetry breaking* would try to compute

$$\langle\Phi\rangle_{numer} \approx \frac{\int_{-\infty}^\infty d\Phi\,|\Phi|\, e^{-\frac{1}{4}g\Phi^4+\frac{1}{2}M^2\Phi^2}}{\int_{-\infty}^\infty d\Phi\, e^{-\frac{1}{4}g\Phi^4+\frac{1}{2}M^2\Phi^2}} \tag{24}$$

This quantity only agrees with (22) for $M^2 \to \infty$,that is, for very deep potential wells. Moreover, it does not possess an asymptotic series in g around the origin (thus disagreeing with the loop expansion solution) because of an essential singularity at $g=0$. Indeed, equation (24) is not even a solution of the original field equations.

The other interesting solution[4] is obtained with the parameter values

$$\alpha = \sqrt{2}, \rho = 0 \tag{25}$$

which corresponds to a pure Z_3^* solution. We find, for the expectation value of the field

$$\langle \Phi \rangle = \frac{1}{M}\sqrt{2}\,\frac{e^{\frac{1}{4}t^2} t^{1/2}}{U(0,t)} \qquad\qquad t = 1/\sqrt{2g} \tag{26}$$

with the integral representation

$$\langle \Phi \rangle = \frac{\int_0^\infty d\Phi\, \Phi\, e^{-\frac{1}{4}g\Phi^4}\left(e^{-\frac{1}{2}M^2\Phi^2} + e^{+\frac{1}{2}M^2\Phi^2}\right)}{\int_0^\infty d\Phi\, e^{-\frac{1}{4}g\Phi^4 - \frac{1}{2}M^2\Phi^2}} \tag{27}$$

Notice that the above expression corresponds fo the formal substitution $M^2 \to -M^2$ in the previously found equation (23). The existence of this solution is a consequence of a symmetry of the ultralocal Schwinger-Dyson equations (14). The expression (27) has an essential singularity at $\lambda = 0$

$$\langle \Phi \rangle \simeq \frac{1}{M}\frac{e^{\frac{1}{4\lambda}}}{\sqrt{\lambda}}\left(1 + \frac{3}{4}\lambda - \frac{87}{32}\lambda^2 + \frac{2889}{128}\lambda^3 - \ldots\right) \qquad\qquad \lambda \to 0^+ \tag{28}$$

Notice also that the **even** Green's functions of this solution are identical to the even Green's functions of the usual path integral solution Z_1^*.

5. Conclusions

We have explicitly shown that there are solutions different than the usual path integral in Quantum Field Theory. A differential formulation, with appropriate boundary conditions, or a less restrictive definition of path integral allows the study of all solutions to the canonical field theory. In particular, we find a new class of solutions which is highly singular at the origin of the bare coupling. These solutions may affect our understanding of the triviality of Φ^4 and mass bounds of the Higgs boson.

Restrictions on the parameters imposed by consistency of a particular expansion method (for instance, $M^2 < 0$ for the loop expansion) are characteristic of the expansion itself, not general properties of a given solution. We have checked above, for the ultralocal model, that solutions with odd Green's functions exist for positive g and either sign of M^2. We do not know of any reason to believe that this is not also possible in the full Φ^4 theory.

In higher dimensions, a purely numerical approach is necessary. The Source Galerkin method, presented elsewhere,[1] is an adequate framework for these calculations.

Acknowledgements This work has been supported in part by NSF Grant ASC-9211072 and DOE Grant DE-FG02-91ER40688—Task D. We thank G. Kilcup, S. Hahn, K. Platt and many colleagues for useful discussions and input in this work. Jim O'Dell of Fort Pond Research provided a site license for the symbolic program ALJABR and considerable help in its application to our ideas.

References

1. S. García, G. Guralnik, and J. Lawson, *Phys. Lett.* **B322** 119 (1994).
2. E.R. Caianello and G.Scarpetta, *Nuovo Cimento* **22A** 454 (1974).
3. C.M. Bender, F. Cooper and L.M. Simmons, Jr, *Phys. Rev.* **D39** 2343 (1989).
4. G. Scarpetta and G. Vilasi, *Nuovo Cimento* **28A** 62 (1975); S. De Filippo and G. Scarpetta, *Nuovo Cimento* **50A** 305 (1979).
5. M. Abramowitz and I. Stegun, *Handbook of Mathematical Functions*, Appl. Math. Series, Vol. 55, Nat. Bureau of Standards (Washington), reprinted by Dover (New York), 1968.
6. S. García and G. Guralnik, to be published.

Infrared Quantum Fluctuations and the Cosmological Constant Problem

Emil Mottola
Los Alamos National Laboratory
Theoretical Div., T-8
Mail Stop B285
Los Alamos, NM 87545 USA

Electronic Mail: emil@pion.lanl.gov

ABSTRACT

Vacuum fluctuations are an essential feature of quantum field theory. Yet, the smallness of the scalar curvature of our universe suggests that the zero-point energy associated with these fluctuations does not curve spacetime. A possible way out of this paradox is suggested by the fact that microscopic fluctuations are generally accompanied by dissipative behavior in macroscopic systems. The intimate relation between the two is expressed by a fluctuation-dissipation theorem which extends to general relativity. The connection between quantum fluctuations and dissipation suggests a mechanism for the conversion of coherent stresses in the curvature of space into ordinary matter or radiation, thereby relaxing the effective cosmological "constant" to zero over very large regions of spacetime.

In classical general relativity, the requirement that the field equations involve no more than two derivatives of the metric tensor allows for the possible addition of a constant term, the cosmological term, to the Einstein equations:

$$R^a_{\ b} - \frac{1}{2}R\delta^a_{\ b} + \Lambda\delta^a_{\ b} = \kappa T^a_{\ b}. \tag{1}$$

If transposed to the right side of the equation this term corresponds to a constant energy density $\frac{\Lambda}{\kappa}$ and isotropic pressure $-\frac{\Lambda}{\kappa}$ permeating all of space uniformly, and independently of any localized matter sources. Hence, even if $T^a_{\ b} = 0$, a cosmological term causes spacetime to become curved with a radius of curvature of order $|\Lambda|^{-\frac{1}{2}}$. From the mere fact that the universe is at least 10^{27} cm. large (and perhaps a great deal larger), we immediately infer an extremely stringent bound on Λ from observations, *viz.*

$$|\Lambda_{obs}| < 10^{-54}\ \text{cm}^{-2} \approx 2 \times 10^{-118}(\hbar\kappa)^{-1}. \tag{2}$$

The fact that this limit is some 118 orders of magnitude smaller than unity in the Planck units, $(\hbar\kappa)^{-1}$ is often called the "cosmological constant problem" (although as I shall argue this name is somewhat misleading).

In purely classical physics there is no natural scale for Λ, and it is perfectly consistent to set it equal to zero. In quantum theory this is no longer the case, because each mode of a field theory has zero point fluctuations which contribute $\frac{1}{2}\hbar\omega$ to the vacuum energy, and the sum over modes is quartically divergent. Renormalization is required, which means that the value of Λ is a free parameter of the quantum theory, which cannot be set equal to zero without fine tuning. Fine tuning is a technically precise term, in the sense that there is no symmetry forbidding the generation of a Λ term of order unity in Planck units. Supersymmetry (or superstrings) would forbid such a Λ term only if the supersymmetry is unbroken. Since it must be broken to agree with particle physics as we know it, the problem of fine tuning remains. As Steven Weinberg has remarked: "Just because something is infinite, does not necessarily mean that it's zero."[1] Clearly, something rather fundamental is missing from our understanding of gravitation, if we cannot explain why the apparent value of the energy of the vacuum is zero in Planck units to one part in 10^{118}.

At accessible distance scales, from centimeters to light years, gravitational phenomena are described quite accurately by the classical Einstein theory. It is very well known that this theory is beset with ultraviolet divergences at the quantum level and is virtually certain to require drastic modification at the Planck scale. A corollary of this severe behavior at ultra-short distances would seem to be mild behavior in the infrared. One does not normally think of quantum fluctuations of the metric field as important at large distances, and indeed in perturbation theory around flat space there is no sign of infrared problems. However, gravitation is a long range force characterized by massless excitations that cannot be shielded. If the classical spacetime is curved on some characteristic distance scale ℓ, fluctuations with wavelengths longer than ℓ need not remain small, a fact long ago pointed out in Newtonian theory by Jeans. Consequently, infrared behavior of fluctuations depends very much on the background geometry, and may force large modifications of the classical background.

The quantum analog of the classical background is the ground state. Prior to and independent of the question of how to tame the ultraviolet divergences bedeviling the quantum theory, the question of what is its correct ground state presents itself. This issue is one that intrinsically involves infrared properties, where we may hope to say something sensible in the language of low energy effective lagrangians, without knowledge of Planck scale phenomena.

The point is brought home most forcefully if one considers a theory with symmetry breaking at a modest energy scale, such as the Weinberg-Salam theory of electroweak interactions. The energy density difference between the symmetric vacuum of the theory, in which it finds itself in the very early universe at temperatures above 1TeV, and the symmetry broken vacuum of the present epoch is of order $1\text{TeV}^4 \approx 10\text{cm}^{-2}$, or some 55 orders of magnitude above the observational bound in eq. (2). Yet the scale of electroweak symmetry breaking 1TeV is still 16 orders of magnitude from the Planck scale. If we cutoff the standard electroweak theory at around 1 Tev (which may be necessary in any case due to the triviality of the Higgs sector) then all phenomena take place at distance scales 16 orders of magnitude larger than the Planck scale. Any mechanism proposed to cancel the energy stored in the electroweak vacuum must work in the range of distance scales where it would be quite surprising if Planck scale physics had not decoupled and become completely irrelevant. Yet this mechanism must be incredibly efficient, since if only a few parts in 10^{55} of the latent heat in the electroweak phase transition did not convert to ordinary matter and radiation but remained as vacuum energy density today, then it would exceed the observational bound. If the universe passed through several such phase transitions at still higher temperatures, the successive fine tunings at very different energy and length scales becomes even more incredible.

Aside from the difficulties that this fine tuning thrusts upon theoretical cosmology and in particular, the inflationary universe scenario, we have direct evidence in the laboratory of the existence of zero point fluctuations and the stresses they cause, in the Casimir effect between two conducting layers[2]. Of course, every success of quantum theory at the microscopic level is indirect evidence for the entire framework, which requires that the problem of vacuum energy be taken seriously.

Since this vacuum energy problem arises at distance scales very much larger than the Planck scale, all the way up to cosmological distance scales, its solution must be independent of the ultraviolet cutoff in general and the physics of the Planck scale in particular. Instead, it is the physics of the correct vacuum state of quantum gravity that is really the issue here, and this is controlled not by ultra high-energy Planckian phenomena but by the long-distance or *infrared* properties of quantum gravity.

An analogy to the vacuum energy problem exists also in electromagnetism. I will call it the "cosmological electric field problem." It consists of the elementary observation that Maxwell's equations *in vacuo* admit a solution with constant, uniform electric field $\mathbf{E}_{cosm}$ of arbitrary magnitude and direction. Why then are all electric fields found in nature always associated with localized electrically charged sources? Why do we not observe some huge uniform electric field pointing in the direction of the magellanic clouds, for example? Classically, we just set $\mathbf{E}_{cosm} = 0$ by a boundary condition, but this explains nothing. Is there a deeper reason in the quantum theory?

The answer is almost certainly yes, because of vacuum polarization effects. Suppose that $\mathbf{E}_{cosm}$ were not zero. Then the quantum vacuum of any charged matter fields interacting with it is polarized. This is described by a polarization tensor,

$$\Pi^{ab}(x, x'; \mathbf{E}) = i\langle T j^a(x) j^b(x')\rangle_{\mathbf{E}}, \tag{3}$$

where j^a is the charge current operator and the expectation value is evaluated in a vacuum state with classical background field $\mathbf{E}$. The time ordering symbol $\mathcal{T}$ enforces the Feynman boundary conditions on the polarization operator, *i. e.* positive frequency modes propagate forward in time, while negative frequency modes propagate backward in time. The polarization operator Π^{ab} contains two pieces, an even and an odd piece under time reversal. The even piece describes the polarizability of the vacuum, since the vacuum fluctuations of virtual charge pairs may be thought of as giving rise to an effective polarizability of the vacuum (dielectric constant). The time reversal odd piece describes the creation of *real* particle anti-particle pairs from the vacuum by the Schwinger mechanism[3]. The creation of real charged pairs means that a real current flows $\mathbf{j} \geq 0$ in the direction of the electric field at later times, even if none existed initially. This implies that the electric field does work at a rate, $\mathbf{j} \cdot \mathbf{E}$. To the extent that this power cannot be recovered because the created particles interact and lose the coherence they may have had in their initial state, this is the rate of energy *dissipation*. The vacuum behaves like a normal conductor with a finite conductivity and resistivity, due to the random, uncorrelated motions of its fluctuating charge carriers. At the same time, the electric field is diminished by the Maxwell equation,

$$\frac{\partial \mathbf{E}}{\partial t} = -\langle \mathbf{j} \rangle, \tag{4}$$

in the case of exact spatial homogeneity of the average current.

An important question is that of the time scale of the dissipation. Does the degradation of the coherent electric field take place rapidly enough to effectively explain why there is no observed $\mathbf{E}_{cosm}$ today? In this connection it must be remembered that the original Schwinger calculation of the decay rate of the vacuum into charged pairs involves a tunneling factor, $\exp(-\pi m^2/eE\hbar)$ for the creation of the first pair from the vacuum. If some charged matter is present initially, the charges are accelerated, radiate, and pair produce without any tunneling suppression factor. Hence one should expect that the time scale for an *induced* cascade of particle pairs to develop and degrade the electric field energy will be very much faster than the Schwinger spontaneous vacuum rate.

The physics of this electromagnetic analogy is quite clear. It is that of a giant capacitor discharging. Any cosmological electric field initially present in the universe eventually shorts itself out, and degrades to zero, when the vacuum polarization effects described by Π^{ab} are taken into account. The real and imaginary parts of Π^{ab} are related by a Kramers-Kronig dispersion relation which is one form of a fluctuation-dissipation theorem for the electrically polarized quantum vacuum. It is simply a consequence of causality, and the existence of a positive Hilbert space of quantum states that quantum vacuum fluctuations and quantum vacuum dissipation are inseparably related. One necessarily implies the other. This is the modern, relativistically covariant formulation for quantum fields of the relation between an equilibrium quantity (mean square displacement) and a time asymmetric dissipative quantity (diffusion coefficient or viscosity) first discussed by Einstein in his theory of Brownian movement[4]. That the "vacuum" of quantum field theory can be described in terms closely parallel to that of a fluid medium has been emphasized by B. Dewitt[5], D. Sciama[6], and by B. L. Hu and coworkers[7].

These observations are of a very general nature, and apply equally well to vacuum fluctuations in a non-abelian gauge field theories or gravitation. Since the geometry of spacetime couples to the energy-momentum stress tensor, it is the fluctuations in this quantity which govern the dynamics of the gravitational field, and we are led to consider the corresponding polarization tensor,

$$\Pi^{abcd}(x,x';\bar{g}) = i\langle TT^{ab}(x)T^{cd}(x')\rangle_{\bar{g}}, \tag{5}$$

where $\bar{g}_{ab}(x)$ represents some classical background metric. This polarization tensor may be handled by exactly the same techniques as (3), and the analogous fluctuation-dissipation theorem relating its real and imaginary parts may be proven. If the background $\bar{g}_{ab}(x)$ possesses a timelike Killing field, and therefore a Euclidean continuation with periodicity β, it is natural to introduce the Fourier transform with respect to the corresponding static coordinate time difference $t-t'$. Defining the Fourier transforms of the symmetric and anti-symmetric parts of (5) by:

$$\begin{aligned} \int_{-\infty}^{\infty} dt\ \langle\{T^{ab}(x),T^{cd}(x')\}_{+}\rangle e^{i\omega(t-t')} &= S^{abcd}(\mathbf{r},\mathbf{r}';\omega), \\ \int_{-\infty}^{\infty} dt\ \langle[T^{ab}(x),T^{cd}(x')]_{-}\rangle e^{i\omega(t-t')} &= D^{abcd}(\mathbf{r},\mathbf{r}';\omega), \end{aligned} \tag{6}$$

we find that the two pieces are related via[8]:

$$D^{abcd}(\mathbf{r},\mathbf{r}';\omega) = \tanh(\frac{\beta\omega}{2})\ S^{abcd}(\mathbf{r},\mathbf{r}';\omega). \tag{7}$$

One such background spacetime is the Schwarzschild metric of a black hole with mass M. Classically, any positive value of M is allowed, and all such spacetimes are stable. Quantum mechanically, the situation is quite different. If initial conditions corresponding to no flux of particles in the past is imposed, then an outward flux of Hawking radiation is observed from the hole at late times[9]. Moreover, this radiation has a thermal spectrum with temperature equal to the Hawking temperature,

$$T_H = \frac{\hbar}{k_B\kappa M}. \tag{8}$$

This Hawking effect in a fixed Schwarzschild background precisely parallels the Schwinger mechanism of e^+e^- pair creation in a fixed electric field background. In each case the temperature is equal to $\frac{\hbar a}{2\pi k_B}$, where a is the proper acceleration of a particle of mass m (and charge e) at the event horizon.

There is one very important difference between the vacuum dissipative mechanism of relativistic systems described here and that of ordinary (non-relativistic) media. In thermodynamics one usually considers systems which are stable to linear perturbations. Then the standard linear response theory [10] gives a response function analogous to the polarization tensors (3) and (5), the time reversal odd piece of which describes the relaxation of the system back to its stable thermodynamic state, after it has been perturbed by a small external force. The perturbation of such a stable system decays with time; it does not grow. Obviously, this cannot be the case for the electric field or the black hole, which is *unstable* to the inclusion of quantum fluctuation and pair creation effects. This is due to the rather different behavior of the polarization tensors (3) and (5) when event horizons are present. One can show that the condition for particle creation effects to occur in the background electric or gravitational field is just the condition that the time asymmetric piece of the polarization function *diverges* for large $t - t'$, in particular that:

$$D^{abcd}(\mathbf{r}, \mathbf{r}'; \omega) \propto \omega^{-1}, \qquad \omega \to 0. \tag{9}$$

In fact it is precisely the residue of this ω^{-1} pole which determines the particle creation rate in the adiabatic limit of slowly varying backgrounds[8]. This *singular* behavior at low frequencies means that the background is unstable to small perturbations, and this is why the inclusion of the effects of quantum fluctuations changes the behavior of the system dramatically over long times, or in the far infrared. The electric field or the black hole, both completely stable classically are not stable quantum mechanically.

Hawking has given a simple thermodynamic argument for the quantum instability of a black hole due to fluctuations of matter or radiation in its vicinity[11]. The key observation is that the relation (8) implies that a black hole has *negative* specific heat. This means that if we place the black hole in a cavity with reflecting boundary condition in equilibrium with its own Hawking radiation, then this equilibrium is unstable. For if in some small time interval δt, the black hole should emit slightly more than its average emittance, due to a quantum/thermal fluctuation, its mass M will suddenly be a tiny bit smaller than its equilibrium value, due to the net energy loss. But from (8) this implies that it will be slightly *hotter* than its equilibrium temperature, and will therefore radiate all the more rapidly, driving it further from equilibrium. The corresponding argument applies if the fluctuation is of the opposite sign.

This behavior is quite different from more familiar systems with positive specific heat, and reflects itself in the singular ω^{-1} behavior of (9), which describes the response of the system to perturbations (or fluctuations) on length scales of order of the horizon size or larger. Notice that this instability could in no way be observed unless we consider small disturbances or fluctuations *away* from the mean equilibrium state. The mathematical existence of the equilibrium configuration (by Euclidean time continuation for example) tells us nothing about the stability of that configuration. For that the detailed behavior of the fluctuation spectrum, as described by the Fourier transform of the polarization tensor (5) is absolutely essential.

With this discussion of the electric field and Schwarzschild backgrounds, we are finally prepared to consider vacuum fluctuations in a cosmological setting, and return to our starting point, the cosmological constant or vacuum energy problem. Consider the Einstein equation for a spatially homogeneous and isotropic spacetime with Robertson-Walker scale factor, $a(t)$:

$$H^2 \equiv \left(\frac{\dot{a}}{a}\right)^2 = \frac{\kappa}{3}\rho, \tag{10}$$

where any cosmological term is included in the right hand side as a vacuum energy contribution with equation of state,

$$\rho_V = -p_V\,. \tag{11}$$

The Einstein equation (10), together with the equation of covariant energy conservation,

$$\dot{\rho} + 3\ H\ (\rho + p) = 0, \tag{12}$$

imply that:

$$\dot{H} = -\frac{\kappa}{2}\ (\rho + p). \tag{13}$$

This equation is to be compared to the Maxwell eq. (4), or the equation for the change of the Schwarzschild mass parameter in Eddington-Finkelstein-Vaidya coordinates[12]. In each case there is a classical static background that solves the Maxwell or Einstein equations, with zero source terms on the right hand side. In the case of (13) it is de Sitter spacetime with $\rho_V = -p_V = \frac{\Lambda}{\kappa}$, constant. In each case this background field is stable to classical charge or matter perturbations. The matter is simply accelerated, and swept out by the electric or gravitational field. In the de Sitter case classical matter (obeying $\rho + p > 0$) is redshifted away by (12). In the first two cases we have seen that there are good reasons to believe that these classically stable backgrounds with event horizons are unstable when quantum fluctuations are considered, and that the evolution of the system is inevitably driven away from the quasi-static initial state towards a final state in which the classical field energy has been dissipated into matter or radiation field modes. The signal of this instability is the singular behavior of the Fourier transform of the polarization tensor (9). In ref. 8 I have argued that the polarization function corresponding to scalar (*i. e.* metric trace) perturbations of the de Sitter background has precisely the required singular ω^{-1} behavior. Particle creation with time asymmetric initial conditions occurs[13], as in the Schwarzschild case. Thus, de Sitter spacetime satisfies the condition for dissipation of curvature stress-energy into matter and radiation modes. Notice also that whereas the Schwinger suppression factor for vacuum tunnelling occurs in electrodynamics because there are no massless charged particles, massless particles do couple gravitationally, and would be expected to dominate the dissipative process.

There is also an analog of the Hawking thermodynamic instability argument in de Sitter spacetime. Consider as the Euclidean defined "vacuum," the Bunch-Davies state which is invariant under the $SO(4,1)$ de Sitter symmetry group[14]. Because of that symmetry, the expectation value of the energy-momentum tensor of any matter fields in this state must itself be of the form $\rho = -p =$ constant. This is the state usually assumed, tacitly or explicitly by inflationary model builders. However, as should be clear from the previous examples, the mere existence of such a state tells us nothing about its stability against small fluctuations. In fact, both the energy within one horizon volume, and the entropy of the de Sitter horizon are *decreasing* functions of the Hawking-de Sitter temperature:

$$\begin{aligned} E_H &= \rho V = \frac{4\pi}{\kappa H} = \frac{2}{\kappa T_H} \\ S_H &= \frac{1}{4} A_H = \frac{\pi}{H^2} = \frac{1}{4\pi T_H^2} \end{aligned} \tag{14}$$

Hence, by considering a small fluctuation in the Hawking temperature of the horizon which (like the black hole case) causes a small net heat exchange between the region interior to the horizon and its surroundings, we find that this interior region behaves like a system with negative specific heat[15]. Since this statement applies to a region centered anywhere in de Sitter space, the entire space is unstable against quantum/thermal fluctuations in its Hawking temperature. The same conclusion is reached by an analysis of the polarization tensor in this background, because of the singular $\frac{1}{\omega}$ dependence in its trace sector. Finally, as if this is not persuasive enough, the Feynman propagator for the linearized metric tensor field in a de Sitter background diverges for large spacelike separations[16]. These divergence will appear in any graph (even tree graphs) involving graviton fluctuations in de Sitter space.

All of these considerations imply that de Sitter spacetime *cannot* be the stable ground state spacetime of the consistent quantum theory, any more than a black hole spacetime can be the ground state of matter with finite total energy in an infinite box, or a state with an arbitrarily large coherent electric field can be the correct stable ground state of QED. In the latter two situations, the system is unstable to the Schwinger/Hawking effect, converting the background field/curvature into matter and radiation which disperse to $\mathcal{I}^+$. If the corresponding evaporation proceeds to completion in the cosmological case, then the universe approaches flat Minkowski spacetime in the asymptotic future, regardless of the value of the cosmological term. One can see the analogy with the electric field discharging by comparing (13) with (4). The creation of real matter with $\rho + p > 0$ causes the scalar curvature parameter H to decrease continuously. In fact, unlike the electric current which is a vector and may change sign, $\rho + p$ is always positive, so we should expect H to decrease monotonically to zero without the plasma oscillations that can occur in the electrodynamic case[17].

It is a very attractive self-consistent picture that emerges from these observations. We began by posing a question, *viz.* why do the zero point energies necessarily associated with quantum fluctuations not curve spacetime? By recognizing that fluctuations are closely connected with dissipation in statistical mechanics, and extending the fluctuation-dissipation analysis to general relativity, we come to the realization that dissipative processes are necessarily associated with the fluctuations of the quantum "vacuum," which behaves in many respects like a reactive medium. Then these dissipative particle creation processes provide a dynamical mechanism to relax the scalar curvature (*i.e.*the effective cosmological "constant," as measured by the time dependent energy density of the universe) to zero. The expansion of the universe may be precisely this relaxation to asymptotically flat space taking place before our eyes, with perhaps some component of vacuum energy (with $p_V = -\rho_V$) of order of the upper bound (2) still present today. If this is the case, it would certainly have consequences for observational cosmology, and imply that searches for non-baryonic dark matter are probably futile. The dark "matter" may be all vacuum fluctuation energy, and so be completely undetectable by ordinary particle detectors.

To see if these ideas really work in detail, I have been following two distinct but parallel lines of thought. The first is to develop the necessary field theory formalism to carry out numerical simulations of the effect (back reaction) of particle creation processes on classical background fields. The main technical difficulties are solved with a systematic $1/N$ approximation[18]. Here N is an artificial, but convenient gauge-invariant parameter, equal to the number of identical copies of the quantum matter field interacting with the classical electric or gravitational field. The expansion of the Schwinger-Dyson equations of field theory in powers of $1/N$ bears a great deal of similarity to the moment expansion in non-equilibrium statistical mechanics. Leading order in $1/N$ exactly reproduces the semi-classical mean-field Maxwell or Einstein equations (after an appropriate rescaling of the coupling constants). However, not until we go beyond this leading mean-field approximation to the next order in $1/N$ have we included the dynamical back reaction and dissipative effects of the polarization tensor on the classical background. As one might expect, just as things become interesting, they also become difficult. Including the effects of the polarization tensor, which is a non-local function of two spacetime points in the correct, causal way requires the closed time path formalism of Schwinger and Keldysh[19]. It also requires numerically solving *integro*-differential equations for the photon or graviton propagator functions in the dynamically evolving classical background. We are just beginning to tackle the details of implementing this in the QED case, and will move on to the gravitational case only after gaining some experience with the QED problem (which incidentally is quite interesting in its own right).

The outstanding question of the time scale for relaxation should be answerable in principle by this direct approach to the problem. However, in many body theory the long time or large distance properties of a system are generally associated with the hydrodynamic regime, which can be notoriously difficult to access by a brute force approach based on the interactions of the underlying microscopic degrees of freedom.

Fortunately, we have at our disposal a quite different line of thought which may be able to bypass the difficult computational issues involved in the direct approach to long time dynamics just sketched. After all, if we are interested only in the asymptotic ground state configuration it should not be necessary to solve for the full evolution needed to reach it. But then one should have identified the correct effective degrees of freedom that dominate at very large spacetime distance scales, and if possible, provide an effective field theory description for the system which is valid in the far infrared. An example of this in particle physics is the non-linear sigma model realization of broken chiral symmetry which provides an effective field theory description of QCD valid at low energies.

In order to construct a low energy effective lagrangian for gravity that incorporates infrared fluctuations correctly, one must decide first of all what is the relevant order parameter field at large distance scales. Here observational cosmology comes to our aid to suggest that the FRW scale factor, or more generally, the conformal factor of the metric tensor is the most likely candidate for this role. In the classical Einstein theory this scalar part of the metric does not propagate. It is determined in terms of the matter sources and has no independent dynamics of its own. At the quantum level there is a trace anomaly in the energy-momentum tensor of conformally coupled matter fields. The existence of the conformal anomaly means that the classical constraints which fix the scalar part of the metric fluctuations in terms of matter sources cannot be maintained upon quantization. In other words, the conformal factor becomes *unconstrained* in the full quantum theory.

In the last few years a great deal of progress has been made in understanding non-critical string theories, which involve the consistent covariant quantization of the conformal part σ of the metric tensor in two dimensions:

$$g_{ab}(x) = e^{2\sigma(x)}\bar{g}_{ab}(x), \tag{15}$$

where $\bar{g}_{ab}(x)$ is a fixed fiducial metric. In two dimensional quantum gravity the effective action induced by the trace anomaly modifies the dynamics of the conformal factor at all distance scales in dramatic fashion. Certainly nothing like the fractal behavior and scaling relations of random surfaces in 2D gravity can arise from the classical two dimensional Einstein-Hilbert action which (being a topological invariant) yields no dynamics whatsoever[20,21]. In four dimensions too, the effective lagrangian must be modified to take account of the trace anomaly due to massless fields and resulting fluctuations of the conformal factor. This modified theory should be analyzed to discover the infrared behavior of the quantum theory of gravity.

It turns out that a great deal of the technology developed to study non-critical string theories (which one may understand as simply 2D quantum gravity coupled to matter) may be carried over directly into four dimensions[22]. The scalar σ field in 4D gravity acquires dynamics through the quantum trace anomaly of matter fields, which is altogether different from the Einstein theory (where σ is constrained). It is not difficult to derive the effective Wess-Zumino action induced by the trace anomaly of conformal matter in four dimensions. Then analysis of this effective theory shows that it possesses a non-trivial, infrared stable fixed point, characterized by certain anomalous scaling relations [23]. Then the radical proposal we make is that this scale invariant phase is the ground state of 4D quantum gravity, approached at distance scales larger than the horizon length of any given classical background. Conformal symmetry, apparently broken by the trace anomaly is restored dynamically by the large fluctuations in the conformal factor at these scales. A consequence of restoration of scale invariance at large distances is the screening of the effective cosmological term as measured by the average scalar curvature in the ground state. In particular, the two-point correlator of Ricci scalars $\langle R(x)R(x')\rangle$ falls to zero with a certain universal power of the invariant distance between x and x' that depends only on the total number of massless fields through the effective central charge of the effective theory. The dynamics of dissipation to effectively zero cosmological "constant" may then be understood as the evolution to a scale invariant fixed point, at which full conformal invariance is restored. There seems to be the exciting possibility that the scaling relations derived by us in the continuum may be verified by the dynamical triangulation approach on simplicial lattices numerically[24].

To conclude, it seems quite possible that the problem of vacuum zero point fluctuations and the curvature of spacetime has within it the seed for its own solution in the dissipative, properties of the "vacuum," necessarily associated with those very fluctuations. In addition to the possibility of a direct verification of this idea, an effective field theory has been proposed in which large quantum fluctuations of the conformal factor drastically modify classical general relativity at cosmological distance scales, resulting in a scale invariant phase of quantum gravity in the far infrared. The result is that any cosmological term in the continuum theory is screened completely at the largest distance scales by these fluctuations. If these ideas are correct (and they may be tested by numerical simulations in the not too distant future), then the *effective* cosmological constant in the correct vacuum of 4D quantum gravity must vanish identically.

Acknowledgement

I wish to thank the organizers of the workshop on Quantum Infrared Physics and particularly H. Fried and B. Müller for inviting me to make this presentation.

References

1. S. Weinberg, *Rev. Mod. Phys.* **61** (1989) 1.

2. H. B. G. Casimir, *Proc. Kon. Ned. Acad. Wetenschap.* **51** (1948) 793;
M. J. Sparnaay,*Physica* **24** (1958) 751;
T. M. Boyer, *Ann. Phys.* **56** (1970) 474.

3. J. Schwinger, *Phys. Rev.* **82** (1951) 664.

4. A. Einstein, *Ann. Phys. (Leipzig)* **17** (1905) 549; Eng. trans. reprinted in *Investigations on the Theory of Brownian Movement*, R. Fürth ed., Dover, New York (1956);
H. B. Callen and T. H. Welton *Phys. Rev.* **83** (1951) 34.

5. B. DeWitt, "Quantum gravity: the new synthesis" in *General Relativity: An Einstein Centenary Survey*, S. W. Hawking and W. Israel eds., Cambridge University Press, Cambridge (1979).

6. D. W. Sciama, P. Candelas, and D. Deutsch, *Adv. Phys.* **30** (1981) 327.

7. B. L. Hu, *Phys. Lett.* **A90** (1982) 375; *Phys. Lett.* **A97** (1983) 368;
L. Parker, *Phys. Rev. Lett.* **50** (1983) 1009.
E. Calzetta and B. L. Hu *Phys. Rev.* **37** (1988) 2878.

8. E. Mottola, *Phys. Rev* **D33** (1986) 2136.

9. S. Hawking, *Nature* **248** (1974) 30;
S. Hawking, *Comm. Math. Phys.* **43** (1975) 199.

10. P. C. Martin and J. Schwinger, *Phys. Rev.* **115** (1959) 1342;
L. P. Kadanoff and G. Baym, *Quantum Statistical Mechanics*, W. A. Benjamin, Menlo Park, Calif. (1962).

11. S. W. Hawking, *Phys. Rev.* **D13** (1976) 191;
J. B. Hartle and S. W. Hawking, *Phys. Rev.* **D13** (1976) 2188.

12. J. M. Bardeen, *Phys. Rev. Lett.* **46** (1981) 382.

13. E. Mottola, *Phys. Rev.* **D31** (1985) 754.

14. E. A. Tagirov, *Ann. Phys. (N. Y.)* **76** (1973) 561;
T. S. Bunch and P. C. W. Davies, *Proc. R. Soc. London* **A360** (1978) 117.

15. E. Mottola, *Phys. Rev.* **D33** (1986) 1616.

16. I. Antoniadis and E. Mottola, *Jour. Math. Phys.* **32** (1991) 1037.

17. F. Cooper, J. M. Eisenberg, Y. Kluger, E. Mottola, and B. Svetitsky, *Phys. Rev. Lett.* **67** (1991) 2427.

18. F. Cooper and E. Mottola, (1989) *Phys. Rev.* **D40** (1989) 456;
F. Cooper, S. Habib, Y. Kluger, E. Mottola, J. P. Paz, and P. R. Anderson, LA-UR-94-783, in press *Phys. Rev.* **D** (1994).

19. J. Schwinger, *Jour. Math. Phys.* **2** (1961) 407;
L. V. Keldysh,*Sov. Phys. JETP* **20** (1965) 1018 [*J. Exptl. Theor. Phys. (U. S. S. R.)* **47** (1964) 1515].

20. A. M. Polyakov, *Phys. Lett.* **B103** (1981) 207;
V. G. Knizhnik, A. M. Polyakov, and A. B. Zamolodchikov, *Mod. Phys. Lett.***A3** (1988) 819.

21. F. David, *Mod. Phys. Lett.* **A3** (1988) 1651;
J. Distler and H. Kawai, *Nucl. Phys.* **B321** (1989) 509.

22. I. Antoniadis and E. Mottola, *Phys. Rev.* **D45** (1992) 2013;
I. Antoniadis, P. O. Mazur and E. Mottola, *Nucl. Phys.* **B388** (1992) 627.

23. I. Antoniadis, P. O. Mazur and E. Mottola, *Phys. Lett.* **B323** (1994) 284.

24. J. Ambjørn, B. Durhuus, J. Fröhlich and P. Orland, *Nucl. Phys.* **B270** (1986) 457;
A. Billoire and F. David, *Nucl. Phys.* **B275** (1986) 617;
D. V. Boulatov, V. A. Kazakov, I. K. Kostov and A. A. Migdal, *Nucl. Phys.* **B275** (1986) 641.

Found State Physics

LONG-RANGE FORCES, INFRARED DIVERGENCES, AND BOUND-STATE QED

J. SUCHER

Center for Theoretical Physics and Department of Physics
University of Maryland, College Park, MD 20742

ABSTRACT

The long-range potential V_2 arising from two-photon exchange between two neutral particles or between a neutral and a charged particle has been studied extensively, in contrast to the case of two charged particles. In this case the two-photon exchange graphs are infrared divergent and a potential can be defined only by careful attention to the role it is required to play in the solution of physical problems, such as the determination of bound state masses. It is vital that the one-photon-exchange potential V_1 be specified in advance; different choices can lead to different asymptotic behavior for large *r*. The role played by such choices in determining the form of the orbit-orbit potential and in the calculation of bound-state energies based on three-dimensional relativistic wave equations is analyzed for the case where the constituents have either spin-0 or spin-1/2. In contrast to the Bethe-Salpeter equation, there are no spurious contributions of order $\alpha^3 m_{red}$ in the one-photon-exchange approximation, even if a form of V_1 is used which is inspired by the use of Feynman gauge; it suffices to include the long-range, spin-independent part of V_2 to get the correct level shift of order $\alpha^4 m_{red}$.

1. Introduction

As you know, this session of our workshop is dedicated to the memory of the late Don Yennie. I only got to know Don well in the mid-80's, when we spent a few months together at the ITP in Santa Barbara. It is there that I learned to appreciate what a splendid person lay behind his quiet demeanor. Among the many lasting contributions which Don made to theoretical physics, his work on bound-state problems in QED and on issues connected with the infrared (IR) divergence in the scattering of charged particles is particularly well known. So I thought it would be appropriate to talk about some work that involves both these topics and which I like to believe he would have enjoyed.

Apart from our common interest in QED there was another element which Don and I shared: We both wrote papers in collaboration with Maurice Lévy. I am happy to see that Maurice is here in the audience.

The research I wish to discuss has its origin in work done long ago with the late Gary Feinberg on the long-range (LR) electromagnetic forces between two particles, *A* and *B*, which are both spinless and neutral. The part of the effective force between such systems which arises simply from the electrostatic interactions between their constituents has a potential which falls off as r^{-6} for large r, as first shown by S.C. Wang for the case of two H-atoms. The forces between such systems were first called dispersion forces by F. London, who generalized the approach of Wang to more complex atoms. Subsequently is was shown by Casimir and Polder that at very large separations the effect of including the exchange of transverse photons is such as to *change* the fall-off to r^{-7}; these forces are therefore variously called Casimir forces or retarded van der Waals forces. Much later Feinberg and I used the techniques of particle theory to show that simply as a consequence of gauge invariance and analyticity properties of scattering amplitudes, independent of particle structure, the potential arising from two-photon exchange, $V_{2\gamma}$, behaves as $-D/r^7$, with D a quadratic polynomial in electric and magnetic polarizabilities. There is also a one-photon exchange potential $V_{1\gamma}$ but it falls off exponentially with r.[1]

If one of the particles, say B, is charged $V_{1\gamma}$ is still short-range while the leading term in $V_{2\gamma}$ is the familiar polarizability potential $V_{pol} = -\alpha_E/2r^4$, which has a well-known classical origin. There are however quantum theoretic retardation corrections to this leading term, which for large r are of order $1/mr$ relative to the leading term; these turn out to be significant in the analysis of the fine structure of the Rydberg states of helium.[2]

The topic I want to discuss today concerns the extension of these results to the case where A and B are both charged and may have spin-1/2. More specifically, one may ask: To what extent can the interaction between two charged particles be described by an effective two-body potential and what are the characteristics of such a potential?

If no nonrelativistic approximations are to be made, then even the computation of $V_{1\gamma}$ is not trivial. If one or both particles have spin-0, useful forms of $V_{1\gamma}$ turn out to be nonlocal. Further, for two Dirac particles naive use of the familiar Breit potential is incorrect, as we shall see. Now what can be said about $V_{2\gamma}$? Of course one expects it to be long-range, as in the other cases, but what is the asymptotic behavior? The zero mass of the photon, which is responsible for the LR nature of electromagnetic forces, now leads to a problem which must be overcome before this question can begin to be addressed: Inspection of the two-photon exchange graphs quickly reveals that, regardless of spin, some have infrared (IR) divergences and these do not cancel among themselves. Thus one seems to be stopped at the gate in defining $V_{2\gamma}$, let alone investigating its LR behavior. It turns out that the solution to this problem sheds new light on alternative methods for studying bound states in QED and on the relation between these methods. So the study of these questions links the topics of the title: *long-range forces, infrared divergences and bound states.*

The resolution of the IR problem lies in the recognition that the potential $V_{2\gamma}$ associated with two-photon exchange is not simply determined from the amplitude $M_{2\gamma}$ computed from the two-photon exchange diagrams: To avoid double counting one must subtract from $M_{2\gamma}$ the amplitude M_I arising from iteration of $V_{1\gamma}$. The key point is that M_I also has an IR divergence and this divergence exactly cancels the one in $M_{2\gamma}$.

I will start by considering the problem for point-like spin-0 particles. i.e. in the context of scalar QED, and then sketch the extension to spin-1/2 particles. This will lead us to a discussion of some aspects of the orbit-orbit interaction and the calculation of bound-state energies.

2. One- and two-photon exchange potentials in scalar QED

2.1 One-photon exchange potential

The potential $V_{1\gamma}$ must of course reduce to the Coulomb potential U_C at large r and low energies,

$$V_{1\gamma} \rightarrow U_C \equiv k/r, \quad k \equiv e_A e_B/4\pi. \tag{2.1}$$

To take into account relativistic corrections to all orders in v/c, U_C must be modified. The nature of this modification is dictated in part by the requirement that when used in lowest-order perturbation theory, $V_{1\gamma}$ is to reproduce the one-photon exchange amplitude $M_{1\gamma}$ exactly. If one uses Feynman gauge in writing down $M_{1\gamma}$ one gets $M_{1\gamma} = e_A e_B(s-u)/t$, where

$$s = (p_A+p_B)^2,\ t = (p_A-p_A')^2,\ u = (p_A-p_B')^2.$$

Since $u = 2m_A^2+2m_B^2-s-t$ we have

$$M_{1\gamma} = e_A e_B(2a+t)/t \qquad (a \equiv s-m_A^2-m_B^2). \tag{2.2}$$

Simple Fourier transformation of (2.2) yields a term proportional to U_C, with an energy-dependent coefficient, plus a contact term proportional to $\delta(\mathbf{r})$. Such a potential is not suitable for use in a Schroedinger type of equation. In second-order perturbation theory it would lead to an ultraviolet (UV) divergence. A potential which is iterable can be obtained by first writing $M_{1\gamma}$ in a different form (which does not change its value on the mass shell) and then finding an equivalent operator in $\mathbf{r}$-space which involves derivative operators.[3] One is thereby led to what can be termed a Feynman-gauge-inspired (FGI) potential $V_{1\gamma}^{FGI}$,

$$V_{1\gamma}^{FGI} = z'_{op}U_C z'_{op} + y_{op}(\mathbf{p}_{op}\cdot U_C\mathbf{p}_{op}/2m_A m_B)y_{op}, \tag{2.3}$$

where $z'_{op} = (1+\mathbf{p}_{op}^2/2E_A^{op}E_B^{op})^{1/2}$, $y_{op} = (m_A m_B/E_A^{op}E_B^{op})^{1/2}$, and $E_i^{op} = (\mathbf{p}_{op}^2+m_i^2)^{1/2}$. The corresponding Coulomb-gauge inspired (CGI) one-photon exchange potential $V_{1\gamma}^{CGI}$ is given by[4]

$$V_{1\gamma}^{CGI} = y_{op}[\{E_A^{op},\{E_B^{op},U_C\}\}$$
$$+(1/2)\{\mathbf{p}_i^{op},\{\mathbf{p}_j^{op},(\delta_{ij}+\hat{r}_i\hat{r}_j)U_C\}\}]y_{op}/4m_A m_B. \tag{2.4}$$

Although these potential operators are clearly different, when taken between plane wave product states in the center-of-momentum system (c.m.s), both reproduce $M_{1\gamma}$. The various square root factors serve the purpose of yielding, to lowest order in k, the correct connection between the invariant Feynman amplitude M and the c.m.s. transition amplitude T_{pot},

$$T_{pot} = M/4E_A E_B. \tag{2.5}$$

Here T_{pot} is to be obtained from a c.m.s. relativistic equation of the natural form

$$(E_A^{op}+E_B^{op}+V)\ \phi = W\phi. \tag{2.6}$$

2.2 Two-photon exchange potential

To compute the field theory amplitude $M^{(4)}$ in scalar QED one must study the integrals associated with the five two-photon-exchange graphs which enter the game: (a) the two-rung ladder graph, (b) the two-rung crossed ladder graph, (c) the two single-seagull graphs, and (e) the double-seagull graph. Both (a) and (b) are UV convergent but IR divergent, whereas (c) and (d) are UV divergent but IR convergent. However, the IR divergence is canceled by the subtraction of the iteration amplitude M_I. One may further make use of the techniques of dispersion theory to both simplify the computation and to finesse the need to deal with the UV divergence when the focus is on the behavior at small t, corresponding to large r. On using $V_{1\gamma}^{FGI}$ to compute M_I one finds that[3]

$$V_{2\gamma}^{FGI} = c_2^{FGI}r^{-2} + c_3^{FGI}r^{-3} + \ \dots \tag{2.7}$$

where

$$c_2^{FGI} = k^2/2(m_A+m_B), \quad c_3^{FGI} = -\ 7k^2/6\pi m_A m_B. \tag{2.8}$$

In contrast, use of $V_{1\gamma}^{CGI}$ yields[4]

$$V_{2\gamma}^{CGI}= c_2^{CGI}r^{-2} + c_3^{CGI}r^{-3} + \dots \tag{2.9}$$

where

$$c_2^{CGI} = 0, \ c_3^{CGI} = -\ 7k^2/6\pi m_A m_B. \tag{2.10}$$

Thus we see that in the case of two charged particles the leading asymptotic behavior of $V_{2\gamma}$ *is either* r^{-2} *or* r^{-3}, *depending on the definition of* $V_{1\gamma}$.

This observation resolves a long-standing puzzle in the literature and shows that in the case of two charged particles not just the concept of potential but even the concept of "asymptotic form" is not without ambiguity. Further, as was noted some time ago by L. Spruch, c_2^{FGI} is classical in character, i.e. if h and c are restored, c_2^{FGI} turns out to be independent of h. One should therefore try to understand the source of this term from classical electrodynamics. It turns out that this is indeed possible, on the basis of a reexamination of the pioneering work of Darwin on the effective Lagrangian describing the interaction of two charged particles to order $1/c^2$.[5]

2.3 Orbit-orbit potential

It is of some interest to study the difference between the two choices of $V_{1\gamma}$. It turns our that this is connected with the form of the so-called orbit-orbit interaction. To see this, note that in the n.r. limit (2.3) yields as the leading correction to U_C an orbit-orbit interaction $U_{o\text{-}o}$ of the form

$$U_{o\text{-}o}^{FGI} = \{p_i^{op},\{p_j^{op},\delta_{ij}U_C\}\}/4m_Am_B,$$

whereas (2.4) yields

$$U_{o\text{-}o}^{CGI} = (1/2)\{p_i^{op},\{p_j^{op},(\delta_{ij}+\hat{r}_i\hat{r}_j)U_C\}\}/4m_Am_B. \tag{2.12}$$

The latter is a manifestly hermitian form of the orbit-orbit interaction $U_{o\text{-}o}$ familiar from atomic physics, usually obtained by reduction of the Breit operator (see below) to n.r. form. This is unfortunate from a pedagogical point of view: *spin has nothing to do with it.*

The difference between these two forms of $U_{o\text{-}o}$ is precisely accounted for by the $1/r^2$ term. In the context of two-body Coulomb bound states, one can show that if one starts with the n.r. Schroedinger equation

$$h_{nr}\phi = (p_{op}^2/2m_{red})\,\phi + U_C\phi = W\phi, \tag{2.13}$$

the leading term $\delta W^{(4)}$ in the level shift δW associated with electromagnetic effects, of order $\alpha^4 m_{red}$, is given by

$$\delta W^{(4)} = \langle\phi|U_{o\text{-}o}^{CGI}|\phi\rangle = \langle\phi|U_{o\text{-}o}^{FGI} + c_2^{FGI}/r^2|\phi\rangle \tag{2.14}$$

The equivalence of these two expressions can be checked most easily by showing that their difference has the form of a commutator $[h_{nr},X]$.

This is an example of the fact that a major source of non-uniqueness of two-body potentials is that one can always add a term of the form $[h_0,X]$ to the lowest-order potential without changing the lowest-order scattering amplitude.[6]

3. One- and two-photon exchange potentials in spinor QED

The inclusion of spin-1/2 particles is straightforward, once one recognizes the main pitfall encountered when dealing with relativistic Dirac-like equations. This is the problem of continuum dissolution.[7] The naive use of local potentials in the case where there are at least two Dirac particles and an external field, e.g. as for the helium atom, leads to an equation which has no normalizable solutions. In the absence of an external field, i.e. in the pure two-body case, the problem is ameliorated because of momentum conservation. However, starting from field theory one again finds that the effective interaction operators come equipped with projection operators, as in the case of He. In particular, one finds that for two spin-1/2 particles the appropriate counterpart of Eq. (2.6) is

$$[h_A^{op}+h_B^{op}+\Lambda_{++}^{op}U\Lambda_{++}^{op}]\psi = E\psi, \tag{3.1}$$

where $h_A^{op} = \boldsymbol{\alpha}_A\cdot\mathbf{p}_{op}+\beta_A m_A$, $h_B^{op} = -\boldsymbol{\alpha}_B\cdot\mathbf{p}_{op}+\beta_B m_B$, and Λ_{++}^{op} is the product of free (Casimir-type) positive-energy projection operators:

$$\Lambda_{++}^{op} = \Lambda_{+;A}^{op}\Lambda_{+;B}^{op}, \qquad \Lambda_{+;i}^{op} = (E_i^{op}+h_i^{op})/2E_i^{op}. \tag{3.2}$$

3.1. One-photon exchange potential

With this understanding, the one-photon exchange potential $V_{1\gamma}$ has the form

$$V_{1\gamma} = \Lambda_{++}{}^{op} U_{1\gamma} \Lambda_{++}{}^{op}, \tag{3.3}$$

where $U_{1\gamma}$ is required to reproduce $M_{1\gamma}$ when sandwiched between on-shell Dirac spinor plane waves. This leads to two natural choices for $U_{1\gamma}$, both of them local in Dirac-spinor space:

$$U_{1\gamma}{}^{FGI} = U_C + U_G, \quad U_{1\gamma}{}^{CGI} = U_C + U_B, \tag{3.4}$$

where U_G and U_B are the Gaunt and Breit potentials defined by

$$U_B = -k(\alpha_A \cdot \alpha_B + \alpha_A \cdot \hat{\mathbf{r}}\, \alpha_B \cdot \hat{\mathbf{r}})/2r, \quad U_G = -k(\alpha_A \cdot \alpha_B)/r. \tag{3.5}$$

The projection operators are vital if the two-body equation is to have a natural origin in field theory and if the equation is to lend itself to obvious generalization to more than two particles. They correspond to the fact that, in contrast to Dirac's one electron theory, in hole theory transitions to the filled negative-energy sea are forbidden by the exclusion principle, a feature which is incorporated by QED.

Reduction of either U_B or U_G to the n.r. limit yields the same familiar spin-dependent interaction operators, while the spin-independent potentials differ in the same way as when both particles has spin zero.

3.2 Two-photon exchange potential

The computation of $V_{2\gamma}$ for the case when both particles have spin-1/2 is a major undertaking. If one uses $V_{1\gamma}{}^{FGI}$ to compute M_I, the spin-independent part (more precisely, the Dirac-matrix independent part) of U_{2c} is the same as that found for two spin-0 particles. For the mixed case of, say, A with spin-0 and B with spin-1/2, the computation of the spin-dependent part has already been carried out.[8] One obtains a correction $V_{2\gamma}{}^{s\text{-}o}$ to one-photon-exchange spin-orbit potential $V_{1\gamma}{}^{s\text{-}o}$. If *A* is a point particle and one uses $V_{1\gamma}{}^{FGI}$ to compute the iteration amplitude one finds that at large *r*

$$V_{2\gamma;pt}{}^{s.o.} \approx -k^2([(3m_A+5m_B)/m_A(m_A+m_B)](\sigma \cdot \ell/4m_B{}^2 r^4). \tag{3.6}$$

If *A* has structure, V_{2c} also contains a spin-orbit polarizability potential

$$V_{2\gamma;pol}{}^{s.o.} \approx (e_B{}^2/4\pi)[(\alpha_E{}^A m_B + \alpha_M{}^B(m_A+m_B))/2m_A m_B{}^2](\sigma \cdot \ell/r^6). \tag{3.7}$$

These terms will also be present in the case of two spin-1/2 particles but in addition there will be a correction $V_{2\gamma}{}^{s\text{-}s}$ to the one-photon-exchange spin-spin potential $V_{2\gamma}{}^{s\text{-}s}$. The calculation of this is in progress.

There are a number of physical situations in which it may be possible to detect the effects of $V_{2\gamma}{}^{s\text{-}o}$. Typically these involve measurements of bound state energies in exotic atoms, where one particle has spin 1/2 and another has spin-0. Examples include anti-protonic atoms with a spin-0 nucleus, such as p-He^4, pionic atoms with a spin-1/2 nucleus, such as pionic hydrogen, and the pi-muon bound state known as pi-muonium. Certain aspects of $V_{2\gamma}{}^{s\text{-}o}$ may be observable in Rydberg states of helium-like ions whose nuclei have spin 1/2. For details see Ref. 8.

It is interesting to compare the use of the FGI potential in the two-body equation such as(2.5) or (3.1), to the use of Feynman gauge in the Bethe-Salpeter equation. One must include with $V_{1\gamma}{}^{FGI}$ at least the LR part of $V_{2\gamma}{}^{FGI}$ to get the correct *spin-independent part* of the level shift of order $\alpha^4 m_{red}$ in bound state masses (the spin-dependent parts are the same as those given by $V_{1\gamma}{}^{CGI}$ alone). However, this is much better than the BS equation, which not only gives the wrong $\alpha^4 m_{red}$ terms but also gives spurious terms, of order $\alpha^3 m_{red}$, in ladder approximation with an FGI kernel.

4. Concluding remarks

The concept of a two-body interaction potential in QED is rather subtle. Extension of methods developed for the cases where at least one of the particles is neutral to the case of two charged particles reveals a new feature: Different choices of the one-photon exchange potential $V_{1\gamma}$ may lead to different results for the leading term in an expansion in powers of r^{-1} of the two-photon exchange potential $V_{2\gamma}$; this feature has a counterpart in classical electrodynamics. Recent extension of the method to spin-1/2 particles, either composite or elementary, has led to formulas for the e^4 corrections to the spin-orbit interaction and work on similar corrections to the spin-spin interaction is in progress. When this is completed one will be able to reanalyze the spin-dependent level structure of a number of two-body physical systems and perhaps gain new insight into some aspects of bound-state QED, at least for states of large orbital angular momentum and relatively large separation between the constituents.

On the conceptual side there are several issues which should be studied. One has to do with the incorporation of radiative corrections to bound states. Because the method is based on the computation of on-shell matrix elements, one encounters IR divergences if both constituents are charged. As long as one restricts attention to generalized ladder graphs this problem is finessed by the computation of the potential, which is IR finite. However, graphs with radiative corrections require the introduction of an IR cutoff and it remains to be seen how this plays itself out in the computation of bound-state energies. Even apart from this, it would be worth exploring the difference between the FGI and CGI potentials in higher orders. In fourth order, the leading spin-independent part of $V_{2\gamma}$ depends on the precise choice of $V_{1\gamma}$ and has a classical character, but the spin-dependent part does not. Do these features persist in higher orders? If so, an interesting theorem about QED may be hidden here. The implications of a particular choice of $V_{1\gamma}$ for the accuracy achievable in the study of many-body bound states also requires study, especially in connection with a search for evidence of three-body forces.

Acknowledgment

This work was supported in part by the U.S. National Science Foundation.

References

1. For a review, with extensive references to the large literature see G. Feinberg, J. Sucher, and C.K. Au, *Phys. Rep.* **180**, 859 (1989).
2. See reviews in *Long-Range Casimir Forces: Theory and Recent Experiments in Atomic Systems*, edited by F.S. Levin and D.A. Micha (Plenum, New York, 1993).
3. G. Feinberg and J. Sucher, *Phys. Rev. D* **38**, 3763 (1988).
4. J. Sucher, *Phys. Rev. D* **49**, 4284 (1994).
5. J. Sucher, *Comm. At. Mol. Phys.* **30,** 129 (1994).
6. For further discussion of this point see J. Sucher, in *Proceedings of the Program on Relativistic, Quantum Electrodynamic and Weak Interaction Effects in Atoms*, ITP, Santa Barbara (AIP Conf. Proc. No. 189, 1989), edited by W.R. Johnson, P. Mohr, and J. Sucher (AIP, New York, 1989), p. 337.
7. J. Sucher, *Phys. Rev. A* **25**, 348 (1980); Phys. Rev. Lett. **55**, 1023 (1985).
8. G. Feinberg and J. Sucher, *Phys. Rev. D* **45**, 2493 (1992).

RECENT PROGRESS IN POSITRONIUM PHYSICS

GREGORY S. ADKINS
Department of Physics and Astronomy
Franklin and Marshall College
Lancaster, Pennsylvania 17604 USA

ABSTRACT

Positronium is arguably the most elementary bound state in quantum field theory. It is interesting because it is accessible both to high precision experiments and to detailed theoretical analysis. In this contribution, I give an introduction to the physics of positronium, and review the status of measurements and calculations of the $n = 1$ and $n = 2$ energy levels and $n = 1$ decay rates.

1. Introduction

Positronium is the electron-positron bound state. The structure of positronium is governed almost completely by QED; strong and weak interaction effects are small compared to electrodynamic effects. However, the effects of relativity, virtual annihilation, radiative corrections, and the details of the binding are all essential to its understanding. Positronium is also accessible to high precision experiments. Thus positronium is an ideal laboratory for the study and use of the bound state formalism in quantum field theory.

The main quantities of physical interest in positronium are energy levels and decay rates. Several energy level splittings of $n = 1$ and $n = 2$ states have been measured with uncertainties of a few MHz. Present theoretical evaluations of these splittings agree with experiment, with terms of $O(m\alpha^6) \approx 19MHz$ still uncalculated. The $n = 1$ decay rates have also been measured. The spin-triplet decay rate is $6-9$ standard deviations away from the present theoretical value, an apparent discrepancy between experiment and theory. The $O(\alpha^2)$ correction to the decay rate will have to be unusually large to account for this difference.

In this contribution I discuss the present status of calculations and measurements of the $n = 1$ and $n = 2$ energy levels, and the $n = 1$ decay rates. I will not be able to give complete references to the original literature in this brief contribution. The reader is referred to the reviews by Stroscio[1], Berko and Pendleton[2], Rich[3], Sapirstein and Yennie[4], Mills and Chu[5], and Dvoeglazov, Faustov, and Tyukhtyaev[6,7].

2. Basic Properties of Positronium

To lowest approximation, positronium is simply a Hydrogen atom with the proton replaced by a positron. The lowest order energy levels are just the Bohr energy levels[8]

$$E_n = -\frac{1}{2}\frac{m}{2}\frac{\alpha^2}{n^2} + O(m\alpha^4) = \frac{-13.6eV}{2} + O(m\alpha^4) \quad , \tag{1}$$

where the reduced mass is $m/2$. The two spin 1/2 factors combine to give $s = 0$ ("para") and $s = 1$ ("ortho") states. These combine with the orbital angular momentum $\ell = 0, 1, 2, \ldots$ to give the total angular momentum J. According to standard arguments, the parity of the positronium state is $P = (-1)^{\ell+1}$, and the charge parity is $C = (-1)^{\ell+s}$. The states of positronium are labeled by their quantum numbers according to $n^{2s+1}\ell_J{}^{PC}$.

Corrections to the Bohr energy levels for positronium start at order $m\alpha^4$. Because the positron and electron have equal mass, the fine and hyperfine structure effects come in at the same order. An energy level diagram, including the effects of these relativistic, spin-orbit, spin-spin, and annihilation contributions, is shown in Fig. 1.

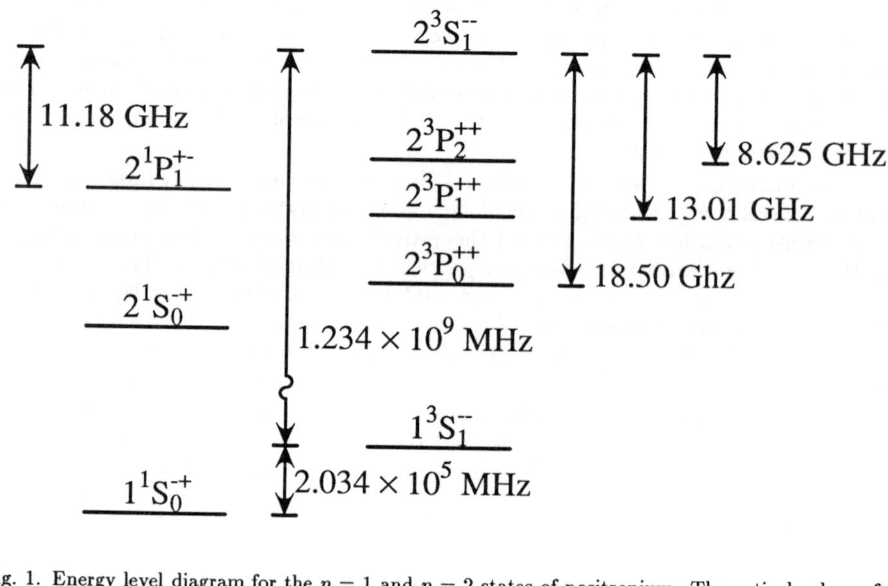

Fig. 1. Energy level diagram for the $n = 1$ and $n = 2$ states of positronium. Theoretical values of transition frequencies for the experimentally interesting transitions are given.

Charge parity conservation in QED imposes selection rules on the decays of positronium to photons. The C-parity of an n-photon state is $(-1)^n$. Consequently, $C = +1$ states of positronium can only decay to even numbers of photons, and $C = -1$ states can only decay to odd numbers of photons. (Decays to a single photon are ruled out by energy-momentum conservation.)

3. Review of Progress in the Calculation of Positronium Energy Levels

3.1. *Theoretical Methods*

Most calculations of positronium properties have used the Bethe-Salpeter equation as a starting point[9]. The inhomogeneous Bethe-Salpeter equation can be written as

$$G' = S' + S'K'G' \quad , \tag{2}$$

where G' is the full e^+e^- to e^+e^- Green's function, S' is the product of two full one-particle electron Green's functions, and K' is the two-particle-irreducible interaction kernel. Of particular use are the related "quasipotential" formulations, which are essentially three-dimensional. The quasipotential bound state equation has the form

$$G = S_0 + S_0 K G \quad , \tag{3}$$

where S_0 is a modified electron propagator, K is the quasipotential, and G is related linearly to G'. A useful quasipotential formulation for positronium was proposed by Caswell and Lepage[10], and a related formalism was developed by Barbieri, Remiddi, and Buchmüller[11,12]. An appropriate approximate kernel K_0 is found which contains the dominant nonrelativistic physics of K and with which the "reference equation"

$$G_0 = S_0 + S_0 K_0 G_0 \tag{4}$$

can be solved exactly. The full Green's function G' can be expressed perturbatively in terms of G_0.

Energy levels are found as the positions of the poles in G' (and also in G), considered as functions of the center-of-mass energy. The lowest order energy levels are given by the positions of the poles of G_0. A perturbation expansion for the true energy levels can be developed as[13]

$$E_n = E_n^0 + (\delta K) + (\delta K \hat{G}_0 \delta K) + (\delta K)(\delta K)' + \cdots \quad , \tag{5}$$

where δK is essentially the difference $K - K_0$ (but for detailed work, δK also involves the difference $S - S_0$), $\hat{G}_0$ is the reference Green's function G_0 with the pole terms at E_n^0 subtracted out, the parentheses represent matrix elements between lowest order wavefunctions, and the prime denotes an energy derivative. A diagrammatic representation of a contribution to (δK) is shown in Fig. 2.

The question of gauge choice is always an issue in bound state work. The noncovariant Coulomb gauge seems to be most natural for photons connecting one fermion

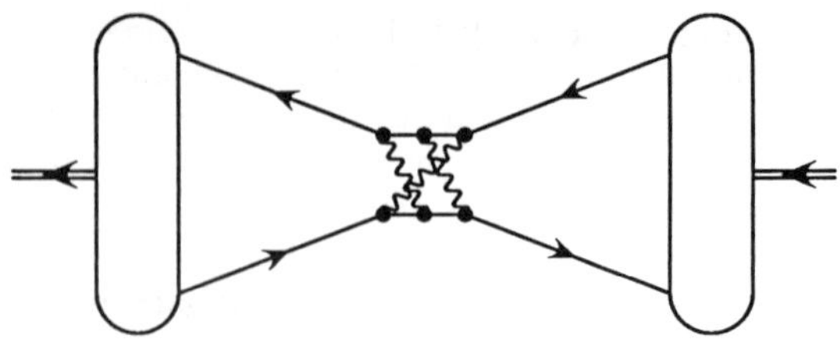

Fig. 2. Example of a contribution to (δK).

line with the other.[14] However, a covariant gauge seems best for photons on a single fermion line, because the covariant gauges are most convenient when dealing with renormalization. (The justification for the use of two different gauges in the same calculation is discussed by Feldman, Fulton, and Heckathorn.[15,16]) In particular, the Yennie gauge is both covariant and infrared safe,[17,18] and is very useful for bound state work.

3.2. *The* $n = 1$ *Hyperfine Splitting*

The ground state hyperfine splitting was the first measured energy level splitting in positronium. It was observed by Deutsch and Dulit in 1951.[19] Several other determinations were made over the years, with the latest being[20]

$$E(1^3S_1) - E(1^1S_0) = 203,389.10 \pm 0.74 MHz \quad . \tag{6}$$

The theoretical result for this splitting is[1-7,21]

$$\begin{aligned} E(1^3S_1) - E(1^1S_0) &= m\alpha^4\Big\{\frac{7}{12} - \frac{\alpha}{\pi}\Big(\frac{8}{9} + \frac{1}{2}\ln 2\Big) + \frac{5}{24}\alpha^2 \ln\Big(\frac{1}{\alpha}\Big) + K\alpha^2 - \frac{7}{8}\frac{\alpha^3}{\pi}\ln^2\Big(\frac{1}{\alpha}\Big) + \cdots\Big\} \\ &= 203,388.1 \pm 0.6 MHz \quad , \end{aligned} \tag{7}$$

where the known contributions to K amount to $K = -0.60(3)$.[22] The uncalculated contributions to K (due to two-loop radiative corrections to the one-photon-annihilation graph) are expected to have order $1/\pi^2$, which corresponds to $1.9 MHz$.

The theoretical and experimental results for the ground state hyperfine splitting are in good agreement.

3.3. *Origin of the* $\ln(1/\alpha)$ *Terms*

The logarithmic terms in the energy levels and decay rates are characteristic of bound states. One source of the logarithms is the standard infrared logarithms of scattering theory. In the Coulomb scattering problem in QED, logarithms of a fictitious photon mass are often encountered. In the bound state formalism, which is best set up without a fictitious photon mass, the same infrared divergences are regulated by the off-shell-ness of the electron and positron. The log of a photon mass naturally becomes a log of α.

I will give an example of another source of $\ln(1/\alpha)$ terms. The Coulomb ground state wavefunction in momentum space is proportional to

$$\frac{8\pi\gamma}{(p^2+\gamma^2)^2} \quad , \tag{8}$$

where $\gamma = m\alpha/2$ is the typical momentum of the positronium state. A logarithmic integral might have the form

$$I = \int \frac{d^3p}{(2\pi)^3} \frac{8\pi\gamma}{(p^2+\gamma^2)^2} \frac{p}{\omega_p} \quad , \tag{9}$$

where $\omega_p = (p^2+m^2)^{1/2}$. Note the two momentum scales, γ and m, that are characteristic of logarithmic integrals. After the trivial angular integrals are done, one finds that

$$I \propto \int_0^\infty dp \frac{p^3}{(p^2+\gamma^2)^2} \frac{1}{(p^2+m^2)^{1/2}} \quad . \tag{10}$$

In the "logarithmic region" $\gamma << p << m$, this has the logarithmic form $\int dp/p$. The wave function provides a natural low-momentum cut-off at momenta of order γ, and the ω_p term provides a high-momentum cut-off at momenta of order m, so the integral becomes

$$I \propto \int_\gamma^m \frac{dp}{p} = \ln(\frac{m}{\gamma}) \approx \ln(1/\alpha) \quad . \tag{11}$$

3.4. *Fine Structure of $n = 2$ States*

Several $n = 2$ fine structure intervals have been measured. The latest results are[23,24]

$$E(2^3S_1) - E(2^3P_2) = 8,624.38 \pm 1.50MHz \quad , \tag{12a}$$
$$E(2^3S_1) - E(2^3P_1) = 13,012.42 \pm 1.68MHz \quad , \tag{12b}$$
$$E(2^3S_1) - E(2^3P_0) = 18,499.65 \pm 4.18MHz \quad , \tag{12c}$$
$$E(2^3S_1) - E(2^1P_1) = 11,181 \pm 13MHz \quad . \tag{12d}$$

The theoretical results for these $n = 2$ intervals are[25]

$$E(2^3S_1) - E(2^3P_2) = \frac{m\alpha^4}{8}\left\{\frac{23}{120} + \frac{\alpha}{\pi}\left[\frac{3}{4}\ln\left(\frac{1}{\alpha}\right) - 1.42148\right] + \frac{1}{12}\alpha^2\ln\left(\frac{1}{\alpha}\right) + O(\alpha^2)\right\}$$
$$= 8,626.2MHz \quad , \tag{13a}$$

$$E(2^3S_1) - E(2^3P_1) = \frac{m\alpha^4}{8}\Big\{\frac{7}{24} + \frac{\alpha}{\pi}\Big[\frac{3}{4}\ln\Big(\frac{1}{\alpha}\Big) - 1.36315\Big] + \frac{1}{12}\alpha^2\ln\Big(\frac{1}{\alpha}\Big) + O(\alpha^2)\Big\}$$
$$= 13,011.9MHz \quad , \tag{13b}$$
$$E(2^3S_1) - E(2^3P_0) = \frac{m\alpha^4}{8}\Big\{\frac{5}{12} + \frac{\alpha}{\pi}\Big[\frac{3}{4}\ln\Big(\frac{1}{\alpha}\Big) - 1.25898\Big] + \frac{1}{12}\alpha^2\ln\Big(\frac{1}{\alpha}\Big) + O(\alpha^2)\Big\}$$
$$= 18,497.1MHz \quad , \tag{13c}$$
$$E(2^3S_1) - E(2^1P_1) = \frac{m\alpha^4}{8}\Big\{\frac{1}{4} + \frac{\alpha}{\pi}\Big[\frac{3}{4}\ln\Big(\frac{1}{\alpha}\Big) - 1.38398\Big] + \frac{1}{12}\alpha^2\ln\Big(\frac{1}{\alpha}\Big) + O(\alpha^2)\Big\}$$
$$= 11,184.9MHz \quad . \tag{13d}$$

The agreement between experiment and theory is good. It should be noted that with a coefficient of unity, the $O(m\alpha^6)$ term would contribute $2.3MHz$.

3.5. *Gross Structure (the $2^3S_1 - 1^3S_1$ Splitting)*

The $E(2^3S_1) - E(1^3S_1)$ "gross structure" interval has been measured to high precision by Fee *et al.*:[26]

$$E(2^3S_1) - E(1^3S_1) = 1,233,607,216.4 \pm 3.2MHz \quad . \tag{14}$$

The theoretical prediction for the gross structure interval is[25]

$$E(2^3S_1) - E(1^3S_1) = m\alpha^2\Big\{\frac{3}{16} - \frac{719}{3072}\alpha^2 + \frac{\alpha^3}{\pi}\Big[-\frac{21}{32}\ln\Big(\frac{1}{\alpha}\Big) + 1.38408\Big] - \frac{7}{96}\alpha^4\ln\Big(\frac{1}{\alpha}\Big) + O(\alpha^4)\Big\}$$
$$= 1,233,607,221.7MHz \quad . \tag{15}$$

This prediction is consistent with the experimental determination. A contribution of $18.7MHz$ would arise from the $O(m\alpha^6)$ term if it had a coefficient of unity.

4. Status of the Decay Rate Calculation

4.1. *Methods*

The energy level of a state calculated through Eq. (5) has an imaginary part related to the decay rate of that state. The formula is

$$\Gamma_n = -2Im(E_n) \quad , \tag{16}$$

which comes from a consideration of the time dependence of the state of a particle when its energy has an imaginary part. An imaginary part occurs when a diagram contributing to E_n can be cut in two through an intermediate state consisting entirely of real particles (here, photons). This is illustrated in Fig. 3, which contains the lowest order contribution to the orthopositronium decay rate.

4.2. *The Parapositronium Decay Rate*

The latest experimental result for the parapositronium decay rate is[27]

$$\Gamma(1^1S_0) = 7.9909(17)\, ns^{-1} \quad . \tag{17}$$

The theoretical result for the parapositronium decay rate is[28]

$$\begin{aligned}\Gamma(1^1S_0) &= \frac{1}{2}m\alpha^5\Big\{1 - \frac{\alpha}{\pi}(5 - \frac{\pi^2}{4}) + 2\alpha^2\ln(1/\alpha) + O(\alpha^2)\Big\} \\ &= 7.9895 ns^{-1} + O(\frac{1}{2}m\alpha^7 \approx 0.0004\, ns^{-1}) \quad .\end{aligned} \tag{18}$$

This result includes the $O(m\alpha^7\ln(1/\alpha))$ contribution, but not the recently obtained $O(m\alpha^8\ln^2(1/\alpha))$ contribution of Karshenbǒim,[21] which is numerically small. The experimental rate is about 2.5σ above the theoretical result.

4.3. *The Orthopositronium Decay Rate*

The latest experimental results for the orthopositronium decay rate are[29,30]

$$\Gamma(1^3S_1) = 7.0514(14)\, \mu s^{-1} \quad , \tag{19a}$$

$$\Gamma(1^3S_1) = 7.0482(16)\, \mu s^{-1} \quad . \tag{19b}$$

The orthopositronium decay rate was first calculated by Ore and Powell.[31] A graph of the lowest order decay process is given in Fig. 3. There are three annihilation photons, and no other virtual photons. The lowest order result is

$$\begin{aligned}\Gamma_{LO} &= \frac{2}{9\pi}(\pi^2 - 9)m\alpha^6 \\ &= 7.2112\, \mu s^{-1} \quad .\end{aligned} \tag{20}$$

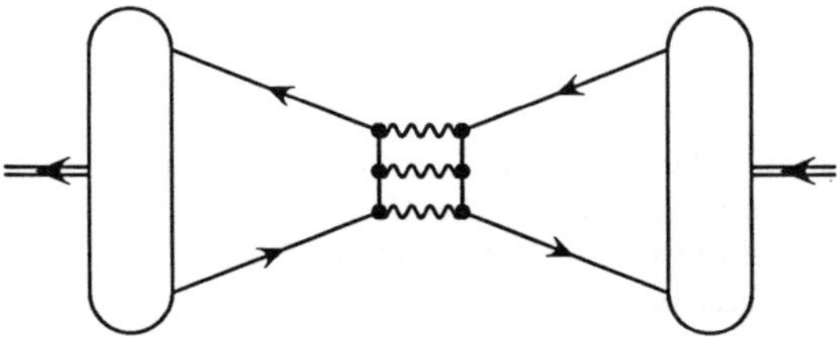

Fig. 3. Diagram for the lowest order contribution to the decay of orthopositronium.

The complete theoretical result is[21,32]

$$\Gamma(1^3S_1) = \Gamma_{LO}\Big\{1 - 10.2866(6)\frac{\alpha}{\pi} - \frac{1}{3}\alpha^2\ln\Big(\frac{1}{\alpha}\Big) + L\alpha^2 - \frac{3}{2}\frac{\alpha^3}{\pi}\ln^2\Big(\frac{1}{\alpha}\Big) + \cdots\Big\}$$
$$= (7.0382 + 0.00038L)\,\mu s^{-1} \quad , \tag{21}$$

where L is the unknown coefficient of $O(\alpha^2)$ corrections.

The orthopositronium decay rate measurement is presently in disagreement with the theoretical rate by $6-9$ standard deviations. A value for L of about 30 would be required to bring theory and experiment into agreement. A value of $L=30$ is unusually large, but it is not impossible to imagine that a complete calculation of L could give such a result. A partial calculation, including only the square of the first order correction amplitude, gives a contribution to L of 2.92(2).[33] The complete calculation of L is the most important task remaining to be done in the area of positronium theory.

5. Conclusion

Bound state physics is an important testing ground for QED. It probes aspects of the theory different from those important in scattering experiments and in the study of anomalous magnetic moments. Positronium physics in particular is interesting because positronium is so fundamental, and is accessible to both detailed theoretical analysis and high precision experimentation.

Most of the current work is at the level of $O(\alpha^2)$ corrections to the basic energy level shifts or decay rates. That is, the order of interest is $O(m\alpha^6)$ for energy levels, $O(m\alpha^7)$ for the parapositronium decay rate, and $O(m\alpha^8)$ for the orthopositronium decay rate. For the most part, experiment is ahead of theory. The $O(\alpha^2)$ corrections that are required for comparison with experiment at this level have not yet been completed.

The agreement between theory and experiment is good for energy levels, and acceptable for the parapositronium decay rate. Theory and experiment are not in good agreement for the orthopositronium decay rate. A complete calculation of the $O(\alpha^2)$ corrections could possibly restore agreement. Work on the calculation of these corrections is in progress.

Acknowledgments

I gratefully acknowledge the support of the NSF through grant No. PHY-9008449, and thank the members and staff of the UCLA Physics Department for their kind hospitality. I am indebted to B. Muller and H. Fried for organizing this Workshop, and to H. Grotch for inviting me to participate.

References

1. M.A. Stroscio, Phys. Rep. **22** (1975) 215.
2. S. Berko and H.N. Pendleton, Ann. Rev. Nucl. Part. Sci. **30** (1980) 543.
3. A. Rich, Rev. Mod. Phys. **53** (1981) 127.
4. J.R. Sapirstein and D.R. Yennie, in *Quantum Electrodynamics*, ed. T. Kinoshita (World Scientific, Singapore, 1990), p. 560.
5. A.P. Mills, Jr., and S. Chu, in *Quantum Electrodynamics*, ed. T. Kinoshita (World Scientific, Singapore, 1990), p. 774.
6. V.V. Dvoeglazov, R.N. Faustov, and Yu.N. Tyukhtyaev, Phys. Part. Nucl. **25** (1994) 58.
7. V.V. Dvoeglazov, R.N. Faustov, and Yu.N. Tyukhtyaev, Mod. Phys. Lett. A **8** (1993) 3263.
8. The conventions and natural units $[\hbar = c = 1, \alpha = e^2/4\pi \approx (137)^{-1}]$ of J.D. Bjorken and S.D. Drell, *Relativistic Quantum Mechanics* (McGraw-Hill, New York, 1964) are used throughout. The symbol m represents the electron mass.
9. E.E. Salpeter and H.A. Bethe, Phys. Rev. **84** (1951) 1232.
10. W.E. Caswell and G.P. Lepage, Phys. Rev. A **18** (1978) 810.
11. R. Barbieri and E. Remiddi, Nucl. Phys. B **141** (1978) 413.
12. W. Buchmüller and E. Remiddi, Nucl. Phys. B **162** (1980) 250.
13. G.P. Lepage, Phys. Rev. A **16** (1977) 863.
14. S. Love, Ann. Phys. (N.Y.) **113** (1978) 153.
15. G. Feldman, T. Fulton, and D.L. Heckathorn, Nucl. Phys. B **167** (1980) 364.
16. G. Feldman, T. Fulton, and D.L. Heckathorn, Nucl. Phys. B **174** (1980) 89.
17. H.M. Fried and D.R. Yennie, Phys. Rev. **112** (1958) 1391.
18. I call this gauge the "Yennie" gauge instead of the "Fried-Yennie" gauge in deference to Herb Fried's generous request that the contribution of Don Yennie be recognized in this way.
19. M. Deutsch and E. Dulit, Phys. Rev. **84** (1951) 601.
20. M.W. Ritter *et al.*, Phys. Rev. A **30** (1984) 1331.
21. S.G. Karshenboǐm, JETP **76** (1993) 541.
22. G.S. Adkins, Y.M. Aksu, and M.H.T. Bui, Phys. Rev. A **47** (1993) 2640.
23. D. Hagena *et al.*, Phys. Rev. Lett. **71** (1993) 2887.
24. R.S. Conti *et al.*, Phys. Lett. A **177** (1993) 43.
25. R.N. Fell, Phys. Rev. Lett. **68** (1992) 25.
26. M.S. Fee *et al.*, Phys. Rev. Lett. **70** (1993) 1397.
27. A.H. Al-Ramadhan and D.W. Gidley, Phys. Rev. Lett. **72** (1994) 1632.
28. I.B. Khriplovich and A.S. Yelkhovsky, Phys. Lett. B **246** (1990) 520.
29. C.I. Westbrook *et al.*, Phys. Rev. A **40** (1989) 5489.
30. J.S. Nico *et al.*, Phys. Rev. Lett. **65** (1990) 1344.
31. A. Ore and J.L. Powell, Phys. Rev. **75** (1949) 1696.
32. G.S. Adkins, A.A. Salahuddin, and K.E. Schalm, Phys. Rev. A **45** (1992) 7774.
33. A.P. Burichenko, Phys. At. Nucl. **56** (1993) 640.

QED BOUND STATE METHODS

D.A. OWEN
Physics Department, Ben Gurion University
Beer Sheva, Israel

ABSTRACT

We describe the main methods that have been used for calculating quantum electrodynamic corrections to two-particle bound states. These include the external field approximation, the quasi-potential approaches, the effective potential approach, the Bethe-Salpeter method as well. Also included is a discussion of the Duffin-Kemmer equation for which new work has just appeared. A brief comparison of these approaches is included as well.

1. Quantum Electrodynamics and Bound Systems

1.1. Weak Coupling Approximation and Spin

The bound states of quantum electrodynamics provide both simplicity and complexity. The simplicity is due to the small coupling constant, α, which governs the strength of the interactions between particles having electric charge. Since $\alpha^{-1}\approx 137$, a perturbative expansion such as energy levels, decay width, etc. in terms of α has not only a range of theoretical validity but, as well known, leads to results consistent with observation to a very high degree of accuracy.[1] This is, of course, in contradistinction from the situation which exists in strong interaction physics in which the coupling constant is of the order unity and for which a perturbation expansion in terms of the coupling constant is not valid.

On the other hand, systems which are most readily available as a test of quantum electrodynamics, contain constituent particles having spin -1/2 which requires a wave function with spinor components. This introduces a degree of complexity that increases the difficulty of evaluating higher-order terms in the perturbative expansion.

It should be noted that we shall not, in the following discussion, refer directly to the electro-weak theory for which quantum electrodynamics is currently viewed as an approximation. For the range of corrections which have been previously considered in quantum electrodynamic calculations, weak corrections are so minute that they can safely be ignored.

In what follows, we shall briefly mention which techniques are useful in calculating radiative corrections for various physical systems and then, starting with section 3, review each of these approaches by themselves and finally, we shall contrast each of them in comparison with the other approaches.

1.2. Hydrogen

In a system such as hydrogen, in which the proton's mass, $m_p\approx 1840\ m_e$ (where m_e is the electron's mass), recoil corrections are relatively small and the largest contribution for to any given order in α may be found by the approximation that $m_p=\infty$. In this approximation (the external field approximation), the energy levels of hydrogen may be found by considering the electron, with its radiative photons, in a fixed external Coulomb field of the proton. For corrections of order α^4th, the lowest order recoil effects are incorporated trivially by replacing, in the relevant formulas, the electron's mass with that of the reduced mass of hydrogen. It was Salpeter[2] who showed that at order $\alpha^3 Ry\frac{m}{M}$ that replacing the electron mass by the reduced mass was not sufficient for including all the recoil corrections to order α^5th.

However, for comparisons with very accurate data, higher order recoil corrections, such as quadratic terms as $\alpha^2\left(\frac{m_e}{m_p}\right)^2$ would be needed. These second order recoil terms are smaller by a factor of $3x10^{-7}$. Thus in magnitude, we see that $\alpha^2\left(\frac{m_e}{m_p}\right)^2 \approx \alpha^5$. In actual practice, its not until order of α^6 that quadratic recoil corrections become important. When this is the case, then one must go beyond the external field approximation even when considering contributions of order α^4th since, as we have just estimated indicates, quadratic recoil corrections to this order cannot be ignored and use the two-particle relativistic equations to evaluation these additional corrections is necessary.

1.3. Positronium and Muonium

For positronium, one cannot use the external field approximation to calculate radiative corrections to its energy levels. Clearly, for a system of two particles of identical mass, such as positronium, each of the particles must be treated on an equal footing. Thus the approximation that one particles has an infinite mass, makes no sense at all. A two-particle wave equation (or another technique which does not does not rely on the mass ratios of the particles comprising the system) must be used from the onset of the calculations. For the non-relativistic energy levels, the Schrödinger equation may be used. However, to calculate relativistic corrections, the Bethe-Salpeter equation (or one of its reduced forms) must be used.

Although muonium, does not strictly fall into the category of a system comprised of equal masses, recoil effects are much greater than those found in hydrogen since the mass ratio of the electron to muon is much larger than that for the electron to proton. In fact, $m_\mu \approx 207\ m_e$, so there is a range of validity of the external field approximation but it is much less than for hydrogen. This requires, that to obtain a reasonable comparison with experiment, recoil corrections must be calculated and the bound state theory pertaining to the relativistic, two-body quantum field theory, invoked.

1.4. Kaonic Hydrogen

The energy levels of exotic systems at present are not sufficiently well observed to determine the radiative corrections to their energy levels. In fact, when it does become possible, the most accessible levels to check the theory for spinless particles will be for non-S states (so strong interaction effects will not be overwhelming). For such systems, such as kaonic hydrogen, pionic hydrogen, etc. the complications of spin are absent but recoil corrections are expected to be large since the mass ratio of the constituents is not small. Thus, again a relativistic two-body formalism would have to be invoked. This requires a modification of the usual Bethe-Salpeter formalism for a system of two particles of spin 1/2 as it relied on the linearity of the momentum in the denominators of the fermion propagators.

1.5. Other Systems

There are many other systems for which radiative corrections have been considered, as for example, many electron atoms, etc. Due to the lack of expertise of the author, the techniques required to treat these systems, shall not be discussed.

2. General Theoretical Considerations

The electro-weak Lagrangian after symmetry breaking is given by

$$\mathcal{L}=\mathcal{L}_F+\mathcal{L}_{GF} \tag{2.1}$$

$$\mathcal{L}_F=\sum_i \bar{\psi}_i(i\gamma\partial - m_i - \frac{gm_i\varphi}{2M_W})\psi_i - \frac{g}{2\sqrt{2}}\sum_i \bar{\psi}_i\gamma^\mu(1-\gamma^5)(T^+W^+{}_\mu + T^-W^-{}_\mu)\psi_i$$
$$- e\sum_i q_i\bar{\psi}_i\gamma^\mu\psi_i A_\mu - \frac{g}{2\cos\theta_W}\bar{\psi}_i\gamma^\mu(V^i - A^i\gamma^5)\psi_i Z_\mu \tag{2.2}$$

$$\mathcal{L}_{GF}= -\frac{1}{4}F_{\mu\nu}F^{\mu\nu} - \frac{1}{2}|D_\mu W^+_\nu - D_\nu W^+_\mu|^2 + \frac{1}{2}g\,F_{\mu\nu}W^-_\rho S^{\mu\nu\rho\sigma}W^+_\sigma$$
$$-\frac{1}{4}g^2 W^-_\rho S^{\mu\nu\rho\delta}W^+_\sigma W^{-\alpha}S_{\mu\nu\alpha\beta}W^{+\beta} \tag{2.3}$$

where φ is the Higgs field, T's are the weak isospin operators. Here $V^i \equiv t_{3L}(i)-2q_i\sin\theta_W$ and $A^i \equiv t_{3L}(i)$ where $t_{3L}(i)$ is the weak isospin of fermion i, and q_i is the charge of ψ_i (in units of e) with $D_\mu=\partial_\mu+igA_\mu$ and $S_{\mu\nu\rho\sigma}=i(g_{\mu\rho}g_{\nu\sigma}-g_{\mu\sigma}g_{\nu\sigma})$.

For most considerations of the energy levels of systems bound primarily by the Coulomb interaction, terms involving W's can be dropped. This is because that terms involving the propagator of a W-gauge boson are 'down' by a factor of M_w^2 (i.e. the mass of the W-boson squared). Thus, the relevant coupling constant for these terms is $\frac{g^2}{M_w^2}$ which is proportional to the Fermi constant $G\approx 1.43\times10^{-49}$erg cm^3.[3] When this is not the case, the terms depending on W can be treated as perturbations and their contributions to the energy levels can be calculated.

3. External Field Approximation

In this approach, that is, of the external field approximation, we consider a two particle system[4] (such as the hydrogen atom) in which one of the particles is relatively much less massive than the other one. The relevant terms comprising the external field Lagrangian, $\mathcal{L}_{EF}$, from Eqs. (2.2) and (2.3) are

$$\mathcal{L}_{EF}= \mathcal{L}_{FD} + \mathcal{L}_{ph} + \mathcal{L}_{int} \tag{3.1}$$

where

$$\mathcal{L}_{FD}= \bar{\psi}_e(i\gamma\partial - m_e)\psi_e \tag{3.2}$$

$$\mathcal{L}_{ph} = -\frac{1}{4}F_{\mu\nu}F^{\mu\nu} \tag{3.3}$$

$$\mathcal{L}_{int} = -e_i\bar{\psi}_e\gamma^{\mu}\psi_e\left(A_{\mu}+A_{\mu}^{(ext)}\right) \tag{3.4}$$

In these expressions, ψ_e, is the field operator of the less massive constituent whose mass is m_e with electric charge e. A_{μ} is the radiative field produced by this charge, e. The only effect of the massive particle is the production of the external field $A_{\mu}^{(ext)}$ in which the other particle moves. In what follows, we take the external field to be that produced by a Coulomb potential, $A_0^{(ext)} = A_0^{(C)}$. Defining the electron propagator (for convenience, we refer to the lighter particle as an electron) in the usual way, i.e.

$$S_F^{(e)\prime}(x,x') = -i\langle 0|T[\psi_e(x)\bar{\psi}_e(x')]|0\rangle \tag{3.5}$$

$S_F^{(e)\prime}$ can be written in the following form[5]

$$S_F^{(e)\prime} = \frac{1}{\not{p} - e\gamma^0 A_0^{(C)} - m - \Sigma^{(e)}} \tag{3.6}$$

where the precise form of $\Sigma^{(e)}$ can be determined by a Wick expansion of the r.h.s. of Eq. (3.5). Furthermore, expanding ψ_e in a complete set of energy eigenstates which we call, φ_K, if follows from Eq. (3.6) that φ_K satisfies the equation

$$\gamma^0\left(K - eA_0^{(C)} - H(\mathbf{p}) - \beta\,\Sigma^{(e)}\right)\varphi_K = 0 \tag{3.7}$$

where $H(\mathbf{p})$ is the free-Dirac Hamiltonian given by $H(\mathbf{p}) = \boldsymbol{\alpha}\cdot\mathbf{p} + \beta m_e$. Equation (3.7) cannot be solved directly and a perturbation strategy is required for calculating the energy eigenvalue, K, to any required order in the fine structure constant, α. Our strategy is to connect, in some way, to be specified, the wave functions for the Dirac equation with a Coulomb potential, which can be solved exactly to Eq. (3.7). We designate the Dirac wave function, $\varphi_{Kc}^{(D)}$ to be the solution of

$$[K_c - eA_0^{(C)} - H(\mathbf{p})]\varphi_{Kc}^{(D)} = 0 \tag{3.8}$$

For our purposes, it is useful to write Eq. (3.8) as follows where K is the eigenvalue found in Eq. (3.7)

$$[K - eA_0^{(C)} - H(\mathbf{p})]\varphi_{Kc}^{(D)} = (K-K_c)\,\varphi_{Kc}^{(D)} \tag{3.9}$$

If Eq. (3.7) is multiplied on the left by $\varphi_{Kc}^{(D)}$ and using Eq. (3.9), we find that[6]

$$\Delta K = K - K_c = \frac{\bar{\varphi}^{(D)}_{Kc}\, \Sigma^e\, \varphi_K}{\bar{\varphi}^{(D)}_{Kc} \varphi_K} \tag{3.10}$$

The wave function φ_K can be found in terms of $\varphi^{(D)}_{Kc}$ from Eq. (3.7) as follows:

$$\left(K - eA^{(C)}_0 - H(\mathbf{p}) - \beta\, \Sigma^{(e)}\right)\left(\varphi^{(D)}_{Kc} + (\varphi_K - \varphi^{(D)}_{Kc})\right) = 0 \tag{3.11}$$

or using Eq. (3.9) again, Eq. (3.11) becomes

$$\Delta K \varphi^{(D)}_{Kc} - \beta\Sigma^{(e)}\, \varphi^{(D)}_{Kc} + \beta G^{-1}_{Fe}\, (\varphi_K - \varphi^{(D)}_{Kc}) = 0 \tag{3.12}$$

where

$$G^{-1}_{Fe} = (K - eA^{(C)}_0)\gamma^0 - \boldsymbol{\gamma}\mathbf{p} - m_e - \Sigma^{(e)} \tag{3.13}$$

Thus, from Eq. (3.12) if follows that

$$\varphi_K = \varphi^{(D)}_{Kc} + G_{Fe}\, \Sigma^{(e)}\, \varphi^{(D)}_{Kc} - \Delta K\, G_{Fe}\, \beta\, \varphi^{(D)}_{Kc} \tag{3.14}$$

Using Eq. (3.14) in Eq. (3.10) we find

$$\Delta K \approx \varphi^{(D)*}_{Kc}\, \Sigma^{(e)} \varphi^{(D)}_{Kc} + \varphi^{(D)*}_{Kc} \Sigma^{(e)} S^{(e)}_F \Sigma^{(e)} \varphi^{(D)}_{Kc} \tag{3.15}$$

Yennie with his co-workers[7,8] succeeded in using this approach and refining it in his calculations of the Lamb shift to calculate higher order corrections in the find structure constant.

4. Quasi-Potential Approach

This approach initially proposed by Logunov and Tavkhelidze[9] and utilized by Rizov[10] and Todorov[11] for calculating quantum electrodynamic corrections as well as corrections to systems with strong interactions. The basic idea behind this approach is to obtain from the scattering matrix amplitudes a "quasi-potential" which can be used with the Schrödinger equation to obtain energy levels.

For example, consider the T-matrix amplitude calculated from the S-matrix for the scattering of two particles, initially with momentum q_1 and q_2 and respectively with spin coordinate η_1 and η_2 and for the final state, the momentum, p_1 and p_2 with the spin coordinates ζ_1 and ζ_2 respectively. Then we may write[12]

$$\begin{aligned}\langle p_1,\zeta_1,p_2,\zeta_2|S|q_1,\eta_1,q_2,\eta_2\rangle &= (2\pi)^6(2p_{10})(2p_{20})\delta_{\eta_1\zeta_1}\delta_{\eta_2\zeta_2}\delta(\mathbf{p}_1-\mathbf{q}_1)\delta(\mathbf{p}_2-\mathbf{q}_2) \\ &\quad + (2\pi)^4 i\, \delta(p_1+p_2-q_1-q_2)\, T(p_1,\zeta_1,p_2,\zeta_2;q_1,\eta_1,q_2,\eta_2)\end{aligned} \tag{4.1}$$

The following Lippmann-Schwinger type equation with the external momenta on the energy shell, $p_1+p_2=q_1+q_2$, (but with the requirement that the off-mass extrapolation should be $p_1{}^2 - p_2{}^2 = m_1{}^2 - m_2{}^2 = q_1{}^2 - q_2{}^2$) is postulated to be

$$T+V+V*GT\equiv T_w(\mathbf{p},\mathbf{q}) + V_W(\mathbf{p},\mathbf{q}) + \int\frac{d^3k}{(2\pi)^3}V_W(\mathbf{p},\mathbf{k})G_W(\mathbf{k})T_w(\mathbf{k},\mathbf{q})=0 \qquad (4.2)$$

where

$$E_1+E_2=w=\sqrt{s}\ ,\ T_W(\mathbf{p},\mathbf{q})\equiv T(p_1,\zeta_1,p_2,\zeta_2;q_1,\eta_1,q_2,\eta_2). \qquad (4.3)$$

If G_W is known, then Eq. (4.2) defines the potential V_W. To see this, consider the weak coupling limit. In this limit, one can expand the scattering amplitude, T (as an expansion in terms of the coupling constant) as $T=T_1+T_2+....$ If we formally write $V=V_1+V_2+...$ as an expansion in terms of the coupling constant, we have from the above Lippmann-Schwinger equation, Eq. (4.2), that

$$V_1=-T_1,\ V_2=-T_2+V_1*GV_1, \quad \text{etc.} \qquad (4.4)$$

Thus, if the Green's function, G, is known, then all the V_i can be determined as in Eq. (4.4).

One of the ways in which the Green's function used in Eq. (4.4) was chosen was by the requirement that the theory satisfy two-body unitarity and that the inverse of the Green's function, $[G_W(\mathbf{k})]^{-1}$, go as $\mathbf{k}^2$. These two requirements determine $G_W(\mathbf{k})$ uniquely. The requirement that the theory satisfy two-body unitarity determines the imaginary part of G_w. To see this, we write the formal solution of Eq. (4.2) as

$$T=-V(1+VG)^{-1} = -(1+VG)^{-1}V \qquad (4.5)$$

This is substituted into the unitarity condition, i.e.

$$T - T^* = i\,T^*\,T \qquad (4.6)$$

However, using Eq. (4.5) in l.h.s. of Eq. (4.6) we have

$$\begin{aligned} T - T^* = V(1+G^*V)^{-1} - (1+VG)^{-1}V &= (1+G^*V)^{-1}V(G-G^*)(1+VG)^{-1}V \\ &= i\,T^*\,\mathrm{Im}G\,T \end{aligned} \qquad (4.7)$$

Comparing this result in Eq. (4.7) with the r.h.s. of Eq. (4.6) we have (where now we have written in the arguments specifically)

$$\mathrm{Im}G = \frac{\pi}{2w}\,\delta(\mathbf{k}^2-b^2) \qquad (4.8)$$

Furthermore, the requirement that G^{-1} is a linear function of $\mathbf{k}^2$ determines G_w:

$$G_W(\mathbf{k}) = \frac{1}{2w}\,\frac{1}{\mathbf{k}^2-b(s)-i\varepsilon} \qquad (4.9)$$

with

$$b(s) = \frac{s^2 - 2s(m_1{}^2+m_2{}^2)+(m_1{}^2-m_2{}^2)^2}{4s}. \qquad (4.10)$$

If we now define the wave function $\phi_q(\mathbf{p})$ by

$$\phi_q(\mathbf{p})=(2\pi)^3\delta(\mathbf{p}-\mathbf{q}) + G_q(\mathbf{p})\,T_q(\mathbf{p},\mathbf{q}) \qquad (4.11)$$

which can be rewritten as by multiplying both sides by G^{-1}

$$G_q(\mathbf{p})^{-1}\,\phi_q(\mathbf{p}) = T_q(\mathbf{p},\mathbf{q}) \tag{4.12}$$

where we have taken q to be the total energy of a bound state. Using Eq.(4.2) in Eq. (4.12) we can write

$$\begin{aligned} G_q(\mathbf{p})^{-1}\,\phi_q(\mathbf{p}) &= -\int\frac{d^3k}{(2\pi)^3}\Big[V_W(\mathbf{p},\mathbf{k})\big((2\pi)^3\delta(\mathbf{k}-\mathbf{q}) + G_W(\mathbf{k})T_W(\mathbf{k},\mathbf{q})\big)\Big] \\ &= -\int\frac{d^3k}{(2\pi)^3}V_q(\mathbf{p},\mathbf{k})\,\phi_q(\mathbf{k}) \end{aligned} \tag{4.13}$$

which can be rewritten as

$$[\mathbf{p}^2 - b^2(w^2)]\phi_W(\mathbf{p}) + \int V_W(\mathbf{p},\mathbf{k})\phi_W(\mathbf{k})\frac{d^3k}{(2\pi)^3} = 0 \tag{4.14}$$

This equation can be used as the starting point for bound state calculations. The use of Eq. (4.14) in this fashion, is one of the quasi-potential approaches for the calculation of bound state energy levels.

5. Relativistic Two-Particle Equations of Quantum Field Theory

The relevant terms in the Lagrangian given in Eqs.(2.2) and (2.3) are given by $\mathcal{L}_{qed}$ where

$$\mathcal{L}_{qed} = \mathcal{L}_{em} + \mathcal{L}_{FD} + \mathcal{L}_{int} \tag{5.1}$$

with

$$\mathcal{L}_{em} = -\tfrac{1}{4}F_{\mu\nu}F^{\mu\nu} \tag{5.2}$$

$$\mathcal{L}_{FD} = \bar{\psi}_a(i\gamma_a\partial - m_a)\psi_a + \bar{\psi}_b(i\gamma_b\partial - m_b)\psi_b \tag{5.3}$$

$$\mathcal{L}_{int} = -e\sum_{1} q_i\bar{\psi}_i\gamma^\mu\psi_i A_\mu \tag{5.4}$$

5.1. Breit Equation

The Breit equation[13] has provided a method for including recoil, radiative corrections for a two particle system. It is conceptually simple to formulate but it is, however, an approximate equation. Essentially one writes the a free-Dirac Hamiltonian for each particle plus the interaction terms between them. That is

$$\left(E - H^{(1)} - H^{(2)} + \frac{\alpha}{r}\right)\varphi = \frac{\alpha}{2r}\left(\boldsymbol{\alpha}_1\cdot\boldsymbol{\alpha}_2 + \frac{(\boldsymbol{\alpha}_1\cdot\mathbf{r})(\boldsymbol{\alpha}_2\cdot\mathbf{r})}{r^2}\right)\varphi \tag{5.5}$$

In this equation $H^{(i)}=\boldsymbol{\alpha}_i\cdot\mathbf{p}_i + \beta m_i$ and r is the distance between the particles. The r.h.s. of Eq.(5.5) is the effective potential for an exchanged, static transverse photon. Energy levels to order α^4 can be calculated by the above equation. A greater physical understanding of this equation is obtained if a Foldy-Wouthuysen transformation is applied to Eq.(5.5). The results first obtained by Barker and Glover.[14]

5.2. Bethe-Salpeter Equation

An exact equation for the two-particle Green's function can be derived directly from field theory. In the most general case which we consider, the Lagrangian given by Eqs. (2.2) and (2.3) can be used. Bethe-Salpeter equation[15] can be written in the following form[16]

$$G(1,2;3,4) = G_0(1,2;3,4) + G_0(1,2;5,6)\, I\, (5,6;7,8)\, G(7,8;3.4) \tag{5.6}$$

In this equation, G is the exact, two particle Green's function, G_0 is the free-two particle Green's function which is given by the product of the two, exact single particle Green's functions and I is the interaction kernel. Equation (5.6) can be used, via the Wick expansion, to display the interaction kernel, explicitly. From Eq. (5.6) the equation for a bound state can be derived using the condition that bound states are localized. One obtains

$$f_K(1,2) = G_0(1,2;3,4)I(3,4;5,6)f_K(5,6) \tag{5.7}$$

where f_K is the two-particle relativistic wave function of energy K (which of course, is less that m_1+m_2). In C.M. coordinates one can write $f_K(1,2)=\phi_K(x)\, e^{-iK\cdot X}$ where X is the center of mass coordinate and $x=x_1-x_2$ is the relative coordinate, then Eq. (5.7) can be written as

$$\phi_K(x) = G_0(x,x')\, I_K(x',x'')\, \phi_K(x'') \tag{5.8}$$

In this form, all the self-energy insertions in G_0 can be associated with the kernel I_K[17] so we can schematically write in place of Eq. (5.8),

$$\phi_K=(S_F^{(1)}\, S_F^{(2)})\tilde{I}_K\, \phi_K \tag{5.9}$$

where

$$(S_F^{(1)}\, S_F^{(2)})(x,x') = \int \frac{d^4p}{(2\pi)^4} e^{-ipx} \frac{1}{(\eta_1 K+p_0)\gamma^0-\boldsymbol{\gamma}\cdot\mathbf{p}-m_1}\, \frac{1}{(\eta_2 K-p_0)\gamma^0-\boldsymbol{\gamma}\cdot(-\mathbf{p})-m_2}$$

$$= \int \frac{d^4p}{(2\pi)^4} e^{-ipx} \frac{1}{K - H^{(1)} - H^{(2)}} \left(\frac{1}{\eta_1 K+p_0 - H^{(1)}(\mathbf{p})} + \frac{1}{\eta_2 K - p_0 - H^{(2)}(-\mathbf{p})} \right)\beta_1\beta_2 \tag{5.10}$$

with $\eta_1+\eta_2 = 1$. In the case of atomic systems, the Coulomb interaction is mainly responsible for most of the binding energy. Designating the corresponding Coulomb kernel by I_C and $I'_K = \tilde{I}_K - I_C$, Eq. (5.9) can be written in a more recognizable form, namely

$$[K - H^{(1)}(\mathbf{p}) - H^{(2)}(-\mathbf{p}) + \frac{\alpha}{r}]\, \phi_K = \Omega(x,\mathbf{x}')\frac{\alpha}{r'}\phi_K(x_0,\mathbf{x}') + \Lambda_K(x,x')I'_K(x'',x''')\, \phi_K(x''') \tag{5.11}$$

where we have assumed that both particles comprising the bound state have unit charge and

$$\Omega(\mathbf{x},\mathbf{x}') = \Lambda(\mathbf{x},\mathbf{x}') - \delta(\mathbf{x},\mathbf{x}') \tag{5.12}$$

i) Bethe-Salpeter perturbation theory

Equation (5.10) can be used as the starting point in developing the perturbation for bound systems. A useful approach is to consider the last term in Eq. (5.11) as a perturbation to the unperturbed equation

$$[K_c - H^{(1)}(\mathbf{p}) - H^{(2)}(-\mathbf{p}) + \frac{\alpha}{r}]\,\phi_{Kc}(\mathbf{x}) = \Omega(\mathbf{x},\mathbf{x}')\frac{\alpha}{r'}\phi_{Kc}(\mathbf{x}') \tag{5.13}$$

By defining a 'hybrid' wave function by

$$\phi'_K(x) = -i\Lambda(x;0,\mathbf{x}')[K_c - H^{(1)}(\mathbf{p}) - H^{(2)}(-\mathbf{p})]^{-1}\frac{\alpha}{r'}\,\phi_{Kc}(\mathbf{x}') \tag{5.14}$$

The choice of the hybrid wave function is dictated by the requirement that

$$\phi'_K(0,\mathbf{x}) = \phi_{Kc}(\mathbf{x}') \tag{5.15}$$

but has the time dependence governed by the eigenvalue K. It follows that[18]

$$\Delta K = K - K_c = -i\,\bar{\phi}'_K(x)I'_K(x,x')[\phi'_K(x') + \bar{G}_K(x',x'')I'_K(x'',x''')\phi'_K(x''')\phi_K(x''')$$

$$-i\bar{G}_K(x',x'')\gamma_0^{(1)}\gamma_0^{(2)}\chi(\mathbf{x}')\Delta K]$$

$$\cong -i\{\bar{\phi}'_K(x)I'_K(x,x')\phi'_K(x') + \bar{\phi}'_K(x)I'_K(x,x')\bar{G}_K(x',x'')I'_K(x'',x''')\phi'_K(x''') \tag{5.16}$$

where

$$\bar{G}_K = \frac{1}{[G_K^c]^{-1} - I'_K} = G_K^c + G_K^c\, I'_K G_K^c + \ldots \tag{5.17}$$

and

$$\chi(\mathbf{x}) \equiv [K_c - H^{(1)}(\mathbf{p}) - H^{(2)}(-\mathbf{p})]^{-1}\frac{\alpha}{r'}\,\phi_{Kc}(\mathbf{x}) \tag{5.18}$$

In principle one can use Eq. (5.16) to calculate to any order thc corrections to the Coulomb energy levels which are given by Eq. (5.13). Although the calculations can be performed and many have been done, the method is tedious and arduous and requires a great deal of care especially since Eq. (5.13) cannot, itself, be solved exactly and must be solved perturbatively.

In the above treatment, we explicitly assumed that the bound state was comprised of two-fermions. The theory can be developed in a similar fashion for systems comprised of a fermion and a scalar particle or comprised of two scalar particles.[19]

ii) On Mass-Shell Various of the Bethe-Salpeter approach

Among those whom have investigated the Bethe-Salpeter approach with one of the particles on mass shell are Grotch and Gorelick[20], Lepage,[21] Blankenbecler and Sugar,[22] F. Gross,[23] etc.[24]

One of the initial approaches was pursued by Grotch & Yennie.[25] They realized that for many bound state calculations, especially those in which one of the masses of the two particle bound state is much heavier than the other, the Bethe-Salpeter formalism is much more complicated than is needed for the results obtained. They obtained an approximate equation whose formal relation to the Bethe-Salpeter equation was later clarified by Grotch and Gorelick.[20]

In particular, Grotch and Yennie by assuming that one of the particles comprising the bound state was much heavier than the other and that this particle could be treated non-relativistically, they were able to reduce the problem to that of a modified Dirac equation for which the solution is known. Key to this work, was to treat the more massive particle as though it were on shell. This work gave impetus to the work of Lepage[21] in which he found the exact connection of this approach to the Bethe-Salpeter equation. One of the main motivations in this approach is to maintain the rigor of the Bethe-Salpeter equation but to bypass some of its complexities. Also this approach bypasses one of the dilemmas in using the Bethe-Salpeter equation, which is that the Bethe-Salpeter perturbation theory takes as its starting point, the non-relativistic Shröedinger equation and thus all relativistic corrections must be calculated perturbatively. On the other hand, in the Grotch and Yennie[25] approach one is able to relate the non-perturbed energy levels to a modified Dirac equation. Hence, the non-perturbed energy levels include not only relativistic corrections to the electron but the relativistic Coulomb corrections as well with the correct infinite mass limit. The latter limit can only be obtained in the Bethe-Salpeter formalism by the inclusion of an infinite number of terms from the interaction kernel (i.e. irreducible two-particle Feynman graphs) in the perturbation expansion. Because of these advantages we shall briefly describe their method.

The initial starting point for Grotch and Yennie[25] is the equation

$$[E-H(\mathbf{x}_e)-H(\mathbf{x}_p)]\psi(\mathbf{x}_e,\mathbf{x}_p) = V(\mathbf{x}_e-\mathbf{x}_p)\psi(\mathbf{x}_e,\mathbf{x}_p) \tag{5.19}$$

Using the Lippmann-Schwinger equations, one may obtain the scattering predicted by Eq. (5.19). The S-matrix becomes ($f\neq i$)

$$-2\pi i\delta(E_1+E_2-E_3-E_4)\ \langle\phi_f|V|\psi_i^+\rangle \tag{5.20}$$

where E_1 and E_2 are the incoming particle energies and E_3 and E_4 are the energies of the outgoing particles. An eigenstate of the free Hamiltonian representing an outgoing state is given by $\langle\phi_f|$ while that of an eigenstate of the full Hamiltonian by $|\psi_i^+\rangle$ having incoming plane waves $|\phi_i\rangle$ and outgoing scattering waves.

By calculating the scattering amplitude directly and with Eq. (5.20), V can be determined as shown explicitly by Grotch and Yennie.[25] For the free Hamiltonian they took

$$H_{free}= \boldsymbol{\alpha}\cdot\mathbf{p} + \beta m + \frac{\mathbf{p}^2}{2M} \tag{5.21}$$

where $m \ll M$. In the particular case studied by Grotch and Yennie, m is the electron mass while M is the mass of the proton.

They found convenient to define the lowest-order potential, V, as

$$V(\mathbf{p}_1-\mathbf{p}_3) \equiv -\frac{e^2}{(\mathbf{p}_1-\mathbf{p}_3)^2}(1-\boldsymbol{\alpha}_\perp\cdot\boldsymbol{\alpha}_p) \tag{5.22}$$

If this potential is evaluated, to order (1/M) between free proton spinors, one finds

$$V_{eff}^{(2)} + \Delta V_{eff}^{(2)} = -\frac{e^2}{\mathbf{q}^2}\left[1 + \boldsymbol{\alpha}_\perp \cdot \left(\frac{\mathbf{p}_1+\mathbf{p}_2}{2M}\right)\right] - \frac{i}{2M}\frac{e^2}{\mathbf{q}^2}\boldsymbol{\alpha}\cdot(\mathbf{q}\mathrm{x}\sigma_p) \qquad (5.23)$$

If one disregards in the lowest approximation $\Delta V_{eff}^{(2)}$, then $V_{eff}^{(2)}$ in position space can be written as

$$V_{eff}^{(2)} = V + \{\boldsymbol{\alpha}\cdot\mathbf{p}/2M, V\} + (1/4M)[\boldsymbol{\alpha}\cdot\mathbf{p},[\mathbf{p}^2,W]] \qquad (5.24)$$

where $V=-\alpha/r$ and $W = -\alpha r$.

Using the free Hamiltonian Eq. (5.21) and the effective potential in Eq (5.24) we arrive at modified Dirac equation

$$\left(\boldsymbol{\alpha}\cdot\mathbf{p} + \beta m + \frac{\mathbf{p}^2}{2M} + V + \{\boldsymbol{\alpha}\cdot\mathbf{p}/2M, V\} + (1/4M)[\boldsymbol{\alpha}\cdot\mathbf{p},[\mathbf{p}^2,W]]\right)\psi(\mathbf{r}) = E\,\psi(\mathbf{r}) \qquad (5.25)$$

At this point, they are able to reduce the problem of find E to solving a further modified Dirac equation given by

$$\left[\boldsymbol{\alpha}\cdot\mathbf{p} + \beta m\left(\frac{1-(E_1/M)}{\sqrt{[1-(m^2/M^2)]}}\right) + \frac{V}{\sqrt{[1-(m^2/M^2)]}}\right]\psi_0 = \frac{E_1[1-(m^2/E_1M)]}{\sqrt{[1-(m^2/M^2)]}}\psi_0 \qquad (5.26)$$

where E_1 is defined as the energy eigenvalue of the Hamiltonian given by

$$H_1 = \boldsymbol{\alpha}\cdot\mathbf{p} + \beta m + V\left(\frac{1-\beta(m/M)}{1-(m^2/M^2)}\right) \qquad (5.27)$$

The Hamiltonian in Eq. (5.25) can be rewritten as

$$H = H_1 + \frac{H_1^2 - m^2}{2M} + (4M)^{-1}[H_1,[\mathbf{p}^2,W]] + \frac{m}{2(M^2 - m^2)}\left\{\boldsymbol{\alpha}\cdot\mathbf{p}, V\beta(1-\beta\frac{m}{M})\right\}$$

$$+\frac{m\beta V}{2M^2}\frac{[1-\beta(m/M)]^2}{[1-(m^2/M^2)]^2} \qquad (5.28)$$

where E is the eigenvalue of H. Using the above equations the reduce mass corrections to the fine structure as well as the hyperfine structure is found by the above method which is relatively conceptually simple. This approach could be thus categorized as lying being between that of quasi-potentials and that of the Bethe-Salpeter equation.

6. Other Approaches

6.1..Duffin-Kemmer Equations

The Duffin-Kemmer[26] equations have been used by Akhiezer and Berestetskii[27] to examine quantum electrodynamic processes involving scalar particles such as bremsstrahlung, annihilation of scalar particles going into e^+e^-, vacuum polarization, etc. One starts with a 'Dirac' type equation, namely

$$\left(i\beta_\mu \frac{\partial}{\partial_\mu} - m\right)\psi = 0 \tag{6.1}$$

where ψ can represent a spin 0 or spin 1 field. The following condition is imposed on the β's (for either spin 0 or spin 1 fields):

$$\beta^\mu\beta^\nu\beta^\lambda + \beta^\lambda\beta^\nu\beta^\mu = g^{\mu\nu}\beta_\lambda + g^{\lambda\nu}\beta_\mu \tag{6.2}$$

For the case of scalar particles, ψ is a five component object while for spin one, they have ten components. For the five component case, the wave function of the scalar particle and the other four, are its derivatives with respect to each of the coordinates. In the ten component case, four components appear to be redundant but can be given physical significance for the massless case.[28] The conditions given by Eq. (6.2) are needed to reproduce the Klein-Gorden equation. To see this we can first show that Eq. (6.1) and Eq. (6.2) implies that

$$\partial_\mu\beta^\mu\beta_\nu\psi = \partial_\nu\psi \tag{6.3}$$

ψ being the solution of Eq. (6.1).

To obtain Eq. (6.3) consider the following expression

$$\frac{1}{2}\left(\partial_\mu\beta^\mu\beta^\nu\beta^\lambda i\partial_\lambda + \partial_\lambda\beta^\lambda\beta^\nu\beta^\mu i\partial_\mu\right)\psi \tag{6.4}$$

which by Eq. (6.1)

$$\begin{aligned}\frac{1}{2}\left(\partial_\mu\beta^\mu\beta^\nu\beta^\lambda i\partial_\lambda + \partial_\lambda\beta^\lambda\beta^\nu\beta^\mu i\partial_\mu\right)\psi &= \frac{1}{2}m(\partial_\mu\beta^\mu\beta^\nu + \partial_\lambda\beta^\lambda\beta^\nu)\psi\ . \\ &= m\partial_\mu\beta^\mu\beta^\nu\psi\end{aligned} \tag{6.5}$$

On the other hand, using Eq. (6.2) we can write for Eq. (6.5)

$$\begin{aligned}\frac{i}{2}\partial_\mu\partial_\lambda\left(\beta^\mu\beta^\nu\beta^\lambda + \beta^\lambda\beta^\nu\beta^\mu\right)\psi &= \frac{i}{2}\partial_\mu\partial_\lambda(g^{\mu\nu}\beta_\lambda + g^{\lambda\nu}\beta_\mu)\psi \\ &= m\partial_\nu\psi\end{aligned} \tag{6.6}$$

So together, Eq. (6.5) and Eq. (6.6) give Eq. (6.3). To obtain the Klein-Gordon equation from Duffin-Kemmer equation we note, using Eq. (6.3) that

$$\partial_\mu\beta^\mu\partial_\nu\beta^\nu\psi = \partial_\nu(\partial_\mu\beta^\mu\beta^\nu\psi) = \partial_\nu\partial^\nu\psi \tag{6.7}$$

However, using the Duffin-Kemmer equation, Eq (6.1) in Eq. (6.7) we have the Klein-Gordon equation

$$\partial_\mu\beta^\mu\partial_\nu\beta^\nu\psi = -\,i\partial_\mu\beta^\mu i\partial_\nu\beta^\nu\psi = -m^2\psi$$

$$= \partial_\nu \partial^\nu \psi$$

That is

$$(\partial_\nu \partial^\nu + m^2)\psi = 0 \qquad (6.8)$$

Until the recent work of Nedjadi and Barrett,[28] little work had been done using the Duffin-Kemmer equation for bound state problems. They obtained the bound state solutions for both a scalar and vector particles in an external central potential.[28] In particular, they considered both Coulomb and harmonic oscillator potentials . For the spin one case, it was not possible to solve the equations analytically, but for the spinless case, there was agreement, as we would expect, to the results of the Klein-Gordon equation.

To proceed to a field theory approach, we would begin by writing the Lagrangian needed to recover Eq. (5.12)

$$L_0 = \frac{1}{2}\left(\bar{\psi} i\beta_\mu \frac{\partial \psi}{\partial x_\mu} - \frac{\partial \bar{\psi}}{\partial x_\mu} i\beta_\mu \psi\right) - m\bar{\psi}\psi \qquad (6.9)$$

The electromagnetic field can be introduced in the usual way, that is, requiring Eq. (6.9) to be invariant under a local U(1) gauge transformation.

At this point, the theory can be development in complete analogy to the case of fermions. The S-matrix can be defined and various amplitudes of interest can be calculated and the location of the poles of the S-matrix can be examined to determine shifts in spectra due to external potentials.

It should be mentioned, that one of the attractive features following from using Duffin-Kemmer equations is that the Duffin-Kemmer propagator can be expressed in covariant form in contrast to the propagator given by the Proca equation. It is this feature which has revived interest in this approach in considerations concerning vector bosons in bound states.

6.2. Tamm-Dancoff Method

Despite its apparent generality, the Tamm-Dancoff method[29,30] this methods has not seen serious applications for calculating radiative corrections to atomic systems. It has been purported that gauge difficulties arise when one tries to invoke this method.[31] The idea behind this method is simple enough. From Eqs. (2.2) and (2.3), one can in the usual fashion, find the corresponding Hamiltonian, H.

i) Pair theory

The eigenvalue equation for the energy can be written symbolically by

$$(E - H_0)|\Psi\rangle = H'|\Psi\rangle \qquad (6.10)$$

where H_0 is the Hamiltonian of the 'bare' particles (in terms of bare mass) and fields. H' is the Hamiltonian for the interaction of the particles with the gauge fields. The Tamm-Dancoff method is to expand $|\Psi\rangle$ in terms of the eigenstates of H_0 which can contain any number of 'bare' particle and photons. We shall try to illustrate this procedure for a number of cases we have considered above, namely, the external field approximation, the two-particle bound state problems.

(i) External field problem

Here, we examine in the context of the Tamm-Dancoff method the case where the Lagrangian is given by $\mathcal{L}_{EF}$ given in Eq. (3.1). The corresponding unperturbed Hamiltonian, H_0 is given by

$$H_0 = \frac{1}{2}(B^2 + E^2) + \psi^\dagger(\boldsymbol{\alpha}\cdot\mathbf{p} + \beta m_e)\psi \tag{6.11}$$

and the state vector $|\Psi\rangle$ we require it to have one net electron. That is,

$$|\Psi\rangle = b^\dagger(p,s)\varphi(p,s) + b^\dagger(k,s)d^\dagger(k',s')b^\dagger(p,s)\phi_{1pair}(p,s)+\ldots \tag{6.12}$$

The interaction Hamiltonian, H' consist of an external field contribution and one from the radiative field. That is,

$$H' = -\frac{\alpha}{r} + H_0^{Coul} + e\bar{\psi}\boldsymbol{\gamma}\psi\cdot\mathbf{A} \tag{6.13}$$

where

$$H_0^{Coul} = \frac{e_0^2}{(8\pi)^2}\int dt\, d^3x\, d^3y\, \frac{:\psi^\dagger(\mathbf{x},t)\psi(\mathbf{x},t)::\psi^\dagger(\mathbf{y},t)\psi(\mathbf{y},t):}{|\mathbf{x} - \mathbf{y}|} \tag{6.14}$$

To be able to proceed further, we must introduce the unitary transformation which carries the interacting fields into their free counter-part. That is[32]

$$\phi(\mathbf{x},t) = U^{-1}\phi_0(\mathbf{x},t)U \tag{6.15}$$

An apparent limitation of the above approach is that our initial equation, Eq. (6.10) is itself, an approximation. This can be seen by comparison of the Bethe-Salpeter equation, which is written in Eq. (5.11) in a similar form. Equation (6.10) is a generalization of the Breit equation.

(ii) Field theory approach

A more rigorous approach to Tamm-Dancoff method has been discussed by Rizov and Todorov.[33] Their objective is to obtain a one-time equation. They consider the following matrix element

$$\Phi_{nB}(x_1,x_2) = \langle n|\psi_1(x_1)\psi_2(x_2)|B\rangle \tag{6.16}$$

where ψ_1 and ψ_2 represent opposite charged fields, $|B\rangle$ is a bound state corresponding to particles 1 and 2 and $|n\rangle$ can be any physical state. Neglecting as we have previously done, all but the electromagnetic interactions we can write the Heisenberg equation for each of the ψ fields. That is

$$\left[i\frac{\partial}{\partial x_a^0} - H_a(\mathbf{p})\right]\psi_a(x_a) = e_a(\alpha^{(a)})_\mu A^\mu(x_a)\,\psi_a(x_a), \qquad \alpha=1,2 \tag{6.17}$$

where H_a is the free Dirac Hamiltonian. At this point it is assumed that x_1-x_2 is space-like and a frame is then chosen such that

$$x_1^0 = x_2^0 = t \tag{6.18}$$

Equation (6.8) yields the following system of equations for the matrix element, Eq.(6.16)

$$\left[i\frac{\partial}{\partial x^0} - H_1(-i\mathbf{\nabla}) - H_2(-i\mathbf{\nabla})\right]\Phi_{nB}(x_1,x_2)=\sum_{n'} A_{nn'}(x_1,x_2)\Phi_{n'B}(x_1,x_2) \tag{6.19}$$

where

$$A_{nn'}(x_1,x_2) = \langle n|e_1\alpha_\mu^{(1)} A^\mu(x_1) + e_2\,\alpha_\mu^{(2)} A^\mu(x_2)|n'\rangle \tag{6.20}$$

Due to the translational invariance of the theory, we can write

$$\left.\begin{aligned}\Phi_{nB}(x_1,x_2) &= \exp[i(P_n-P_B)X]\mathcal{F}_{nB}(\mathbf{x})\\ A_{nn'}(x_1,x_2) &= \exp[i(P_n-P_{n'})X]\mathcal{A}_{nn'}(\mathbf{x})\end{aligned}\right\} \tag{6.21}$$

where $X=\eta_1x_1+\eta_2x_2$ and $x=x_1-x_2$ with $\eta_1+\eta_2=1$. P_n and P_B are the respective four-vector momenta of the states $|n\rangle$ and $|B\rangle$. In terms of these C.M. coordinates, Eq. (6.19) can be written as

$$[P_B^0 - P_n^0 - H_1(\eta_1(\mathbf{P}_B-\mathbf{P}_n)-i\mathbf{\nabla}) - H_2(\eta_2(\mathbf{P}_B-\mathbf{P}_n)+i\mathbf{\nabla})]\mathcal{F}_{nB}(\mathbf{x}) =\sum_{n'} \mathcal{A}_{nn'}(\mathbf{x})\mathcal{F}_{n'B}(\mathbf{x}) \tag{6.22}$$

Introducing the retarded Green's function for the two particle system by

$$[H_1(\eta_1(\mathbf{P}_B-\mathbf{P}_n)-i\mathbf{\nabla}) - H_2(\eta_2(\mathbf{P}_B-\mathbf{P}_n)+i\mathbf{\nabla}) - P_0]\mathcal{G}^R{}_0\ (P,\mathbf{x}) = \delta(\mathbf{x}) \tag{6.23}$$

and defining K by

$$K(P_B;\mathbf{x},\mathbf{y})_{ns}=\sum_{n'} \mathcal{A}_{nn'}(\mathbf{x})\mathcal{G}_0^R(P_B-P_{n'},\mathbf{x-y})\mathcal{A}_{nn'}(\mathbf{y}) \tag{6.24}$$

we can rewrite symbolically, Eq. (6.22) as

$$\left\{[\mathcal{G}_0^R]^{-1} + K\right\}\mathcal{F}_B = 0 \tag{6.25}$$

If we define the projection operator $\Pi_0=|0\rangle\langle 0|$ (i.e. the projection on the $|0\rangle$ state), Eq. (6.25) can now be written as

$$\Pi_0\left\{[\mathcal{G}_0^R]^{-1} + K\right\}\Pi_0\mathcal{F}_B + \Pi_0K(1-\Pi_0)\mathcal{F}_B = 0 \tag{6.26}$$

$$(1-\Pi_0)\left\{[\mathcal{G}_0^R]^{-1} + K\right\}(1-\Pi_0)\mathcal{F}_B + (1-\Pi_0)K\Pi_0\mathcal{F}_B = 0 \tag{6.27}$$

At this point it is assumed that $(1-\Pi_0)\left\{[\mathcal{G}_0^R]^{-1} + K\right\}(1-\Pi_0)$ is invertible in the subspace $(1-\Pi_0)\mathcal{H}_\mathcal{F}$ of the entire Hilbert space $\mathcal{H}_\mathcal{F}$ and we designate this inverse by Q. Thus

$$Q\,(1-\Pi_0)\Big\{[\mathcal{G}_0^R]^{-1} + K\Big\}(1-\Pi_0) = (1-\Pi_0);\;\; Q\Pi_0=\Pi_0 Q \qquad (6.28)$$

Letting $\mathcal{F}_{0B} = \Pi_0\mathcal{F}_B$ we can now write

$$\Big\{[H_1(\eta_1\mathbf{P}-i\nabla) - H_2(\eta_2\mathbf{P}+i\nabla) - P_0] - K'_{00}\Big\}\mathcal{F}_{0B} = 0 \qquad (6.29)$$

where

$$K'_{00} = \Pi_0 K'\Pi_0 \qquad (6.30a)$$

$$K' = K - Q(1-\Pi_0)K \qquad (6.30b)$$

Further the non-trivial part $\mathcal{V}$ can be separated from K'_{00} by removing the self-energy part, Σ. That is

$$K'_{00} = \Sigma + \mathcal{V} \qquad (6.31)$$

Designating $[H_1(\eta_1\mathbf{P}-i\nabla) - H_2(\eta_2\mathbf{P}+i\nabla) +\Sigma - P_0]$ by $[\mathcal{G}^R]^{-1}$ we can write Eq. (6.29) by

$$\Big\{ [\mathcal{G}^R]^{-1} - \mathcal{V} \Big\}\mathcal{F}_{0B} = 0 \qquad (6.32)$$

K'_{00} can, by Eq. (6.30b), be expanded in powers of α and the usual perturbation techniques can be employed to find the energy eigenvalues.

7. Conclusions

In the above we have considered various approaches to the bound state problem for QED. Some of these have yet to be exploited such as the Duffin-Kemmer approach. Others, such as the quasi-potential approach, may be conceptually simpler than the Bethe-Salpeter equation, in the sense, that they are 'single-time' equations. However, since they involve solving an equation which although satisfies a number of reasonable physical requirements such as two-body unitarity, and an off mass shell requirement, the equation obtained is an approximate one whose relation to the exact Bethe-Salpeter equation is not clear. Thus, how good an approximation the energy levels calculated from the quasi-potential approach is difficult to gauge. Furthermore, the single time approach excludes the possibility of including the retardation effects properly. By always going back to the scattering amplitude in defining higher-order corrections to the quasi-potential, to some extent, there is a self-correcting mechanism.

However, in principle, the Bethe-Salpeter equation is exact. It does not include any assumptions (as the external field approximation does) concerning the relative size of the particles' masses and all recoil corrections, in principle, can be calculated. Furthermore, the perturbation theory allows one to calculate the terms neglected in any approximation used. The main draw back using the Bethe-Salpeter formalism is that it very complicated and its generality is frequently more than is needed. To some extent, to find a method which would allow some reasonable approximations more easily than the Bethe-Salpeter formalism, may have motivated Grotch and Yennie in their effective potential approach. One of the great advantages of their approach is that their unperturbed equation is a modified Dirac equation which can be solved exactly. To obtain from the Bethe-Salpeter approach, the relativistic corrections that Grotch and Yennie obtained from their exact solution, requires a tedious perturbative calculation.

8. Acknowledgments

I wish to thank Professor Howard Grotch who suggested the usefulness and the timeliness to speak on the methods used in bound state QED as well as his many insights in this subject as well. I also wish to thank Dr. Misha Eides for many useful discussions as well and Drs. Roger Barrett and Youcef Nedjadi for acquainting me with their work using Duffin-Kemmer equations for bound state considerations.

9. References

1. The decay width of the triplet state of positronium still remains a problem as there is a discrepancy between the calculated life times and those measured by experiment.
2. E. Salpeter, *Phys. Rev.* **87**, 328 (1952).
3. Particle Properties Data Booklet, 1988 (North-Holland & Amsterdam).
4. In most of the general cases, we will consider both particles to be fermions as these are the most common systems that one encounters plus the possibility that all scalar particles, in fact, may be composite systems of fermions.
5. J. Bjorken & S. Drell, *Relativistic Quantum Fields* (McGraw Hill Book Co., 1965) (cf. ch. 19).
6. In this express and those that follow, we use matrix notation. Thus, in both the numerator and denominator, there are implied integrations.
7. G. Erickson & D. Yennie, *Ann. Phys.*(N.Y.), **35**, 271, 447 (1965).
8. J. Fox & D. Yennie, *Ann. Phys.* (N.Y.), **81**, (1973).
9. A. Logunov & A. Tavkhelidze, *Nuovo Cim.* **29**, 380 (1963); A Logunov, et. al., *Nuovo Cim.* **30**, 134 (1963).
10. V.A. Rizov, *Ann.Phys.***165**, 59 (1985).
11. I.T. Todorov, *Rep. Math. Phys.* **24**, 95 (1986).
12. V.A. Rizov, I.T. Todorov & B.L. Aneva, *Nucl. Phys.* **B98**, 447 (1975).
13. G. Breit, *Phys. Rev.* **34**, 553 (1929); **36**, 383 (1930); **39**, (1932).
14. W. Barker & F. Glover, *Phys. Rev.* **99**, 317 (1955).
15. E. Salpeter & H. Bethe, *Phys. Rev.* **84**, 1232 (1951); J. Schwinger, *Proc. Natl. Acad. Sci.* (U.S.A.) **37**, 452 (1951); M. Gell-Mann & F. Low, *Phys. Rev.* **84**, 350 (1951).
16. P. Roman, *Introduction to Quantum Field Theory* (John Wiley and Sons, New York 1969)
17. D.A. Owen, *Corrections to the Positronium Hyperfine Structure of Order* $\alpha^2 \ln\alpha^{-1}$ (JHU thesis, 1970 (unpublished)).
18. R. Karplus & A. Klein, *Phys. Rev.* **87**. 848 (1952); T.Fulton & P.Martin, *Phys. Rev.* **95**, 811 (1954); For further details and references, see D. A. Owen, *Topical Seminar on Electromagnetic Interactions* (ICTP Trieste IC/71/140).
19. D. A. Owen, *Phys. Rev.* **D42**, 2534 (1990); M. Halpert & D. A. Owen, *J. Phys.* **G20**, 51 (1994); D. A. Owen, *Found. Phys.* **20**, 273 (1994)
20. J. Gorelick & H. Grotch, *J. Phys.*, **G3**, 751 (1977).
21. G. Lepage, *Phys. Rev.* **A16**, 863, (1977).
22. R. Blankenbecler & R. Sugar, *Phys. Rev.* **142**, 1051 (1966).
23. F. Gross, *Phys. Rev.* **186**, 1448 (1969).

24. A. Klein & B. Lee, *Phys. Rev.* **D10**, 4308 (1974), A. Logunov & A. Tavkhelidze, *Nuovo Cim.* **29**, 380 (1963).
25. H. Grotch & D. Yennie, *Rev. Mod. Phys.* **41**, 350 (1969).
26. R. Duffin, *Phys. Rev.* **54**, 1114 (1938), N. Kemmer, *Proc. of Roy. Soc.* (London) **A173**, 91 (1939)
27. A. Akhiezer & V. Berestetskii, *Quantum Electrodynamics* (State Technico-Theoretical Literature Press (1953).
28. Y. Nedjadi & R.C. Barrett, *J. Physics,* **G19**, 87 (1993); preprint, University of Surrey "Solutions of the Central Field Problem for a Duffin-Kemmer-Petiau Vector Boson"
29. I. Tamm, *J. Phys.* (USSR), **9**, 449 (1945); S.M. Dancoff, *Phys. Rev.* **78**, 382 (1950).
30. H. Bethe & E. Salpeter,*Quantum Mechanics of One- and Two- Electron Systems* (Springer-Verlag,Berlin 1957).
31. Private communication, M. Eides.
32. The form of this transformation is found in most field theory books, for example, J. Bjorken & S. Drell, *Relativistic Quantum Fields* (McGraw-Hill Book Company 1965).
33. V. Rizov & I Todorov, *Sov. J. Part. Nucl.* **6**, 269 (1976)

New Developments in the Theory of Muonium Hyperfine Splitting

M. I. Eides*
Petersburg Nuclear Physics Institute,
Gatchina, St.Petersburg 188350, Russia

Abstract

The present status of the investigation of the muonium hyperfine splitting with emphasis on the recent theoretical achievements is reviewed.

1. Introduction and Experimental Data

The hyperfine splitting (HFS) in one-electron atoms can be qualitatively understood in the framework of classical physics. It originates from the interaction of the magnetic moments of the electron and the nucleus. According to classical electrodynamics parallel currents attract each other and antiparallel ones repel. Hence, it is clear that there is a splitting in energy between states with different mutual orientation of magnetic moments and the state with antiparallel spins and parallel magnetic moments is the state with lowest energy (see Fig.1).

The HFS in the ground state of hydrogen is one of the best measured quantities in modern physics [1, 2]

$$\Delta\nu_H(n=1) = 1\ 420\ 405.751\ 766\ 7\ (9)\ \text{kHz} \qquad \delta = 6\cdot 10^{-13}. \tag{1}$$

Unfortunately, theoretical calculation of this quantity is impeded by our poor knowledge of spin-dependent proton structure functions and by our inability to perform precise low-energy QCD calculations. There was no significant theoretical progress in the investigation of the hydrogen HFS recently and we will not discuss hydrogen any more below.

The situation is much more promising with the hyperfine splitting in the ground state of muonium. It is also measured with high precision [3]

*E-mail address: eides@lnpi.spb.su

$$\Delta\nu_{MU}(n=1) = 4\ 463\ 302.88\ (16)\ \text{kHz} \qquad \delta = 3.6\cdot 10^{-8} \tag{2}$$

and there are further opportunities for improving the accuracy. The lifetime of the state with higher energy in respect to the $M1$-transition to the lower level ($F = 1 \rightarrow F = 0$) is extremely large $\tau = 1\cdot 10^{13}$ s and gives negligible contribution to the linewidth. The linewidth is completely defined by the muon lifetime $\tau_\mu \approx 2.2\cdot 10^{-6}$ s which leads to the width $\Gamma_\mu/h = 72.3$ kHz.The experimental accuracy attained thus far corresponds to measuring the energy splitting with an experimental error of $\Delta\nu_{exp}/(\Gamma_\mu/h) \approx 2.2\cdot 10^{-3}$ of the natural linewidth.

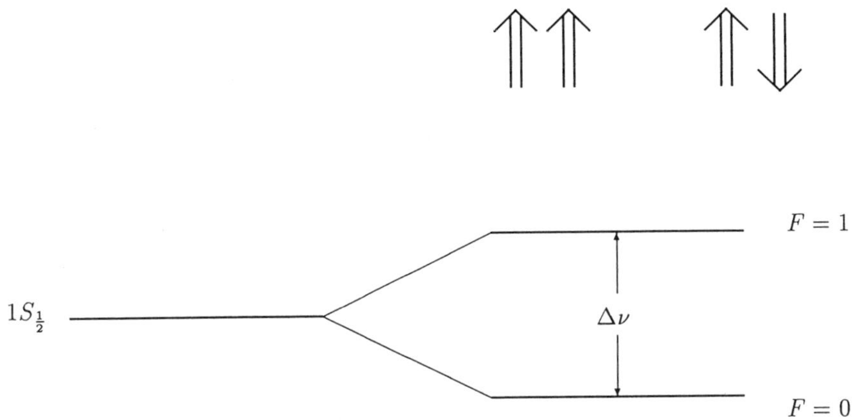

Fig.1. Scheme of hyperfine energy levels in the ground state.

A new experiment is being undertaken at Los Alamos now [4], with the goal of measuring $\Delta\nu_{MU}(n=1)$ with precision $1\cdot 10^{-8}$.

For comparison with theory one also needs to know the ratio m_μ/m_e, which enters as a factor in the theoretical expression for hyperfine splitting. It is measured now with significantly less precision than the hyperfine splitting itself [3, 5]

$$\frac{m_\mu}{m_e} = 206.\ 768\ 262\ (62) \qquad \delta = 3\cdot 10^{-7}. \tag{3}$$

A new measurement of the electron-muon mass ratio (or of the ratio of the muon and proton magnetic moments, which may be readily translated to the mass ratio above) with precision about $5\cdot 10^{-8}$ is planned at the LAMPF experiment mentioned above [4].

2. Theory of Hyperfine Splitting

2.1. Leading Order Contribution

The main contribution to hyperfine splitting was already explained above as due to the interaction of magnetic moments. It was first obtained in the framework of nonrelativistic quantum mechanics by E. Fermi a long time ago [6]

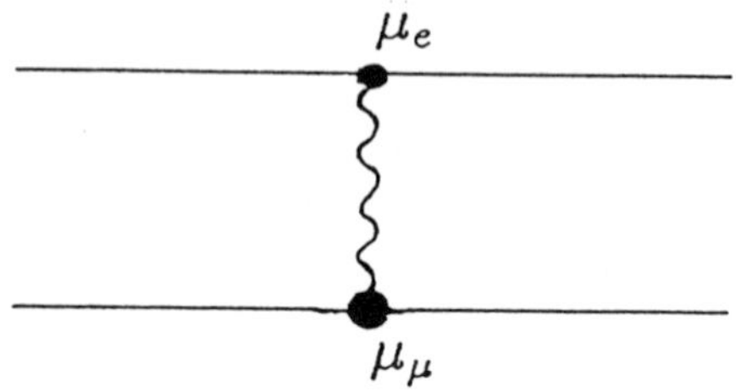

Fig.2. Transverse photon exchange, producing the main contribution to HFS.

$$\begin{aligned} E_F &= \frac{8}{3}(Z\alpha)^4 \frac{m}{M} \cdot mc^2 \cdot (1+\frac{m}{M})^{-3}(1+a_\mu) \\ &= \frac{16}{3} Z^4\alpha^2 \cdot R_\infty \cdot ch \cdot \frac{m}{M} \cdot (1+\frac{m}{M})^{-3}(1+a_\mu), \end{aligned} \tag{4}$$

where m and M are the electron and muon masses, respectively, Z is the charge of the muon in terms of the proton charge[1], c is the velocity of light, a_μ is the muon anomalous magnetic moment, R_∞ is the Rydberg constant and h is the Planck constant.

2.2. Classification of Corrections

Quantum Electrodynamics provides one with a tool for systematic calculation of numerous corrections to the Fermi formula for hyperfine splitting

$$\Delta\nu_{HFS} = \widetilde{\nu}_F \cdot (1 + corrections), \tag{5}$$

where[2]

$$\widetilde{\nu}_F = \frac{16}{3} Z^4\alpha^2 \cdot R_\infty \cdot c \cdot \frac{m}{M} \cdot (1+\frac{m}{M})^{-3} \tag{6}$$

[1]Of course, $Z = 1$ for muon, but the Fermi formula is valid for any heavy nucleus with arbitrary Z. Even in the case of muonium it is useful to preserve Z as a parameter in all expressions below, since it helps to clarify the origin of different corrections.

[2]Note that the expression for $\widetilde{\nu}_F$ does not include, unlike Eq.(4) muon anomalous magnetic moment a_μ which we now consider on the same grounds as other corrections to hyperfine splitting.

There are three different small parameters in the problem under consideration: the fine structure constant α, $Z\alpha$ and the small electron-muon mass ratio m/M.

Respectively corrections to the Fermi hyperfine energy are classified as:

a) *Relativistic corrections* – expansion over $Z\alpha$,

b) *Radiative corrections* – expansion over α and $Z\alpha$,

c) *Recoil corrections* – expansion over $\frac{m}{M}$ and $Z\alpha$,

d) *Radiative-Recoil corrections* – expansion over α, $\frac{m}{M}$ and $Z\alpha$,

e) *Correction induced by Weak Interaction* – of order $G_F \cdot m^2 \cdot (Z\alpha) \cdot R_\infty$.

We are going to discuss these corrections in more detail below. However, we would like to emphasize at this stage that it is now a very appropriate time for a new review of the theory of muonium hyperfine splitting. Until recently certain corrections of order $\alpha(Z\alpha)/\pi \cdot (m/M) \cdot \tilde{E}_F$ and $\alpha^2(Z\alpha)/\pi \cdot E_F$ which could have been as large as 1 kHz, an order of magnitude larger than the current experimental precision, remained uncalculated. The new developments are connected with calculation of all these corrections and will now be described below. *Thus the theory of muonium HFS has successfully passed the* 1 *kHz barrier and achieved a new stage. Theoretical uncertainty of the ground state hyperfine splitting is reduced below* 0.2 *kHz* ($5 \cdot 10^{-8}$).

2.3. Relativistic and Radiative Corrections

Relativistic and radiative corrections may be calculated in the framework of the external field approximation and, hence, the relativistic two-body equations are not needed for their consideration.

Let us note first, that the heavy particle magnetic moment factorizes for these corrections and one has to substitute in Eq.(5) the original Fermi expression Eq.(4) for the main contribution to hyperfine splitting. This occurs because the virtual momenta involved in calculation of the relativistic and radiative corrections are small in comparison with the muon mass, which sets the natural momenta scale for the corrections to the muon magnetic moment.

2.3. 1. Relativistic Corrections

Relativistic corrections are the simplest corrections to hyperfine splitting. They correspond to the nonrelativistic expansion of the classical relativistic square root expression for the energy of the light particle

$$\sqrt{\mathbf{p}^2+m^2} \approx m + \frac{\mathbf{p}^2}{2m} - \frac{\mathbf{p}^4}{8m^3} + \dots .$$

The quantity $\mathbf{p}^2/m^2$ in the loosely bound two-body system corresponds to $(Z\alpha)^2$ and the whole expansion in $(Z\alpha)^2$ may be easily obtained from the exact solution of the Dirac equation in the external Coulomb field. Even quantization of the electromagnetic field is not necessary for consideration of these corrections. We would like to mention the absence of the factor π in the denominators in this expansion (see explicit expression for the Breit correction [10] in Table 1 and contribution of the next order [28] in Table 2). This is characteristic for the Coulomb problem and emphasizes the nonradiative nature of the relativistic corrections!

Table 1. Relativistic and Radiative Corrections

	E_F	kHz
Fermi (1930)[6]	1	4 459 032.22 (1.33)
Schwinger (1948)[7]	$\alpha/(2\pi)$	5 178.76
Sommerfield (1957)[8] Peterman (1957)[9]	$-0.328 \cdot (\alpha/\pi)^2$	−7.89
Breit (1930)[10]	$3/2 \cdot (Z\alpha)^2$	356.17
Kroll, Pollock (1951)[25] Karplus, Klein, Schwinger (1951)[12]	$\alpha(Z\alpha)(\ln 2 - \frac{5}{2})$	−429.04
Zwanziger (1964)[13] Layzer (1964)[14]	$-8\alpha(Z\alpha)^2/(3\pi) \cdot \ln Z\alpha$ $\times(\ln Z\alpha - \ln 4 + \frac{281}{480})$	−41.40
Sapirstein (1983)[15]	$\alpha(Z\alpha)^2/\pi \cdot 15.38(29)$	8.48(16)
Eides, Karshenboim, Shelyuto (1989)[16]	$36/35 \cdot \alpha^2(Z\alpha)/\pi$	0.57
Eides, Karshenboim, Shelyuto (1989)[16]	$(224/15 \ln 2 - 38\pi/15$ $-118/225) \cdot \alpha^2(Z\alpha)/\pi$	1.03
Eides, Karshenboim, Shelyuto (1989)[16]	$\{-4/3 \ln^2[(1+\sqrt{5})/2]$ $-20\sqrt{5}/9 \cdot \ln[(1+\sqrt{5})/2]$ $-64/45 \ln 2 + \pi^2/9 + 3/8$ $+1043/675\} \cdot \alpha^2(Z\alpha)/\pi$	−0.37
Eides, Karshenboim, Shelyuto (1990)[17]	$-0.310\ 742 \cdot \alpha^2(Z\alpha)/\pi$	−0.17
Eides, Karshenboim, Shelyuto (1991,1993)[18] Kinoshita, Nio (1993)[19, 20]	$-0.472\ 51 \cdot \alpha^2(Z\alpha)/\pi$	−0.26
Kinoshita, Nio (1993)[19, 20]	$-0.63(4) \cdot \alpha^2(Z\alpha)/\pi$	−0.35(2)

2.3. 2. Radiative Corrections

Contribution to the hyperfine splitting corresponding to the electron anomalous magnetic moment is the simplest kind of *radiative correction* [7, 8, 9]. The respective expansion goes over α/π and is independent of the binding parameter $Z\alpha$. This means that all such corrections contribute only to the effective electron vertex in Fig.2 and are completely independent of the bound-state nature of muonium. Hence, the value of the electron anomalous magnetic moment entering in the expression for HFS coincides with the one for the free electron.

Nontrivial interplay between radiative corrections and binding effects first emerges in calculation of combined expansion over α and $Z\alpha$. The simplest contribution of this type is of order $\alpha(Z\alpha)\cdot E_F$ and was calculated a long time ago in classical papers [25, 12].

Let us explain, by the example of corrections of order $\alpha(Z\alpha)\cdot E_F$ a general method of calculation radiative corrections which was recently applied for calculation of corrections of order $\alpha^2(Z\alpha)/\pi \cdot E_F$ [16].

All diagrams generating corrections of order $\alpha(Z\alpha)\cdot E_F$ may be obtained by one-loop radiative insertions in the skeleton diagram in Fig.3 with two external photons attached to the electron line.

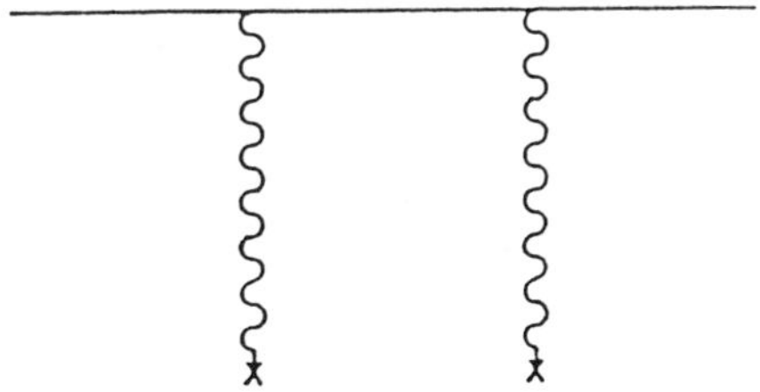

Fig.3. Skeleton diagram for radiative corrections.

Polarization insertions in the photon lines produce graphs in Fig.4, while radiative photon insertions in the electron line generate diagrams in Fig.5.

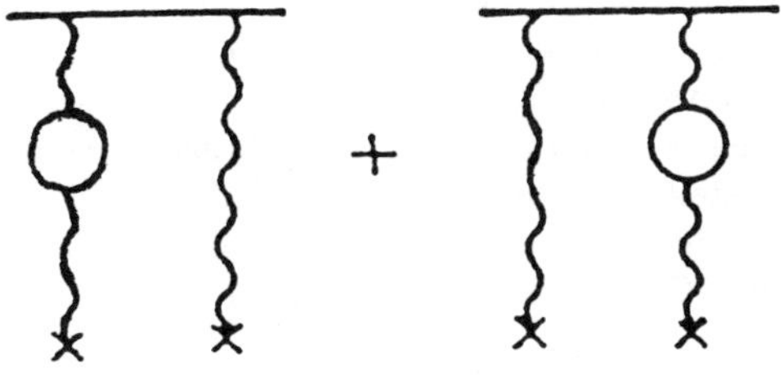

Fig.4. Diagrams with one-loop polarization insertions.

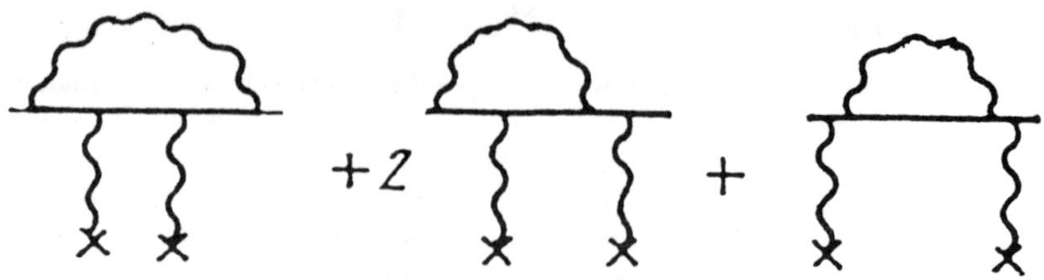

Fig.5 Diagrams with radiative photon insertions.

It is evident that due to polarization operator insertions low integration momenta (of atomic order $mZ\alpha$) are suppressed in the exchange loops in Fig.4 and effective integration momenta are of order m. The same is valid for the diagrams in Fig.5 (see, e.g. [21], let us note in passing that use of the Fried-Yennie gauge greatly facilitates the proof of the correctness of this statement). Hence, one may neglect small virtuality of external fermion lines and calculate contributions generated by the diagrams under discussion using on-mass-shell external momenta. The contribution to HFS is given by the product of the square of the Schrödinger-Coulomb wave function at the origin $|\psi(0)|^2$ and the matrix element corresponding to the diagram. In these conditions the diagrams above comprise a gauge invariant set and may be easily calculated. It is not difficult to present this calculation in a universal way. Contribution of each of the diagrams in Figs. 4,5 to HFS is given by the expression

$$\frac{8Z\alpha}{\pi n^3}\int_0^\infty d|\mathbf{k}|\frac{F(\mathbf{k})}{\mathbf{k}^2}, \tag{7}$$

where $d|\mathbf{k}|/|\mathbf{k}|^2$ corresponds to the skeleton integrand in Fig.3, while the factor $F(\mathbf{k})$ describes radiative corrections and turns infrared divergent skeleton integral into a convergent one. Hence, the problem of calculating contributions to HFS generated by the diagrams in Fig.5 turns into the problem of calculating the electron factor describing radiative insertions in the electron line. This last problem admits a straightforward solution.

It is very easy now to describe the latest theoretical achievement in the field, namely, calculation of all contributions of order $\alpha^2(Z\alpha)/\pi \cdot E_F$. Once again only diagrams with mass-shell external electron lines are relevant for calculation of these corrections. As was shown in [16] six gauge invariant sets of graphs in Fig.6 produce corrections of order $\alpha^2(Z\alpha)/\pi \cdot E_F$ to HFS. Respective contributions once again may be readily calculated with the help of the integral in Eq.(7). Contributions corresponding to the diagrams in Figs.6a-6c were calculated analytically in [16] in the Fried-Yennie gauge. Contribution of the diagrams in Fig.6d was reduced to the form of an one-dimensional integral, whose integrand is an elliptic function. This integral was then calculated numerically [17]. Explicit expression for the light-by-light scattering contribution was reduced to a three-dimensional integral which was also calculated numerically [18]. All these results were confirmed by an independent numerical calculation in the Feynman gauge in [19, 20][3]

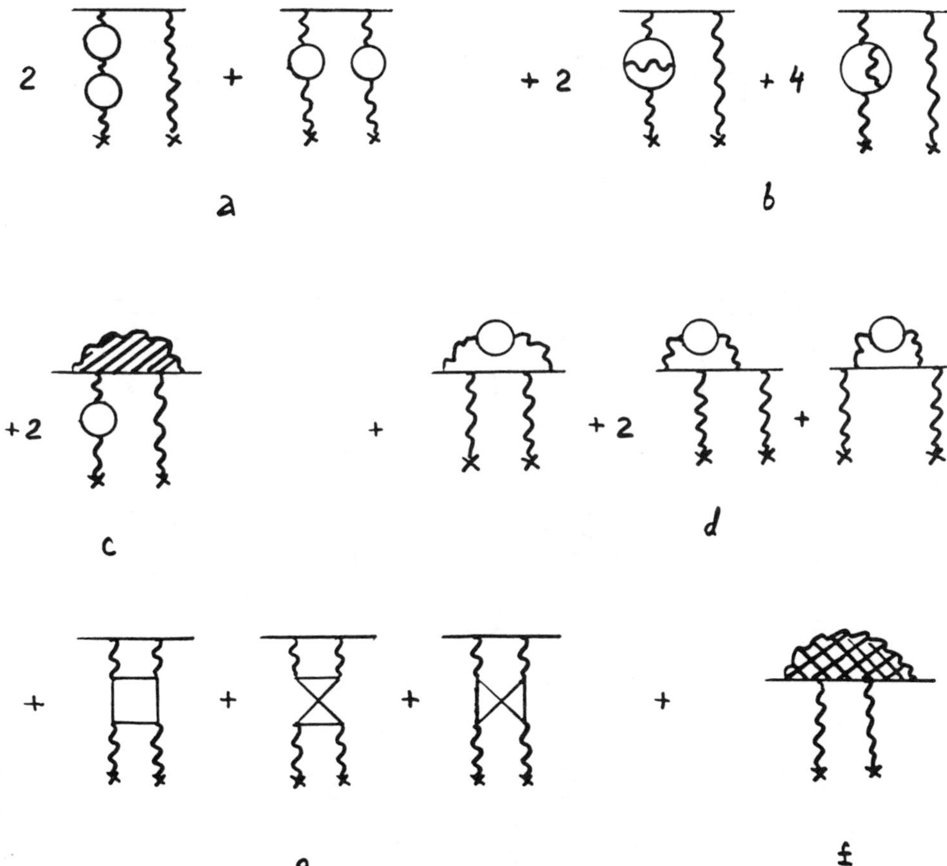

Fig.6. Six gauge-invariant sets of diagrams for corrections of order $\alpha^2(Z\alpha)/\pi \cdot E_F$.

[3]Work [19] revealed a 2% discrepancy between the value of the light-by-light contribution obtained in this work and in the first of references [18]. After intensive reexamination of all analytic calculations in this reference we found a minor arithmetic error at one of the last stages of calculations. Corrected result (see the last of references [18]) is in perfect agreement with the result in [19, 20].

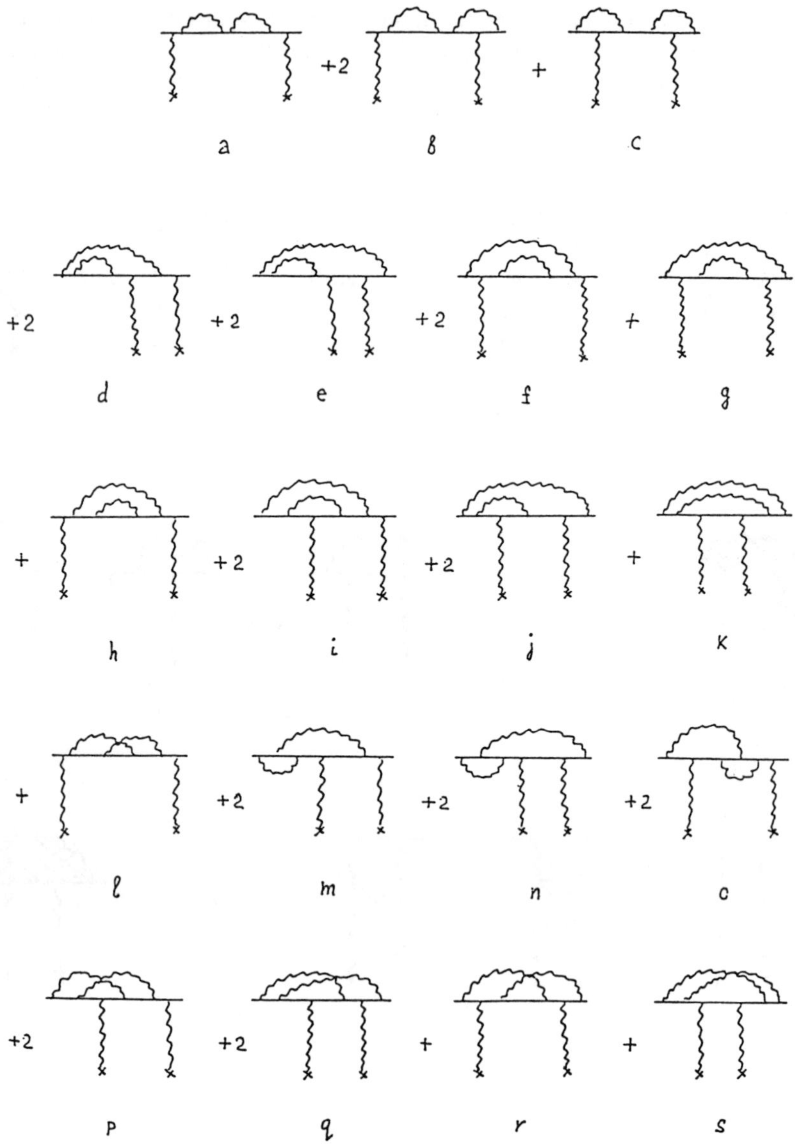

Fig.7. 19 topologically different diagrams with two radiative photon insertions for corrections of order $\alpha^2(Z\alpha)/\pi \cdot E_F$.

Last of the six sets of diagrams presented in Fig.6 consists of 19 topologically different graphs which are presented in Fig.7. These are all diagrams which may be obtained from the skeleton graph in Fig.3 by insertions of two radiative photons in the electron line. Respective contribution to HFS was recently calculated numerically in the Feynman gauge in [19, 20] and the result is presented in Table 1. Contributions generated by ten diagrams in Fig.7 were also obtained in the Fried-Yennie gauge in [22, 23],and results for the remaining diagrams will be available in the near future [24].

2.3. 3. Radiative Corrections of Order $\alpha(Z\alpha)^2/\pi \cdot E_F$

We have not yet discussed radiative corrections of order $\alpha(Z\alpha)^2/\pi \cdot E_F$ also present in Table 1. Binding effects play the crucial role in calculation of this contribution to HFS. Unlike all cases considered above, exchanged photon loops with low (of order $\sim mZ\alpha$) exchanged momenta give significant contribution in this case and moreover, external wave functions at the origin do not factorize. The factor before $\alpha(Z\alpha)^2/\pi \cdot E_F$ does not reduce to a constant but contains low-frequency logarithmic terms $\log^2 Z\alpha$ and $\log Z\alpha$. Hence, the situation with this calculation resembles the classical case of the main contribution to the Lamb shift, namely, while factors before logarithmic terms themselves depend only on diagrams with three and two exchanged photons and may be easily calculated analytically [13, 14], for calculation of the nonlogarithmic factor summation of an infinite sequence of diagrams is necessary [15]. In view of the significant theoretical progress achieved recently in calculation of all other contributions to HFS and described in this review, insufficient accuracy of the calculation in [15] (about 0.2 kHz) became the main source of the uncertainty in the theoretical expression for muonium HFS. A new calculation of this correction is being performed now by M. Nio [25].

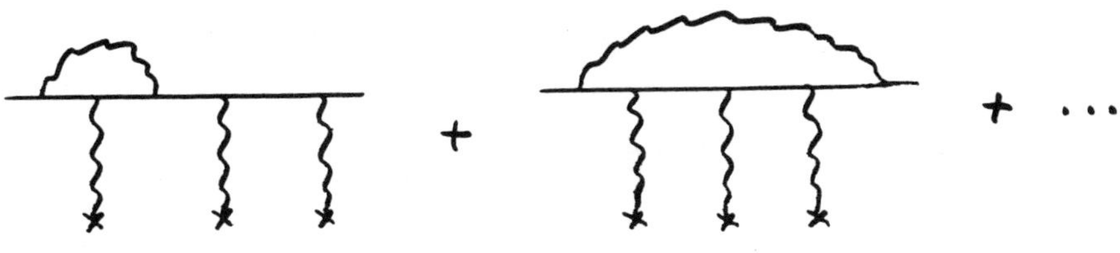

Fig.8. Some diagrams for corrections of order $\alpha(Z\alpha)^2/\pi \cdot E_F$.

2.3. 4. Higher Order Purely Radiative Logarithmic Corrections

The normal scale of corrections containing the fourth power of α or $Z\alpha$ is of order 0.02 kHz. However, some of them are enhanced by logarithms and may become important at the current level of theoretical accuracy. It is not difficult to calculate all such logarithmic contributions. We present all already known higher order radiative contributions in Table 2, but more systematic study of all such corrections is still necessary. It is highly probable that all as yet uncalculated terms are smaller than 0.1 kHz.

Table 2. Relativistic and Radiative Corrections(sub 1 kHz)

	E_F	kHz
Lepage (1994)[26]	$c_2 \cdot \alpha(Z\alpha)^3 \cdot \ln Z\alpha$	−0.381
Kinoshita (1973)[27]	$1.176 \cdot (\alpha/\pi)^3$	0.07
Karshenboim (1993)[28]	$-4/3 \cdot \alpha^2(Z\alpha)^2/\pi^2 \cdot \ln^2 Z\alpha$	−0.04
Kinoshita, Nio (1994)[20]	$c_1 \cdot \alpha^2(Z\alpha)^2 \cdot \ln Z\alpha$	−0.007
Kinoshita, Nio (1994)[20]	$c_2 \cdot \alpha^2(Z\alpha)^2 \cdot \ln Z\alpha$	0.004
Karshenboim (1993)[28]	$17/8 \cdot (Z\alpha)^4$	0.03

2.4. Recoil corrections

Let us consider now recoil corrections to HFS. In this case the external field approach is inadequate and, in principle for calculation of recoil corrections one needs the complete machinery of two-particle equations. However, many results may be understood without a cumbersome formalism.

2.4. 1. Leading Recoil Correction

The leading recoil correction of order $Z\alpha \cdot (m/M) \cdot \tilde{E}_F$ is produced by the graphs in Fig.9 and was calculated a long time ago [29]. Subtraction of the heavy particle pole contribution is necessary here to get rid of the main non-recoil contribution to HFS. Since this main contribution is connected with small (atomic scale $\sim m(Z\alpha)$) momenta of exchanged photons, it is not difficult to realize that after subtraction only high loop momenta k ($m < k < M$) are relevant for calculation.

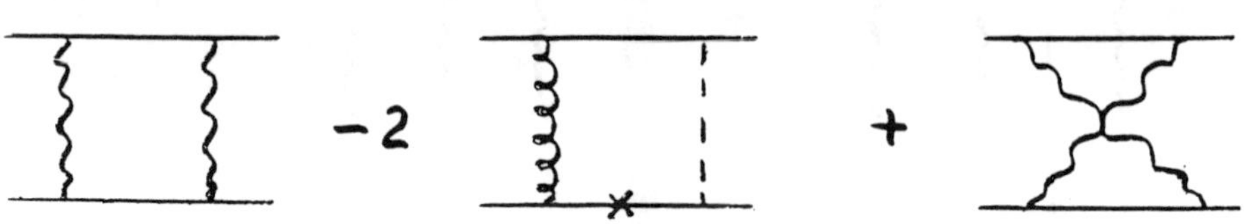

Fig.9. Leading recoil correction.

Then two main features of the leading recoil correction immediately emerge as a consequence of the large magnitude of loop momenta. First, the virtual momenta in the exchanged photons are of the same scale as virtual momenta relevant for calculation of the muon anomalous magnetic moment. Hence, the anomalous magnetic moment of the muon does not enter the expression for the recoil corrections as a factor, as is the case for the expression for the radiative corrections. Second, due to the large magnitude of loop momenta the wave function at the origin factorizes in the calculation of the leading recoil correction, this situation being similar to the one discussed above in connection with several cases of radiative corrections. Let us mention that the explicit expression for the leading recoil correction (see Table 3) contains logarithm of the muon-electron mass ratio, connected with the high momenta integration region, unlike typical logarithms of $Z\alpha$ in radiative corrections which are always connected with low momenta integration region.

2.4. 2. Recoil Correction of Order $(Z\alpha)^2 \cdot (m/M) \cdot \widetilde{E}_F$

Pure recoil correction of order $(Z\alpha)^2 \cdot (m/M) \cdot \widetilde{E}_F$ is induced by the graphs with two and three exchanged photons in Fig.10. As for the case of radiative corrections of order $\alpha(Z\alpha)^2/\pi \cdot E_F$ the factor before $(Z\alpha)^2 \cdot (m/M \cdot \widetilde{E}_F$ does not reduce to a constant but contains also low-frequency logarithmic terms $\log^2 Z\alpha$ and $\log Z\alpha$. Once again this means that low momenta are important in the exchanged loops in Fig.10, despite the subtraction of nonrecoil contributions.

Fig.10. Pure recoil correction of order $(Z\alpha)^2 \cdot (m/M) \cdot \widetilde{E}_F$.

Due to the importance of low virtual momenta wave functions do not decouple here and the exact calculation is quite intricate. It is interesting to emphasize one significant and general difference between recoil and nonrecoil low momenta contributions. While nonrecoil contributions admit only numerical calculation of the complete contribution containing low energy logarithms (compare correction $\alpha(Z\alpha)^2/\pi \cdot E_F$ above, main contribution to the Lamb shift, etc.), recoil corrections with low energy logarithmic enhancement admit exact analytic calculation as in the case under consideration [32] (compare, e.g. with the similar situation for the recoil corrections to the Lamb shift [33]).

Table 3. Recoil Corrections

	$\widetilde{E}_F$	kHz
Arnowitt (1953)[29]	$-\frac{3(Z\alpha)}{\pi}\cdot\frac{mM}{M^2-m^2}\cdot\ln\frac{M}{m}$	−800.30
Lepage (1977)[30] Bodwin, Yennie (1978)[31]	$(Z\alpha)^2\cdot\frac{m_r^2}{mM}\cdot 2\ln\frac{1}{2Z\alpha}$	9.60
Bodwin, Yennie, Gregorio (1982)[32]	$(Z\alpha)^2\cdot\frac{m_r^2}{mM}\cdot(-6\ln 2+\frac{65}{18})$	−0.62
Karshenboim (1993)[28]	$-\frac{2}{3}\cdot\frac{(Z\alpha)^3}{\pi}\cdot\frac{m_r^2}{mM}\cdot\ln^2 Z\alpha$	−0.04
Kinoshita, Nio (1994)[20]	$c_1\cdot(Z\alpha)^3\cdot\frac{m}{M}\cdot\ln Z\alpha$	−0.189
Karshenboim (1994)[28] Kinoshita, Nio (1994)[20]	$3\cdot\frac{(Z\alpha)^3}{\pi}\cdot\frac{m}{M}\cdot\ln Z\alpha\cdot\ln\frac{M}{m}$	−0.21

In the same way as for the case of radiative corrections after recent achievements in the theory of hyperfine splitting it is now reasonable to investigate logarithmically enhanced corrections of the next order, namely of order $(Z\alpha)^3\cdot(M/m)$. A number of such corrections is also presented in the Table 3 below double line.

2.5. *Radiative-Recoil corrections*

As in the case of purely radiative corrections of order $\alpha(Z\alpha)\cdot E_F$ considered above all diagrams relevant for calculation of radiative-recoil corrections of order $\alpha(Z\alpha)(m/M)\cdot\widetilde{E}_F$ may be obtained by radiative insertions in the skeleton diagrams, the only difference being the nature of the skeleton diagrams themselves. The skeleton diagrams for contributions of order $\alpha(Z\alpha)(m/M)\cdot\widetilde{E}_F$ are presented in Fig.11. It is easy to see that the skeleton diagram in Fig.3 corresponds to the residue in the heavy particle pole in Fig.11.

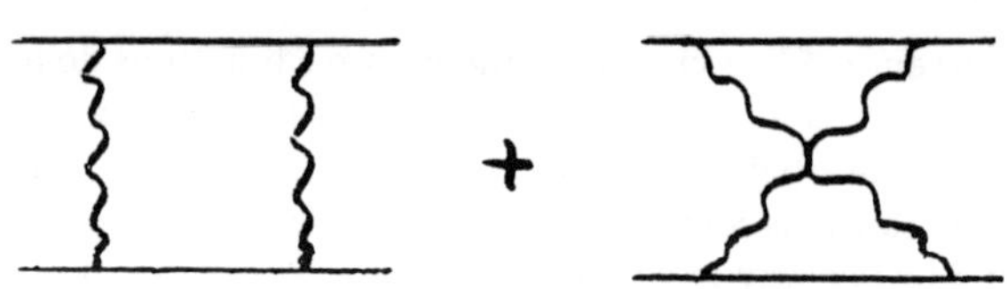

Fig.11. Skeleton diagrams for contributions of order $\alpha(Z\alpha)\cdot(m/M)\cdot\widetilde{E}_F$.

Diagrams generating contributions to HFS of order $\alpha(Z\alpha)(m/M)\cdot\widetilde{E}_F$ (or $(Z^2\alpha)\cdot(Z\alpha)(m/M)\cdot\widetilde{E}_F$ in case of insertions in the muon line) and obtained by inserting a polarization operator in the exchanged photon lines, by inserting a radiative photon in the electron line and by inserting a radiative photon in the muon line are presented in Figs. 12, 13 and 14, respectively.

Once again, due to radiative insertions the low (order $mZ\alpha$) exchanged momentum region is suppressed in the diagrams in Figs. 12-14, wave functions decouple and one may calculate matrix elements of diagrams in the scattering approximation, when all external electron lines are on mass-shell and diagrams comprise gauge invariant sets. It turns out that all corrections in Figs. 12-14 admit complete analytic calculation [34, 35, 37, 38] and respective results are presented in Table 4. The only nonanalytic contribution of the order under consideration [36] is connected with the hadron vacuum polarization insertion in the exchanged photon and also admits very precise calculation.

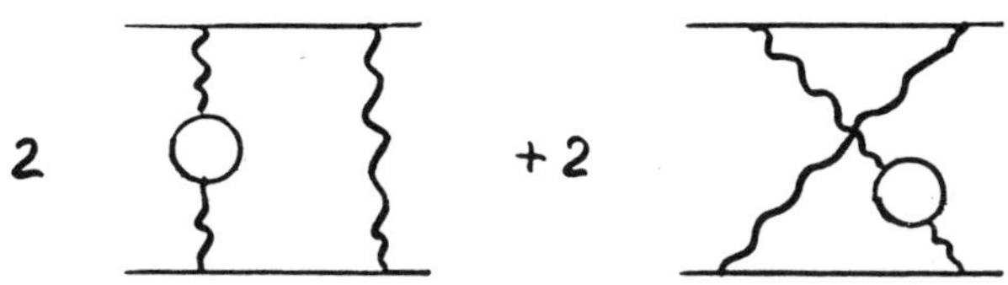

Fig.12. Diagrams with polarization operator insertions.

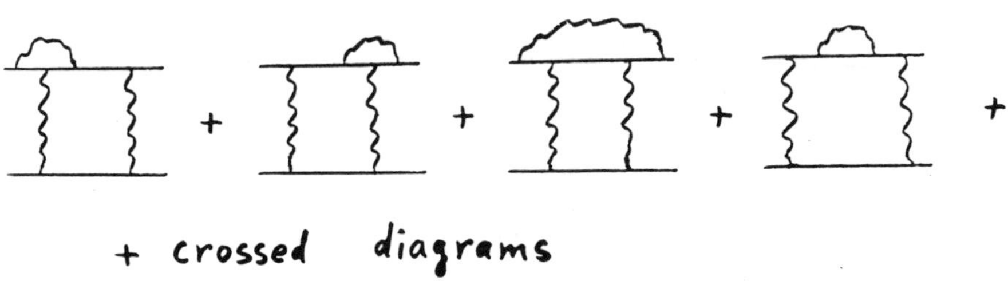

Fig.13. Diagrams with radiative photon insertions in the electron line.

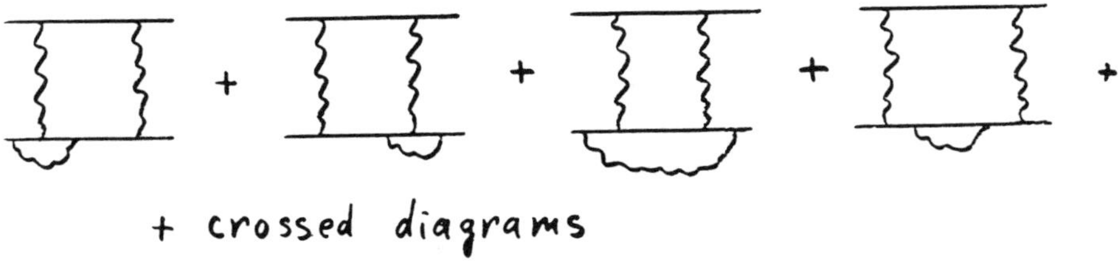

Fig.14. Diagrams with radiative photon insertions in the muon line.

Recent progress in the theory of HFS has made further investigation of small higher order radiative-recoil corrections to HFS more relevant. Some logarithmic contributions of higher order are presented in Table 4. Further analysis of all such contributions is needed.

Table 4. Radiative-Recoil Corrections

	$\tilde{E}_F$	kHz
Caswell, Lepage (1978)[34]	$-2 \cdot \ln^2 \frac{M}{m} \cdot \frac{\alpha(Z\alpha)}{\pi^2} \cdot \frac{m}{M}$	-6.61
Terray, Yennie (1982)[35]	$(\frac{13}{12} \ln \frac{M}{m} - \frac{\pi^2}{3} - \frac{28}{9}) \times \frac{\alpha(Z\alpha)}{\pi^2} \cdot \frac{m}{M}$	-0.07
Karimkhodzhaev, Faustov(1991)[36]	$2.15(14) \cdot \frac{\alpha(Z\alpha)}{\pi^2} \cdot \frac{m}{M}$	0.25(2)
Eides, Karshenboim, Shelyuto (1986)[37]	$(6\zeta(3) + 3\pi^2 \ln 2 + \frac{\pi^2}{2} + \frac{17}{8}) \cdot \frac{\alpha(Z\alpha)}{\pi^2} \cdot \frac{m}{M}$	4.04
Eides, Karshenboim, Shelyuto (1988)[38]	$(\frac{9}{2}\zeta(3) - 3\pi^2 \ln 2 + \frac{39}{8}) \times \frac{Z^2\alpha(Z\alpha)}{\pi^2} \cdot \frac{m}{M}$	-1.19
Karshenboim (1993)[28]	$\frac{16}{3} \cdot \frac{\alpha(Z\alpha)^2}{\pi} \cdot \frac{m}{M} \cdot \ln^2 Z\alpha$	0.34
Eides, Shelyuto (1984)[39]	$-\frac{4}{3} \cdot \frac{\alpha^2(Z\alpha)}{\pi^3} \cdot \frac{m}{M} \cdot \ln^3 \frac{M}{m}$	-0.05
Eides, Karshenboim, Shelyuto (1989)[40]	$\frac{4}{3} \cdot \frac{\alpha^2(Z\alpha)}{\pi^3} \cdot \frac{m}{M} \cdot \ln^2 \frac{M}{m}$	0.01
Eides, Li, Samuel (1993)[41]	$0.6455 \cdot \frac{\alpha^2(Z\alpha)}{\pi} \cdot \frac{m}{M} \cdot \ln \frac{M}{m}$	0.01

2.6. *Weak Interaction contribution*

The weak interaction contribution is induced by the four fermion interaction Lagrangian

$$H_{int} = -\frac{G}{\sqrt{2}} j_\mu j^\mu, \tag{8}$$

where

$$j_\mu = (-\frac{1}{2} + 2\sin 2\theta_W)\bar{e}\gamma_\mu e - \frac{1}{2}\bar{e}\gamma_\mu\gamma_5 e \tag{9}$$
$$+(-\frac{1}{2} + 2\sin 2\theta_W)\bar{\mu}\gamma_\mu\mu - \frac{1}{2}\bar{\mu}\gamma_\mu\gamma_5\mu.$$

It is easy to obtain [42]

$$\delta E = -\frac{G}{\sqrt{2}} \cdot \frac{3mM}{4\pi Z\alpha} \cdot E_F \approx 0.063\text{kHz}. \tag{10}$$

3. Phenomenological Implications

As was already mentioned above the error of the theoretical expression for HFS (0.18 kHz or $4 \cdot 10^{-8}$) is dominated by the inaccuracy of the correction to HFS of order $\alpha(Z\alpha)^2 \cdot E_F$. All other theoretical uncertainties are smaller by at least an order of magnitude and, hence, after recalculation of this correction the accuracy of the theoretical formula for HFS would be of order 10^{-9}.

A precise value of the fine structure constant α is necessary for calculation of the HFS in the ground state. Current values of the fine structure constant, obtained by measuring the electron anomalous magnetic moment, with the help of the devices based on alternating current Josephson effect, and by measuring the quantum Hall effect resistance, are respectively, equal to

$$\alpha_{a_e}^{-1} = 137.035\,992\,22\,(94) \qquad \delta = 7 \cdot 10^{-9}, \tag{11}$$

$$\alpha_{acJ}^{-1} = 137.035\,977\,0\,(77) \qquad \delta = 6 \cdot 10^{-8}, \tag{12}$$

$$\alpha_{H}^{-1} = 137.035\,997\,9\,(32) \qquad \delta = 2 \cdot 10^{-8}. \tag{13}$$

Respective theoretical values for the HFS in the ground state of muonium are

$$\Delta\nu_{th,a_e} = 4\;463\;302.94\;(0.06)\;(0.18)\;(1.33)\;\text{kHz}. \tag{14}$$

$$\Delta\nu_{th,acJ} = 4\;463\;302.59\;(0.50)\;(0.18)\;(1.33)\;\text{kHz}. \tag{15}$$

$$\Delta\nu_{th,H} = 4\;463\;302.57\;(0.18)\;(0.18)\;(1.33)\;\text{kHz}. \tag{16}$$

The first error in parenthesis reflects the uncertainty of the fine structure constant itself and the second is induced by the uncertainty of the contribution of order $\alpha(Z\alpha)^2 \cdot E_F$.

The third, and by far the largest contribution to the error in the theoretical value of HFS is defined by the experimental error in measuring electron-muon mass ratio m/M.

The agreement between theory and experiment is excellent for all values of the fine structure α above. For example,

$$\Delta\nu_{th,a_e} - \Delta\nu_{exp} = 0.08\;(0.16)\;(0.06)\;(0.18)\;(1.33)\;\text{kHz}. \tag{17}$$

One may extract electron-muon mass ratio from the experimental value of HFS and the most precise value of α

$$\frac{M}{m} = 206.768\,265\,(15), \tag{18}$$

this result being at least four times more precise than the direct experimental value in Eq.(3). The accuracy of this result is better than $5 \cdot 10^{-8}$, planned at LAMPF.

In conclusion it is appropriate to emphasize once more that the rather complicated calculation of order $\alpha(Z\alpha)/\pi\cdot(m/M)\cdot\widetilde{E}_F$ and $\alpha^2(Z\alpha)/\pi\cdot E_F$ contributions has recently been completed and now the theory of muonium HFS has successfully overcome the 1 kHz barrier. The current theoretical uncertainty is thus reduced to about 0.2 kHz $(5\cdot 10^{-8})$ and is defined by the uncertainty in the correction of order $\alpha(Z\alpha)^2/\pi\cdot E_F$. The natural scale of all other contributions is about an order of magnitude smaller than 1 kHz and the problem of obtaining of all logarithmically enhanced terms is definitely less complicated than the calculations described above. Hence, pending the results of the LAMPF experiment, one may clearly envisage a new stage in the investigation of the muonium ground state splitting.

I am deeply grateful to S. Karshenboim and V. Shelyuto for longtime collaboration on the theory of hyperfine splitting. I would also like to thank H. Grotch for reading the manuscript and for useful remarks. This work was completed during my visit to the Penn State University. I am deeply grateful to H. Grotch for his invitation to visit Penn State and for the kind hospitality extended to me there. I wish to thank the organizers of the workshop on Infrared Physics for the invitation to take part in its work and the International Science Foundation for providing travel support to attend the meeting. This work was made possible in part by grant #R2E000 from the International Science Foundation and was also supported by the Russian Foundation for Fundamental Research under grant #93-02-3853.

References

[1] H. Hellwig, R. F. Vessot, M. Levine et al, IEEE Trans. **IM-19** (1970) 200.

[2] L. Essen, R. W. Donaldson et al, Nature **229** (1971) 110.

[3] F. G. Mariam, W. Beer, P.R. Bolton et al, Phys. Rev. Lett. **49** (1982) 993.

[4] V. W. Hughes, Z. Phys. C **56** (1992) S35.

[5] E. Klempt, R. Schulze, H. Wolf et al, Phys. Rev. **D25** (1982) 652.

[6] E. Fermi, Z. Phys. **60** (1930) 320.

[7] J. Schwinger, Phys. Rev. **73** (1948) 416.

[8] C. M. Sommerfeld, Phys. Rev. **107** (1957) 328; Ann. Phys. (NY) **5** (1958) 26.

[9] A. Peterman, Helv. Phys. Acta **30** (1957) 407.

[10] G. Breit, Phys. Rev. **35** (1930) 1477.

[11] N. Kroll and F. Pollock, Phys. Rev. **84** (1951) 594; ibid. **86** (1952) 876.

[12] R. Karplus, A. Klein, and J. Schwinger, Phys. Rev. **84** (1951) 597; R. Karplus and A. Klein, Phys. Rev. **85** (1952) 972.

[13] D. Zwanziger, Nuovo Cim. **34** (1964) 77.

[14] A. Layzer, Nuovo Cim. **33** (1964) 1538.

[15] J. R. Sapirstein, Phys. Rev. Lett. **51** (1983) 985.

[16] M. I. Eides, S. G. Karshenboim, and V. A. Shelyuto, Phys. Lett. **229B** (1989) 285.

[17] M. I. Eides, S. G. Karshenboim, and V. A. Shelyuto, Phys. Lett. **249B** (1990) 519.

[18] M. I. Eides, S. G. Karshenboim, and V. A. Shelyuto, Phys. Lett. **268B** (1991) 433; **316B** (1993) 631 (E); **319B** (1993) 545 (E).

[19] T. Kinoshita, Preprint CLNS 93/1219, May 1993.

[20] T. Kinoshita and M. Nio, Phys. Rev. Lett. **72** (1994) 3803.

[21] M. I. Eides, S. G. Karshenboim, and V. A. Shelyuto, Ann. Phys. (NY) **205** (1991) 231, 291.

[22] M. I. Eides, S. G. Karshenboim, and V. A. Shelyuto, Phys. Lett. **312B** (1993) 358; Penn State preprint PSU/TH/129/hep-ph@lanl.gov - 9306266, June 1993, to be published in the Russian Journal of Nuclear Physics.

[23] M. I. Eides, S. G. Karshenboim, and V. A. Shelyuto, Preprint PNPI-1942/hep-ph@xxx.lanl.gov - 9312337, December 1993; to be published in the Russian Journal of Nuclear Physics.

[24] M. I. Eides and V. A. Shelyuto, work in progress.

[25] T. Kinoshita, private communication.

[26] P. Lepage, unpublished, as cited in [20].

[27] T. Kinoshita, in *Quantum Electrodynamics*, ed. T. Kinoshita (World Scientific, Singapore, 1990).

[28] S. Karshenboim, Sov. Phys. JETP **76** (1993) 541.

[29] R. Arnowitt, Phys. Rev. **92** (1953) 1002.

[30] G. P. Lepage, Phys. Rev. **A16** (1977) 863.

[31] G. T. Bodwin and D. R. Yennie, Phys. Rep. **43C** (1978) 267.

[32] G. T. Bodwin, D. R. Yennie, and M. A. Gregorio, Phys. Rev. Lett. **48** (1982) 1799.

[33] H. Grotch and K. Pachucki, submitted to Physical Review A.

[34] W. E. Caswell and G. P. Lepage, Phys. Rev. Lett. **41** (1978) 1092.

[35] E. A. Terray and D. R. Yennie, Phys. Rev. Lett. **48** (1982) 1803.

[36] A. Karimkhodzhaev and R. N. Faustov, Sov. J. Nucl. Phys. **53** (1991) 1012.

[37] M. I. Eides, S. G. Karshenboim, and V. A. Shelyuto, Phys. Lett. **177B** (1986) 425.

[38] M. I. Eides, S. G. Karshenboim, and V. A. Shelyuto, Phys. Lett. **202B** (1988) 572.

[39] M. I. Eides and V. A. Shelyuto, Phys. Lett. **146B** (1984) 241.

[40] M. I. Eides, S. G. Karshenboim, and V. A. Shelyuto, Phys. Lett. **216B** (1989) 405.

[41] G. Li, M. A. Samuel, and M. I. Eides, Phys. Rev. **A47** (1993) 876.

[42] M. A. Beg and G. Feinberg, Phys. Rev. Lett. **33** (1974) 606.

Status of the Theory of the Hydrogen Lamb Shift

Howard Grotch

Department of Physics
Penn State University
104 Davey Laboratory
University Park, PA USA 16802

Abstract

The theory of the hydrogen Lamb shift is reviewed and discussed in relation to experiment. Corrections from radiative recoil of order $\alpha(Z\alpha)^5 m^2/M$, pure recoil of order $(Z\alpha)^6 m^2/M$, and nonrecoil of order $\alpha^2(Z\alpha)^5 m$ are emphasized and remaining uncertainties are mentioned.

The splitting between the $2S_{\frac{1}{2}}$ and $2P_{\frac{1}{2}}$ energy levels in hydrogen is called the Lamb shift. The theory of this has been developed over nearly five decades in terms of power expansions in $\alpha, Z\alpha, m/M$, and R/a_0 (see Ref. (1) for a recent review including references to previous work).

The experiments on the hydrogen Lamb shift provide the following results for the $2S_{\frac{1}{2}} - 2P_{\frac{1}{2}}$ splitting:

1057.845(9) MHz	Lundeen and Pipkin (1981)[2]
1057.851(2) MHz	Pal'chikov, et al. (1983)[3]
1057.839(12) MHz	Hagley and Pipkin (1994)[4]

Prior to 1984 (see review of Kinoshita and Sapirstein)[5], the theoretical numbers were

$$1057.857(11)\, MHz \qquad \text{if } \sqrt{r_p^2} = 0.805(11)\, fm[6]$$

and

$$1057.875(11)\, MHz \qquad \text{if } \sqrt{r_p^2} = 0.862(12)\, fm[7]$$

depending on which value one took for the proton radius, and consequently there appeared to be a discrepancy with the result for the larger radius. We present below an up-to-date expression for the theory. Terms below the dotted line represent contributions added to the Lamb shift formula after the 1984 review article previously mentioned.

	MHz
$\delta E = \Delta E_{2S_{\frac{1}{2}}} - \Delta E_{2P_{\frac{1}{2}}}$	
$= \frac{\alpha(Z\alpha)^4 m}{6\pi}\Big\{(\frac{\mu}{m})^3\Big[\frac{1}{8}\frac{m}{\mu} + \ell n(Z\alpha)^{-2} - 2.207909$	1050.560
$+Z\alpha\pi(\frac{427}{128} - \frac{3}{2}\ell n 2)$	7.129
$+(Z\alpha)^2\left(-\frac{3}{4}\ell n^2(Z\alpha)^{-2} + (4\ell n 2 + \frac{55}{48})\ell n(Z\alpha)^{-2}\right)$	−0.246
$+(Z\alpha)^2(-23.81)$	−0.172
$+\alpha(\frac{0.323}{\pi})\Big]$	0.101
$+\frac{Zm}{M}(\frac{1}{4}\ell n(Z\alpha)^{-2} + 2.39977)\Big\}$	0.359
$+\frac{1}{12}(Z\alpha)^4 m \cdot m^2 < r_p^2 >$	0.145
	if $r_p = .862(12) fm$
	(0.127
	if $r_p = .805(11) fm$
– –	
$-\frac{1}{48}(Z\alpha)^4 m \cdot \frac{m^2}{M^2}$	−0.0022
$+\frac{\alpha(Z\alpha)^5 m^2}{8M}\Big[(\frac{35}{4}\ell n 2 - \frac{7333}{960}) - 0.415\Big]$	−0.0025
$+\frac{(Z\alpha)^6 m^2}{8M}\Big[4\ell n 2 - \frac{7}{2}\Big]$	−0.0009
$+\frac{\alpha^2(Z\alpha)^5 m}{8\pi}[-0.06 + 0.51 + 0.61 - 0.07 - 0.12$	0.0047
$-7.61]$	−0.0411
	1057.834
	if $r_p = .862(12) fm$
	1057.816
	if $r_p = .805(11) fm$

(1)

These new results include quadratic recoil of $-2.2kHz$, radiative recoil of $-2.5kHz$, pure recoil of $-0.9kHz$, as well as higher order nonrecoil effects of $4.7kHz$ and $-41.1kHz$. Finally there is a small $(1kHz)$ change above the dotted line due to an improved numerical calculation of the $\alpha(Z\alpha)^6 m$ term. We will now discuss these new corrections.

The quadratic recoil correction of $-2.2kHz$ is contained in the Breit Hamiltonian treatment of hydrogen[8]. When the Hamiltonian is expanded in inverse powers of the electron and the proton mass an extra piece

$$\Delta H = -\frac{Ze^2}{4\mu^2 r^3}\vec{\sigma}\cdot\vec{L}\left(\frac{m}{m+M}\right)^2 \tag{2}$$

is present which breaks the degeneracy of $2S_{\frac{1}{2}}$ and $2P_{\frac{1}{2}}$ levels.

Radiative recoil corrections of order $\alpha(Z\alpha)^5 m^2/M$ can be understood as corrections beyond those associated with appropriate reduced mass factors. If we consider the Dirac equation, including nuclear kinetic energy and the convection potential, as done by Grotch and Yennie[9], we obtain

$$\left[\vec{\alpha}\cdot(\vec{p}-e\vec{A})+\beta m+V+\frac{p^2}{2M}\right]\mid n\rangle = E_n \mid n\rangle$$

or

$$(\not{\pi} - m)\mid n\rangle = 0 \tag{3}$$

with $\pi_o = E_n - V - \dfrac{p^2}{2M}$ and $\vec{\pi} = \vec{p} - e\vec{A}$.

In this treatment the proton is on its positive energy mass shell. We now consider the Lamb shift expression

$$\Delta E_n = \frac{\alpha}{4\pi^3}\int\frac{d^4k/i}{k^2+i\varepsilon}\langle n\mid\gamma_\mu\frac{1}{\not{\pi}-\not{k}-m+i\varepsilon}\gamma^\mu\mid n\rangle - \delta m < n \mid n > \tag{4}$$

in the same manner as it was calculated by Erickson and Yennie[10]. We obtain the standard Lamb shift expression with reduced mass factors but there are additional terms of order $\alpha(Z\alpha)^5 m^2/M$ which give

$$\frac{\alpha(Z\alpha)^5 m^2}{n^3 M}\left[\frac{35}{4}\ell n2 - \frac{7333}{960}\right] [11] \tag{5}$$

for S states and a negligible result for P states. For hydrogen this gives $-2kHz$ for the $2S$ state. This correction comes from an external field approximation in which the Dirac electron experiences the nuclear convection potential and is also altered by inclusion of nuclear kinetic energy, but there are additional contributions from Feynman diagrams not included in Eq. (3). These diagrams are shown in Fig. (1) and after a long calculation they contribute an extra $-0.5kHz$.

The pure recoil correction of $-0.9kHz$ is a new result[12] which corrects earlier incomplete work[13]. The corrected work changes the old calculation by $4kHz$. The new result, of order $(Z\alpha)^6 m^2/M$, arises from three distinct sources. There are contributions from Coulomb exchange graphs. Specifically, these come from two and three Coulomb exchange graphs. Next there is a contribution which involves a single transverse photon. It is necessary to include graphs which explicitly bring in one or two Coulomb exchanges along with the transverse photon exchange. Finally there is also a term which involves a seagull correction, as well as a seagull corrected by one Coulomb exchange. The correction has now been evaluated both analytically and numerically and the results are in agreement. The analytic result is

$$\frac{(Z\alpha)^6 m^2}{8M}\left[4\ell n2 - \frac{7}{2}\right] \tag{6}$$

for the $n = 2$ state.

The next set of graphs, shown in Fig. (2), produce contributions of order $\alpha^2(Z\alpha)^5 m$. The diagrams a, b, c, d, and e, all of which have closed electron loops, have been separately calculated and confirmed by two groups[14, 15]. These yield $4.7kHz$ for the $n = 2$ state of hydrogen. A sixth set of diagrams, f, involve many graphs with two photons on the electron line. A result

$$\Delta E_f = -7.61\frac{1}{n^3}\frac{\alpha^2(Z\alpha)^5 m}{\pi} \tag{7}$$

has been obtained by Pachucki[16]. This gives a huge answer of $-41.1kHz$ for the $n = 2$ state of hydrogen. It should be noted that the corresponding lower order contribution, with one less power of $Z\alpha$, is[17]

$$\frac{\alpha^2(Z\alpha)^4 m}{6\pi^2}(.323) \quad \longrightarrow \quad 101kHz. \tag{8}$$

We see that Eq. (7) is a 40% correction to Eq. (8). This seems unusually large. On the other hand, it might be argued that the coefficient in Eq. (8) is anomalously small. It should also be mentioned that in Eq. (1) above there is a term[18]

$$\frac{\alpha(Z\alpha)^6 m}{6\pi}[-23.81] \tag{9}$$

which clearly has a very large coefficient. This illustrates that terms with large coefficients do arise and hence it may also be possible to have a large coefficient in Eq. (7).

We also want to mention that the number presented in Eq. (9) above is due to Pachucki[18] who has redone the calculation of Mohr[19] by a different method. His answer adds $1kHz$ to the Lamb shift and reduces the error associated with the $\alpha(Z\alpha)^6 m$ correction.

If all calculations have been done correctly, then theory and experiment agree reasonably well if the larger value of the proton radius is used. It is clearly essential to have a verification of the $-41.1 kHz$ correction. Several groups are now working towards this goal.

Finally, it should be noted that other corrections also need to be calculated to be sure all theoretical effects at the level of $1 kHz$ have been accounted for. Still to be calculated are terms of order $\alpha^3(Z\alpha)^4 m$. These will involve the derivative at the origin of the sixth order contribution to the electron form factor. In addition, there are also terms of the type $\alpha^2(Z\alpha)^6 m(\ell n Z\alpha)^p$. Recently the $p = 3$ term has been worked out and found to give $-3.6 kHz$ for the $2S$ state[20]. We have not included this in the Lamb shift formula since other terms of order $\alpha^2(Z\alpha)^6 m$ are not known. It might be argued, however, that the $p = 3$ term should dominate other corrections.

The experimental accuracy of the measured hydrogen Lamb shift is limited by the P state lifetime. However, very accurate measurements of the $2S - 1S$ energy separation are now being carried out[21] and consequently there is a need to have very accurate theoretical results for radiative level shifts, even beyond the $1 kHz$ level.

Acknowledgement

I am grateful for the support of the National Science Foundation which has funded this research under Grant No. PHY-9120102. My discussions and collaboration with M. Eides, K. Pachucki, and Y. Khriplovich have been extremely rewarding.

References

[1] H. Grotch, *Foundations of Physics* **24**, 249 (1994).

[2] S. R. Lundeen and F. M. Pipkin, *Phys. Rev. Lett.* **46**, 232 (1981).

[3] V. G. Pal'chikov, Yu L. Sokolov, and V. P. Yakovlev, *JETP Lett.* **38**, 418 (1983).

[4] E. W. Hagley and F. M. Pipkin, *Phys. Rev. Lett.* **72**, 1172 (1994).

[5] T. Kinoshita and J. Sapirstein, in *Atomic Physics* **9**, eds. Robert S. Van Dyck, Jr., and E. Norval Fortson, World Scientific (Singapore, 1984).

[6] L. N. Hand, D. J. Miller and R. Wilson *Rev. Mod. Phys.* **35**, 335 (1963).

[7] G. G. Simon, Ch. Schmitt, F. Borkowski and V. H. Walther, *Nucl. Phys. A* **333**, 381 (1981).

[8] W. A. Barker and F. N. Glover, *Phys. Rev.* **99**, 317 (1955).

[9] H. Grotch and D. R. Yennie, *Rev. Mod. Phys.* **41**, 350 (1969).

[10] G. W. Erickson and D. R. Yennie, *Ann. Phys. (N.Y.)* **35**, 217 (1965); **35**, 447 (1965).

[11] G. Bhatt and H. Grotch, *Phys. Rev. A* **31**, 2794 (1985); *Phys. Rev. Lett* **58**, 471 (1981); *Ann. Phys.***178**, 1 (1987).

[12] K. Pachucki and H. Grotch, submitted to *Phys. Rev. A*.

[13] M. A. Doncheski, H. Grotch and G. W. Erickson, *Phys. Rev. A***43**, 2152 (1991).

[14] M. I. Eides, H. Grotch and D. A. Owen, *Phys. Lett.* **294B**, 115 (1992); M. I. Eides and H. Grotch, *Phys. Lett.* **301B**, 127 (1993); M. I. Eides and H. Grotch, *Phys. Lett.* **308B**, 389 (1993); M. I. Eides, H. Grotch and P. Pebler, *Phys. Lett.* **326B**, 197 (1994); M. I. Eides, H. Grotch and P. Pebler, *Phys. Rev. A* (1994).

[15] K. Pachucki, *Phys. Rev.* **A48**, 2609 (1993).

[16] K. Pachucki, *Phys. Rev. Lett.* **72**, 3154 (1994).

[17] T. Appelquist and S. J. Brodsky, *Phys. Rev. A***2**, 2293 (1970); R. Barbieri, J. A. Mignaco, and E. Remiddi, *Nuovo Cim. Lett.***3**, 588 (1970).

[18] K. Pachucki, *Ann. Phys. (N.Y.)* **226**, 1 (1993).

[19] P. J. Mohr, *Ann. Phys.* **88**, 26, 52 (1974).

[20] S. G. Karshenboim, *JETP* **76**, 541 (19930.

[21] F. Schmidt-Kaler, D. Leibfried, M. Weitz and T. W. Hänsch *Phys. Rev. Lett.* **70**, 2261 (1993); M. Weitz, A. Huber, F. Schmidt-Kaler, D. Leibfried, and T. W. Hänsch, *Phys. Rev. Lett.* **72**, 328 (1994).

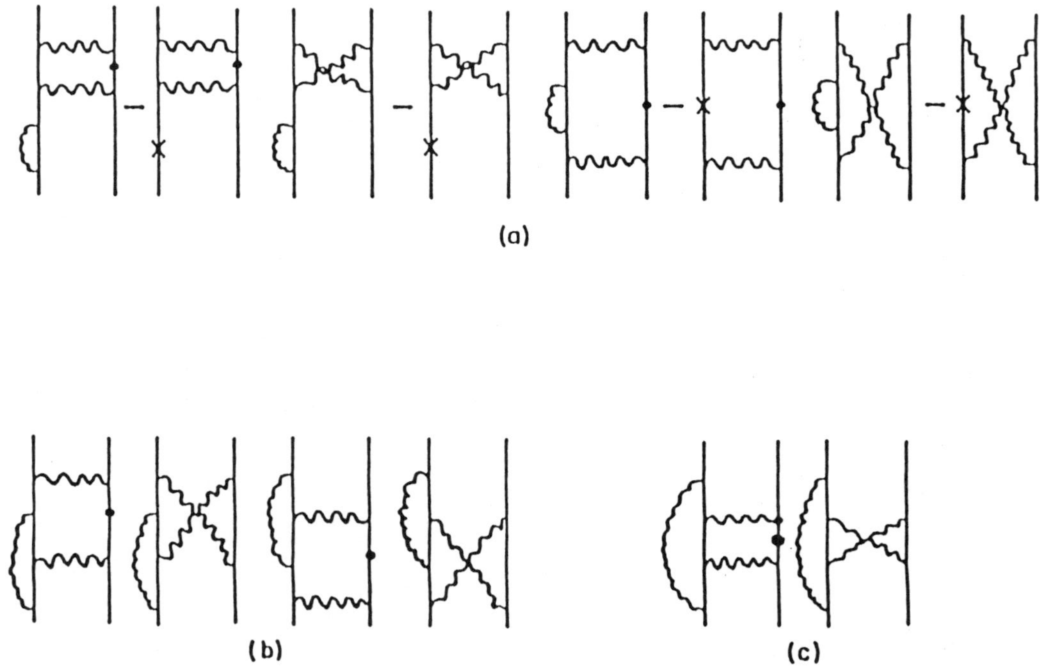

Figure 1: Corrections to EFA. Two-photon exchange self-energy (a), vertex (b), and the spanning photons (c) diagrams. The "dot" on the proton propagator in the ladder graphs implies the subtraction of EFA, while the "cross" represents the mass renormalization.

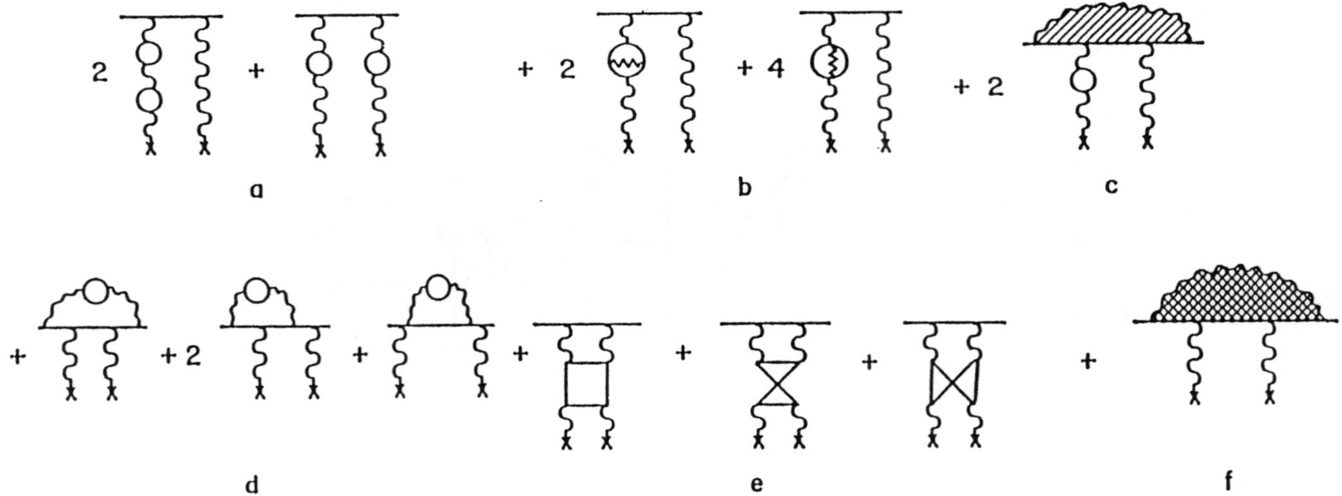

Figure 2: All gauge invariant sets of diagrams that generate corrections of order $\alpha^2(Z\alpha)^5m$ to the Lamb shift.

Eikonal Scattering Phenomenology

BEHAVIOR OF TOTAL AND ELASTIC CROSS SECTIONS AT VERY HIGH ENERGIES

TAI TSUN WU
Gordon McKay Laboratory, Harvard University, Cambridge, MA 02138, USA, and Theoretical Physics Division, CERN, Geneva, Switzerland

1. Introduction

In this talk, I would like to present an overview of the original theoretical prediction of the increasing total cross section on the basis of gauge quantum field theory, phenomenological predictions from this theoretical understanding, comparison with later experimental data, and expected future data from HERA.

The discussion will be limited to elastic scattering, since it is the simplest scattering process. Through the optical theorem, the total cross section—which gives the overall strength of the scattering process—is proportional to the imaginary part of the elastic scattering amplitude in the forward direction. Therefore, the theoretical understanding of elastic scattering is often of primary importance.

Some of these developments can be found in an excellent book by Herbert Fried, the organizer of this workshop: Herbert M. Fried, *Functional Methods and Eikonal Models*, Editions Frontieres (1990). A few years earlier, Hung Cheng and I also wrote a book on this subject: Hung Cheng and Tai Tsun Wu, *Expanding Protons: Scattering at High Energies*, The MIT Press (1987).

A recent topic of current interest, of course not covered in either of these books which are a few years old, is the data from the electron-proton colliding accelerator HERA at DESY, Hamburg, Germany. This is discussed in some detail in Section 5.

2. The Beginning: 1967-1970

In 1967, Hung Cheng and I learned that the United States Atomic Energy Commission (predecessor of the Department of Energy) had decided to fund the Fermi National Accelerator Laboratory, Batavia, Illinois. This was for the construction and operation of a 200-GeV proton accelerator, later upgraded to 400 GeV. When a 200-GeV proton hits a proton at rest, the energy of each proton in the center-of-mass system is about 10 GeV. If we say that a particle is extremely relativistic if its energy is at least ten times its rest mass, then for the first time it would be possible to observe the interaction of two extremely relativistic protons.

Because of this exciting news, we asked ourselves what we could say about such scattering processes at very high energies. It seemed reasonable to expect some simple features in such a limit, perhaps analogous to optics as a limiting case of electromagnetic theory where the wavelength is small. How could we get hold of these simple features?

The approach we decided to take was to study the high-energy behavior of quantum field theory. Cheng was an expert on Regge-pole theory, and I had some experience in getting Regge poles and cuts from ϕ^3 field theory. However, we both felt that, in order to be able to make theoretical predictions for the scattering processes at very high energies, we needed take the best quantum field theory and study its high-energy behavior without preconceived prejudices as to the outcome. At that time, the only definitive piece of information was the Froissart-Martin bound.[1]

In the sixties, it was easy to decide which quantum field theory was the best: the only well-understood one was quantum electrodynamics. Since our interest was in using QED as a clue for high-energy hadron scattering, we made two modifications to QED: We permitted a nonzero photon mass, and we allowed the coupling constant to be larger than the fine-structure constant. Neither changed the calculation in any appreciable way. Perhaps influenced by the optical limit of Maxwell's equations, we decided to investigate first the high-energy behavior of Compton scattering, i.e., the scattering of a vector gauge particle by a fermion.

To the lowest order, the total Compton scattering cross section approaches zero as the energy increases without bound. This is also true to the next, or fourth, order. Therefore, Cheng and I concentrated on calculating the sixth-order cross sections. This was one of the longest calculations that I have engaged in, and it took us over a year to complete.

The result of this sixth-order asymptotic calculation gave a total cross section that no longer goes to zero at infinite energy.[2] Thus at high energies, the sixth-order matrix element is much larger than the second- and fourth-order ones. We found this result to be very satisfying: for example, unlike the second- and fourth-order results, this sixth-order total cross section at least bears some resemblance to that of the optical scattering from a sphere.[3]

Of course this calculation also gives the elastic differential cross section. The high-energy behavior of this cross section is somewhat complicated and is most compactly expressed in terms of what we call "impact factors." The same impact factor for the vector gauge particle also appears in the eighth-order scattering of two such vector particles.

Once the sixth-order Compton scattering cross section at high energies was understood, we gradually became more and more efficient in extracting the high-energy behavior of various processes. Our calculations became much more exciting when we found extra factors of $\ln s$ in the higher-order diagrams, giving the first indication that the total cross section may increase with energy rather than approaching a constant value. Figure 1 shows the orders of magnitude of some of these higher-order terms. Last year at the Blois conference, Cheng explained this development in detail and very well.[4] I therefore refer the reader to his presentation.

About sixteen months after obtaining the sixth-order result for Compton scattering, Hung Cheng and I were able to predict that at very high energies the total cross sections increase without bound for all hadronic scattering. In our 1970 paper in Physical Review Letters,[5] the abstract says:

> "At infinite energy, we predict that: (1) σ_{tot} approaches infinity; (2) the ratio of the real part to the imaginary part of the forward elastic amplitude approaches zero; (3) $\sigma_{\text{el}}/\sigma_{\text{tot}}$ approaches 1/2; (4) the width of the diffraction peak approaches zero; its product with σ_{tot} is a constant. We give theoretical evidences based on massive quantum electrodynamics as well as experimental evidence in support of these predictions, and a physical picture for high-energy scattering."

By the time this volume comes out next year, it will be the twenty-fifth anniversary of this first prediction of increasing total cross sections. I am happy to say that these predictions have fared well after all these years.

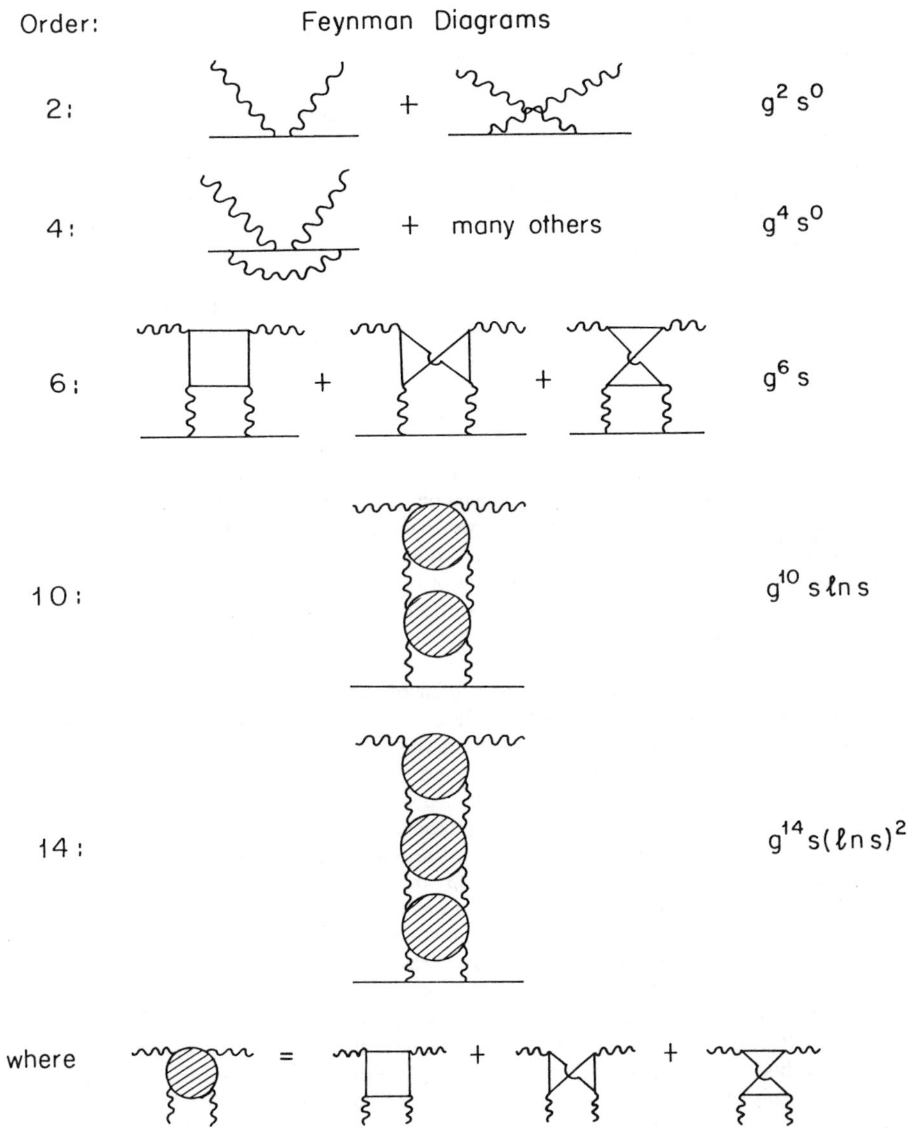

Fig. 1. The orders of magnitude of the matrix elements for Compton scattering in an Abelian gauge theory when s is large but t is fixed. [s is the square of the center-of-mass energy and $-t$ is the square of the momentum transfer.]

3. The Early Phenomenology: 1972-1974

As mentioned above, the original motivation for our investigation was provided by the approval of the Fermi National Accelerator Laboratory by the U. S. Atomic Energy Commission. However, before 1970, we did not expect such a dramatic prediction of the increasing total cross section. For the purpose of obtaining experimental verification for this prediction, the Intersecting Storage Rings (ISR) at CERN, Geneva, Switzerland, was the most promising accelerator. We therefore discussed with the experimentalists, especially Giorgio Bellettini, about measuring the total proton-proton cross section at ISR. Naturally he raised the good question as to how much increase was to be expected in this total cross section.

On the basis of the so-called *tower diagrams* in quantum field theory, we had learned that the effective interaction strength increases with energy in the form of

$$\frac{s^{1+c}}{(\ln s)^{c'}}. \tag{1}$$

To answer the question of Bellettini, we needed to be able to find an approximate determination of the value of c, since the dependence on c' is only of secondary importance.

In order to determine the value of c, we asked ourselves whether there were any experimental data that we could use. At that time, the best experimental evidence for the increasing cross section was the measurement of the ratio of the real to the imaginary parts of the proton-proton forward elastic cross section: it showed a distinct tendency to cross over to positive values.[6] However, it was difficult to get an estimate for Bellettini of the increase in total cross section from these data.

Since the values of c and c' are independent of the hadronic scattering process under consideration, we could use any hadronic data. Before the Fermilab accelerator and the CERN ISR, the highest-energy accelerator in the world was that at Serpukhov, Russia. That accelerator produced 70 GeV protons, and there were secondary π and K beams of both charges, up to nearly the same energy. Thus there were data for pp, $\bar{p}p$, π^+p, π^-p, K^+p, and K^-p elastic scattering from Serpukhov.[7] Of these six total cross sections, five are decreasing and only one is increasing in the Serpukhov energy range, the increasing one being that of K^+p. Moreover, the K^+p total cross section is below that for K^-p. While the almost universal belief at that time was that these two total cross sections would approach the same constant value at very high energies due to the Pomeranchuk theorem,[8] we took the opposite view that this K^+p total cross section as measured at Serpukhov can be used to estimate the increase in all total cross sections, including the pp cross section to be measured at the CERN ISR and the Fermilab accelerator.

Figure 2 shows this original estimate[9] of the increase in pp total cross section obtained by this method and given to Bellettini for the planning of the experiment at the CERN ISR. The result of this rough estimate was that the increase of the total pp cross section from its minimum value to the ISR energy of 53 GeV is almost 3 mb. The year after this phenomenological estimate was obtained, the total cross section at this energy was measured by both the Pisa-Stony Brook Collaboration,[10] of which Bellettini was a most prominent member, and the CERN-Rome Collaboration,[11] and it was found that this estimate was too low by about 35%. In view of the crudeness of the assumptions involved in this estimate, the result of Fig. 2 was considered to be a theoretical triumph.

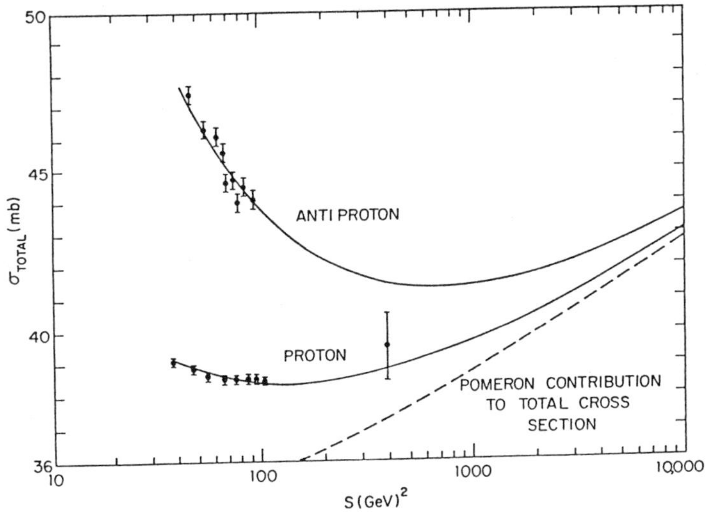

Fig. 2. An early rough prediction of rising pp and $\bar{p}p$ total cross sections, carried out before the first experimental observation of the rise in 1973.

The natural next step was to use these ISR data[10,11] as an input in addition to the Serpukhov data.[7] However, there was still no way to determine the value of c'. For this reason, two sets of phenomenological parameters were obtained at this second stage, one set with $c' = 0$ and the other with $c' = 1$. For these two sets, the corresponding values of c were of course different:[12]

$$\text{For } c' = 0, \quad c = 0.08; \qquad \text{For } c' = 1, \quad c = 0.2. \tag{2}$$

Once these parameters were obtained, it was possible to predict not only the pp and the $\bar{p}p$ total cross sections, but also those for π^+p, π^-p, K^+p, and K^-p. The result of this early phenomenology is shown in Fig. 3 together with the experimental data available in 1973.[12] In this energy range, there is virtually no difference between the two sets of parameters.

Various other quantities can also be obtained easily besides the total cross sections. One example, the ratio of the real and imaginary parts of the elastic amplitude in the forward direction, is shown in Fig. 4 together with the experimental data available at that time. Diffraction scattering was also studied.

It should perhaps be mentioned that the value of $c = 0.08$ as given in Eq. (2) for $c' = 0$ was later used extensively by many authors on increasing total cross sections.[15]

Because of the important contributions of Pomeranchuk[8] to the theoretical study of total cross sections, the word "pomeranchukon," later shortened to "pomeron," is used to describe the exchanged object that is responsible for elastic scattering. The theoretical prediction and later experimental verification of rising total cross sections mean that the pomeron is a more complicated and hence more interesting object than originally visualized. This pomeron is of course also responsible for diffraction scattering.[16]

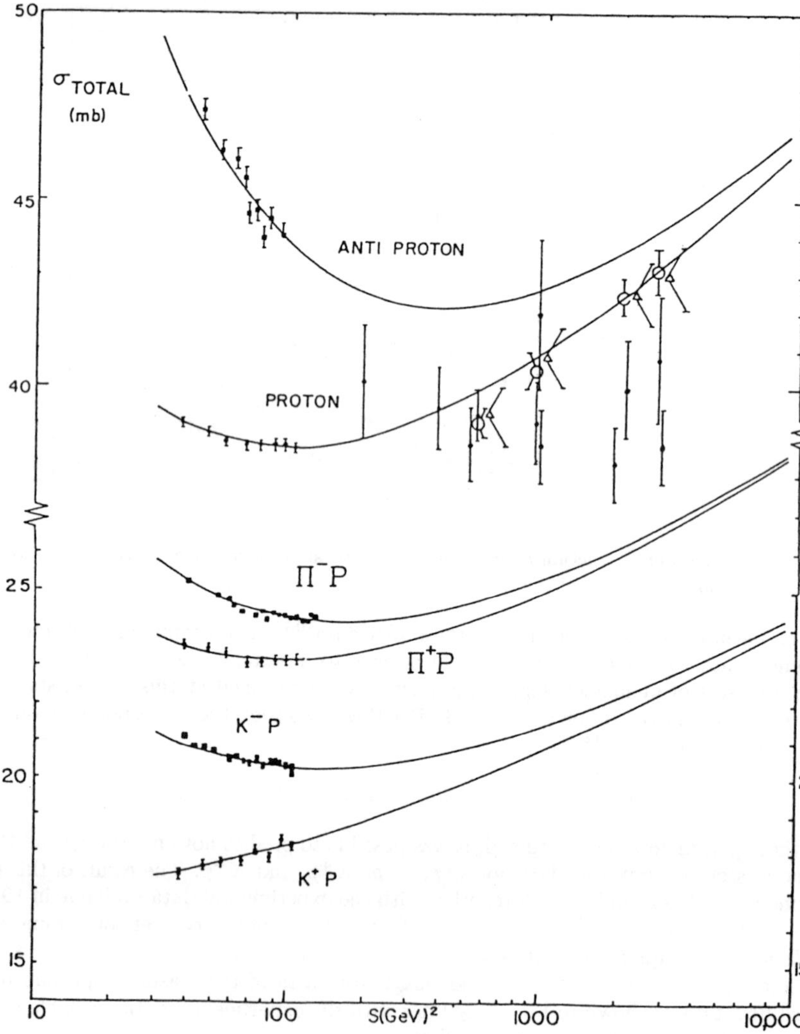

Fig. 3. Comparison of the phenomenology with the parameters of Eq. (2) with the experimental data from Serpukhov,[7] Fermilab,[13] and ISR.[10,11,14]

It should be emphasized that there is only one pomeron. In particular, the so-called soft pomeron and the hard pomeron are merely aspects of the same object. For example, with reference to Eq. (2) which gives two different parametrizations of this pomeron, the value of 0.08 (1.08 when 1 is added) is sometimes considered to pertain to the soft pomeron, while the value of 0.2 (similarly 1.2 when 1 is added) to the hard pomeron.

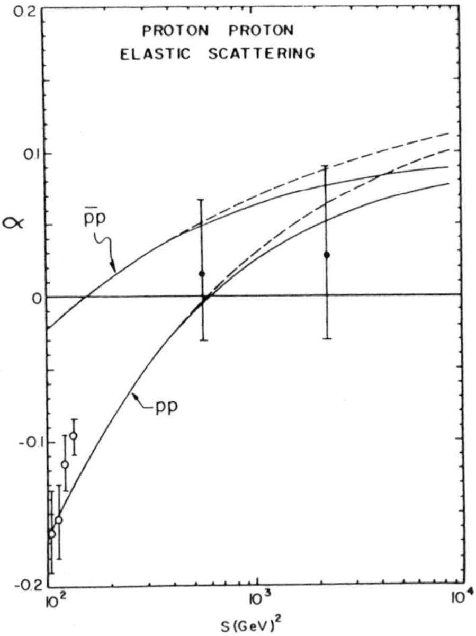

Fig. 4. Theoretical values of α, the ratio of real to imaginary parts of the pp and $\bar{p}p$ forward scattering amplitude, together with the experimental data available up to 1973.

4. Theoretical Predictions and Experimental Verifications: 1977-1993

After obtaining the results of Figs. 3 and 4 from the early phenomenological analyses in 1973 and 1974, we were not able to make any further progress on this topic. The major problems were clear: (a) The value of c' remained to be determined; and (b) we needed to take better account of the exchange of other objects, such as the ρ Regge pole. This problem remained dormant until I was fortunate to meet Claude Bourrely and Jacques Soffer of Marseille, France. They were great experts on a number of subjects, including spin physics and phenomenology in general. Since they found the problem of increasing total cross sections both interesting and challenging, we decided to collaborate on working out explicit theoretical predictions to be compared with future experimental data. As of now, I have been collaborating with them for nearly twenty years.

In order to solve problem (a) above, Bourrely, Soffer and I carried out a systematic study of the experimental data available up to 1979. This lead unambiguously to the conclusion that c' is positive:[17]

$$c' = 0.756. \tag{3}$$

This result has the following important implication: The pomeron is *not* a Regge pole (for which c' must be zero), but is instead a Regge cut above 1. A few years later, this systematic study was repeated[18] using additional data and this value of c' was revised to 0.748 (see below)—a reduction of an insignificant amount—meaning that the determination of c' is quite stable.

The present model (sometimes referred to as the impact picture) for proton-proton and antiproton-proton elastic differential cross section is given by the following simple formulas, when Coulomb contributions are included:[18]

$$a(s,t) = a^N(s,t) \pm a^c(t), \tag{4}$$

where the upper sign is for $\bar{p}p$ while the lower one is for pp, $a^N(s,t)$ is the hadronic amplitude and $a^c(t)$ is the $\bar{p}p$ Coulomb amplitude. The Coulomb amplitude is that of West and Yennie,[19] and the hadronic amplitude at high energies is given by

$$a^N(s,t) = is\int_0^\infty J_0(b\sqrt{-t})(1 - e^{-\Omega_0(s,b)})b\,db, \tag{5}$$

where

$$\Omega_0(s,b) = S_0(s)F(b^2), \tag{6}$$

when the Regge background is neglected. The energy dependence is given by the crossing symmetric version of Eq. (1),

$$S_0(s) = \frac{s^c}{(\ln s)^{c'}} + \frac{u^c}{(\ln u)^{c'}}. \tag{7}$$

In Eq. (7), u is the third Mandelstam variable and both s and u are expressed in $(\mathrm{GeV})^2$. The t-dependence of $a^N(s,t)$ is controlled by $F(b^2)$ whose Fourier transform is taken to be

$$\tilde{F}(t) = f[G(t)]^2[(a^2+t)/(a^2-t)], \tag{8}$$

where $G(t)$ is given by

$$G(t) = [(1 - t/m_1^2)(1 - t/m_2^2)]^{-1}. \tag{9}$$

The values of the parameters are

$$\begin{aligned} c &= 0.167, & c' &= 0.748, \\ m_1 &= 0.586\ \mathrm{GeV}, & m_2 &= 1.704\ \mathrm{GeV}, \\ a &= 1.953\ \mathrm{GeV}, & f &= 7.115\ \mathrm{GeV}^{-2}. \end{aligned} \tag{10}$$

These values have not changed at all during the last ten years. In fact, they have changed only slightly over the previous five years.

Since these formulas are rather simple, anyone with a computer can calculate the theoretical predictions and compare with experimental results. The inclusion of ρ and ω exchanges makes the formulas only slightly more complicated. The most extensive tabulation of the results from these formulas was published several years ago.[20] Such comparisons between theoretical predictions and later experimental results have been carried out on many occasions, and the reader is referred to the literature. Instead, I would like to mention only a couple of examples.

First, as already discussed in Sec. 2, the third prediction of the 1970 paper[5] was that the ratio $\sigma_{\mathrm{el}}/\sigma_{\mathrm{tot}}$ of the integrated elastic cross section and the total cross section approaches 1/2 as the center-of-mass energy approaches infinity. At the time of this prediction, this ratio for pp scattering was less than 0.2 and decreasing. The predicted increase of this ratio was not verified experimentally until over a decade later by the UA4 Collaboration.[21]

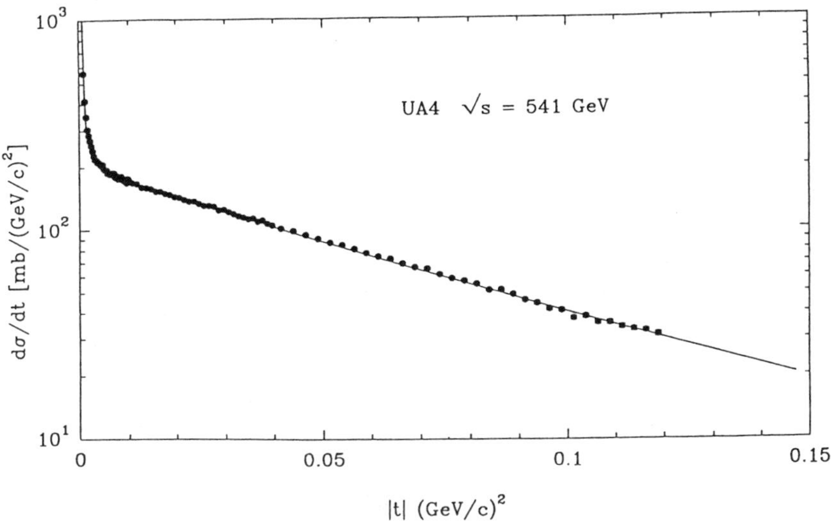

Fig. 5. Comparison of theoretical prediction and new experimental measurement by UA4[24] at $\sqrt{s} = 541$ GeV in the small-t region.

As a second example, I would like to relate the story of the ratio of the real and imaginary parts of the elastic scattering amplitude in the forward direction. The experimental data on this quantity[6] provided the first indication that the total cross section may not be approaching a constant[8] at infinite energy. Later experimental data up to the early eighties gave good confirmation of the theoretical calculations. However, in 1987, the UA4 Collaboration published an experimental value for this ratio of 0.24 ± 0.04,[22] much above the prediction of 0.13. This discrepancy led to a flurry of theoretical activities claiming new physics. Not believing this new physics, Bourrely, Soffer and I analyzed the UA4 data[22] in detail and concluded that their value of 0.24 ± 0.04 was questionable.[23] It was nevertheless gratifying when this same Collaboration repeated their experiment with more extensive data and found that, instead of 0.24 ± 0.04, this ratio is 0.135 ± 0.015.[24] A comparison of the theoretical prediction and the new data is shown in Fig. 5; a more detailed comparison is in Ref. 25. Some further comparisons between our theoretical predictions and the recent data from the CERN Collider and the Fermilab Tevatron Collider will be presented by André Martin.[26]

5. Photon-Proton Scattering at HERA Energies

So far, I have described only pp and $\bar{p}p$ scattering, since these have been the most reliable source of experimental information on the rising total cross sections, first predicted in 1970.[5] With the operation of HERA, there will soon be a second source of reliable information about rising total cross sections.

Although the theoretical development was initially based on U(1) gauge theory, the phenomenology has been applied almost exclusively to hadronic scattering. In the seventies, we were not sufficiently far-sighted to realize the possibility of constructing electron-proton

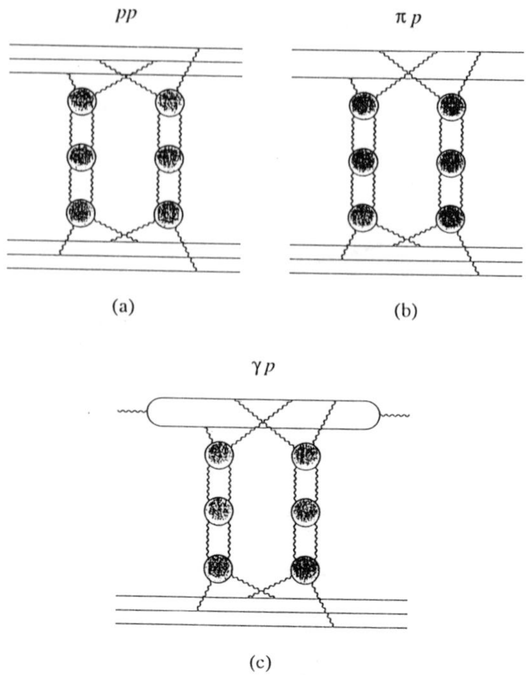

Fig. 6. Multi-tower diagrams for (a) pp and $\bar{p}p$, (b) πp, and (c) γp scattering.

colliding accelerators to study the scattering of photons and protons at very high energies. This remarkable feat has now been accomplished at the Deutsches Electronen-Synchrotron in Hamburg, Germany, under the leadership of Bjorn Wiik and Gus Voss. This new accelerator is called HERA, with an initial proton energy of about 800 GeV and an electron energy of about 30 GeV, leading to a center-of-mass energy of about 300 GeV. This makes it possible to study very high-energy γp scattering.

The theory of rising total cross sections is most easily obtained by studying the multi-tower diagrams. These are shown in Fig. 6a for pp and $\bar{p}p$ and in Fig. 6b for π^+p and π^-p. Clearly the corresponding multi-tower diagrams for γp are those in Fig. 6c. The similarity of the diagrams in Fig. 6b and Fig. 6c are striking, the major differences being: i) the electromagnetic coupling e appears twice in Fig. 6c, and ii) the photon is its own anti-particle. Because of these two differences, it is natural to expect that

$$\sigma_{\text{tot}}(\gamma p) = (\text{const})\alpha[\sigma_{\text{tot}}(\pi^+ p) + \sigma_{\text{tot}}(\pi^- p)], \tag{11}$$

where α is the fine-structure constant. The constant is determined by comparing the low-energy data of the γp and πp total cross sections; it was found to be $1/3$.

Figure 7 shows the theoretical prediction on the basis of Eq. (11) together with the π^+p and π^-p total cross sections of Fig. 3.[27] The HERA data[28] from the ZEUS and H1

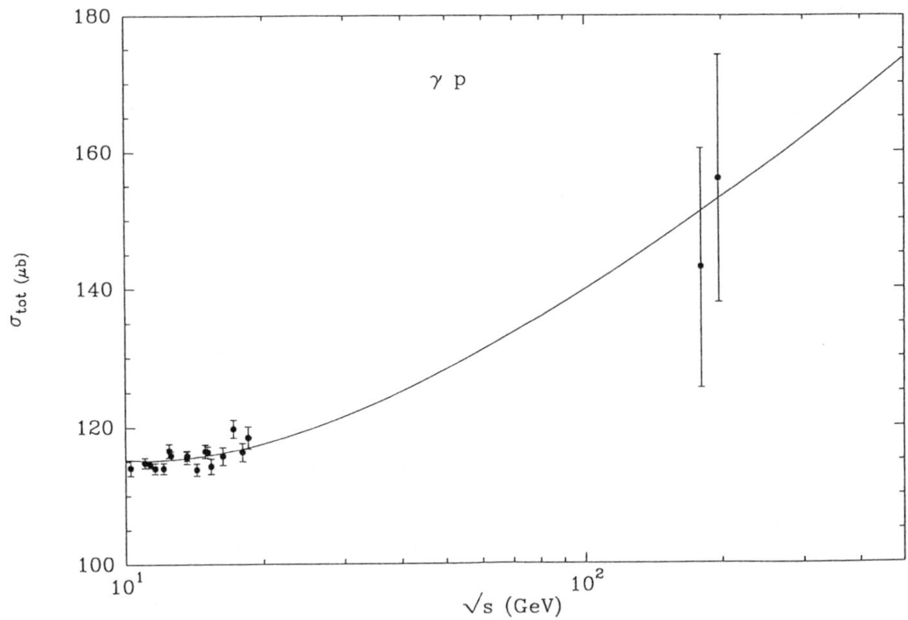

Fig. 7. γp total cross section as a function of the center-of-mass energy.

Collaborations are also shown together with the previous data at lower energies.[29] The HERA data are surely going to improve rapidly, and we are eager to see a much more accurate determination of the γp cross sections. In fact, while the points shown in Fig. 7 are the most recent ones published by the H1 and ZEUS Collaborations, they are based on their 1992 data. They have not yet finished analyzing the 1993 data, but this is to be expected any time now. These more accurate data will provide a stringent test of the accuracy of Eq. (11).

Thus, we have encountered the expanding proton unambiguously in two different situations, first the rising total cross sections of pp and $\bar{p}p$, and secondly that of γp. To us this is especially satisfying: we began with the study of Compton scattering at high energies twenty-five years ago, and now HERA is providing data on Compton scattering at high energies. The cycle is complete.

Acknowledgments

I am grateful to the organizers of this Workshop, especially to Herbert Fried, for such an interesting and pleasant meeting. I also wish to thank the Theoretical Physics Division of CERN for its kind hospitality. This work was supported in part by the the U.S. Department of Energy under Grant No. DE-FG02-84ER40158 with Harvard University.

References

1. M. Froissart, *Phys. Rev.* **123** (1961) 1053; A. Martin, *Nuovo Cim.* **42** (1966) 930; **44** (1966) 1219.
2. H. Cheng and T. T. Wu, *Phys. Rev. Letts.* **22** (1969) 666; *Phys. Rev.* **182** (1969) 1852, 1868, 1873, 1899.
3. T. T. Wu, *Phys. Rev.* **104** (1956) 1201.
4. H. Cheng, *Proc. of the Vth Blois Workshop*, eds. H. M. Fried, K. Kang, and C.-I. Tan (World Scientific, Singapore, 1994), p. 236.
5. H. Cheng and T. T. Wu, *Phys. Rev. Letts.* **24** (1970) 1456.
6. K. J. Foley et al., *Phys. Rev. Letts.* **19** (1967) 857.
7. S. P. Denisov et al., *Phys. Letts.* **B36** (1971) 415.
8. I. Ya. Pomeranchuk, *Sov. Phys. JETP* **7** (1958) 499.
9. H. Cheng, J. K. Walker, and T. T. Wu, Paper 524, 16th Int. Conf. on High Energy Physics, 1972 (unpublished).
10. S. R. Amendolia et al., *Phys. Letts.* **B44** (1973) 119.
11. U. Amaldi et al., *Phys. Letts.* **B44** (1973) 112.
12. H. Cheng, J. K. Walker, and T. T. Wu, *Phys. Letts.* **B44** (1973) 97.
13. G. Charlton et al., *Phys. Rev. Letts.* **29** (1972) 515; F. T. Dao et al., *Phys. Rev. Letts.* **29** (1972) 1627; J. W. Chapman et al., *Phys. Rev. Letts.* **29** (1972) 1686.
14. Aachen-CERN-Harvard-Geneva-Torino Collaboration, paper presented by C. Rubbia at 16th Int. Conf. on High Energy Physics, 1972 (unpublished).
15. See, for example, A. Donnachie and P. V. Landshoff, *Nucl. Phys.* **B231** (1983) 189.
16. H. Cheng, J. K. Walker, and T. T. Wu, *Phys. Rev.* **D9** (1974) 749.
17. C. Bourrely, J. Soffer, and T. T. Wu, *Phys. Rev.* **D19** (1979) 3249.
18. C. Bourrely, J. Soffer, and T. T. Wu, *Nucl. Phys.* **B247** (1984) 15.
19. G. B West and D. R. Yennie, *Phys. Rev.* **172** (1968) 1413.
20. C. Bourrely, J. Soffer, and T. T. Wu, *Z. Phys. C–Particles & Fields* **37** (1988) 369.
21. UA4 Collaboration, R. Battiston et al., *Phys. Letts.* **B115** (1982) 333; **B117** (1982) 126; and **B127** (1983) 472.
22. UA4 Collaboration, D. Bernard et al., *Phys. Letts.* **B198** (1987) 583.
23. C. Bourrely, J. Soffer, and T. T. Wu, *Mod. Phys. Letts.* **A6** (1991) 2973.
24. UA4.2 Collaboration, C. Augier et al., *Phys. Letts.* **B316** (1993) 448.
25. C. Bourrely, J. Soffer, and T. T. Wu, *Phys. Letts.* **B315** (1993) 195.
26. A. Martin, this Workshop.
27. C. Bourrely, J. Soffer, and T. T. Wu, CERN Preprint TH-7387 (1994), *Phys. Letts. B* (to be published).
28. H1 Collaboration, T. Ahmed et al., *Phys. Letts.* **B299** (1994) 374; ZEUS Collaboration, M. Derrick et al., DESY Preprint 94-032 (1994).
29. D. O. Caldwell et al., *Phys. Rev. Letts.* **40** (1978) 1222.

DO WE UNDERSTAND TOTAL CROSS-SECTIONS?

ANDRE MARTIN
Theoretical Physics Division, CERN
CH 1211 Geneva 23

ABSTRACT

We show that we have at least a qualitative and even semi-quantitative understanding of total cross-sections based on general principles plus various models which even if they look superficially different have a lot in common. We discuss the problem of existence or non-existence of the odderon and in spite of our prejudice against it insist on the need of direct fixed target experiments which would clarify the matter and at the same time would give previous information on the relation between cross-sections and the quark content of the colliding particles at high energies.

The reason why the title of this talk has an interrogative form is maybe that one of my grand-parents was from French Normandy. In this part of France people never answer a question by Yes or No. They prefer to say in a non-committal way "P'têt ben qu'oui, p'têt ben qu'non".

Indeed in the problem of total cross-sections we can answer YES and NO.

NO because we cannot predict total cross-sections in an absolute way like the mass of the Z_0 particle can be predicted from the Glashow-Weinberg angle. It is clearly out of question to make accurate predictions like the ones we see in Z_0 physics, like, besides the masses of Z_0, and W, the total and partial widths, which agree so well with experiment.

On the other hand, I do not feel too much of an optimist if I say that we have a qualitative and even semi-quantitative understanding of total cross-sections in terms of various – superficially very different – models, depending on a few parameters. I can quote the most well-known ones:

- the Cheng-Wu model [1], which became the Cheng-Wu-Walker model [2], and was eventually fully exploited by Wu, Bourrely and Soffer [3];
- the Donnachie-Landshoff model [4]

- the Kaidalov-Ter-Martyrosian model [5] based on a supercritical pomeron.

The authors of these models claim to have completely different basis. Yet their conclusions, as we shall see, are very similar.

I would like to begin with a historical perspective. Let me remind you that up to 1971 total cross-sections were believed to approach constants at high energies. It was thought that around 30 GeV laboratory energy the asymptotic regime was reached.

The most naïve model to understand this, namely a nucleon which is an object with a finite size, which we could call "geometric", would tell you that the scattering amplitude is approximately

$$F(s,t) \cong \mathrm{i}\ \ \mathrm{s}\ f(t)\ , \tag{1}$$

the total cross-section being proportional to Im $f(0)$. This model was "killed" already in 1961 by V. Gribov [6] who proved that such an assumption leads to a contradiction with analyticity and unitarity in the s and t channels.

However, the constancy of the total cross-sections was "saved" (from the theoretical point of view) by the appearance of Regge poles. In 1962 everybody was happy with the idea that high-energy scattering was dominated by one special Regge pole, the "Pomeron". Then the scattering amplitude became essentially

$$\begin{aligned} &F(s,t) = f(t)\ s^{\alpha(t)} \\ &\alpha'(t) \neq 0 \ \text{ for } \ t = 0\ , \quad \alpha(0) = 1 \end{aligned} \tag{2}$$

Then we still have a constant total cross-section – but a shrinking diffraction peak, with a slope $\alpha'(0) \log s$, which is what was observed at least from proton-proton scattering. I shall not inflict on you the story of Regge cuts which seem unavoidable but complicate the picture.

However, some time around 1972 it was discovered by two groups at the ISR [7,8] that total cross-sections were rising. Over the total energy range of the ISR, from 30 to 60 GeV cm energy, it was only 4 millibarns, i.e., only 10 % of the so-called asymptotic value of 40 millibarns. Later with the Sp$\bar{\text{p}}$S collider the increase became 50 %, and more with the Tevatron.

Even these initial 10 % were a great surprise. I am insisting on that because some of our younger colleagues might realize it. Yet at least two theoreticians were prepared to that and had predicted it, H. Cheng and T.T. Wu [1], but even their friends admit that they did not believe them [9].

I shall return to the Cheng-Wu model later and to its developments. I apologize for the unavoidable overlap with the talk of Tai Wu. Now I would like to mention more general rigorous considerations. In 1961 Marcel Froissart was visiting the

Berkeley laboratory where Geoff Chew was preaching that everything could be obtained from analyticity and unitarity. Froissart [10] took this program seriously and indeed succeeded to prove a most remarkable result (which influenced a lot my own career as a theoretician), namely that the total cross-section cannot increase faster than $(\log s)^2$:

$$\sigma_t < C(\log s)^2 \ . \tag{3}$$

The assumption of Froissart was that there is a strongly interacting particle with a minimum mass – the pion – and the validity of Mandelstam representation. Mandelstam representation could never be proved, but fortunately, a few years later, I could prove that the validity of the Froissart bound is a direct consequence of the axioms of local field theory [11].

Now if one looks into the details of the proof of the Froissart bound, one sees that basically one starts with an amplitude which grows like a power of s but that unitarity forbids that the partial wave amplitudes be larger than unity. On the other hand, the partial wave amplitudes should eventually decrease because of the finite range of the interactions. Taking the best of the two bounds gives the Froissart bound.

Froissart in his paper [10] has a heuristic derivation which was later rediscovered by Feynman at a conference in Stony Brook in 1966. One can view the interaction between particles as due to an energy-dependent potential

$$V = g_{\text{AB}} s^{\nu} \exp -\mu r \ , \tag{4}$$

where g can be complex. μ is the <u>intrinsic</u> range of the interaction. The effective range will be where $V = 1$, i.e.,

$$R = \frac{1}{\mu}\left[\log g_{\text{AB}} + \nu \log s\right] \ . \tag{5}$$

Then σ_T is a finite fraction of $2\pi R^2$, and hence behaves like $(\log s)^2$.

At the beginning, many people thought that the $(\log s)^2$ bound could be improved (in spite of the fact that the Cheng-Wu model predicted precisely that!), but all efforts were vain in the end. Now, hopes to improve the Froissart bound in a qualitative way are very small especially since, in 1982, Joachim Kupsch [12] constructed a completely crossing symmetric scattering amplitude, satisfying what can be called "inelastic unitarity" in all three channels, namely, with a convenient normalization:

$$\text{Im}\, f_\ell \geq |f_\ell|^2 \ . \tag{6}$$

The Kupsch scattering amplitude saturates the Froissart bound. For the "connoisseurs" this was a remarkable achievement because all sorts of things happen to a scattering amplitude saturating the Froissart bound, in particular a curious scaling property in the high-energy limit, predicted long ago by Auberson, Kinoshita and myself [13]. Anyway it was the end of reasonable hopes to improve the bound, though it will never be possible to have a complete certainty.

In general there is, for instance in all models we have mentionned, a mechanism which tends to make the total cross-section rise very rapidly – like a power of the energy – but unitarity prevents this, but not to the point of producing a constant total cross-section. A total cross-section approaching a constant would be the result of a kind of miraculous cancellation which I do not expect.

My prejudice, to which I will stick to day, is that such a miraculous cancellation does not occur. Some people argue that it is anyway an irrelevant question because the energies at which saturation occurs are inaccessible and then one can work with amplitudes increasing like a power of s larger than one [4]. Others claim to see already visible signs of the Froissart bound saturation, as we shall see. Anyhow I believe that it is eventually saturated.

I have another prejudice, much less well founded, which is that there is no odderon. An odderon is a contribution to the odd signature amplitude of the same order of magnitude as the Pomeron. I must admit that there is no fundamental principle which forbids the odderon, and nothing excludes the extreme model of Łukaszuk and Nicolescu [14]:

$$\begin{aligned} F_{\text{even}}(t=0) &\sim is(\log s - i\frac{\pi}{2})^2 \\ F_{\text{odd}}(t=0) &\sim Cs(\log s - i\frac{\pi}{2})^2 , \end{aligned} \tag{7}$$

which is called the "Maximal odderon" F_{even} is dominantly imaginary, F_{odd} is dominantly real and this is compatible with s channel unitarity and the Froissart bound, and more explicitly with the "New Pomeranchuk Theorem" of Kinoshita and Eden [15] which says that $\sigma_{\bar{p}p}/\sigma_{pp} \to 1$ as $s \to \infty$ if anyone of the two cross-sections goes to infinity. If (7) holds, we have $\sigma_{pp} \sim \sigma_{p\bar{p}} \sim (\log s)^2$ but $\sigma_{pp} - \sigma_{p\bar{p}} \sim \text{const} \times \log s$, but also instead of having

$$\rho = \frac{\text{Re}F}{\text{im}F} \sim \frac{\pi}{\log s} \quad \text{if} \quad C = 0 , \tag{8}$$

we have

$$\rho_{p\bar{p}} \to +\pi C \qquad \rho_{pp} \to -\pi C .$$

This odderon, as you know, became fashionable after the first UA4 experiment which gave at 546 GeV c.m. energy [16]

$$\rho = 0.24 \pm 0.04$$

The odderon was one way out, at the price of having $\sigma_{pp} > \sigma_{p\bar{p}}$ in the high-energy limit, i.e., having σ_{pp} and $\sigma_{p\bar{p}}$ crossing at some energy [17]. One would tend to believe that $\sigma_{p\bar{p}} > \sigma_{pp}$, because of all the annihilation channels of $p\bar{p}$, but this argument cannot be quantified.

Now we have a new measurement of ρ, by UA4.2, using a more refined detector and going to much smaller t to investigate fully the region of interference between the nuclear and Coulomb amplitudes. This measurement gives [18]:

$$\rho = 0.135 \pm 0.015 \qquad \text{at} \qquad \sqrt{s} = 550 \text{ GeV}$$

compatible with all expectations from a smooth fit of cross-sections without odderon. The D710 measurement

$$\rho = 0.1 \pm 0.1 \qquad \text{at} \qquad \sqrt{s} = 1800 \text{ GeV}$$

is also compatible with all standard fits.

For instance the fit made in 1984 by Bourrely and myself [19] predicted $0.120 < \rho < 0.155$ at $\sqrt{s} = 550$ GeV, and a very refined fit by Kroll and Schweger [20] is in perfect agreement with the data

x LEAR PS172 P. Schiaron et al. Nucl.Phys.A505 (1989) 595

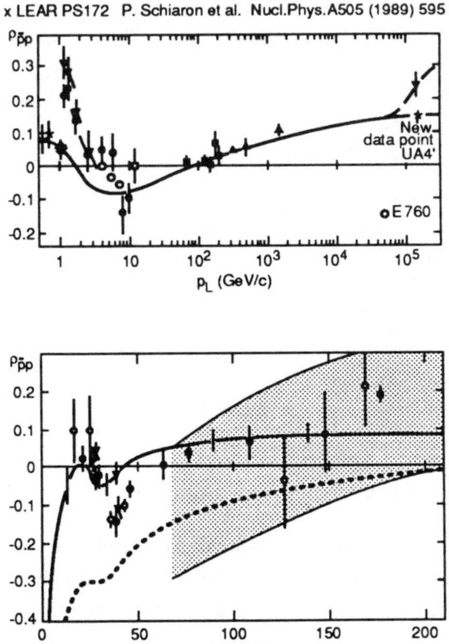

P. Kroll, W. Schweiger Nucl. phys. A503 (1989) 865

Figure 1

The main reason for me to disbelieve the odderon is that when one looks directly at total cross-section differences there is absolutely no indication in that direction. All cross-section differences, within the limited range of enerties where experiments have been performed, up to 300 GeV/c lab momentum, can be fitted by a power law $s^{-\nu}, \nu$ being close to 1/2. This is shown in Fig. 2, taken from the experiment of Caroll et al. in 1976 [21]

However, I must admit that we have now the theoretical problem that a perturbative odderon with intercept larger than unity has been obtained [22]. This does not mean that the physical odderon has an intercept at one.

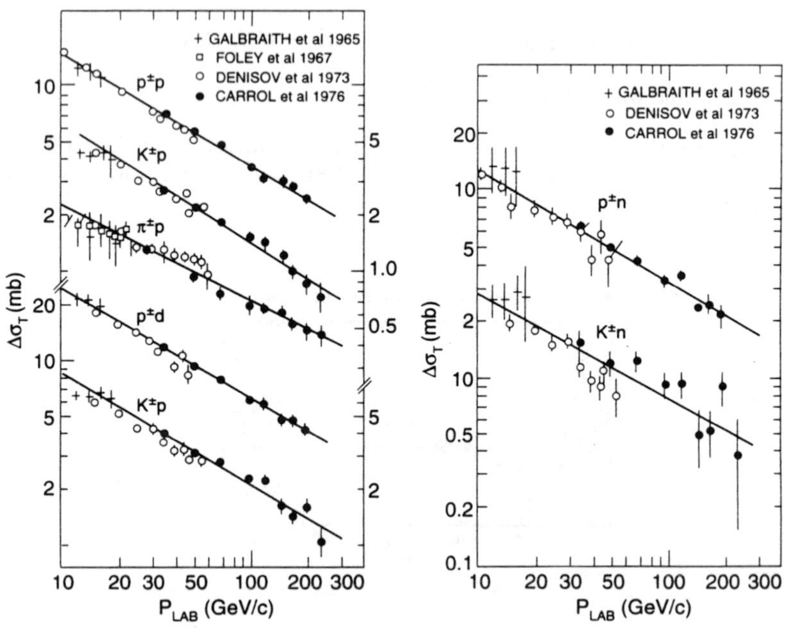

Figure 2

Now I would like to spend some time on one of the models which fits and predicts the data, that of Cheng, Wu and their collaborators, Walker and Bourrely-Soffer:

1) it predicts correctly the cross-section at Sp$\bar{p}$S energies (Fig. 3); concerning the Tevatron it gives a value which is between the E710 value of 72.8 ± 3.1 [23] and the CDF value which is 80.6 ± 2.3 [23]; these two numbers are at the limit of compatibility;
2) it predicts correctly ρ, which is not a complete surprise since dispersion relations are satisfied, but indicates that the model has reasonable cross-sections at very high energies;

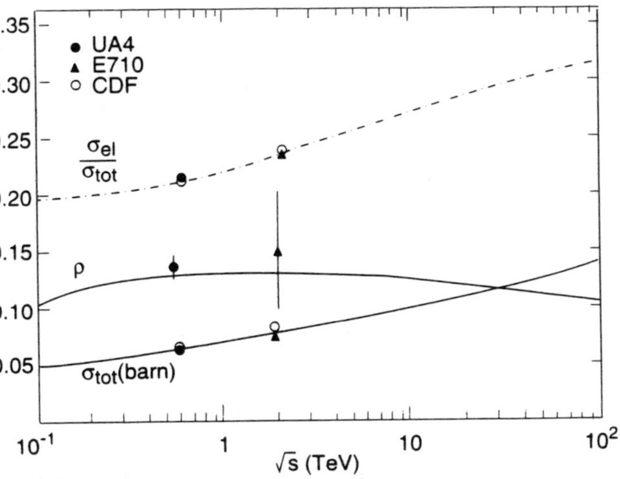

Figure 3

3) it predicts correctly the ratio σ_{el}/σ_{tot} which increases, in the model, to the asymptoticvalue 1/2 at extremely high energies; the values [23] of 546 GeV $\sigma_{el}/\sigma_{tot} = 0.215 \pm 0.005$ (UA4), 0.210 ± 0.002 (CDF) and at 1.8 GeV, 0.23 $\pm$ 0.012 (E710), 0.246 ± 0.004 (CDF), were both predicted, at a time when σ_{el}/σ_{tot} looked constant at the ISR energies and supported the belief of "geometrical scaling" [19] (the Ter Martyrosyan-Kaidalov model also predicted that geometrical scaling was only a transient phenomenon);
4) it predicts correctly the variation of the slope of the diffraction peak, as well as the general shape of the peak (Fig. 4).

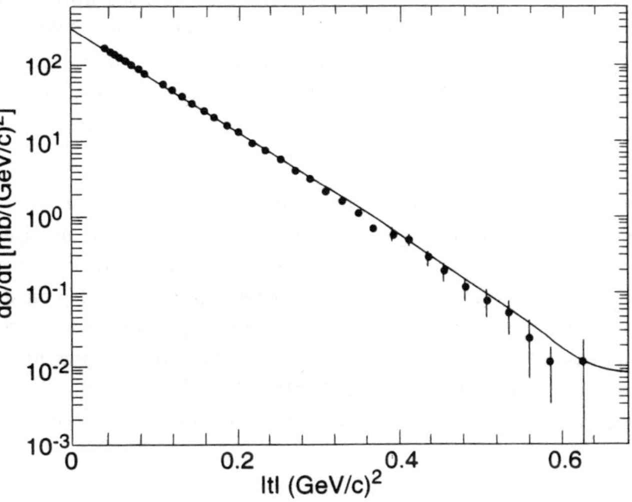

Figure 4

This is what I call "understanding". Now returning to my prejudice that the Froissart bound is saturated, let me present to you the exercize done by the UA4.2 group which is the repetition of what was called the "Amaldi fit" [24] which led to the successful prediction of $\sigma_t \sim 65$ millibarns at the Sp$\bar{\text{p}}$S.

Again, the new fit is [25]

$$\sigma_{\pm} = A_1 E^{-N_1} + A_2 E^{-N_2} + \sigma_\infty$$

with

$$\begin{aligned} \sigma_\infty &= C_0 + C_2(\log s/s_0)^\gamma \\ s_0 &= 1 \text{ GeV} \quad \text{(this is inessential)} \end{aligned}$$

The best fit leads to

$$\gamma = 2.25^{+0.35}_{-0.31}$$

supporting the saturation of the Froissart bound. The fit (like the Wu-Bourrely-Soffer model) predicts a cross-section of 76.5 ± 2.5 millibarn at the Tevatron, between the E710 and CDF measurements, and a cross-section at the LHC of 111 ± 8 mb, and a ρ parameter of 0.133 ± 0.018.

To finish I would like to come back to future experiments [26]. Naturally one must perform elastic scattering at the LHC to measure cross-sections and real parts (a real part which according to Nick Khuri might give surprises as he will explain himself).

There is, however, another very important issue which is to consolidate our belief that there is no odderon. Real part measurements are very indirect. A large real part might be due to other sources. What is really missing is a systematic knowledge of total cross-section differences. Now it goes up only to 300 GeV/c lab momentum, while in the Tevatron fixed target mode of 800 GeV/c protons one could produce secondary beams of up to 600 GeV/c, doubling the range where $\sigma_{\pi^-p} - \sigma_{\pi^+p}$, $\sigma_{K^-p} - \sigma_{K^+p}$, etc., could be measured. This, also, should not be forgotten if one day UNK is constructed, and, above all, at the LHC where one can hope to extract beams with bent crystals.

For those who think that cross-section differences are not interesting, let me stress that cross-sections in themselves are interesting. At low energies, where cross-sections are all flat, the rules of the additive quark model work with 10 % accuracy (not only the famous 3/2 ratio between nucleon-nucleon and pion-nucleon but relations involving hyperon nucleon, K nucleon and pion-nucleon cross-sections). At higher energies we do not know what happens. We do not know first of all if cross-sections are rising (except a mild indication for K^+p) at all. We do not know if flavour dependence remains, i.e., $\sigma_{kp} < \sigma_{\pi p}$, and we do not know if quark content dependence remains.

In the naïve model presented at the beginning with an energy dependent potential $g_{AB}s^{\nu}\exp -\mu r$, we expect only g to depend on the particles in collision, but ν and μ to be universal. Then the range of the interaction

$$R = \frac{1}{\mu}\left[\log g_{AB} + \nu \log s\right]$$

has a leading term independent of A and B. If it is so you expect to have $\sigma_{nn}/\sigma_{\pi n} \to 1$ at very high energies. Is this true?: only experiment can tell, and we should try to know because

"the truth shall make you free" [27].

1. H. Cheng and T.T. Wu, *Phys.Rev.Lett.* **24** (1969) 666.
2. H. Cheng, J.K. Walker and T.T. Wu, *Phys.Lett.* **B44** (1973) 112.
3. C. Bourrely, J. Soffer and T.T. Wu, *Nucl.Phys.* **B247** (1984) 15, *Z.Phys.* **C37** (1988) 369, *Phys.Lett.* **B252** (1990) 287.
4. A. Donnachie and P.V. Landshoff, *Z.Phys.* **C2** (1979) 55, *Nucl.Phys.* **B231** (1983) 189, *Nucl.Phys.* **B267** (1986) 690.
5. See, for instance: A.B. Kaidalov, *Physics Reports* **50** (1977) 157.
6. V.N. Gribov, Proceedings of the 1960 International Conference on High Energy Physics, Rochester (1960), p. 340.
7. U. Amaldi et al., *Phys.Lett.* **B44** (1973) 112.
8. S.R. Amendolia et al., *Phys.Lett.* **B44** (1973) 119.
9. C.N. Yang, Talk given on the occasion of T.T. Wu's 60th birthday.
10. M. Froissart, *Phys.Rev.* **123** (1961) 1053.
11. A. Martin, *Nuovo Cimento* **42** (1966) 930.
12. J. Kupsch, *Nuovo Cimento* **70B** (1982) 85.
13. G. Auberson, T. Kinoshita and A. Martin, *Phys.Rev.* **D3** (1973) 3185.
14. L. Lukaszuk and B. Nicolescu, *Nuovo Cimento Lett.* **8** (1973) 405.
15. R.J. Eden, *Phys.Rev.Lett.* **16** (1966) 39;
    T. Kinoshita, in Perspectives in Modern Physics, ed. R.E. Marshak, New York (1966) p. 211;
    See also: G. Grunberg and T.N. Truong, *Phys.Rev.Lett.* **B31** (1973) 63.
16. D. Bernard et al., *Phys.Lett.* **B198** (1987) 583.
17. P. Gauron, E. Leader and B. Nicolescu, *Phys.Rev.Lett.* **B54** (1985) 2656 and **55** (1985) 639;
    D. Bernard, P. Gauron and B. Nicolescu, *Phys.Lett.* **B199** (1987) 125.
18. C. Augier et al., *Phys.Lett.* **B316** (1993) 448.
19. C. Bourrely and A. Martin, Proceedings of the ECFA-CERN Workshop on LHC, CERN (1984) p. 323.
20. P. Kroll and W. Schweiger, *Nucl.Phys.* **A503** (1989) 865.
21. A.S. Carroll et al., *Phys.Lett.* **61B** (1976) 303.
22. P. Gauron, L. Lipatov and B. Nicolescu, *Phys.Lett.* **B304** (1993) 334.
23. To simplify our life we just refer to the recent review by G. Matthiae, Hadron-Nucleon Cross-Sections at High Energies, Preprint ROM2F/94/04, to be published in "Reports on Progress in Physics". Some results are also private communications from K. Goulianos.
24. U. Amaldi et al., *Phys.Lett.* **B66** (1977) 390.

25. C. Augier et al., *Phys.Lett.* **B315** (1993) 503.
26. For more details, see:
A. Martin, in "Elastic and Diffractive Scattering", eds. H.M. Fried, K. Kang and L.I. Tan, Proceedings of the 5th "Blois" Workshop, World Scientific, Singapour (1994), p. 1.
27. John, 32

HIGH ENERGY HADRONIC COLLISION AS DIFFUSION PROCESS AND HETEROTIC POMERON

CHUNG-I TAN
Physics Department, Brown University, Providence, RI 02912, USA

ABSTRACT

The two-dimensional transverse nature of high energy scattering is emphasized. In particular, we explain the fact that, while a hard Pomeron can be understand in terms of a diffusion process in the log of virtuality, a soft Pomeron corrsponds to a diffusion in the impact parameter space. The large-N_c cylinder structure, *i.e.,* closed-string like structure, of the Lipatov Pomeron is pointed out next. Since the soft-Pomeron also represents effective multiperipheral ladder graphs with cylinder topology, it provides the basis for a unified treatment of high enrgy hadronic collisions in QCD, leading to a "Heterotic Pomeron".

1. Introduction

There are currently two seemingly conflicting interpretations of Pomeron in QCD: One based on perturbative leading log approximation (LLA) and another based on nonperturbative (large-N_c and/or phenomenological) considerations. Thus one of the main puzzles for our understanding of high energy hadronic collisions in QCD is the relation, if any, between the perturbative (Lipatov) Pomeron[1] and the nonperturbative (soft) Pomeron.[2,3] Two recent theoretical studies in QCD at high energies have heightened interests in this area of research. One is an attempt by Verlinde and Verlinde in trying to derive an effective two-dimensional field theory for QCD.[4,5] The second is due to Mueller[6,7], who provided an intuitively attractive dipole picture for the high energy hard processes.

In this brief report, we would like to bring out the two-dimensional transverse nature of high energy scattering. In particular, we explain the fact that, while a hard Pomeron can be understand in terms of a diffusion process in the log of virtuality, a soft Pomeron corrsponds to a diffusion in the impact parameter space. We next point out the large-N_c cylinder structure, *i.e.,* closed-string like structure, of the Lipatov Pomeron. Since the soft-Pomeron is also supposed to represent the effective multiperipheral ladder graphs emerging from a cylinder topology, it provides the basis for a unified treatment of high enrgy hadronic collisions in QCD, leading to a "Heterotic Pomeron".[8]

Let us recall certain common features of Lipatov and soft Pomerons. In spite of their apparently different dynamical origins, the hard and the soft Pomerons share a structural similarity since they are both generated by summing ladder graphs. For Lipatov Pomeron, one deals with the hard gluon ladders, whereas the soft Pomeron involves multiperipheral ladders. In a ladder sum, one encounters amplitudes which satisfy a recursion relation

$$\mathcal{A}_n(p_1,p_3;p_2,p_4) = \int d^4p_1' d^4p_3' K(p_1,p_3;p_1',p_3')\mathcal{A}_{n-1}(p_1',p_3';p_2,p_4), \qquad (1)$$

where $\mathcal{A}_n(p_1,p_3;p_2,p_4)$ corresponds to the n-rung contribution to the absorptive part of a non-forward two-to-two amplitude, $p_1+p_2 \to p_3+p_4$.

Formally, a high energy hadronic collision in the near-forward limit corresponds to the mixing of a longitudinal "short-distance" phenomenon with a transverse "long-distance" phenomenon. By treating the longitudinal and transverse degrees of freedom separately, one could hope to achieve a "dimensional reduction", reducing QCD at high energies to an effective two-dimensional theory. This is indeed the case under either a hard Lipatov or a soft multiperipheral approximation. At high energies where $s >> |t|$, ($s \equiv (p_1+p_2)^2$ and $t \equiv (p_3-p_1)^2$), the longitudinal structure for either case can indeed be factorized, leading to a two dimensional transverse recursion equation. It is often more convenient to parametrize momenta in Eq. (1) in terms of four-vectors p, p', p_0 and a four-momentum transfer vector q where $p_{1/3} = p \pm q/2$, $p'_{1/3} = p' \pm q/2$ $p_{2/4} = p_0 \mp q/2$, so that $s = (p+p_0)^2$ and $t = q^2$. It can then be shown that the sum $\mathcal{A}(p,p_0;q) = \sum_n \mathcal{A}_n(p,p_0;q)$ is power-behaved in s, leading to the respective Pomerons.

The detail of the longitudinal structure of the kernel K depends on the nature of exchanged spins. Nevertheless, both pictures can readily lead to increasing total cross sections.* Instead of concentrating on their difference in the longitudinal direction, we shall instead bring out in this note their key difference in the structure in the two-dimensional transverse direction. Two most important features we would like to concentrate on are: (i) The production mechanism for a hard Pomeron leads to an increase in "virtuality", *i.e.*, the average transverse momentum squared, $\langle p_T^2 \rangle$ increases with $\log s$, whereas a soft Pomeron exchange corresponds to processes with limited $\langle p_T^2 \rangle$. (ii) The soft Pomeron has a relatively large slope at $t=0$, whereas the hard Pomeron has a much weaker t-dependence.

$\star$ We assume that, for current collider energies, with a Pomeron intercept greater than 1, (either the hard or the soft), an eikonal representation is operating for restoring unitarity.

2. Transverse Structure of High Energy Collisions

(a) Tranverse Momemtum Representation:

To be more precise, it is possible to choose a frame in which q is transverse, and after performing a Laplace transform with respect to the longitudinal rapidity, Eq. (1) becomes

$$\mathcal{A}_n(\vec{p},\vec{p}_0;\vec{q};\omega) = \int d^2p' K(\vec{p},\vec{p}';\vec{q};\omega)\mathcal{A}_{n-1}(\vec{p}',\vec{p}_0;\vec{q};\omega), \tag{2}$$

where ω plays the role of a complex angular momentum. As pointed out above, the ω-dependence of the kernel $K(\vec{p},\vec{p}';\vec{q};\omega)$ depends on the nature of the spin exchanges. We shall instead concentrate in what follows on the transverse dependence only. We can define a four-point function $\mathcal{A}(\vec{p},\vec{p}_0;\vec{q};\omega) = \sum_n \mathcal{A}_n(\vec{p},\vec{p}_0;\vec{q};\omega)$, which then satisfies an integral equation. We note that $\mathcal{A}(\vec{p},\vec{p}_0;\vec{q};\omega)$ depends on a set of two-vectors. In particular, the phase space integration in Eq. (2) involves a two-vector.

(b) Coordinate Representation:

Instead of working in the momentum space, it is instructive to introduce two-dimensional impact parameter coordinates via Fourier transforms. In terms of two-vectors $\vec{p}_1 = \vec{p}+\vec{q}/2$, $\vec{p}_3 = \vec{p}-\vec{q}/2$, $\vec{p}_2 = \vec{p}_0-\vec{q}/2$, $\vec{p}_4 = \vec{p}_0+\vec{q}/2$, we can define a four-point function via

$$G_c(\vec{x}_1,\vec{x}_3;\vec{x}_2,\vec{x}_4;\omega) = \int \{\prod_i d^2\vec{p}_i\}\delta^{(2)}(p_1+p_2-p_3-p_4) \\ e^{ip_1x_1-ip_3x_3+ip_2x_2-ip_4x_4}\mathcal{A}(\vec{p},\vec{p}_0;\vec{q};\omega).$$

Because of translational invariance, G_c can only depend on relative coordinates. Introducing a pair of ralative coordinates $x_{13} = \vec{x}_1 - \vec{x}_3$, $x_{24} = \vec{x}_2 - \vec{x}_4$ and a pair of center of mass coordinates $\bar{x}_{13} = (\vec{x}_1+\vec{x}_3)/2$, $\bar{x}_{24} = (\vec{x}_2+\vec{x}_4)/2$, one readily verifies that

$$\mathcal{A}(\vec{p};\vec{p}_0;\vec{q};\omega) = \int\int\int dx_{13}dx_{24}d\bar{x}_{13} \\ e^{-ipx_{13}-ip_0x_{24}-iq(\bar{x}_{13}-\bar{x}_{24})}G_c(x_{13},x_{24};\bar{x}_{13}-\bar{x}_{24};\omega).$$

We shall refer to the difference $\vec{b} = \bar{x}_{13} - \bar{x}_{24}$ as the (relative) impact parameter for a near-forward scattering.

In a standard ladder treatment, q^2 is the overall momentum transfer squared and $|\vec{p}|^2$ is referred to as the "virtuality". We now see that, from a two-dimensional spatial view-point, $\vec{q}$ is conjugate to the relative separation $\vec{b}$ of the respective center of mass of the two sources and $\vec{p}$ and $\vec{p}_0$ are conjugate to their transverse sizes. Kinematically, it is clearly equivalent to use either the momentum space or the coordinate representation.

(c) Mixed Representation:

We shall be interested in the behavior of the correlation function as the impact parameter $\vec{b}$ becomes large with either virtualities p and p' or relative sizes x_{13} and x_{24} held fixed. Tranditional analysis for ladder sum has been carried out using the momentum coordinates, *e.g.*, Eq. (2). More recently, in 6 and 7, the coordinate representation was used. In what follows, we shall adopt a mixed representation: $(\vec{p};\vec{b})$, *i.e.*, we use virtuality together with impact parameter. (We shall set $\bar{x}_{24}=0$ for convenience.)

Performing an inverse Fourier transform to Eq. (2) with respect to $\vec{q}$, one arrives at

$$\mathcal{A}_n(\vec{p},\vec{p}_0;\vec{b};\omega) = \int d^2\vec{p}' \int d^2\vec{b}' K_\omega(\vec{p},\vec{p}';\vec{b}-\vec{b}')\mathcal{A}_{n-1}(\vec{p}',\vec{p}_0;\vec{b}';\omega). \tag{3}$$

We shall use Eq. (3) as the starting point for our discussion of high energy scattering as a diffusion process.

3. Random Walk Picture

(a) Hard Pomeron as Diffusion in Virtuality:

In this section, we concentrate on the Lipatov equation and explain why its key feature corresponds to a diffusion in the space of the logarithm of the virtuality. We shall not going into the detail structure of the Lipatov integral equation except by noting the following key features: (1) Due the spin-1 nature of gluon, the kernel has a pole at $\omega = 0$, (2) The kernel receive contributions from two pieces (gluon Reggeizing and two gluon exchange); each contribution separately is infrared singular at $\vec{p}=\vec{p}'$. However, the sum is infrared finite! The LLA approximation for Reggeized two-gluon ladder corresponds to retaining this residual finite contribution. (3) Furthermore, since the dominant contribution now comes from the region where $\vec{p}\simeq\vec{p}'$, and since it can also be shown that the average magnitude for $\vec{p}'$ is much larger than $|\vec{q}|$, it follows that the kernel in Eq. (2) has a very weak

t-dependence. In what follows, we shall approximate the kernel by setting $\vec{q} \to 0$. In a mixed representation, *i.e.*, Eq. (3), the kernel is proportional to $\delta(\vec{b}-\vec{b}')$.

Because of scale invariant, after the angular variation is integrated out, the Lipatov kernel can depend only on the ratio $|\vec{p}|/|\vec{p}'|$. In terms of the logarithm of virtualities, $r \equiv \log p^2$ and $r' \equiv \log p'^2$, it corresponds to the situation where the kernel depends only on the difference $r-r'$. We can therefore write for the recursive Lipatov equation as

$$\mathcal{A}_n(r,\vec{b};\omega) \simeq \int_{-\infty}^{\infty} dr' \int d^2\vec{b}' K_L(r-r';\omega)\delta(\vec{b}-\vec{b}')\mathcal{A}_{n-1}(r';\vec{b}';\omega). \tag{4}$$

The normalization for the kernel is such that, when Eq. (4) is iterated, $\mathcal{A}_n(y;r) \sim e^{-r/2}$ for r large, up to powers of r. With this normalization, the kernel is symmetric in r and r', *i.e.*, it depends only on $|r-r'|$. It follows that, both for $r \to \pm\infty$, $\mathcal{A}_n(y;r) \sim e^{-|r|/2}$, up to a logarithmic factor.

Eq. (4) belongs to a class of recursion relation where the kernel depends only on differences, *i.e.*, $K_\omega(r-r';\vec{b}-\vec{b}')$. It thus truly corresponds to a random walk and the kernel has $e^{i\nu r}e^{i\vec{q}\cdot\vec{b}}$ as its eigenfunctions. The leading eigenfunction corresponds to $\nu = 0 = \vec{q}$, However, because of the weak t-dependence, it is sufficient to treat as if the motion in the $\vec{b}$-space as frozen. It thus reduces to a random walk in the r-space only. Therefore, properties of the hard Pomeron can be understood in terms of a random walk picture in r.

The Lipatove Pomeron intercept is determined by an eigenvalue condition, $\int dr K_L(r;\omega^*) = 1$. With a K_L symmetric, its eigenvalues are real, and they can be expressed in terms of $\chi(\nu) = 2\pi[2\Psi(1) - \Psi(\frac{1}{2}+i\nu) - \Psi(\frac{1}{2}-i\nu)]$, where $\Psi(z)$ is the logarithmic derivative of the gamma function. This function $\chi(\nu)$ takes on a maximum at $\nu = 0$, which leads to the celebrated formula for the Lipatov Pomeron intercept $\omega^* = N_c\alpha_s\chi(0)/\pi = (\alpha_s N_c/\pi)4\ln 2$.

To bring out the diffusion nature, let us go back to Eq. (4). Denote $\Psi(n,r,\vec{b}) = \mathcal{A}_n(r,\vec{b};\omega^*)$, and assume that its dependence on n is smooth for n large. We can expand the integrand about n,r, *i.e.*, $\Psi(n-1,r',\vec{b}) \simeq \Psi(n,r,\vec{b}) - \frac{\partial\Psi(n,\vec{b},r)}{\partial n} + \frac{\partial\Psi(n,\vec{b},r)}{\partial r}(r'-r) + \frac{1}{2}\frac{\partial^2\Psi(n,\vec{b},r)}{\partial r^2}(r'-r)^2 + \cdots$, so that

$$\frac{\partial\Psi(n,\vec{b},r)}{\partial n} = D_r\frac{\partial^2\Psi(n,\vec{b},r)}{\partial^2 r}, \tag{5}$$

where $D_r \equiv < r^2 > \equiv \int_{-\infty}^{\infty} dr r^2 K_L(r;\omega^*)$. In arriving at Eq. (5), we have made use of the fact that $< r > = 0$ and the fact that $K_L(r;\omega^*)$ is normalized to unity.

It is well-known from a longitudinal analysis that the average number of steps $< n >$ is proportional to the longitudinal rapidity Y. Eq. (5) leads to a spread in virtuality with increasing rapidity: $\langle p_t^2 \rangle \sim e^{\text{const.}\sqrt{Y}}$. We stress that a hard Pomeron does not lead to diffusion in the impact parameter space.

(b) Soft Pomeron as Diffusion in Impact parameter:

We will be more sketchy for the case of a soft Pomeron. Note that, by definition, a soft ladder structure also involves a strong cutoff in p_t^2, and, for simplicity, we assume that all p_t^2's are of the order p_0^2. The kernel must also be cutoff in t, thus leading at high energies to a recursion relation

$$\mathcal{A}_n^{(s)}(r_0, \vec{b}; \omega) \simeq \int d^2\vec{b}' K_s(\vec{b} - \vec{b}'; \omega) \mathcal{A}_{n-1}^{(s)}(r_0; \vec{b}'; \omega), \tag{6}$$

where $r_0 \simeq 0$. That is, a soft Pomeron corresponds to a random walk in the impact parameter space. Under a factorizable approximation, one has $K_s(\vec{b} - \vec{b}'; \omega) \simeq \frac{I_s(\vec{b}-\vec{b}')}{\omega - \omega_s}$, where $\omega_s \sim -1/2$ and $I_s(\vec{b})$ decreases rapidly for $\vec{b}^2$ large.

The soft Pomeron intercept is now fixed by the leading eigenvalue condition: $\int d^2b K_s(\vec{b}; \omega_s^*) = 1$. Denote $\Psi(n, r, \vec{b}) = \mathcal{A}_n(r, \vec{b}; \omega_s^*)$. For n large, one again arrives at a diffusion equation

$$\frac{\partial \Psi(n, \vec{b}, r_0)}{\partial n} = D_b \nabla^2 \Psi(n, \vec{b}, r_0), \tag{7}$$

where the diffusion coefficient D_b is related to the one-step fluctuation in b^2. This leads to $\langle b^2 \rangle \propto \log s$, which, when translated back into t, corresponds to the well-known shrinkage of the forward peak due to the exchange of a soft Pomeron.

4. Large-N_c Structure and Heterotic Pomeron in QCD

In a perturbative treatment, a Lipatov Pomeron loosely corresponds to the color-singlet bound state of two (reggeized) gluons. In a nonperturbative treatment, the soft Pomeron is thought to correspond to nonperturbative gluon exchanges having the topology of a closed string. It has been emphasized in Ref. 8 that a "two-gluon ladder" also has the topological structure of a "cylinder" in the color space. This fact offers the possibility of a unified approach to both perturbative and nonperturbative Pomerons in a systematic large-N_c treatment.

Large-N_c expansion leads to a topological expansion for QCD, which provides a theoretical basis for discussing the nonperturbative structure of QCD. In the high energy limit, it is generally accepted that the dominant topological structure for near-forward scattering corresponds to that of a closed-string exchange[2], figure

1. The cylindrical surface in figure 1 is understood to have been populated with arbitrary number of gluons, provided that they are connected in a "planar" fashion in the color space. This can best be represented if each gluon is labeled in a "double-line" representation, *i.e.*, as far as color is concerned, a gluon can be thought of as a pair of quark and antiquark. In this fashion, the cylindrical surface is "tiled" by polygons whose edges are gluon lines. The color structure is such that the boundary of each polygon corresponds to a closed "quark loop", formed by identifying a quark line from each gluon of the polygon edges.

In a soft approach, as discussed in Ref. 2, such a topological structure leads to a unique "two-chain" feature in multiparticle production. When supplemented with generally accepted theoretical principles like duality, unitary, etc., it provides the basis underlying the dual parton model (DPM), which has been recently reviewed in a Physics Report volume.[2]

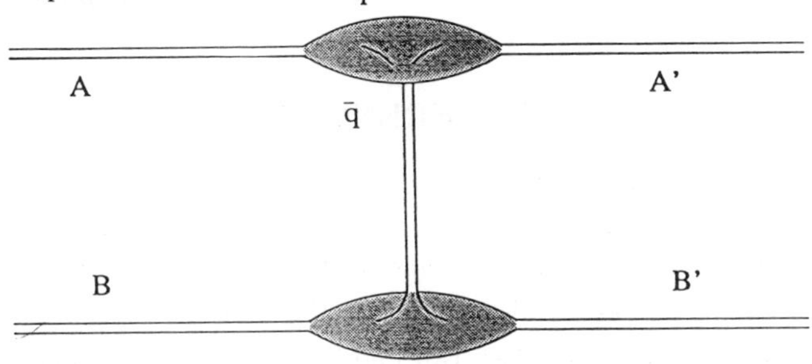

What about the hard processes? We emphasize that, most significantly, a "two-gluon ladder" also has the topology of a "cylinder" in the color-space, it therefore survives in the leading large-N_c limit. This can be seen as follows. Note that, figure 1 can be obtained by joining of a closed-string propagator, *i.e.*, the long neck, to wave functions at both ends of the graph. By construction, this closed-string propagator corresponds to a color-neutral exchange. (In what follows, it will be represented by the color-neutral component of two-gluon exchange.) Concentrate next on the topological structure of the wave-functions at each end. Each wave function, before the closed-string exchange, topologically has the structure of a disk. In order to join to the closed-string, a "window", or more precisely, a "quark-line" boundary must be cut out. In a light-cone wave function approach,[6] this "quark-loop" represents the propagation of a color-dipole. Finding the distribution of this quark-loop corresponds to finding the desired "dipole density"[6] *.

* Indeed, we have recently carried a study for QCD in 2+1 dimension using Mueller's dipole picture. This study, because of its simpler analytic calculation, provides a clearer picture for the physics of hard processes. Even more importantly, it allows us to clarify and contrast the relation between a perturbative approach and that of a nonperturbative soft Pomeron.[7]

Since the soft-Pomeron is also supposed to represent the effective ladder graphs emerging from the cylinder graphs, it suggests that, instead of simply adding, they should be treated as different manifestations of a more general "structure" which truly represents the asymptotic behavior of the cylinder graphs in QCD. We have indeed constructed a Heterotic Pomeron, incorporating both these features.

Details of our constructure can be found in a forthcoming publication[9] and a brief outline for our proposal has previously been given in Ref. 8. We simply state here that, based on the Lipatov equation and on the large-N_c expansion, a probabilistic approach can be formulated in which the perturbative (hard) and non-perturbative (soft) components are merged into a single "Heterotic Pomeron". As a Regge trajectory, it is nonlinear, approaching 1 in the limit $t \to -\infty$. In this framework, "diffusion" processes in both the (logarithm of) virtuality and the impact parameters take place. This significantly modifies the physical picture of high energy collisions both near the forward limit and in the small-x region in DIS.

REFERENCES

1. L. N. Lipatov, review in *Perturbative QCD*, ed. A. H. Mueller (World Scientific, Singapore, 1989), and references therein. For phenomenological applications: L. V. Gribov, E. M. Levin, and M. G. Ryskin, *Phys. Rep.* **100C** (1983) 1; E. M. Levin and M. G. Ryskin, *Phys. Rep.* **189C** (1990) 267.
2. A. Capella, U. Sukhatme, C-I Tan and Tran T. V., Phys. Lett. B81 (1979) 68; *Phys. Rep.* 236 (1994) 225.
3. P. V. Landshoff, Proc. of 3rd Int. Conference on Elastic and Diffractive Scattering, *Nucl. Phys.* **B12** (1990) 397.
4. H. Verlinde and E. Verlinde, preprint PUPT-1319, revised version.
5. Miao Li and C-I Tan,, *Phys. Rev. D* **50** (1994) 1140.
6. A.H. Mueller, Nucl. Phys. B415 (1994) 373; A.H. Mueller and B. Patel, Columbia preprint, March (1994). See also the contribution by Mueller. For related work, see N.N. Nikolaev, B.G. Zakharov and V.R. Zoller, KFA-IKP(Th)-1993-34, November (1993).
7. M. Li and C-I Tan, "High Energy Scattering in 2+1 QCD: A Dipole Picture", Brown-HET-943, 1994.
8. See:C-I Tan, Moriond Proceedings, 1993, edited by J. Tran Thanh Van.
9. E.M. Levin and C-I Tan, "Heterotic Pomeron", to appear.

Screening Corrections in High Energy Hadron and Photoproduction Reactions

Uri Maor

School of Physics and Astronomy
Raymond and Beverly Sackler Faculty of Exact Sciences
Tel Aviv University, Tel Aviv 69978 *

and

LAFEX, Centro Brasileiro de Pesquisas Físicas (CNPq)
Rua Dr. Xavier Sigaud 150, 22290 - 180, Rio de Janiero, RJ, BRASIL

Invited talk presented at the Workshop on Quantum Infrared Physics
Paris, June 6-10,1994

Abstract

The role of screening corrections, calculated using the eikonal approximation, is investigated for elastic and diffractive hadron-hadron and photon-hadron scattering in the high energy limit. We examine the differences between our results and those obtained from the supercritical Pomeron-Reggeon model with no unitarity corrections. It is argued that the saturation of cross sections is attained at different scales for different channels. A review of the relevant data and its implications is presented.

---

*Permanent address

Over the past few years phenomenological investigations of Pomeron exchange reactions have been confined mostly to the study of forward elastic scattering and total cross sections. In particular, Donnachie and Landshoff (DL) have promoted [1] an appealing and very simple Regge picture in which

$$\sigma_{tot} = X(\frac{s}{s_0})^{\Delta} + Y(\frac{s}{s_0})^{-\eta} \tag{1}$$

with universal $\Delta = 0.0808$ and $\eta = 0.4525$. This corresponds to a supercritical Pomeron trajectory $\alpha_P(t) = 1+\Delta+\alpha'_P t$ and a Regge trajectory $\alpha_R(t) \simeq 0.5+t$. Indeed,the analysis of Block, Kang and White [2], who examine the nuclear slope of high energy pp and $\bar{p}p$ elastic scattering, yields an excellent reproduction of the data with $B = b_0 + 2\alpha' ln(\frac{s}{s_0})$ where $\alpha'_P = 0.25\ GeV^{-2}$. Even tough DL offer a global fit to all available hadron-hadron and photon-hadron total cross sections, it should be noted that in reality only $\bar{p}p$ and γp reactions have attained high enough energies in, which the Pomeron parameters can be unambiguously tested, provided experimental errors are small enough. Clearly, such a simple model is bound, eventually, to violate s-channel unitarity. Nevertheless, as we shall see, DL have introduced a P-P cut correction which extends appreciably the domain of applicability of their model.

Elastic scattering and diffraction dissociation are similar processes which have predominantly forward imaginary amplitudes corresponding to the exchange of vacuum quantum numbers in the t-channel. As such, both are dominated in the high energy domain by Pomeron exchange and are expected, in a simple Regge model, to exhibit rather similar dependences on the kinematic variables. In this review I wish to address some of the apparent differences between these processes and their interpretation. In particular, I wish to examine the role of screening corrections in the calculation of these reactions and show that they saturate the above cross sections at very different scales. I shall add a few comments on the new data coming from HERA and its relevance to this discussion. Both H1 and ZEUS report [3,4)] that in the HERA energy domain $(\sigma_{el} + \sigma_{diff})$ is about 35-40% of the total cross section. This is similar to the ratio observed in $\bar{p}p$ scattering in the ISR-Tevatron energy range, where the need for some unitary corrections is apparent [1,5].

In order to catalog the differences between an uncorrected supercritical Pomeron model and a similar model which includes screening corrections, let us specify their features side by side. I wish to present a calculation which is reasonably realistic but also as simple as possible. To this end I present the Pomeron amplitude in an exponential form and calculate the screening corrections utilizing the eikonal approximation. This approximation accounts for elastic rescatterings which are the leading contribution to the screening process.

Our amplitude is normalized so that

$$\frac{d\sigma}{dt} = \pi |f(s,t)|^2 \tag{2}$$

$$\sigma_{tot} = 4\pi Im f(s,0) \tag{3}$$

The scattering amplitude in b-space is defined as

$$a(s,b) = \frac{1}{2\pi} \int d\mathbf{q} \;\; e^{-i\mathbf{q}\cdot\mathbf{b}} f(s,t) \tag{4}$$

where $t = -q^2$.
In this representation

$$\sigma_{tot} = 2 \int d\mathbf{b} \;\; Im a(s,b) \tag{5}$$

$$\sigma_{el} = \int d\mathbf{b} \;\; |a(s,b)|^2 \tag{6}$$

The introduction of screening rescattering corrections is greatly simplified in the eikonal approximation where at high energy $a(s,b)$ is assumed to be pure imaginary, and can be written in the simple form

$$a(s,b) = i(1 - e^{-\Omega(s,b)}) \tag{7}$$

where the opacity $\Omega(s,b)$ is a real function. As we shall utilize Regge parameterizations, analyticity and crossing symmetry are easily restored by substituting $s^{\alpha} \rightarrow s^{\alpha} e^{-i\pi\alpha/2}$, where α denotes the exchanged Regge trajectory.

In previous publications [6,7] we have shown that the eikonal approximation can be summed analytically for a Gaussian input

$$\Omega(s,b) = \nu(s) e^{-\frac{b^2}{R^2(s)}} \tag{8}$$

which corresponds to an exponential representation in t space. This is a good approximation [5,6] for Regge type amplitudes, where

$$Im f(s,t) = C e^{R_0^2 t} (\frac{s}{s_0})^{\alpha(t)-1} sin[\frac{\pi\alpha(t)}{2}] \simeq C(\frac{s}{s_0})^{\Delta} e^{\frac{1}{4}R^2(s)t} \tag{9}$$

With this amplitude we obtain

$$\sigma_{tot} = 4\pi C(\frac{s}{s_0})^{\Delta} \tag{10}$$

$$\sigma_{el} = \frac{\pi C^2 (\frac{s}{s_0})^{2\Delta}}{B_{el}} \tag{11}$$

$$\sigma_{in} = \sigma_{tot} - \sigma_{el} \tag{12}$$

On transforming to b-space we get

$$\nu(s) = \frac{\sigma_0}{2\pi R^2(s)} (\frac{s}{s_0})^{\Delta} \simeq \frac{\sigma_{tot}}{4\pi B_{el}} \tag{13}$$

$$R^2(s) = 4[R_0^2 + \alpha' ln(\frac{s}{s_0})] \tag{14}$$

where $\sigma_0 = \sigma(s_0)$ and $B_{el} = \frac{1}{2}R^2(s)$. With this input, we obtain in the eikonal approximation

$$\sigma_{tot} = 2\pi R^2(s)[ln\nu(s) + C - Ei(-\nu(s))] \tag{15}$$

$$\sigma_{in} = \pi R^2(s)[ln2\nu(s) + C - Ei(-2\nu(s))] \tag{16}$$

$$\sigma_{el} = \sigma_{tot} - \sigma_{in} \tag{17}$$

where $Ei(x) = \int_{-\infty}^{x} \frac{e^t}{t} dt$, and C = 0.5773 is the Euler constant.

The above parameterization also allows one to obtain a closed expression for single diffraction dissociation , where in the triple Regge limit with no screening corrections we have

$$\frac{M^2 d\sigma_{sd}}{dM^2 dt} = \sigma_0^2 (\frac{s}{M^2})^{2\Delta + 2\alpha' t} [G_{PPP}(\frac{M^2}{s_0})^{\Delta} + G_{PPR}(\frac{M^2}{s_0})^{-\frac{1}{2}}] \tag{18}$$

With the introduction of screening corrections [7] we obtain

$$\frac{M^2 d\sigma_{sd}}{dM^2} = \frac{\sigma_0^2}{2\pi \bar{R}_1^2(\frac{s}{M^2})} (\frac{s}{M^2})^{2\Delta} \cdot [G_{PPP}(\frac{M^2}{s_0})^{\Delta} a_1 \frac{1}{(2\nu(s))^{a_1}} \gamma(a_1, 2\nu(s))$$

$$+ G_{PPR}(\frac{M^2}{s_0})^{-\frac{1}{2}} a_2 \frac{1}{(2\nu(s))^{a_2}} \gamma(a_2, 2\nu(s))] \tag{19}$$

where G_{PPP} and G_{PPR} are the triple Regge couplings corresponding to single diffraction dissociation,

$$\bar{R}_i^2(\frac{s}{M^2}) = 2R_{0i}^2 + r_{0i}^2 + 4\alpha' ln(\frac{s}{M^2}) \tag{20}$$

$r_{0i} \leq 1 GeV^{-2}$ denotes the radius of the triple vertex and can safely be neglected.

$$a_i = \frac{2R^2(s)}{\bar{R}_1^2(\frac{s}{M^2}) + 2\bar{R}_i^2(\frac{M^2}{s_0})} \tag{21}$$

The indices i =1,2 corresponds to P (Pomeron) and R (Reggeon) exchanges. $\gamma(a,2\nu)$ denotes the incomplete Euler gamma function $\gamma(a,2\nu) = \int_0^{2\nu} z^{a-1}e^{-z}dz$. In the high energy limit the above expression simplifies to

$$\frac{M^2 d\sigma_{sd}}{dM^2} = \pi R^2(s)[G_{PPP}(\frac{M^2}{s_0})^{\Delta} + G_{PPR}(\frac{M^2}{s_0})^{-\frac{1}{2}}] \tag{22}$$

Table I. Asymptotic predictions of the Supercritical Pomeron and Eikonal models

	Supercritical Pomeron	Eikonal model
σ_{tot}	s^{Δ}	$a + b\ ln^2(\frac{s}{s_0})$
σ_{el}	$\frac{s^{2\Delta}}{R_0^2+\alpha' ln(\frac{s}{s_0})}$	$\frac{1}{2}[a + b\ ln^2(\frac{s}{s_0})]$
σ_{sd}	$\frac{s^{2\Delta}}{R_0^2+2\alpha' ln(\frac{s}{<M^2>})}$	$R_0^2 + \alpha' ln(\frac{s}{s_0})$

The remarkable differences between the non eikonalized and eikonalized versions of the supercritical Pomeron model are best illustrated in the asymptotic region and are summarized in Table I. As can be readily seen, the most dramatic change takes place for single diffraction where a $s^{2\Delta}$ divided by a ln(s) term is replaced by $R^2(s)$ which behaves as ln(s). Problem is that the information summarized in Table I is of no practical use as long as we do not specify the appropriate energy scales at which the screening corrections become appreciable. To this end we need to asign some numerical values to the free parameters of our equations. It is an easy excersise to see that if we insert the DL parameters into Eqs. (9) and (13) our amplitude will violate s-channel unitarity for small b at c.m. energies which are as low as 2.5 TeV, just above the Tevatron range. Indeed, the CDF group [8] finds that a(s,b=0) = 0.96 at 1.8 TeV. This problem is avoided in the DL model by their introduction of a weak P-P cut, its strength is fixed so as to reproduce the high t dip observed in elasic differential cross sections. This correction extends considerably the energy range at which the DL model can be applied provided we consider Δ as an effective parameter which should go down eventually. The net practical result is that by confining our study only to total cross sections in the available energy range, we are in no position to assess the importance of the screening corrections and the scales at which they become appreciable. To attain this wider scope of investigation let us turn to a discussion of some elastic and diffractive data and its interpratation.

Much of the information we are trying to get can be deduced from a close examination of available data on $\frac{\sigma_{el}}{\sigma_{tot}}$ and $\frac{\sigma_{sd}}{\sigma_{tot}}$. Without screening corrections we expect both ratioes to grow indefenitely like s^{Δ} devided by the appropriate slopes which behave as ln(s). This is, obviously, not a physical picture. In a rather general formulation, Pumplin [9] has suggested that $\frac{\sigma_{el}+\sigma_{diff}}{\sigma_{tot}}$ is bounded by $\frac{1}{2}$. Indeed, once we eikonalize we get $\frac{\sigma_{el}}{\sigma_{tot}} \rightarrow \frac{1}{2}$ wheras $\frac{\sigma_{sd}}{\sigma_{tot}} \rightarrow 0$. The emerging picture is that we expact $\frac{\sigma_{el}}{\sigma_{tot}}$ to start its rise behaving approximately as s^{Δ} and at the energy scale where screening becomes important to temper its rise approaching $\frac{1}{2}$ from below at exeedingly high energies. $\frac{\sigma_{sd}}{\sigma_{tot}}$, on the other hand, is supposed to reach a maximum at its screening scale after which it will start going down becoming diminishingly small in the high energy limit. The relevant data is shown in Fig. 1 with the highest data points at c.m. energy of 1.8 TeV [8,10]. The remarkable conclusion we make, after inspecting these data, is that wheras we have not reached, as yet, the screening scale corresponding to elasic $\bar{p}p$ scattering, the single diffraction scale is considerably smaller at c.m. energies of about 50-100 GeV.

An additional source of interesting information is available by inspecting the ratio between the forward single diffraction cross section at fixed M^2 and the corresponding forward elastic cross section $\frac{(\frac{d\sigma_{sd}}{dM^2dt})_{t=0}}{(\frac{d\sigma_{el}}{dt})_{t=0}}$. By virtue of the factorization theorem we expact in a simple Regge model that this ratio, being just the ratio of the relevant couplings to aproach rather quickly a constant value. This constant behaviour should change into a moderate increase if we chose to integrate over the diffractive mass. The introduction of screening corrections diminishes the forward single diffraction cross section and as a result we expact once again that the above ratio should reach a maximum at about the diffraction screening scale after which it should go down rather rapidely. The data which is shown in Fig. 2 supports this point of view. Moreover, the ratioes presented in Fig. 1 and Fig. 2 peak at the same energy of 50-100 GeV as expacted.

The CDF experiment [8], which was carried out at c.m. energies of 546 and 1800 GeV, conviniently enables us to examine the ratioes of the total, elastic and single diffraction cross sections. These are presented in Table II together with the theoretical expectations. By using the data of a single experiment we avoid the need to compare the relative normalisations and other algorithms used in the analysis of different experiments. The data, as compared with a simple supercritical Pomeron model, showes again the inability of the simple model to reproduce the energy dependence of the diffractive cross sections. On the other hand, once screening corrections are included (through the eikonal sum) a very reasonable reproduction of the experimental ratioes is attained. Some caution should be applied when regarding these data points. There is a clear discrapency between the 1800 GeV total cross section reported by the CDF [8] and the one reported by E710 [10] which is considerably lower. We have tried to overcome this difficulty by examining the ratioes rather than the magnitudes of the cross sections of interest. Nevertheless, we see that a $\Delta = 0.11$ value is preffered on the DL value of $\Delta = 0.0808$. Regardless of this difficulty, I consider the diffractive cross section ratioes and magnitudes to be a very serious problem for supercritical Pomeron models without a severe unitary correction. DL do not discuss diffraction dissociation in their publications. However, considering the fact that their unitarity correction is very weak and thus their Δ is universal and fixed, I do not beleive that in a model like theirs it is possible to ammend the problems we have just encountered.

Table II. CDF cross section ratioes $[\frac{\sigma(1800)}{\sigma(546)}]$ against the theoretical predictions

	CDF	Supercritical Pomeron $\Delta = 0.0808$	Supercritical Pomeron $\Delta = 0.11$	Eikonal model $\Delta = 0.0808$	Eikonal model $\Delta = 0.11$
σ_{tot}	1.31 ± 0.04	1.21	1.30	1.18	1.26
σ_{el}	1.53 ± 0.075	1.36	1.57	1.30	1.48
σ_{sd}	1.20 ± 0.075	1.43	1.65	1.21	1.33

A very exiting new window of information, through which we can re-assess our ideas on Pomeron physics, has been opened recently when the second generation of HERA results became available. Actually, we have two sources of new knowledge. To begin with, and as we have already noted, real photoproduction is the only channel, in addition to $\bar{p}p$, where our c.m. energies exceed 100 GeV. On a more profound level, a carefull study of DIS and its Q^2 dependence reveals new information on the Pomeron and in particular the dependence of $\alpha(0)$ on Q^2.

Clearly, a very important element in the DL model is the universality of Δ. As is evident from Ref. [1], this is perfectly compatible with the available data on πp and Kp total cross sections. Problem is that this data is confined to relatively low energies where the Pomeron parameters can be checked for consistency but not be determined independently. The analysis of the global data on $\sigma_{tot}(\gamma p)$, including the HERA points [3,4], are presently consistent with a wide range of choices. In particular, we specify the DL value of $\Delta = 0.0808$, as well as the excelent ALLM [11] reconstruction of the potoproduction and DIS data which has $\Delta = 0.045$. Obviously, this is just a reflection of the fact that at present the systematic error on $\sigma_{tot}(\gamma p)$ in the HERA measurements is too big. This experimental deficiency should be ammended within the next year or so and finally enable us yet another independent look at the behaviour of total cross sections.

Our discussion, thus far, was confined to soft processes, within the domains of conventional Regge physics. In perturbative QCD a hard Pomeron can be percieved [12] through the summation of gluon ladder diagrams. As a result one obtains a series of poles in the complex j plane above unity. These poles sum to an effective $\Delta = \frac{12}{\pi}\alpha_s ln2$. We shall refer to this as the BFKL Pomeron. One of the more exciting observations established by the HERA groups is that the proton structure function $F_2(x, Q^2)$ behaves in the exceedingly small limit of x like x^{-J} with $J \simeq 0.30$. This is remarkably close to the BFKL Pomeron and very different from the $Q^2 = 0$ photoproduction results which are reproduced with the soft Pomeron parameters. The transition from $Q^2 = 0$ to the high Q^2 domain has been carefully studied by ALLM [11] in a dual model. Their results, which were subsequently corroborated also by other studies, indicate a fairly rapid transition from the soft to the hard domain taking place at Q^2 values of a few GeV^2. This is shown in Fig. 3. To further study the exact nature of this transition we need better data at moderate Q^2 which is not available as yet. Nevertheless, the above observation, as such, invites some speculations. We shall mention here just two possible explenations to this behaviour:
1) We have actually two Pomerons, a soft one and a hard one. Wheras the small Q^2 data is dominated by the soft Pomeron the higher Q^2 data is dominated by the BFKL Pomeron. The net effect of the two contributions is shown in Fig. 3.
2) We have only a hard BFKL Pomeron which, due to its high Δ, receives very strong unitarity corrections. These result in an effective low Δ observed in hadron-hadron and photon-hadron reactions as we have seen above. However, for highly virtual photons those corrections start to diminish at some moderate Q^2 values. Only preliminary attempts in developing such a model have been undertaken thus far [5] and the idea seems to me intriguing enough to be persued furthermore.

Acknowledgements: I wish to thank my colleagues at LAFEX/CBPF for their kind hospitality and help, and the CNPq for partial financial support. Special thanks to my friends E. Gotsman and E.M. Levin for a very constractive and enjoyable collaboration.

References

[1] A. Donnachie and P.V. Landshoff, *Nucl. Phys.* **B231**, (1984) 189: *Phys. Lett.* **B296** (1992) 227.

[2] M.M. Block, K. Kang and A.R. White, Northwestern University Preprint No. 172 (1992).

[3] H1 Collaboration, *Phys. Lett.* **B299**, (1993) 376; DESY Red Report 93-077; presentations at the Eilat Conference, January 1994.

[4] ZEUS Collaboration, *Phys. Lett.* **B293**,(1992) 465; DESY Red Report 94-032; presentations at the Eilat Conference, January 1994.

[5] E. Gotsman, E.M. Levin and U. Maor, *Zeit. für Phys.* **C57** (1993) 667.

[6] E. Gotsman, E.M. Levin and U. Maor, *Phys. Lett* **B309** (1993) 109.

[7] E. Gotsman, E.M. Levin and U. Maor, *Phys. Rev.* **D49** (1994) R4321.

[8] F. Abe et. al., CDF Preprint July 1993; S. Belforte et. al., CDF Preprint July 1993.

[9] J.D. Pumplin, *Phys. Rev.* **D8** (1973) 2849.

[10] E710 collaboration,N.A. Amos et. al., *Phys. Lett.* **B243** (1990) 158; *Phys. Rev. Lett.* **68** (1992) 2443; *Phys. Lett.* **B301** (1993) 313.

[11] H. Abramowicz, E.M. Levin, A. Levy and U. Maor, *Phys. Lett.* **B269** (1991) 465.

[12] E.A. Kuraev, L.N. Lipatov and V.S. Fadin, *Sov. Phys. JETP* **44** (1976) 443; ibid **45** (1977) 199; Ya.Ya. Balitskii and L.N. Lipatov, *Sov. J. Nucl. Phys.* **28** (1978) 822.

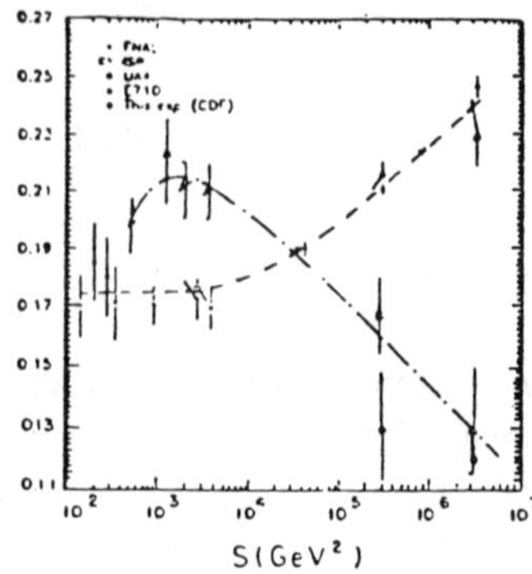

Fig. 1: s **dependence of** $\frac{\sigma_{el}}{\sigma_{tot}}$ **(dashed line) and** $\frac{\sigma_{sd}}{\sigma_{tot}}$ **(dashed dotted line)**

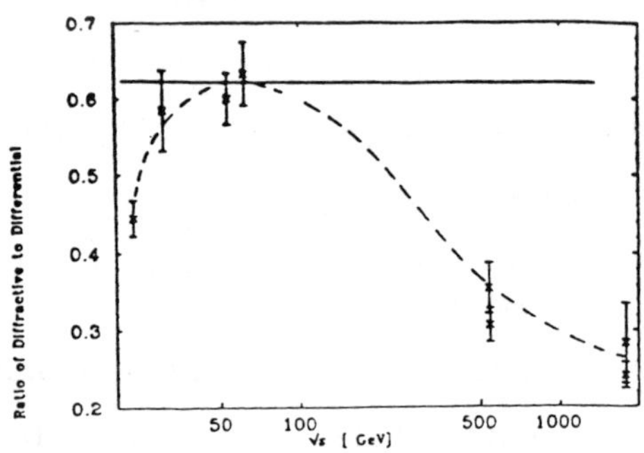

Fig. 2: $\sqrt{s}$ dependence of the ratio between $(\frac{d\sigma_{sd}}{dt})_{t=0}$ and $(\frac{d\sigma_{el}}{dt})_{t=0}$

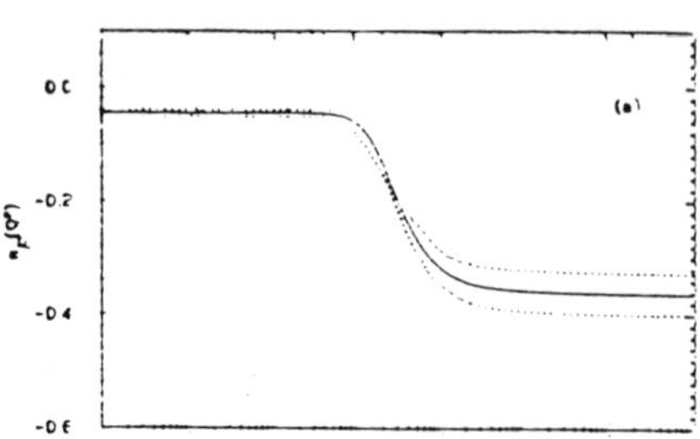

Fig. 3: Q^2 dependence of $a_P = \alpha_P(0) - 1$

SUDAKOV CORRECTIONS IN HARD EXCLUSIVE PROCESSES

Thierry GOUSSET[1,2] and Bernard PIRE[2]

1. *CEA/DAPNIA/Service de Physique Nucléaire, CE Saclay 91191 Gif sur Yvette, France*
2. *Centre de Physique Théorique, Ecole Polytechnique 91128 Palaiseau, France*

ABSTRACT

We give a short review of the long history of theoretical progress in the understanding of Sudakov form factors. We emphasize their recent implementation in the study of hard exclusive reactions and examine some phenomenology related to them.

1. Sudakov form factor

Whereas infrared divergencies subtletly cancel in inclusive hard reactions[1], exclusive reactions, which do not allow real soft processes to compete with virtual ones, turn out to be rather sensitive to infrared regions of integration. The well known Sudakov effect has been recently shown to play an important role in the physics of this domain which covers electromagnetic form factors, exclusive scattering at fixed angle and heavy meson exclusive decays.

1.1. Leading Log approximation

Let us begin with the original study of Sudakov[2], namely the off-shell electron vertex function in QED. The one loop contribution, Fig. 1-a, to the vertex function in Feynman gauge is:

$$\Gamma^\mu(p,p') = -ie^2 \int \frac{d^4k}{(2\pi)^4} \frac{\gamma^\alpha(\not p' - \not k + m)\gamma^\mu(\not p - \not k + m)\gamma_\alpha}{(k^2+i\varepsilon)[(p-k)^2 - m^2 + i\varepsilon][(p'-k)^2 - m^2 + i\varepsilon]}, \quad (1)$$

and we want to examine the leading behavior in the kinematical region ($q = p' - p$):

$$|q^2| >> |p^2|,\ |p'^2| >> m^2, \quad (2)$$

that is leading in power of $Q = \sqrt{|q^2|}$ (Leading Twist analysis) and leading in logarithms (Leading Log). This second requirement corresponds to searching for terms $\sim (e^2 L_1 L_2)^n$ at n-loops level, where L_1 and L_2 are two large logarithms.

The vertex function is divergent in the Ultra-Violet region, and renormalization is supposed to be carried out whenever it is necessary. In the UV region the remaining

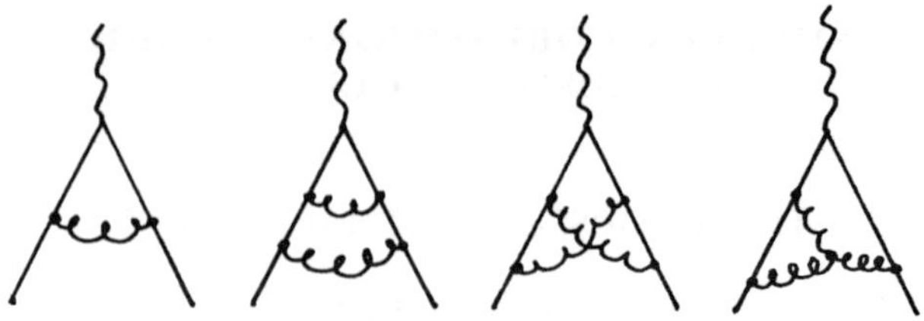

(a) (b) (c) (d)

Figure 1: 1-loop (a) and 2-loop (b,c,d) graphs which contribute, in Feynman gauge, to LL approximation (graph d is absent in QED).

term after subtraction is a single logarithm and is ignored in LL computation (self-energy corrections are dropped for the same reason).

Following Sudakov, one introduces the parametrization: $k = up + vp' + \mathbf{k}_T$ and identifies the domain of integration which gives the leading contribution:

$$|p'^2/q^2| \ll |u| \ll 1,\ |p^2/q^2| \ll |v| \ll 1,\ 0 \leq x = \mathbf{k}_T^2 \ll \min(|uq^2|, |vq^2|). \tag{3}$$

The denominator in Eq. (1) is approximately $(-x - uvq^2 + i\varepsilon)\, vq^2\, uq^2$, while in the numerator, after transverse angular summation, all terms except $\gamma^\alpha \not{p}' \gamma^\mu \not{p} \gamma_\alpha$ are combinations of small components u, v or x which would convert, upon integration, one logarithm to a non logarithmic term. The further projection onto external propagators $S_F \approx \not{p}/p^2$, dropping Higher Twist terms, leads to*:

$$\Gamma^\mu(p,p') = \frac{-ie^2}{16\pi^3}\gamma^\mu \mathrm{sign}(q^2) \int \frac{du}{u} \int \frac{dv}{v} \int \frac{dx}{x + uvq^2 - i\varepsilon}. \tag{4}$$

At this stage, the integral is apparently triply-logarithmic. However, once the x integration is performed, one finds, comparing the symmetries of the (u, v) domain and of the integrand, that the summation of the Principal Part term vanishes. Only the imaginary term survives and one gets:

$$\Gamma^\mu(p,p') = -\frac{e^2}{8\pi^2}\gamma^\mu \ln|q^2/p^2| \ln|q^2/p'^2|. \tag{5}$$

The same calculation may be carried out in a non abelian gauge theory like QCD, changing the internal photon to a gluon. One gets for the electromagnetic (*i.e.* color neutral) form factor the same expression with an overall color factor $C_F = 4/3$. However, because non abelian gauge theories are not infrared free, there is a major change when one includes charge renormalization effects in such a calculation[3]. Formally, in the above expression, one replaces $g^2/4\pi$ by the effective coupling $\alpha_S(x) = \pi(\beta_1 \ln x/\Lambda^2)^{-1}$, the one-loop $\overline{\mathrm{MS}}$ running coupling. To preserve the

*At Leading Twist, only the γ^μ term contributes so one defines the *form factor* F: $\Gamma^\mu = F.\gamma^\mu$.

coherence of the perturbative calculation, the additionnal condition $|p^2||p'^2| \gg |q^2|\Lambda^2$ should be fullfilled.

So far, we got large one-loop correction to the tree level expression and we have to examine higher orders. The relevant two loops diagrams in QED are, according to Sudakov's study, the two diagrams where the photon loops overlap the vertex μ (see Fig. 1-b,c). Again, leading logarithms arise from a given region:

$$\begin{aligned}
&|p'^2/q^2| \ll |u_1| \ll |u_2| \ll 1, \\
&|p^2/q^2| \ll |v_{i1}| \ll |v_{i2}| \ll 1, \qquad (i1,i2) = (1,2),(2,1), \\
&0 \le x_i \ll \min(|u_i q^2|, |v_i q^2|), \qquad i = 1,2,
\end{aligned} \tag{6}$$

and repeating the various steps of the one loop calculation, one gets the tree graph value γ^μ times the square of the one loop integral divided by a factor $(2!)^2$ due to the ordering of the u's and the v's in the definition of the domain of integration. Taking into account the 2! diagrams, one gets the beginning of the exponential form derived by Sudakov:

$$\Gamma^\mu(p,p') = \gamma^\mu \exp\left(-\frac{e^2}{8\pi^2}\ln|q^2/p^2|\ln|q^2/p'^2|\right). \tag{7}$$

In QCD, two other graphs have to be considered[4]: Fig. 1-d and its mirror diagram. These new graphs, where the 3-gluon coupling appears, are due to the non abelian nature of the gauge group and lead to question the exponentiation. However, to this order the four graphs combine to give[4]:

$$(b)+(c)+2(d) = \frac{1}{2!}\left(\frac{1}{2}C_F^2 + \frac{1}{2}C_F(C_F - \frac{1}{2}C_A) + \frac{1}{4}C_F C_A\right)\left(\frac{g^2}{8\pi^2}\ln|\frac{q^2}{p^2}|\ln|\frac{q^2}{p'^2}|\right)^2 \gamma^\mu, \tag{8}$$

i.e. the beginning of the exponential expansion.

Sudakov has shown the exponentiation of LL to all orders in QED, while this has been done in a non-abelian gauge theory by Dahmen and Steiner[5] and by Belokurov and Ussyukina[6].

1.2. Beyond Leading Logarithms

The LL Sudakov form factor rapidly decreases to 0, as $|q^2| \to \infty$. As a consequence, the non-leading logarithms, ignored until now, may play an important role and spoil the LL result. It is apparent from the above presentation, that a direct evaluation of all the large logarithms $\sim g^{2n}L^{2\nu}$ is a tremendous task and efficient technical methods have been developped to overcome the various difficulties (for a recent review, see Collins[7]). In the following, we shall only enumerate the principal steps of the method derived by Sen[8].

The first ingredient is the power counting method[9] which is necessary to identify, among all singular points in a Feynman amplitude, those (the *pinch singularities*) which effectively lead to infrared divergencies (*i.e.* large logarithms in any regulated form). Sen carries his work in axial gauge for technical reasons. The choice of gauge determines the kind of configurations, in Feynman graphs, which have collinear divergencies. For example, the double-log region in the graph of Fig. 1-a is due to gluons with longitudinal polarizations. In axial gauge, they are absent and only one logarithm remains, while double-logs appear in self energy corrections of the vertex function.

The second problem is to find an algorithm to disentangle soft from collinear divergencies. It turns out that the simplest object to study is not the form factor itself, but its derivative with respect to the large scale Q. The graphs contributing to this derivative at leading power are rearranged using algebraical techniques and Ward identities: this results in a suitable set of new Feynman rules.

The evolution equation looks like:

$$\frac{\partial}{\partial \ln Q}\Gamma^{\mu} = \left(G(\frac{Q}{\mu}, g) + \cdots\right)\Gamma^{\mu}, \tag{9}$$

where in the factor $(G + \cdots)$ the large scale dependence has been isolated, in G, from the low scale dependence, so that an improved perturbative expansion, *i.e.* free from large logarithms, can be performed. One can write a Renormalization Group equation for G:

$$G\left(\frac{Q}{\mu}, g(\mu)\right) = G\left(1, g(Q)\right) - \int_{\mu}^{Q} \frac{du}{u}\gamma_G(g(u)), \text{ with } \gamma_G = 2C_F\frac{\alpha_S}{\pi}. \tag{10}$$

The double log comes from the γ_G integral:

$$\Gamma^{\mu} = \exp\left(-\frac{2C_F}{\beta_1}\ln Q \ln\ln Q + \cdots\right)\gamma^{\mu},\ \beta_1 = \frac{33 - 2n_F}{12}, \tag{11}$$

where the $\cdots$ represent non leading logs. Note that the $\ln Q \ln\ln Q$, instead of $\ln^2 Q$ in Eq. (7), comes from the implementation of the running coupling.

2. Hard exclusive reactions

2.1. Physical picture

Before turning to a rigourous definition of the formalism, we shall give some indications of how the Sudakov form factors appear in the study of exclusive processes.

We have seen above that, in QCD, the elastic scattering of a quark is suppressed at high energy Q, because of the radiation of long wavelength quanta (long with respect to $1/Q$), and this may occur even for wavelengths shorter than some infrared cut-off or confining scale. If we now consider the elastic scattering implying a quark antiquark dipole (*i.e.* a color neutral object), the size of the dipole b enters the game in the following way. Gluons with wavelength longer than b cannot resolve the system and hence do not couple to it: one gets a natural infrared cut-off for corrections of

Sudakov type. For short wavelength quanta, on the other hand, the system does not respond coherently and the elastic scattering of the system is suppressed. As a consequence, neutral systems with $b \lesssim 1/Q$ are not affected by Sudakov corrections.

As we will see below, in some observables (form factors, Compton scattering cross section ...), the elementary process between "incoming" and "outgoing" quark systems entirely take place in a small spacetime region and Sudakov corrections act only as an extra rule for mini configuration selection in the reaction. However for more complicated processes, like hadron hadron elastic scattering, quarks are kinematically allowed to interact at distant spacetime points (Landshoff multiple independent scattering processes) and Sudakov corrections play an important role[10].

2.2. Perturbative pion form factor

To be specific, let us concentrate on the spacelike form factor of the pion, one of the simplest observable in exclusive scattering. The spacelike form factor measures the ability of a pion of momentum p to absorb a virtual photon (carrying a momentum q with $q^2 = -Q^2 < 0$) while remaining intact. It is defined by the formula:

$$< \pi(p+q)|J^\mu|\pi(p) >= e_\pi.F(Q^2).(2p+q)^\mu, \tag{12}$$

where e_π is the pion electric charge.

In the hard scattering regime, that is when Q is very large with respect to the low energy scales of the theory (the QCD scale Λ and the pion mass), Brodsky and Lepage have motivated the following three step picture for the process, valid in the light-front formalism[11]:

- the pion exhibits a valence quark-antiquark state,
- which interacts with the hard photon leading to another $q\bar{q}$ state,
- which forms the final pion.

Higher light-front Fock state component $q\bar{q}g$, $q\bar{q}q\bar{q}$ ... also contribute to the process, but these contributions are suppressed by powers of the large energy scale Q and may be ignored in the computation.

The form factor is written in the convolution form:

$$\begin{aligned} F &= \int_0^1 dx \int_0^1 dy \int d\mathbf{k} \int d\mathbf{l}\; \psi(x,\mathbf{k},p)\; \psi^*(y,\mathbf{l},p+q)\; T_H(\mathbf{k},\mathbf{l},x,y,Q), \\ T_H &= 16\pi\alpha_S C_F \frac{xQ^2}{xQ^2+\mathbf{k}^2-i\varepsilon}\, \frac{1}{xyQ^2+(\mathbf{k}-\mathbf{l})^2-i\varepsilon}, \end{aligned} \tag{13}$$

where T_H is a sum of QCD tree graph amplitudes and ψ, ψ^* are respectively the incoming and outgoing pion wave functions.

2.3. Factorization of Infra Red logarithms

As in the case of the quark form factor, one has to examine the effect of loop corrections to the above expression. It turns out[10,12] that a study similar to the one depicted in subsection *1.2* can be performed. For this purpose, however, it is simpler to take the transverse Fourier transform of the wave function:

$$\widehat{\psi}(x, \mathbf{b}) = \int d^2\mathbf{k} e^{i\mathbf{k}\mathbf{b}} \psi(x, \mathbf{k}), \tag{14}$$

and, accordingly, of the hard scattering. The leading corrections can then be factorized into the wave function $\widehat{\psi}$ with an evolution equation of the form:

$$\frac{\partial}{\partial \ln Q}\widehat{\psi} = \left(\frac{1}{2}G(\frac{xQ}{\mu}, g) + \frac{1}{2}G(\frac{(1-x)Q}{\mu}, g) + K(b\mu, g)\right)\widehat{\psi}, \tag{15}$$

where G organises the Q dependence of the corrections while K contains the infrared one, through $b = |\mathbf{b}|$.

The solution is derived as for the Sudakov form factor:

$$\widehat{\psi}(x, \mathbf{b}; Q, \mu) = \exp - \left(\frac{C_F}{2\beta_1} \ln xQ \ln \frac{\ln xQ}{\ln 1/b} + (x \to 1 - x) + \cdots\right) \widehat{\psi}^0(x, \mathbf{b}), \tag{16}$$

where $\widehat{\psi}^0(x, \mathbf{b})$ is an integration constant which is governed by confinement dynamics not accessible to such a perturbative treatment. The important property of the effective wave function amplitude is that it is exponentially suppressed as $b \to \Lambda^{-1}$, whatever the fraction x. This is crucial when one examines the consistency of the perturbative expansion at intermediate Q. The analysis of radiative corrections to the hard scattering amplitude T_H in Eq. (13) leads to the choice $\mu = \max(\sqrt{xy}Q, b^{-1})$ for the argument of α_S and this scale is small only when the effective wave function is suppressed. Li and Sterman[12] have quantified this statement showing that the perturbative computation of the pion form factor becomes safe for Q as low as $2 - 4\,GeV$ for a standard x dependence of the wave function.

3. A phenomenological implication: comparison of timelike and spacelike form factors.

The timelike form factor is the observable in the crossed channel, $s = q^2$:

$$< \pi(q-p)\pi(p)|J^\mu|0 >= e_\pi.F(s).(q-2p)^\mu. \tag{17}$$

Experimentally, around $10\,GeV^2$, the ratio of the timelike over spacelike form factors is about 2 for the proton. The same ratio seems to occur for the pion, even though the experimental extraction of the spacelike value is a matter of controversy and prevents one from giving a definite expertimental result for the ratio. At leading order in perturbative QCD, the ratio is 1. Therefore if a perturbative calculation is reliable for energy range where data exist, one is lead to wonder where the discrepancy comes from. The possible corrections to a leading order evaluation are of two kinds. One may expect that higher order terms in the perturbative expansion, do not behave in the same way in the two regions. A simple guess follows from the analytic continuation of the one-loop running coupling itself $\ln^{-1}(-Q^2) \approx \ln^{-1}(Q^2)\left(1 + i\pi \ln^{-1}(Q^2)\right)$.

Another possibility arises from the different behaviors of power suppressed terms. In the formalism described above some of them have been essential to ensure the reliability of the perturbative treatment at intermediate Q^2. Moreover, it is apparent from the expression of T_H in Eq. (13), changing $Q^2 \to -s$, that the $\mathbf{k}_T^2$ poles which were far from the domain of integration in the spacelike expression, are close to it in the timelike case. This feature effectively results in different behaviors[13]. For the pion, we show, in Fig. (2), the predicted ratio for different x dependences of the wave function: one can see that the obtained ratio is around 2 for various wave functions and on the other hand that the evolution to 1 is rather slow.

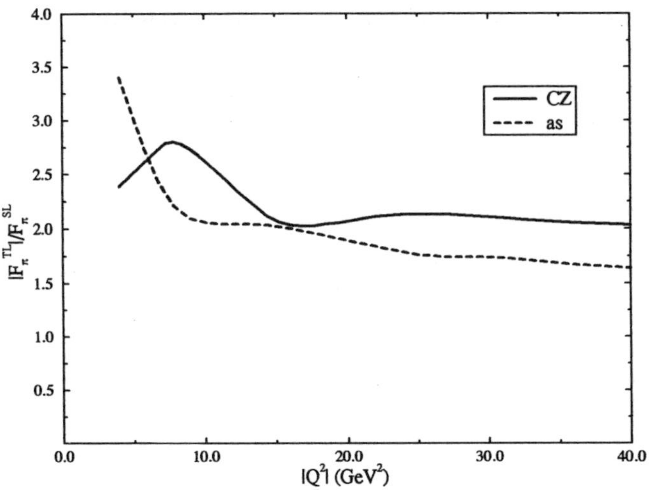

Figure 2: Timelike over spacelike ratio for various wave functions.

4. Conclusions.

Although substantial progress in the implementation of infra-red Sudakov effect in exclusive processes has been made, much work remains to understand the phenomenology of hard exclusive reactions. The importance of the intrinsic transverse behavior of $\widehat{\psi}^{(0)}$ has been considered[14]. The case of large $|t|$ exclusive scattering at small angle has also been tackled recently[15].

5. References

1. D. Kossover, this meeting.

2. V. V. Sudakov, *JETP* **3** (1956) 65.
3. A. H. Mueller, *Phys. Rep.* **73** (1981) 237.
4. J. M. Cornwall and G. Tiktopoulos, *Phys. Rev.* **D13** (1976) 3370.
5. H. D. Dahmen and F. Steiner, *Zeit. Phys* **C 11** (1981) 247.
6. V. V. Belokurov and N. J. Ussyukina, *Phys. Lett.* **94B** (1980) 251.
7. J. C. Collins, in *Perturbative Quantum Chromodynamics*, ed. A. H. Mueller (World Scientific, Singapore, 1989).
8. A. Sen, *Phys. Rev.* **D24** (1981) 3281.
9. G. Sterman, *Phys. Rev.* **D17** (1978) 2773.
10. J. Botts and G. Sterman, *Nucl. Phys. B* **325** (1989) 62.
11. S. J. Brodsky and G. P. Lepage, *Phys. Rev.* **D22** (1980) 2157.
12. H-N. Li and G. Sterman, *Nucl. Phys. B* **381** (1992) 129.
13. T. Gousset and B. Pire, *preprint* DAPNIA/SPhN **94-09**.
14. R. Jakob and P. Kroll, *Phys. Lett.* **B315** (1993) 463.
15. M. G. Sotiropoulos and G. Sterman, *preprint* ITP-SB **93-83**.

Semiclassical QCD

Yang Millsian Field Theory and Riemannian Geometry*

Ken Johnson

Center for Theoretical Physics
Laboratory for Nuclear Science and
Department of Physics
Massachusetts Institute of Technology
Cambridge, MA 02139

I. Introduction

A study will be made of the Yang-Mills field theory in a conventional quantum mechanical framework, that is, in a setting with a Hamiltonian expressed as a function of canonical variables. This is the case in so-called "temporal" or "Weyl" gauges where $A_0 = 0$. The plural rather than singular is used to describe such gauges since there are an infinite number of them, which differ classically by time independent gauge transformations. Since here the study shall be done in the conventional quantum framework of the fixed time or "Schrödinger" picture, these are the only relevant gauge transformations. As an alternative to "fixing" the gauge irrelevant variables, here the goal shall be to find new variables which are independent of gauge fixing, that is, to introduce usable variables which are gauge invariant. In this the pioneering work of Goldstone and Jackiw[7] is followed. The difference from them will be that other new variables for the gauge fields shall be sought. It will be shown first that there is a natural framework for the definition of such variables in the Hamiltonian formulation of the field theory. This is because the canonical variables and the description of local gauge transformations in terms of them are covariant under arbitrary transformations of the space coordinates, that is, in three dimensions the group GL(3). This is the group based upon the linear transformations of tensors using the matrix $\frac{\partial x'^i}{\partial x^j}$, where $x'^i = x'^i(x)$ are arbitrary functions of the space coordinates. This is not a symmetry of the conventional Yang-Mills field theory. However, it is a symmetry of a closely related local gauge field theory which may at least plausibly share many of the dynamical features present in the conventional Yang-Mills field theory. It should be stressed that anything motivated by or necessarily related to gravity, that is "general relativity," is not being pursued here. The GL(3) symmetry is already present in the canonical variables of any non-Abelian gauge theory in three space dimensions. However, there is a natural way to "find" a conventional Riemannian geometry associated with it by focusing on finding a gauge invariant set of variables. (It will have a different

*This work is supported in part by funds provided by the U. S. Department of Energy (D.O.E.) under cooperative agreement DE-FC02-94ER40818.

metric than that recently introduced by Bauer, Freedman and Haagensen[2].) It is simplest to "find" the Riemannian geometry associated with the GL(3) coordinate transformations for the SU(2) gauge group, but the same geometry, with the addition of new independent tensor variables, can also be "found" in the Yang-Mills field theory based on SU(N), and perhaps others as well, although this last possibility has not yet been investigated. This report will be developed as follows.

In section II it shall be shown why there is a GL symmetry for any spacial metric. It will then be shown that finding a way to make a change of variables which makes the basic coordinate transform as a vector rather than as a gauge "connection" introduces a metric. In section III, the various Yang-Mills tensors will be related to Riemannian geometric tensors based on the metric. The Bianchi and Gauss law identities will be discussed as well as the form that the Yang-Mills Hamiltonian takes in terms of the geometric variables. The "closely related" GL(3) symmetric Yang-Mills theory will be defined. In section IV singularities in the functional energy density will be discussed. It will be shown that these occur in general when the Einstein tensor has zero eigenvalues, unless the gauge invariant wave functional satisfies additional constraints. In section V it will be shown how the formalism can be extended to bigger gauge groups. Finally, in VI, the results and conclusions will be presented.

II. The Canonical Version of the Yang-Mills Field Theory

The conventional quantum version of the Yang-Mills field theory is expressed in terms of canonical variables A_i^a and E^{ai}, a vector potential and the "electric" field. They are both three vectors where $i, j \ldots$ indicate the space components and $a, b \ldots$ denote the gauged components. They obey the canonical commutation relations (or classical Poisson brackets),

$$\left[A_i^a(x), E^{bj}(y)\right] = i\delta^{ab}\delta_i^j\delta(x-y) \ . \tag{1}$$

In terms of the vector potential there is a "magnetic" field,

$$B^{ak} = \epsilon^{k\ell m}\left(\partial_l A_m^a + \frac{1}{2}f^{abc}A_l^b A_m^c\right) \ . \tag{2}$$

There is a group of local gauge transformations under which E and B transform as vectors and A transforms as a "gauge connection." The infinitesimal elements of the group are generated by the operator G where

$$G^a(x) = D_i E^{ai} \equiv \partial_i E^{ai} + f^{abc}A_i^b E^{ci} \ , \tag{3}$$

thus,

$$\begin{aligned} \left[A_i^a(x), G^b(y)\right] &= -i\delta^{ab}\partial_i\delta(x-y) + if^{abc}A_i^c(x)\delta(x-y) \ , \\ \left[E^{ai}(x), G^b(y)\right] &= if^{abc}E^{ci}(x)\delta(x-y) \ , \end{aligned} \tag{4}$$

and

$$\left[G^a(x), G^b(y)\right] = if^{abc}G^c(x)\delta(x-y) \ . \tag{5}$$

The finite gauge transformations are

$$\Omega^{-1}A_i^a(x)\Omega = U^{ab}(x)A_i^b(x) + \frac{1}{N}f^{abc}U^{b'b}\partial_i U^{b'c} \ , \tag{6}$$

and

$$\Omega^{-1}E^{ai}(x)\Omega = U^{ab}(x)E^{bi}(x) \ , \tag{7}$$

where f^{abc} are the structure constants of the Lie group and U is an arbitrary element of the group. The feature that shall now be exploited is the covariance of all of the above equations under the group of general local coordinate transformations,

$$\dot{x}^i = \dot{x}^i(x) \qquad ; \qquad \frac{\partial \dot{x}^i}{\partial x^j} \ . \tag{8}$$

The invariance requires that A_i^a transforms as the components of a covariant vector and E^{ai} and B^{ai} as the components of a contravariant vector density. Hence the notation used in the preceding formulas with upper and lower indices. The Yang-Mills Hamiltonian:

$$\begin{aligned} H &= \int dx\, T^{00}(x) \ , \\ T^{00}(x) &= \delta_{ij}\left(\frac{1}{2}\frac{B^{ai}B^{aj}}{g^2} + \frac{1}{2}g^2E^{ai}E^{aj}\right) \ , \end{aligned} \tag{9}$$

is gauge invariant:

$$[H, G^a(x)] = 0 \ , \tag{10}$$

but is not GL(3) invariant since the GL(3) tensors are contracted in (9) with a fixed metric. Therefore there is no GL(3) symmetry in the Yang-Mills field theory. However, it shall be shown later that there is a GL(3) invariant Hamiltonian which is closely related to the conventional Yang-Mills Hamiltonian.

If one looks naïvely at the operators which appear in the Hamiltonian, namely E and B, they transform as vectors under local gauge transformations and the Hamiltonian is a gauge scalar. However, the gauge connection A transforms more like a cartesian coordinate. Trying to get an insight into the dynamics of the Yang-Mills field is somewhat like trying to get an insight into the quantum mechanics of a particle moving in a rotationally invariant potential by expressing the Hamiltonian in cartesian coordinates. Thus many attempts have been made to find a natural way to introduce a "polar" version of the vector potential, that is, a coordinate which transforms as a vector under local gauge transformations. Another such attempt will now be made. Here the local GL(3) symmetry of the variables of the gauge theory will be taken as the constraint on the introduction of the new coordinate. The simplest way to make a close relation between a vector variable and the gauge connection is by using the covariant derivative, so the gauge covariant derivative is used in the transformation equation,

$$\epsilon^{ijk}\left(\partial_j u_k^a + f^{abc} A_j^b u_k^c\right) = 0 \ . \tag{11}$$

Further, because of the presence of the epsilon , this equation is GL(3) covariant as well as gauge covariant. To convince oneself that the differential equations which appear in (11) are mutually compatible one may derive them from a function of u,

$$W(u) = -\frac{1}{2}\int dx\, \epsilon^{ijk} u_i^a \partial_j u_k^a \ . \tag{12}$$

Here the sum goes over the dimension of the adjoint representation, $a = 1$ to $N^2 - 1$. When the independent variable undergoes a local gauge transformation,

$$U^{ab} u_i^b(x) \ , \tag{13}$$

it is found that

$$W\left(U^{ab} u_i^b\right) = W(u_i^a) + \frac{1}{2}\int f^{abc} U^{b'b} \partial_j U^{b'c} \bar{u}^{cj}(x) dx \ . \tag{14}$$

The quantity $\bar{u}$ is given by

$$\bar{u}^{ai} = \frac{1}{2} f^{abc} \epsilon^{ijk} u_j^b u_k^c \ . \tag{15}$$

Consequently one may see that if the vector potential is defined by

$$A_i^a = \frac{\delta W}{\delta \bar{u}^{ai}} \ , \tag{16}$$

then when u undergoes a vectorial gauge transformation. A as defined as a functional of u by (16) transforms as a gauge connection according to (6). This expression is therefore equivalent to (11). In this simplest case of SU(2), A can be explicitly calculated as a functional of u. In the next simplest case, for SU(3), it has not been possible up to now to determine A explicitly in terms of u, and it is probable that even if an explicit formula can be given, it will be fairly complicated and therefore may not be very useful. In the SU(2) case,

$$A_i^a = \frac{\epsilon^{nml} \partial_m u_l^b \left(u_n^a u_i^b - \frac{1}{2} u_n^b u_i^a\right)}{\det u} \ . \tag{17}$$

Although the above formulas all are covariant on GL(3) transformations and there is no specific expression which involves a metric, a metric has nevertheless been implicitly introduced. From this point forward, all of the considerations will apply to the simplest case of the SU(2) Lie group. The generalization to larger values of N will be briefly indicated in section V. To see how the metric comes about a formula is written which follows from (11) and which specifies all of the first coordinate derivatives of u,

$$\partial_j u_k^a + \epsilon^{abc} A_j^b u_k^c = \left\{ {s \atop jk} \right\} u_s^a \ . \tag{18}$$

The expression in the curly brackets is symmetric in j and k, and is invariant under the SU(2) gauge transformations since the gauge covariant derivative of u on the left side of (18) transforms as a vector as does the coefficient of the curly brackets on the right side of (18). If (18) is multiplied by u_l^a and one takes the part of the resulting equation which is symmetric in k, l one finds that

$$\partial_j g_{kl} - \left\{ {s \atop jk} \right\} g_{sl} - \left\{ {s \atop jl} \right\} g_{sk} = 0 \ , \tag{19}$$

where,

$$g_{kl} = u_k^a u_l^a \tag{20}$$

The unique solution to (19) is well known and is the symmetric affine connection (or "Christoffel" symbol) associated with the metric g_{ij}. Thus:

$$\left\{ {k \atop ij} \right\} = -\frac{1}{2} g^{km} \left(\partial_m g_{ij} - \partial_i g_{mj} - \partial_j g_{mi} \right) \ . \tag{21}$$

Thus a metric has been introduced as a consequence of the transformation (11) as well as an accompanying ordinary Riemannian geometry. Once there is a metric, g_{ij} it is clear how to make a GL(3) invariant gauge theory: use g_{ij} instead of δ_{ij} to form a Hamiltonian. Before concluding this section, a few brief comments will be made which concern the maps $u = u(A)$ and its inverse $A = A(u)$. Since equation (18) is a linear homogeneous equation for A in terms of u and its derivative, it is clear that there is a unique A for given u up to the global scale of u on which A does not depend (as long as $\det u$ is not zero). Therefore one must form wave functions of u which are globally homogenous. This situation is very different from our earlier attempt to use B as a vector variable where several A's can yield the same B. (="Wu/Yang" ambiguity). In the case of the direct map $u = u(A)$ one should like there to be at least one u for every A. This depends on the linear operator

$$\epsilon^{ijk} \left(\partial_j \delta^{ab} + \epsilon^{acb} A_j^c \right) \ ,$$

having at least one zero eigenvalue for any A. It is believed that this is true although a complete proof of it up to now does not exist. Clearly when A has some symmetry there may be more than one. Thus in this case the space of u is larger than that of

A, in contrast to the case studied before where the space of B was smaller than that of A. This property of the map shall be referred to in what follows.

Further, since we have defined the "gauge connection" A in terms of u, there will be no difficulty in constructing non-local gauge invariant quantities such as "Wilson loops" as functionals of u. Finally we should not fail to mention that the metric g_{ij} is locally gauge invariant, and its determinant is $\det g = (\det u)^2$. In what follows we shall often find $\sqrt{\det g} = \sqrt{g} \equiv \det u$, since $\det u$ can have both signs.

III. Geometric Expressions for the Yang-Mills Tensors

Now it will be shown how the various gauge theory tensors look in terms of the geometric tensors defined in terms of the metric tensor. In the case of the group SU(2) one may anticipate that the gauge invariant metric tensor which has six components will be sufficient to express completely the gauge invariant content of the gauge invariant Yang-Mills tensors. This is because locally nine field components, $A(x)$ or $u(x)$, describe a Schrödinger wave functional, and local gauge invariance imposes three constraints leaving six local degrees of freedom. One starts with the basic relation (18) which enforces the A to u transformation,

$$D_j u_k^a = \left\{ \begin{matrix} s \\ jk \end{matrix} \right\} u_s^a \,, \tag{22}$$

where we use the gauge covariant derivative notation. If the gauge Ricci identity on the operator is used one finds from (22)

$$[D_i, D_j] u_k^a = \epsilon^{abc} F_{ij}^b u_k^c = R_{kij}^s u_s^a \,. \tag{23}$$

The function of the geometric connection which appears is just the ordinary Riemann curvature tensor specified by the metric tensor g. In three space dimensions the Riemann curvature is completely specified by the Ricci or Einstein tensors, it is the latter which gives the simplest form,

$$R_{skij} = -\frac{\epsilon_{sku}\epsilon_{iju}}{g} G^{uv} \,. \tag{24}$$

If this is used one can rewrite (23) in the form,

$$B^{ai} = \sqrt{g} G^{ik} u_k^a \tag{25}$$

(In general by $\sqrt{g}$ we shall mean $\det u$). It is also simple to see that the gauge invariant Yang-Mills magnetic tensor density is given by

$$B^{ai} B^{aj} = g G^{ik} G^{jl} g_{kl} \,. \tag{26}$$

As promised, this locally gauge invariant quantity is specified now in terms of the gauge invariant variable g_{ij}.

It is next possible to work out the form that the Yang-Mills Bianchi identity takes in terms of the geometric variables.

One has, for the identity,

$$D_i B^{ai} = 0 \ , \tag{27}$$

and if one uses equation (25) together with (22) one quickly finds that

$$D_i B^{ai} = \sqrt{g} d_i G^{ij} u^a_j \ . \tag{28}$$

Thus, the equivalence of the geometric $(d_i G^{ij}) = 0$ and Yang-Mills identities has been established. The next step is to consider the quantum variable associated with the color electric field which is a functional operator in the Schrödinger picture. It is convenient to define an electric analog of the "magnetic" Einstein tensor,

$$\frac{\delta \Psi}{\delta A^a_i} = \sqrt{g} u^a_j \left(e^{ij} \Psi \right) \ . \tag{29}$$

In terms of the tensor operator e, in the same way the magnetic Bianchi identity was worked out, one can work out how the generator of local gauge transformations appears when it acts through the tensor operator,

$$i G_a \Psi = D_i \left(\frac{\delta \Psi}{\delta A^a_i} \right) = \sqrt{g} u^a_j d_i (e^{ij} \Psi) \ . \tag{30}$$

However, in contrast to working within the gauge frame where the condition that G acting on a functional of A vanishes is a constraint on the wave function, the situation when using u as a variable should be viewed differently. If the wave functional depends only on the gauge invariant variable (the metric) then any such wave functional will be allowed. The above formula is then not a constraint on the wave functional, but one of a set of equations needed to compute the form that the electric field operator takes when acting on such a state. Since the electric tensor operator has nine components, six additional equations are needed in addition to the gauss law equation to find the state which results when the electric field operator acts on a gauge invariant state. If (22) is used one can work out the additional equations for e which will determine its form completely. Thus if for convenience

$$\tilde{e}^i_j = e^i_j - \frac{1}{2} \delta^i_j e^s_s \tag{31}$$

where the inverse is

$$e^i_j = \tilde{e}^i_j - \delta^i_j e^s_s \tag{32}$$

one finds (after some algebra)

$$\epsilon^{inm} d_n (\tilde{e}^j_m) = 2 \frac{\delta}{\delta g_{ij}} + \frac{\epsilon^{ijk}}{2\sqrt{g}} u^a_k \left(\epsilon^{abc} u^b_s \frac{\delta}{\delta u^c_s} \right) \tag{33}$$

The second operator on the right side of (33) vanishes on gauge invariant states which are those for which it is both necessary and sufficient to depend only on the metric g. For such states the electric operator must be obtained from the equation

$$\epsilon^{inm} d_n(\tilde{e}^j_m)\Psi = 2\frac{\delta\Psi}{\delta g_{ij}} \tag{34}$$

IV. Singularities

To determine the electric field, one must invert the linear operator which appears in (34). The resulting expression will be a non-local relation between the quantity $\tilde{e}^j_m\Psi$ and $\delta\Psi/\delta g_{ij}(x)$. However, without an explicit inversion one finds that the quantity $\tilde{e}^j_m\Psi$ will not be finite at points where the curvature vanishes. This is because the covariant divergence of (34) takes the form

$$2d_i\left(\frac{\delta\Psi}{\delta g_{ij}(x)}\right) = -\frac{1}{2}\epsilon^{ist}\left(e_{sp}\Psi\right)G^p_t \tag{35}$$

Therefore, unless the left side vanishes at any point where $G = 0$, the electric field will diverge. This is a local gauge invariant constraint which a physical state vector must obey. It is easy enough to show that a similar constraint is imposed if only two of the three principal curvatures vanish at a point. One can write the formal solution to (34) in terms of the inverse of the linear operator,

$$\tilde{e}^{jn}\Psi = 2\int dy\, H^{nj}_{rs}(x,y)\frac{\delta\Psi}{\delta g_{rs}(y)} \tag{36}$$

where H is the inverse. This inverse is not defined when acting on vectors which do not obey the constraint

$$0 = \int dy\, g_{ij}\frac{\delta\Psi}{\delta g_{ij}} \tag{37}$$

which one recognizes as the condition, previously mentioned, of homogeneity in g. Formally g_{ij} is a zero eigenvector of the linear operator. This property of the operator will be discussed more completely in our paper on this subject. Here it will simply be asserted that the operator has a spectrum with accumulation points both at infinity and at zero. The former are associated with the transverse modes of the field and the later with the longitudinal modes. From the equation for H one can easily infer that

$$\frac{\epsilon^i_{mn}}{\sqrt{g}}G^m_p H^{pn}_{k\ell} = \frac{1}{\sqrt{g}}\delta^i_\ell d_k\delta(x-y) \tag{38}$$

which is the form that the local constraint already mentioned values where expressed in terms of H. This property of the inverse operator H implies that there is an infinite sequence of eigenvalues of the linear operator which asymptote to zero.

V. Beyond SU(2)

Here it will be shown that although the group SU(2) and the gauge field in three space dimensions are particularly closely connection, the geometric formalism is also a property of the groups $SU(N > 2)$. If one uses the same functional (12) which is GL(3) invariant it has already been shown that the equations

$$\epsilon^{ijk}\left(\partial_j u_k^a + f^{abc} A_j^b u_k^c\right) = 0 \tag{39}$$

can be derived. There are $(N^2-1)\cdot 3$ equations to express A_k^a as a function of u_k^a. The next step must be to define a basis u_x^a which includes u_k^a, but allows general gauge tensors to be expanded in terms of u_x^a. The lower index must be of dimension N^2-1. Since SU(N) has an adjoint subgroup, O(3), we use the tensors which belong to the irreducible representations of this subgroup, which are u_i^a, u_{ij}^a, u_{ijh}^a, etc. where the tensors $u_{ij}^a \ldots$ are traceless (using a metric) symmetric tensors. These have dimensions $3+5+7\cdots+2N-1 = N^2-1$. One can form these tensors using u_i^a as the basis. Now one proceeds as in the SU(2) case, starting from (39),

$$\partial_j u_k^a + f^{abc} A_j^b u_k^c = \begin{Bmatrix} s \\ jk \end{Bmatrix} u_s^a + K_{jk}^{st} u_{st}^a + \cdots \tag{40}$$

where u_s^a, u_{st}^a, ... are the basis tensors and $\{\}$, K, etc. are symmetric in jk. We choose the basis tensors to be orthogonal so

$$u_s^a u_{ij}^a = 0\,,\text{etc.} \tag{41}$$

In this case if one multiplies (??) on the left by u_ℓ^a and takes the symmetric part in kl one again finds that $\left\{ {s \atop jk} \right\}$ is the symmetric affine connection of a Riemannian geometry with the metric tensor

$$g_{ij} = u_i^a u_j^a \tag{42}$$

Now one must use higher rank tensors to form a complete set of gauge invariant variables. For example, in SU(3) one supplements the metric (6 variables) with the invariant tensor $d^{abc}u_i^a u_j^a u_k^a$ (10 variables) for a total of 16 variables which is 24–8, the number needed to span the Gauss law obeying subspace of the Hilbert space. We shall say no more here.

VI. Conclusions

In conclusion, in this brief report an attempt has been made to show that a non-abelian gauge theory can be closely associated with a Riemannian geometry. In this framework, the gauge constraints on wave functionals can be very simply imposed. We have not had time to indicate in detail how the geometric approach can be extended beyond the group SU(2) to SU(N) gauge theories. We have also not had the opportunity to show more than that wave functionals of the theory which have finite energy must obey conditions which go beyond those of local gauge invariance.

Acknowledgments

This work is the result of a collaboration of the author with P. E. Haagensen. I would also like to thank D. Z. Freedman for many discussions on these matters.

References

1. J. Goldstone and R. Jackiw, *Phys. Lett.* **74B** (1978) 81.
2. M. Bauer, D. Z. Freedman and P. E. Haagensen, "Spatial Geometry of the Electric Field Representation of Non-Abelian Gauge Theories," preprint CERN-TH 7238/94, to be published in *Nucl. Phys.* **B**.
3. See P. E. Haagensen, K. Johnson, to appear, for more references to earlier work.

Effective Lagrangian for Constituent Quarks

M. Baker*
University of Washington, Seattle, WA 98105

ABSTRACT

Using dual potentials to describe long distance QCD, we find the color field distribution surrounding a quark anti-quark pair to first order in their velocities. Using these distributions we eliminate the dual potentials and obtain an effective interaction Lagrangian $L_I(\vec{x}_1, \vec{x}_2; \vec{v}_1, \vec{v}_2)$ depending only upon the quark and anti-quark coordinates and velocities, valid to second order in their velocities. We propose L_I as the Lagrangian describing the long distance interaction of constituent quarks.

1. Introduction

In 1964 Gell-Mann and Zweig proposed that the constituents of hadrons are quarks u, d, and s which have masses of order of 300 MeV and which carry color degrees of freedom. Elementary calculations analogous to those used in atomic physics give a good qualitative description of the hadronic spectrum. We know that the forces in an atom are mediated by the electromagnetic field A_μ. One might ask what is the field outside constituent quarks which mediates their interaction? From the point of view of QCD constituent quarks are superpositions of interacting gluons A_μ^a and fundamental spin 1/2 quark fields ψ_α. At short distances where the coupling is weak (asymptotic freedom) one can use the Yang Mills potentials A_μ^a and carry out perturbation calculations. At long distances the coupling becomes strong (anti-screening) and there is no procedure available to use the potentials A_μ^a to calculate the force between quarks.

In this talk we review the proposal that dual potentials C_μ^a mediate the long distance interactions between constituent quarks. We construct an effective Lagrangian density $\mathcal{L}_{\text{eff}}$ to describe the C_μ^a quark interactions, treating the quarks as classical point particles having coordinates $\vec{x}_i(t)$ and velocities $\vec{v}_i(t)$. We solve the resulting classical field equations for C_μ^a in the presence of a quark anti-quark pair to first order in their velocities. Using these solutions we eliminate the dual potentials C_μ^a from $\mathcal{L}_{\text{eff}}$ and obtain an effective interaction Lagrangian $L_I(\vec{x}_1, \vec{x}_2; \vec{v}_1, \vec{v}_2) \equiv \int d\vec{x} \mathcal{L}_{\text{eff}}(\vec{x})$ depending upon the quark and anti-quark positions and velocities, valid to second order in these velocities. It contains two parameters α_s and the string tension σ. We then use L_I and the canonical formalism to construct the quark anti-quark Hamiltonian. Finally, treating L_I at the classical level we determine the leading Regge trajectory. For the sake of simplicity of this presentation we do not treat the spin degrees of freedom here.†

*Work supported in part by the U.S. Dept. of Energy under Contract No. DE-FG06-91ER40614

†see Ref. [1] for a discussion of spin.

2. Dual Potentials In Electrodynamics

We first review the use of dual potentials in electrodynamics. Consider the inhomogeneous Maxwell equations,

$$\vec{\nabla} \cdot \vec{D} = \rho , \quad \vec{\nabla} \times \vec{H} = \vec{J} + \frac{\partial \vec{D}}{\partial t} , \tag{1}$$

for a pair of oppositely charged particles. Then

$$\rho(\vec{x}, t) = e[\delta(\vec{x} - \vec{x}_1(t)) - \delta(\vec{x} - \vec{x}_2(t))] , \tag{2}$$

$$\vec{J}(\vec{x}, t) = e[\vec{v}_1(t)\delta(\vec{x} - \vec{x}_1(t)) - \vec{v}_2(t)\delta(\vec{x} - \vec{x}_2(t))] . \tag{3}$$

Introduce a line of polarization $\vec{P}_s(\vec{x}, t)$ connecting the particles (a Dirac string[2]):

$$\vec{P}_s(\vec{x}, t) = e \int_{\vec{x}_2(t)}^{\vec{x}_1(t)} d\vec{y} \delta(\vec{x} - \vec{y}(t)) , \tag{4}$$

where the integral $d\vec{y}$ is along any path $L(t)$ connecting $\vec{x}_1(t)$ and $\vec{x}_2(t)$. (See Fig. 1.)

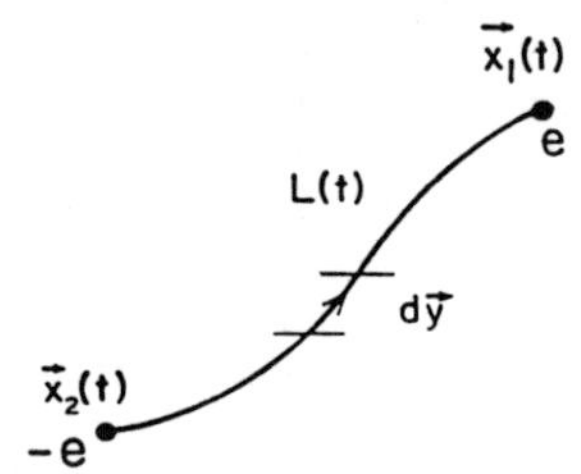

Fig. 1. Dirac string connecting oppositely charged particles

As the particles move, so does the Dirac string. There results a magnetization $\vec{M}_s(\vec{x}, t)$,

$$\vec{M}_s(\vec{x}, t) = \int_{\vec{x}_2(t)}^{\vec{x}_1(t)} e d\vec{y} \times \frac{\partial \vec{y}}{\partial t} \delta(\vec{x} - \vec{y}(t)) . \tag{5}$$

It is readily checked that

$$\rho = -\vec{\nabla} \cdot \vec{P}_s , \quad \text{and} \quad \vec{J} = \vec{\nabla} \times \vec{M}_s + \frac{\partial \vec{P}_s}{\partial t} . \tag{6}$$

Inserting eqs. (6) into (1), we obtain the solution for $\vec{D}$ and $\vec{H}$ of the form:

$$\vec{D} = -\vec{\nabla} \times \vec{C} - \vec{P}_s \quad , \quad \vec{H} = -\vec{\nabla} C_0 - \frac{\partial \vec{C}}{\partial t} + \vec{M}_s . \tag{7}$$

Thus, Maxwell's equations (1) have become kinematic equations whose solution eq. (7) defines dual potentials C_μ.

The remaining Maxwell equations,

$$\vec{\nabla} \cdot \vec{B} = 0\,, \quad \text{and} \quad \vec{\nabla} \times \vec{E} = -\frac{\partial \vec{B}}{\partial t}\,, \tag{8}$$

are dynamical equations determining C_μ. In the vacuum $\vec{E} = \vec{D}$, $\vec{B} = \vec{H}$, and eqs. (8) are obtained from the Lagrangian density:

$$\mathcal{L}_C = -\frac{1}{4}[\partial_\mu C_\nu - \partial_\nu C_\mu + G^s_{\mu\nu}]^2\,, \tag{9}$$

where the components of the covariant string polarization tensor $G^s_{\mu\nu}$ are:

$$G^s_{0k} \equiv \vec{M}_{s,k}\,, \quad G^s_{ij} = -\epsilon_{ijk}\vec{P}_{s,k}\,. \tag{10}$$

The Lagrangian $\mathcal{L}_C$ gives a formulation of electrodynamics equivalent to the usual formulation defined by the Lagrangian $\mathcal{L}_A$ expressed in terms of the ordinary vector potentials A_μ:

$$\mathcal{L}_A = -\frac{1}{4}[\partial_\mu A_\nu - \partial_\nu A_\mu]^2 + J_\mu A^\mu\,. \tag{11}$$

In a dielectric medium with dielectric constant ϵ the Lagrangian $\mathcal{L}_C$ is multiplied by a factor $\mu = \frac{1}{\epsilon}$. This means that in a medium with antiscreening, that is, where $\epsilon < 1$ and $\mu > 1$, the dual potentials are screened at long distances: they are thus the natural variables to use in such circumstances.

3. Non-Abelian Theory

Mandelstam[3] has shown that we also have the freedom to use dual potentials C^a_μ in non-Abelian gauge theory. 'tHooft[4] showed that if the Wilson loop satisfies an area law then the dual Wilson loop (defined in terms of dual potentials) satisfies a perimeter law. This is the non-Abelian analogue of the fact that in a medium with $\epsilon < 1$, dual potentials C^a_μ are screened at large distances. What is the Lagrangian density $\mathcal{L}_{\text{eff}}$ describing long-distance QCD in terms of these dual potentials? Although in non-Abelian gauge theory there is no direct relation between ordinary Yang Mills potentials A^a_μ and dual potentials C^a_μ, we can determine the form of $\mathcal{L}_{\text{eff}}$ by imposing the following requirements:

1. $\mathcal{L}_{\text{eff}}$ must be invariant under non-Abelian gauge transformations of the C^a_μ:

$$C_\mu = \Omega^{-1} C_\mu \Omega + \frac{i}{g}\Omega^{-1}\partial_\mu \Omega\,, \tag{12}$$

where $C_\mu = \sum_{a=1}^{8} C^a_\mu \frac{1}{2}\lambda_a$, Ω is an $SU(3)$ matrix and $g = \frac{2\pi}{e}$ where e is the Yang Mills coupling constant, i.e. $\alpha_s = \frac{e^2}{4\pi}$.

2. $\mathcal{L}_{\text{eff}}$ generates a mass for the C_μ field via a Higgs mechanism coupling C_μ to 3 scalar octets $B_i (i = 1, 2, 3)$. The C_μ are then screened at long distances.

3. We write a minimal $\mathcal{L}_{\text{eff}}$ satisfying (1) and (2).

4. We assume higher dimension operators not included in $\mathcal{L}_{\text{eff}}$ are not quantitatively important at large distances.

The minimal $\mathcal{L}_{\text{eff}}$ then has the form:

$$\mathcal{L}_{\text{eff}} = 2\,\mathrm{tr}\left\{-\frac{1}{4}G^{\mu\nu}G_{\mu\nu} + \frac{1}{2}(\mathcal{D}_\mu B_i)^2\right\} - W\,, \tag{13}$$

where

$$\mathcal{D}_\mu B_i = \partial_\mu B_i - ig[C_\mu, B_i]\,, \tag{14}$$

$$G_{\mu\nu} = \partial_\mu C_\nu - \partial_\nu C_\mu - ig[C_\mu, C_\nu]\,, \tag{15}$$

and the "Higgs potential" W has the explicit color structure[5]

$$W = 3\mu^2\,\mathrm{tr}\,B_i^2 + \frac{2}{3}\lambda\{3\,\mathrm{tr}\,B_i^2 B_j^2 + (\mathrm{tr}\,B_i^2)^2 + 2(\mathrm{tr}\,B_i B_j)^2\}\,, \tag{16}$$

with $\lambda > 0$ and $\mu^2 < 0$, so that W has a minimum at a non vanishing value B_0^2 of B^2. Then the mass squared M_C^2 of the C_μ field is of order $M_C^2 \simeq g^2 B_0^2$. Note that C_μ couples to B_i like A_μ couples to electrically charged fields, i.e. the B_i carry magnetic color charge.

To couple C_μ to a quark anti-quark pair we introduce a Dirac string polarization tensor $G^s_{\mu\nu}$ given by eqs. (4), (5) and (10) multiplied by the hypercharge matrix Y. This guarantees that one unit of Z_3 flux flows from the anti-quark to the quark. We then replace $G_{\mu\nu}$ by $G_{\mu\nu} + G^s_{\mu\nu}$ in $\mathcal{L}_{\text{eff}}$. The dual potential C_μ is proportional to Y so that the non-Abelian contribution to eq. (15) vanishes. (A continuous deformation in $SU(3)$ of our particular choice of C_μ into a non-Abelian configuration will leave unchanged $P\exp(ig\int \vec{C}\cdot d\vec{l}) = e^{\frac{2\pi i}{3}}$). We then obtain the Lagrangian $\mathcal{L}_{\text{eff}}$ coupling dual potentials to a quark anti-quark pair by making the replacement

$$G_{\mu\nu} \rightarrow (\partial_\mu C_\nu - \partial_\nu C_\mu + G^s_{\mu\nu})\,, \tag{17}$$

in eq. (13) so that

$$\mathcal{L}_{\text{eff}} = 2\,\mathrm{tr}\{-\frac{1}{4}(\partial_\mu C_\nu - \partial_\nu C_\mu + G^s_{\mu\nu})^2 + \frac{1}{2}(\mathcal{D}_\mu B_i)^2\} - W\,. \tag{18}$$

We shall see that this Lagrangian density describes a dual superconductor and provides a concrete realization of the Mandelstam 'tHooft dual superconductor mechanism of confinement[6,7]

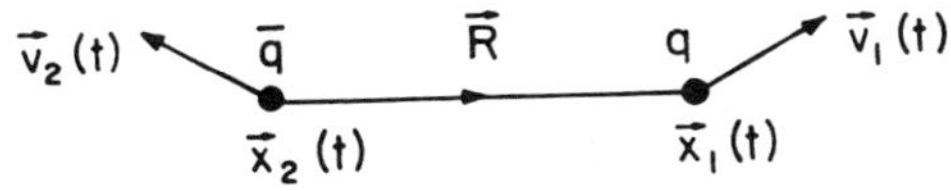

Fig. 2. Quark anti-quark pair moving along trajectories $\vec{x}_1(t)$ and $\vec{x}_2(t)$, respectively.

4. Effective Lagrangian for Constituent Quarks

Consider a quark (anti-quark) at $\vec{x}_1(t)(\vec{x}_2(t))$ moving with velocities $\vec{v}_1(t)(\vec{v}_2(t))$ (see Fig. 2). To find the effective interaction Lagrangian $L_I(\vec{x}_1\,,\vec{x}_2\,;\vec{v}_1\,,\vec{v}_2)$ we solve the classical field equations for C_μ and B_i generated by $\mathcal{L}_{\text{eff}}$, (eq. (18)), to first order in $\vec{v}_1$ and $\vec{v}_2$. We obtain

$$C_\mu = C_\mu(\vec{x}\,,\vec{R}(t)\,,\vec{v}_1(t)\,,\vec{v}_2(t)) \equiv \bar{C}_\mu\,, \tag{19}$$

$$B_i = B_i(\vec{x}\,,\vec{R}(t)\,,\vec{v}_1(t)\,,\vec{v}_2(t) \equiv \bar{B}_i\,, \tag{20}$$

where $\vec{R} \equiv \vec{x}_1 - \vec{x}_2$. For $\vec{v}_1 = \vec{v}_2 = 0$, the scalar potential $\bar{C}_0 = 0$ and $\vec{\bar{C}}$ and $\bar{B}_i$ reduce to static flux tube solutions $\vec{C}^s(\vec{x}\,,\vec{R})$, $B_i^s(\vec{x}\,,R)$, describing a configuration[8] where the color flux connecting the quark anti-quark pair is confined to a tube of radius $R_{FT} \sim \frac{1}{M_C^2}$ (see Fig. 3). The Higgs fields B_i^s vanish at the center of the flux tube

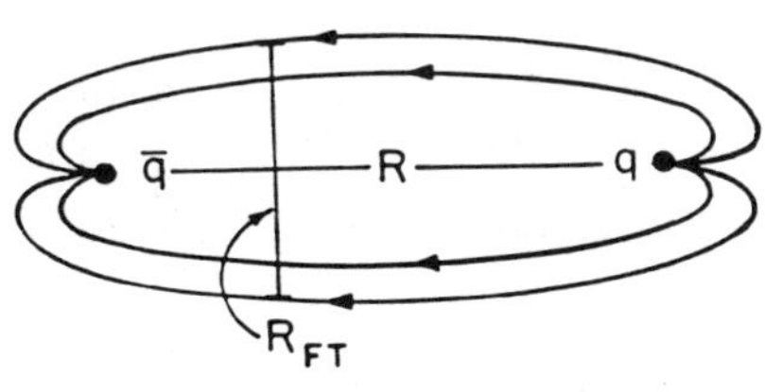

Fig. 3. Static flux tube configuration

and approach their vacuum value B_0 at large distances. Inserting this static solution

$\vec{C}^s$, B_i^s into $\mathcal{L}_{\text{eff}}$ (C_μ, B_i) and integrating over all space gives the static potential $V_0(R)$, i.e.,

$$V_0(R) = -\int d\vec{x}\mathcal{L}_{\text{eff}}(\vec{C} = \vec{C}^s, C_0 = 0, B_i = B_i^s), \tag{21}$$

At large R, $V_0(R) \to \sigma R$, where the string tension $\sigma \sim (5B_0)^2$.

Having solved the static equations, the first order solutions $\bar{C}_\mu, \bar{B}_i$, eqs. (19) and (20) are readily calculated since, to first order in $\vec{v}_1(t)$ and $\vec{v}_2(t)$, the static distribution $\vec{C}^s(\vec{x}, \vec{R})$ and $B_i^s(\vec{x}, \vec{R})$ follow the quark motion adiabatically. The time dependence of $\vec{C}$ and B_i arise only from the explicit time dependence of $\vec{R}(t)$ in $\vec{C}^s$ and B_i^s. The time varying color electric field along with the moving Dirac string then generate a scalar potential $\bar{C}_0$ and color magnetic fields which, when inserted into $\mathcal{L}_{\text{eff}}$, produce the velocity dependent terms in the quark anti-quark effective Lagrangian L_I. That is: evaluating $\mathcal{L}_{\text{eff}}(C_\mu, B_i)$ at $C_\mu = \bar{C}_\mu, B_i = \bar{B}_i$ and integrating all space gives

$$L_I(\vec{x}_1, \vec{x}_2; \vec{v}_1, \vec{v}_2) = \int d\vec{x}\mathcal{L}_{\text{eff}}(C_\mu = \bar{C}_\mu, B_i = \bar{B}_i) \equiv \int d\vec{x}\mathcal{L}_{\text{eff}}(\vec{x}, \vec{R}; \vec{v}_1, \vec{v}_2), \tag{22}$$

valid to second order in $\vec{v}_1$ and $\vec{v}_2$. L_I is the QCD analogue of the Darwin Lagrangian[9] describing the interaction of electrons and positrons after eliminating the electromagnetic field. The concept of an effective Lagrangian depending only upon quark anti-quark variables has no meaning to order higher than second because of the appearance of radiation.

Separating L_I into its color electric and color magnetic components, we have

$$-L_I(\vec{R}, \vec{v}_1, \vec{v}_2) = V_0(R) + V_2(\vec{R}, \vec{v}_1, \vec{v}_2), \tag{23}$$

where the static potential $V_0(R)$ is determined by eq. (21), and where the color magnetic component V_2 has the form,

$$\begin{aligned} V_2(\vec{R}, \vec{v}_1, \vec{v}_2) &= \frac{[\vec{R} \times (\vec{v}_1 + \vec{v}_2)]^2}{4R^2} V_-(R) + \frac{[\vec{R} \cdot (\vec{v}_1 + \vec{v}_2)]^2}{4R^2} V_\|(R) \\ &+ \frac{[\vec{R} \times \frac{d\vec{R}}{dt}]^2}{4R^2} V_+(R) + \frac{(\vec{R} \cdot \frac{d\vec{R}}{dt})^2}{4R^2} V_L(R) + V_{\text{spin}}, \end{aligned} \tag{24}$$

where the potentials $V_-(R), V_\|(R), V_+(R)$, and $V_L(R)$ are moments of the field distributions determined by the integrand in eq. (22). The potentials $V_-(R)$ and $V_\|(R)$ are related to $V_0(R)$ by Lorentz invariance:

$$V_-(R) = -\frac{1}{2}V_0(R), \quad V_\|(R) = -\frac{1}{2}V_0(R) + \frac{R}{2}\frac{dV_0}{dR}. \tag{25}$$

The function $V_+(R)$ determines the moment of inertia $I(R)$ of the rotating flux tube according to the equation:

$$I(R) = -\frac{1}{2}R^2 V_+(R). \tag{26}$$

At large R, $V_+(R) \rightarrow -AR$, where $A \sim .21\sigma$. The function $V_L(R)$ determines the energy associated with longitudinal oscillations $\hat{R} \cdot \frac{d\vec{R}}{dt}$. The spin dependent potential V_{spin}, which includes both spin-spin and spin orbit contributions,[1] has the qualitative features of Monte Carlo data obtained from lattice gauge theory.[10]

For $g = 0$ the Lagrangian L_I, eq. (22), reduces to

$$L_I = \frac{4}{3}\frac{\alpha_s}{R} - \frac{2}{3}\frac{\alpha_s}{R}\left\{\vec{v}_1 \cdot \vec{v}_2 + \frac{(\vec{v}_1 \cdot \vec{R})(\vec{v}_2 \cdot \vec{R})}{R^2}\right\} \equiv L_D\,, \tag{27}$$

where L_D is the Darwin Lagrangian[9] (multiplied by the color factor $\frac{4}{3} = 2\,\mathrm{tr}\,Y^2$). Furthermore,

$$L_I \rightarrow L_D\,, \qquad \text{as} \quad R \rightarrow 0\,. \tag{28}$$

The Lagrangian L_I describes the quark anti-quark interaction at distances R greater than the flux tube radius R_{FT}. (See Section 5.) At shorter distances this interaction is described by a Lagrangian L_{PT} obtained from perturbation theory. The Lagrangian L_{PT} is essentially the Darwin Lagrangian L_D, eq. (27), with α_s replaced by the running coupling constant. Since at $R = R_{FT}\,, L_I \approx L_D$, one can obtain a Lagrangian applicable at all distances by using eq. (27) for $R < R_{FT}$ with α_s replaced by the running coupling constant. Therefore the elimination of the gauge fields involved in the construction of L_I makes it possible to connect the long distance confinement domain described by C_μ^a to the short distance perturbative domain described by A_μ^a.

To determine R_{FT} we have calculated the energy density $T_{00}(\vec{x})$ surrounding a static quark anti-quark pair separated by a distance R. The energy density on the median plane $z = 0$ falls off exponentially with the distance ρ from the flux tube axis, which is compatible with the results of the calculation of analogous quantities in lattice gauge theory.[11] Although the correspondence between $T_{00}(\vec{x})$ and Monte Carlo data is not completely clear and the comparison is to some extent qualitative, it provides a check on the physical picture underlying L_I. Such tests are not even in principle possible for purely phenomenological quark anti-quark potentials.

The Monte Carlo calculations[11] were carried out in $SU(2)$ gauge theory for an interquark spacing $R \sim 1.5fm$ and gave a root mean squared flux tube radius $R_{FT} \sim .2fm$. The energy distribution calculated from $\mathcal{L}_{\text{eff}}$ broadens as a function of R. For $R = .7fm$. $R_{FT} = .3fm$, for $R = 1.3fm$, $R_{FT} = .4fm$, and for $R = 2.6fm$, $R_{FT} = .45fm$, at which point the "flux tube" is close to its maximum width.

5. The Long Distance Hamiltonian

To treat mesons containing light quarks it is essential to use relativistic kinematics. The Lagrangian describing a quark anti-quark pair of masses m_1 and m_2 respectively is

$$L = -m_1\sqrt{1 - \vec{v}_1^2} - m_2\sqrt{1 - \vec{v}_2^2} - V_0 - V_2\,. \tag{29}$$

The Lagrangian (29) accounts for the kinematic effects of relativity but not for the emission of radiation. In fact, as in electrodynamics, the whole concept of a potential which depends only upon particle positions and velocities has a meaning only when radiation can be neglected. To estimate when radiation is important, consider a quark anti-quark pair separated by a distance R and moving on a circular orbit with frequency ω. Then radiation occurs if $\hbar\omega > Mc^2$, where M is the mass either of the C_μ field or of the Higgs field. (These masses are of order 500 MeV).[12] Since $\omega \sim \frac{c}{R}$, this means $\hbar c/R > Mc^2$, or $R < \hbar/Mc$. That is, radiation occurs when the distance R between the quarks is less than the flux tube radius $R_{FT} \sim \frac{\hbar}{Mc}$. Thus, for $R > R_{FT} \sim .4fm$, the dominant relativistic effects are kinematic and are included in the Lagrangian (29).

The Hamiltonian H obtained from the Lagrangian (29) is

$$H = \vec{p}_1 \cdot \vec{v}_1 + \vec{p}_2 \cdot \vec{v}_2 - L\,, \tag{30}$$

where

$$\vec{p}_i = \frac{\partial L}{\partial \vec{v}_i}\,. \tag{31}$$

From eqs. (29),(30), and (31) we find,

$$H = H_0 + V_2(\vec{v}_i\,, \vec{R})\,, \tag{32}$$

where

$$H_0 = \sqrt{\vec{p}_1^{\,2} + m_1^2} + \sqrt{\vec{p}_2^{\,2} + m_2^2} + V_0(R)\,, \tag{33}$$

and

$$\vec{v}_i = \frac{\vec{p}_i}{\sqrt{\vec{p}_i^{\,2} + m_i^2}}\,, \qquad i = 1, 2\,. \tag{34}$$

In obtaining the result (32), we have retained terms only to first order in V_2, consistent with the calculation of V_2. The result (32) also follows immediately from the general rule[13] for finding the change in the Hamiltonian due to a small change in the Lagrangian. Finally, we quantize the theory by imposing canonical commutation relations and suitably symmetrizing V_2 so that it is a hermitian operator.

In collaboration with Lewis Fulcher we are now using the Hamiltonian (32) to calculate the energy levels of mesons containing one light quark. Since the relevant distance scale is larger than in mesons containing only heavy quarks, the energy levels should be more sensitive to the long distance regime where the Hamiltonian (32) is applicable, and we feel that these calculations may provide a more sensitive test of the theory. The only new parameters will be the constituent masses of the light quarks.

One cannot readily justify the use of a two particle wave equation such as that generated by the Hamiltonian (32) to calculate the lowest lying bound state energies of mesons containing two light quarks, such as the π meson or the ρ meson. On the other hand, the energy levels of excited states composed of a light quark anti-quark pair receive substantial contributions from the long distance regime where (32) is applicable. Furthermore, the semi-classical description is appropriate for describing the physics of this smooth long distance confinement regime, particularly for calculating the orbital excitations of these mesons. We are now in the process of calculating the Regge trajectories predicted by (32) in the WKB approximation.

6. The Leading Regge Trajectory

To calculate the leading Regge trajectory $\alpha(E^2)$ having the largest angular momentum α for a given bound state energy, the classical approximation is applicable. Furthermore, we can consider circular orbits so that $\vec{v}_1 \cdot \vec{R} = \vec{v}_2 \cdot \vec{R} = 0$. Then the Lagrangian L, eqs. (29) and (24), reduces to

$$L = -m_1\sqrt{1-\vec{v}_1^2} - m_2\sqrt{1-\vec{v}_2^2} - V_0 - \left(\frac{\vec{v}_1+\vec{v}_2}{2}\right)^2 V_- - \left(\frac{\vec{v}_1-\vec{v}_2}{2}\right)^2 V_+ \,. \tag{35}$$

From eq.(35) we calculate the canonical momenta $\vec{p}_i = \frac{\partial L}{\partial \vec{v}_i}\,, i = 1,2$, the energy $E = \vec{p}_1 \cdot \vec{v}_1 + \vec{p}_2 \cdot \vec{v}_2 - L$ and the angular momentum $\vec{J} = \vec{x}_1 \times \vec{p}_1 + \vec{x}_2 \times \vec{p}_2$. We obtain

$$E = \frac{m_1}{\sqrt{1-\vec{v}_1^2}} + \frac{m_2}{\sqrt{1-\vec{v}_2^2}} + \left[1 - \frac{1}{2}(\frac{\vec{v}_1+\vec{v}_2}{2})^2\right] V_0 + \frac{1}{2}I\omega^2\,, \tag{36}$$

$$\vec{J} = \frac{\vec{R}}{2} \times \left\{\frac{m_1\vec{v}_1}{\sqrt{1-\vec{v}_2^2}} - \frac{m_2\vec{v}_2}{\sqrt{1-\vec{v}_2^2}}\right\} + I\vec{\omega}\,, \tag{37}$$

where the moment of inertia I is given by eq. (26) and $\vec{\omega}$ is the angular velocity of the rotating quark anti-quark pair. In obtaining eq. (37) we have made use of the fact that we are working in the center of mass system, $\vec{p}_1 = -\vec{p}_2$.

Using Hamilton's principle, the Lagrangian (35), and the fact that we have circular orbits so that $\vec{v}_1 \cdot \dot{\vec{v}}_1 = \vec{v}_2 \cdot \dot{\vec{v}}_2 = 0$, we obtain the equation of motion

$$\frac{m_1\dot{\vec{v}}_1}{\sqrt{1-\vec{v}_1^2}} - \frac{(V_+ + V_-)}{2}\dot{\vec{v}}_1 - \frac{(V_- - V_+)}{2}\dot{\vec{v}}_2 = -\frac{\vec{R}}{R}\left[\frac{\partial V_0}{\partial R} + \frac{(\vec{v}_1-\vec{v}_2)^2}{4}\frac{\partial V_+}{\partial R} + \frac{(\vec{v}_1+\vec{v}_2)^2}{4}\frac{\partial V_-}{\partial R}\right], \tag{38}$$

along with a corresponding equation obtained by replacing m_1 by m_2 and interchanging $\vec{v}_1$ and $\vec{v}_2$ in eq. (38). For circular motion of frequency $\vec{\omega}$, $\vec{v}_1 = \vec{\omega} \times \vec{R}_1$ and $\vec{v}_2 = \vec{\omega} \times \vec{R}_2$ where $\vec{R}_1 = R_1\hat{R}\,, \vec{R}_2 = -R_2\hat{R}$, and $R_1 + R_2 = R$. We can then write eq. (38) in the form

$$\omega^2\left[\frac{m_1R_1}{\sqrt{1-\omega^2R_1^2}} - \frac{(V_+ + V_-)R_1}{2} + \frac{(V_- - V_+)R_2}{2}\right] = \frac{\partial V_0}{\partial R} + \frac{\partial V_+}{\partial R}\left(\frac{\omega^2R^2}{4}\right) + \omega^2\frac{\partial V_-}{\partial R}\frac{(R_1-R_2)^2}{4}, \tag{39}$$

along with the corresponding equation obtained from eq. (39) by replacing m_1 by m_2 and interchanging R_1 and R_2. Eqs. (39) determine R_1 and R_2 in terms of ω. Substituting these expressions for R_1 and R_2 into eqs. (36) and (37) give $\vec{J}$ and E as functions of ω, which upon elimination of ω yield the leading Regge trajectory.

For the equal mass case, $m_1 = m_2 = m\,, R_1 = R_2 = \frac{R}{2}$, and eqs. (39), (36) and (37) reduce to

$$\frac{m}{\sqrt{1-\frac{\omega^2R^2}{4}}} = \frac{2}{\omega^2R^2}R\frac{\partial V_0}{\partial R} + V_+ + \frac{\partial V_+}{\partial R}\frac{R}{2}\,, \tag{40}$$

$$E = \frac{2m}{\sqrt{1-\frac{\omega^2 R^2}{4}}} + V_0 + \frac{1}{2} I\omega^2 , \tag{41}$$

$$\vec{J} = \frac{mR^2\vec{\omega}}{2\sqrt{1-\frac{\omega^2 R^2}{4}}} + I\vec{\omega} . \tag{42}$$

The Regge trajectory J as a function of E^2 obtained from these equations is plotted in Fig. 4 for the cases of $m = 100\text{MeV}$ and $m = 300\text{MeV}$.

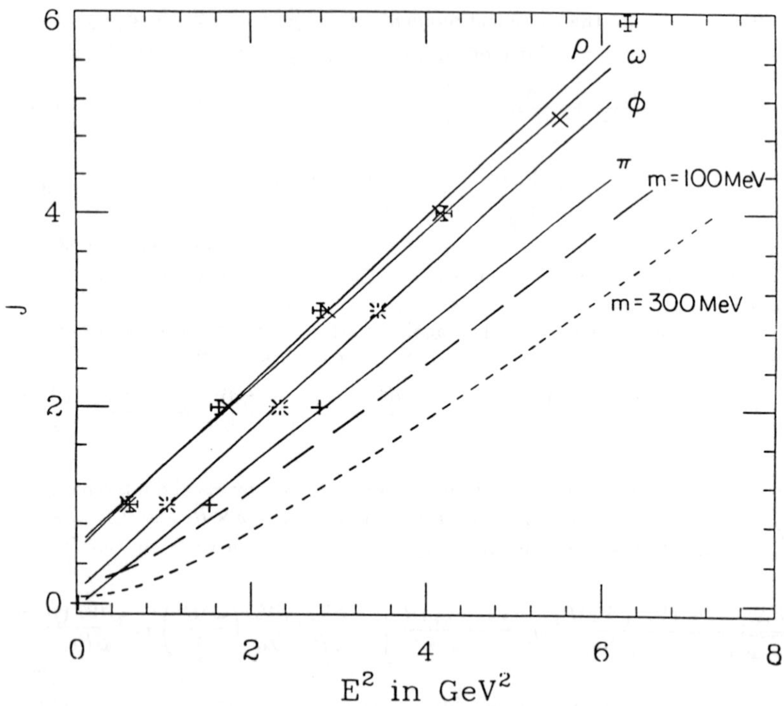

Fig. 4. The points represent the particles on the π, ρ, ω, and ϕ Regge trajectories, and the solid lines are the corresponding linear fits. Dashed lines are Regge trajectories calculated from eqs. (40), (41) and (42): $m = 100\text{MeV}$ (long dashes) and $m = 300\text{MeV}$ (short dashes)

To calculate the large E^2 limit of the slope of the Regge trajectory, the quark mass is irrelevant and V_0 and V_+ can be replaced by their values at large R,

$$V_0 \to \sigma R\,, \quad V_+ \to -AR\,, \quad \text{where} \quad A \approx .21\sigma\,. \tag{43}$$

For large $R\,, \frac{\omega^2 R^2}{4} \sim 1$ and eq. (39) becomes

$$\begin{aligned} \frac{m}{\sqrt{1-\frac{\omega^2R^2}{4}}} &= \frac{1}{2}\sigma R - AR - \frac{AR}{2} = \frac{1}{2}(\sigma - 3A)R\,. \\ R &\to \infty \end{aligned} \tag{44}$$

Note that the presence of the velocity dependent potential V_+ tends to destabilize the system, and there are solutions to eq. (44) for ω only for $A < \sigma/3$. This condition is satisfied by the result (43) of our numerical calculations.[12]

Inserting the equilibrium condition eq. (44) into eqs. (41) and (42) yields the following expressions for E and $\vec{J}$ valid at large R:

$$E = 2R(\sigma - A)\,, \quad \vec{J} = \hat{\omega}\frac{R^2}{2}(\sigma - A)\,. \tag{45}$$

Eq. (45) then yields a linear Regge trajectory $|\vec{J}| = \alpha' E^2$ with

$$\alpha' = \frac{1}{8\sigma(1-\frac{A}{\sigma})}\,, \tag{46}$$

which, from (43), gives

$$\alpha' \approx \frac{1}{6.3\sigma}\,. \tag{47}$$

Had we not included V_2 in H, we would have obtained,

$$\alpha' = \frac{1}{8\sigma}\,. \tag{48}$$

Thus the angular momentum associated with the rotating color fields determined by V_+ gives an additional contribution to the total angular momentum, thereby increasing α' from the value (48) obtained by neglecting the color fields produced by the moving quarks[14] to the value (47). With $\sigma \simeq .21\text{GeV}^2$, eq. (47) gives,

$$\alpha' \approx .75\text{GeV}^{-2}\,, \tag{49}$$

which is very close to the observed slope. (Note that to the accuracy of our numerical evaluation of $\frac{A}{\sigma}$ the expression (47) for α' agrees with the string model prediction, $\alpha' = \frac{1}{2\pi\sigma}$.)

7. Summary and Comments

The original quark model hypothesized the existence of constituent quarks. It left the nature of their interaction an open question. We have argued that dual potentials are the natural variables describing long distance QCD and have written down an effective Lagrangian $\mathcal{L}_{\text{eff}}$ determining their interaction with quarks. Since dual potentials describe only long distance QCD they cannot have a local interaction with the fundamental quarks of QCD. We therefore assume that the dual potentials in $\mathcal{L}_{\text{eff}}$ interact with constituent quarks. The electromagnetic interaction can be used to define electrons and positrons as the source of photons. Similarly the interaction mediated by dual potentials can be used to define constituent quarks as the source of dual gluons. Solving the field equations for these dual potentials in the presence of a quark anti-quark pair leads to an effective Lagrangian for constituent quarks inside mesons, containing α_s and σ as parameters. Physically this Lagrangian is the non-Abelian analogue of the effective Lagrangian describing the interaction of a monopole anti-monopole pair in a superconductor. It thus provides a concrete realization of the Mandelstam 'tHooft dual superconducting mechanism for confinement which can now be confronted by experiment.

Acknowledgements

I would like to acknowledge the contributions of my collaborators James S. Ball and F. Zachariasen during all stages of this work. I would also like to thank Steve Sutlief and Stefan Gössling for their contributions toward the calculation of Regge trajectories.

References

1. M. Baker, James S. Ball and F. Zachariasen, Phys. Rev. **D44** (1991) 3949; M. Baker, James S. Ball and F. Zachariasen, Phys. Lett. **B283** (1992) 360.
2. P. A. M. Dirac, Phys. Rev. **74** (1948) 817.
3. S. Mandelstam, Phys. Rev. **D19** (1979) 2391.
4. G. 'tHooft, Nucl. Phys. **B153** (1979) 141.
5. M. Baker, James S. Ball and F. Zachariasen, Phys. Rev. **D37** (1988) 1036.
6. S. Mandelstam, Phys. Rep. **23C** (1976) 245.
7. G. t'Hooft, in Proceedings of the European Physical Society Conference on High Energy Physics (1976) p. 1225.
8. M. Baker, James S. Ball and F. Zachariasen, Phys. Rev. **D44** (1991) 3328; S. Maedan, Y. Matsubara and T. Suzuki, Prog. Theor. Phys. **84** (1990) 130.
9. C. G. Darwin, Phil. Mag. 39 (1920) 537.
10. A. Huntley and C. Michael, Nucl. Phys. **B286** (1987) 211.
11. R. Sommer, Nucl. Phys. **B306** (1988) 181.
12. M. Baker, James S. Ball and F. Zachariasen, Phys. Rev. **D12** (1993) 3021.
13. Landau and Lifshitz, Mechanics, Pergamon Press (1976) p. 132.
14. W. Lucha, F. Shöberl and D. Gromes, Phys. Rep. 200 (1990) 127.

DUAL QCD AND THE DETERMINATION OF QUARKONIUM MASSES

James S. Ball
Physics Department, University of Utah
Salt Lake City, Utah 84112, USA

ABSTRACT

An effective potential to order (quark mass)$^{-2}$ for a constituent quark and anti-quark has been derived from Dual QCD. The resulting potentials - central, spin-spin, spin-orbit and spinless velocity dependent are completely determined in terms of two parameters. Two additional parameters, the two quark masses, are required to calculate $q\bar{q}$ energy levels. We have made a "best" fit to all of the observed $c\bar{c}$ and $b\bar{b}$ bound states using the Schrödinger equation and the Salpeter equation which incorporates some relativistic effects. Once the four parameters are determined we can then predict the masses of a number of states that have not yet been observed. The quality of the fit obtained with the Schrödinger equation is very good, fitting the 17 observed states with an average error of .36%. Finally, we find that the Schrödinger fit is improved by using a running coupling constant at small R (one new parameter). Using the Salpeter equation to treat the quark kinematics relativistically requires some modification of the methods used to calculate the s-wave mass splitings. This is because the wave functions are singular at $R = 0$ and δ-function that appears in our hyperfine (spin-spin) potential requires a more sophisticated treatment. Examining the origin of the δ-function, which arises when one takes the nonrelativistic limit of the Dirac equation, provides a prescription for the hyperfine calculation in relativistic wave equations. When the Salpeter equation is used to fit the spectrum the results are significantly poorer than the Schrödinger fit. Finally, using our central potential in the Dirac equation for the heavy-light quark systems of $D - D^*$ and $B - B^*$ we obtain encouraging results.

1. Introduction

This talk will be organized as follows: The next section will discuss the potential obtained from dual QCD as described in the previous talk (M. Baker) and it will present empirical formulas suitable for use in the Schrödinger equation or its relativistic generalizations. We will also consider a simple form which incorporates a coupling constant that runs at small R (dual QCD should not be applicable to this region).

The next section will describe our fit to the energy levels of the known $c\bar{c}$ and $b\bar{b}$ states and predictions for the as yet unobserved states. We also predict the masses of the $^1S_0$ and $^3S_1$ states in $c\bar{s}$ and $b\bar{s}$ systems, though, because the s quark is very light, we do not expect our predictions to be very reliable.

Section 4 will discuss the method of solution of the Salpeter equation and problems associated with the behavior of the wave functions at the origin. Our proposed resolution of these problems is then used to fit the existing data.

The final section will describle our preliminary investigations of heavy-light systems employing our potential in the Dirac equation.

2. Dual QCD Potentials

The entire heavy quark potential to order $(mass)^{-2}$ can in general be written in the form[1]

$$\begin{aligned} V = \;& V_0(R) + \left(\tfrac{\vec{\sigma}_1\cdot\vec{L}}{4m_1^2} + \tfrac{\vec{\sigma}_2\cdot\vec{L}}{4m_2^2}\right)\tfrac{1}{R}[V_0'(R) + 2V_1'(R)] + \tfrac{(\vec{\sigma}_1+\vec{\sigma}_2)\cdot\vec{L}}{4m_1m_2}\,\tfrac{1}{R}V_2'(R) \\ & + \tfrac{1}{4m_1m_2}\left(\tfrac{\vec{\sigma}_1\cdot\vec{R}\vec{\sigma}_2\cdot\vec{R}}{R^2} - \tfrac{\vec{\sigma}_1\cdot\vec{\sigma}_2}{3}\right)V_3(R) + \tfrac{\vec{\sigma}_1\cdot\vec{\sigma}_2}{12m_1m_2}V_4(R) \\ & + \tfrac{L^2}{4R^2}\left[\left(\tfrac{1}{m_1}+\tfrac{1}{m_2}\right)^2 V_+(R) + \left(\tfrac{1}{m_1}-\tfrac{1}{m_2}\right)^2 V_-(R)\right] \\ & + \tfrac{[\vec{R}\cdot(\vec{v}_1+\vec{v}_2)]^2}{4R^2}V_{\|}(R) + \tfrac{[\vec{R}\cdot(\vec{v}_1-\vec{v}_2)]^2}{4R^2}\left(-\tfrac{4\alpha_s}{3R}\right) . \end{aligned} \tag{1}$$

Within the classical approximation to dual QCD, we[2] have evaluated the seven potentials appearing in eq. (5.1) as functions of R, the quark-antiquark separation, by first evaluating them numerically from dual QCD differential equations and then by constructing analytic fits to these data. The potentials have an overall power of R (determined by dimensions) and are otherwise functions of λ and $\tilde{F}_0^2$. We have fixed $g^2/\lambda = 5$ and as a result $\alpha_s = \frac{\pi}{5\lambda}$. Analytic fits are

$$\begin{aligned}
V_0(R) &= -\frac{4}{3}\frac{\alpha_s}{R}e^{-.655x} + 1.0324(-\tilde{F}_0^2)R - 0.520\left(-\frac{\tilde{F}_0^2}{\lambda}\right)^{1/2}, && (2)\\
V_1'(R) &= \frac{4}{3}\frac{\alpha_s}{R}\sqrt{-\lambda\tilde{F}_0^2}\;0.655(1-e^{-0.655x}) + 1.0324\tilde{F}_0^2, && (3)\\
V_2'(R) &= \frac{4}{3}\frac{\alpha_s}{R}\sqrt{-\lambda\tilde{F}_0^2}\left(0.655 + \frac{e^{-0.655x}}{x}\right), && (4)\\
V_3(R) &= 12\pi\alpha_s(-\lambda\tilde{F}_0^2)^{3/2}(0.1018x + 0.0534x - 0.007x^3)\cdot\frac{e^{-0.360x}}{x^3}, && (5)\\
V_4(R) &= \frac{32\pi\alpha_s}{3}\delta^3(\vec{R}) + 12\pi\alpha_s(-\lambda\tilde{F}_0^2)^{3/2}(0.0311 - 0.00361x)\frac{e^{-0.530x}}{x}, && (6)\\
V_+(R) &= -\frac{\alpha_s}{R}(\frac{2}{3}e^{-1.458x} - 1.433x + 0.342x^2), && (7)\\
V_-(R) &= -\frac{1}{2}V_0(R), && (8)\\
V_{\|}(R) &= V_-(R) + \frac{R}{2}\frac{\partial V_0(R)}{\partial R}, && (9)
\end{aligned}$$

where the dimensionless distance $x = R\sqrt{-\lambda\tilde{F}_0^2}$.

The relativistic invariance of the Lagrangian density imposes several exact relations between the various terms in Eq.(1). The first is the Gromes' relation[3] for the spin-orbit potentials. This requires that

$$V_1'(R) = V_2'(R) - \frac{\partial V_0(R)}{\partial R} . \tag{10}$$

Our numerical results agree with this to the accuracy of our numerical calculation, and the analytic fits to V_1', V_2' and V_0 satisfy it exactly. Two additional relations for V_- and $V_\parallel$ have been derived by Barchielli, Brambilla and Prosperi[4]. These are

$$V_- = -\frac{1}{2}V_0 \tag{11}$$

and

$$V_\parallel = V_- + \frac{R}{2}\frac{\partial V_0}{\partial R}\,. \tag{12}$$

These too are satisfied by our numerical results although $V_\parallel$ is rather small in magnitude and is therefore not as accurately determed as our other potentials. It should also be noted that neither of these potentials are necessary for our quarkonium fits with equal mass quarks.

Another quark-antiquark potential that we have calculated via dual QCD is the potential associated with the creation of a pair within a flux tube and the subsequent separation of the pair to form closed ends on each of the divided flux tubes. No fields remain in the original region of pair creation. In this calculation one considers a fixed length L of an infinitly long flux tube. Initially, the energy is simply σL. When the quarks have been created but are close together, they interact with a Coulomb-like potential with all of the quark flux absorbed by the antiquark. The flux-tube flux remains nearly unchanged. As the pair separates further, an increasing amount of quark flux is absorbed by the flux tube in which it resides and less flux passes through the region between the pair. This is the result of the screening properties of the QCD vacuum state. At large separations the only change in the energy is due to the decreasing amount of flux tubes remaining in the length L. If we define the potential $V_b(R)$ as the change in energy in going from the initial state to a configuration in which the quarks are separated a distance R, we find that it is very well fit by the following expression:

$$V_b(R) = -\frac{4}{3}\frac{\alpha_s}{R}e^{-.655x} - 1.0324(-\tilde{F}_0^2)R. \tag{13}$$

This is essentially $V_0(R)$ with the opposite sign for the string tension because the flux tube length decreases with increasing R. From this potential and knowing the distribution of color fields in a flux-tube, it should be possible to calculate the probability of pair production per unit length in a long flux tube.

3. Fits and Predictions for $c\bar{c}$ and $b\bar{b}$ Energy Levels

The procedure for obtaining a best fit to the energy levels of the known $c\bar{c}$ and $b\bar{b}$ states is as follows. We define an effective χ^2 to be:

$$\chi^2 = \sum \left(\frac{(experiment - theory)}{(.01 \times experiment)}\right)\,. \tag{14}$$

This would be the actual χ^2 if the experimental statistical error was in fact 1% or equivalently what might be expected to be equal to the number of degrees of freedom if the theory was good to 1%. Our four parameters, λ, $\sqrt{-\tilde{F}_0^2}$, and the two-quark masses m_c and m_b are then varied to minimize the effective χ^2. Our procedure is the following: Using our central potential we solve the Schrödinger equation to determine the eigenvalues and the wave functions for the necessary orbital angular momentum states. The spin and angular momentum dependent potentials are then used perturbatively to calculate the energies of the individual states and the χ^2 is evaluated. The four parameters are then varied to minimize χ^2. It should be emphasized that these are our only parameters and the dependence of the potentials on these parameters is completely determined by dual QCD.

Once the best fit parameters are determined we can predict the unobserved energy levels. Our best fit to the seventeen observed states is given in Table 1. The resulting χ^2 is 1.70, corresponding to an average least-square error of .36%.

Table 1. Schrödinger equation fit masses of all observed $c\bar{c}$ and $b\bar{b}$ states below threshold[5]. Parameters are $\lambda = 1.770$, $\sqrt{-\tilde{F}_0^2} = 455\text{MeV}$, $m_c = 1.320\text{GeV}$ and $m_b = 4.750\text{GeV}$.

State	Pred. Mass (GeV)	Exper. Mass (GeV)
$\eta_c(1S)$	2.969	2.980
$\psi(1S)$	3.114	3.097
$\psi(2S)$	3.687	3.686
$\chi_{c_0}(1P)$	3.440	3.415
$\chi_{c_1}(1P)$	3.502	3.511
$\chi_{c_2}(1P)$	3.542	3.556
$h_c(1P)$	3.514	3.526
$\Upsilon(1S)$	9.464	9.460
$\Upsilon(2S)$	9.994	10.023
$\Upsilon(3S)$	10.344	10.355
$\Upsilon(4S)$	10.635	10.580
$\chi_{b_0}(1P)$	9.861	9.860
$\chi_{b_1}(1P)$	9.891	9.892
$\chi_{b_2}(1P)$	9.914	9.913
$\chi_{b_0}(2P)$	10.223	10.232
$\chi_{b_1}(2P)$	10.248	10.255
$\chi_{b_2}(2P)$	10.268	10.268

The predicted energy levels for the as yet unobserved (nearly) stable states of these systems are shown in Table 2.

Table 2. Predicted masses of the unobserved $c\bar{c}$ and $b\bar{b}$ states

State	Pred. Mass (GeV)
$\eta_c(2S)$	3.581
$(^1D_2)c\bar{c}$ $(n=1)$	3.840
$(^3D_1)c\bar{c}$ $(n=1)$	3.830
$\eta_b(1S)$	9.330
$\eta_b(2S)$	9.931
$\eta_b(3S)$	10.294
$\eta_b(4S)$	10.591
$(^1P_1)b\bar{b}$ $(n=1)$	9.900
$(^1P_1)b\bar{b}$ $(n=2)$	10.256
$(^1D_2)b\bar{b}$ $(n=1)$	10.153
$(^1D_2)b\bar{b}$ $(n=2)$	10.456
$\chi_{b_0}(3P)$	10.519
$\chi_{b_1}(3P)$	10.542
$\chi_{b_2}(3P)$	10.561
$(^3D_1)b\bar{b}$ $(n=1)$	10.144
$(^3D_2)b\bar{b}$ $(n=1)$	10.152
$(^3D_3)b\bar{b}$ $(n=1)$	10.159
$(^3D_1)b\bar{b}$ $(n=2)$	10.447
$(^3D_2)b\bar{b}$ $(n=2)$	10.455
$(^3D_3)b\bar{b}$ $(n=2)$	10.462

Finally, we have also computed the $^1S_0$ and $^3S_1$ states of the $c\bar{s}$ and $b\bar{s}$ systems. (These are known as D_s, D_s^*, B_s and B_s^* respectively). Our results for these fits are shown in Table 3, and use a value $m_s = 430$MeV.

Table 3. Predicted masses of observed $c\bar{s}$ and $b\bar{s}$ states[5]

	$c\bar{s}$		$b\bar{s}$	
	Theory	**Expt.**	**Theory**	**Expt.**
$^1S_0$	2.385	1.969	5.820	$5.359 \rightarrow 5.409$
$^3S_1$	2.542	2.536	5.877	$5.406 \rightarrow 5.456$

This mass is very low, and it is not surprising that the fit in this case is much poorer than for the $c\bar{c}$ and $b\bar{b}$ systems.

The dual QCD region of applicability is in the IR or long range behavior and therefore provides no information about short range effects. From what is known about QCD, one would certainly expect the coupling constant to run at very short distances reflecting the fact that the theory is assymptotically free. We have investigated what such a modification might do to our fits to the data by using a modified formula for the coupling constant for the central potential.

$$\alpha_s(R) = \begin{cases} \alpha_s & R > R_0 \\ \frac{\alpha_s}{1+\frac{R-R_0}{R_0}\beta \ln(R/R_0)} & R < R_0 \end{cases} \tag{15}$$

β was calculated from the β-function for QCD for no flavors and R_0 was arbitarily fixed at 1/GeV. The resulting χ^2, obtained from fitting the data with this modification, was 1.2, certainly a noticeable improvement over our fixed α_s results. This shows that the data certainly favors a running coupling constant.

4. Fits to the $c\bar{c}$ and $b\bar{b}$ Energy Levels Using the Salpeter Equation

There are several important reasons for seeking a relativistic treatment of quark-antiquark boundstate. First of all, for quarks moving in a coulomb potential $v/c \simeq \alpha$ where α is the coeficient of the $1/R$ singularity, this number is about .5 for our potential, making even the $b\bar{b}$ states quite relativistic. This certainly causes some doubt as to the reliability of calculations that employ the Schrödinger equation, and quantities that depend on short range behavior are particularly suspect. A second reason is the need to treat lower mass quarks where relativistic effects will be even more important.

The Salpeter equation

$$(2\sqrt{p^2 + m^2} + V)\Psi = E\Psi \tag{16}$$

represents a simple generalization to treat the quark kinematics relativistically. We have used the method proposed by Fulcher[6], employing a finite number of basis functions to obtain a matrix representation of the operator on the left of Eq.(16). The eigenvalues and eigenvectors are easily calculated for our central potential.

The next step in the calculation of quarkonium energy levels is the perturbative calculation of the splitting produced by the potentials given in Eqs.(3-9). Because of the δ-functions that appear in V_4 and in the last term in Eq.(1) this is not a trivial step.

It has been shown by Durand et al[7] that the Salpeter wave function behaves like $1/R^{1-\gamma}$ for a potential that behaves like α/R near the origin.

$$\gamma = \frac{\alpha}{2}\tan(\frac{1}{2}\pi\gamma) \tag{17}$$

Note that as $\alpha \to 4/\pi$, γ goes to zero. For larger values of α, the solution of Eq.(17) shifts to another branch of the tangent and no normalizable wave function exists. Fortunately, our value of $\frac{4}{3}\alpha_s$ is about .5 so that a solution exists but $\gamma \simeq .8$. If one uses a running coupling constant to weaken the singularity at the origin, the power behavior disappears, but Durand[8] has shown that a logarthimic singularity remains and $\Psi(0)$ is still infinite.

We have developed a method to regularized the δ-function terms. Recall that the δ-function in V_4 is the result of taking the nonrelativistic limit of the Dirac equation in which the vector potential of the nucleus is treated as a perturbation. The hyperfine matrix element is

$$\Delta E \sim \int d^3 r \bar{\Psi} \alpha \cdot \vec{A} \Psi \tag{18}$$

where

$$\vec{A} = \frac{\vec{\mu} \times \vec{r}}{r^3} \tag{19}$$

where $\vec{\mu}$ is the nuclear magnetic moment. Writing this matrix element in terms of the large g and small f components of the spin we obtain:

$$\Delta E \sim \int dr f(r) g(r). \tag{20}$$

Note that this is finite even though, like the Salpeter wave function, the Dirac wave functions are singular at the origin. If we eliminate the small component, we find

$$\Delta E \sim -\int dr \frac{2m}{(E+m-V)} g(r) \frac{dg}{dr}, \tag{21}$$

where V is the Coulomb potential. In this form it is clear that V in the denominator is regulating the singularity at the origin. Taking the nonrelativistic limit we obtain

$$\Delta E \sim -\int dr g(r) \frac{dg}{dr} = \frac{1}{2} g(0)^2. \tag{22}$$

This would be a problem if it were not for the fact that g now is the Schrödinger wave function which is well behaved at the origin. This is exactly the same term that the δ-function potential produces. Evidently the correct replacement for the δ-function for a relativistic wave function is

$$\delta^3(\vec{r}) \rightarrow -\frac{4m}{(E+m-V)} \Psi(r) \frac{d\Psi}{dr}. \tag{23}$$

This prescription has the further advantage that no new parameters have been introduced.

The fit to the data using the Salpeter eq. and the potential with a running coupling yielded χ^2=2.7, significantly worse that our earlier results (χ^2=1.2). Qualitatively, the Salpeter fit was somewhat better for the $c-\bar{c}$ states and considerably worse for the $b-\bar{b}$ states. The parameters for this fit are $\lambda = 1.642$, $\sqrt{-\tilde{F}_0^2} = 493\text{MeV}$, $m_c = 1.255\text{GeV}$, and $m_b = 4.662\text{GeV}$. Again R_0 was fixed at $1/\text{GeV}$.

Aside from the fact that the Salpeter eq. doesn't help the fitting of the data, it has several other defects. First of all, it only applies to scalar particles and provides no guidance about how spin should be introduced. Secondly, we have simply assumed that our potential is the fourth component of a four-vector. If part of our V_0 was a scalar rather than a vector (indistinguishable for the Schrödinger eq. and in our derivation of the potential) it should have been added to m in Eq. (16) with only the vector term in V. Finally, since the Salpeter eq. doesn't come from any correct relativistic equation it provides no guidance as to recoil corrections to the potential such as factors of m/E.

5. Dirac Equation

In contrast to the Salpeter eq., the Dirac eq. does provide a correct relativistic treatment for atomic systems. Although the mass ratios are not as large for heavy-light systems, they probably are large enough to keep the heavy quark in the non-relativistic regime. In a recent paper, Mur et al[9] give theoretical arguments for the use of the Dirac equation to describe heavy-light quark-antiquark systems in QCD. We begin with the investigation states of a light quark bound to a c or b quark, and determine the eigenvalues and eigenvectors of the Dirac eq. in the limit that the heavy mass is infinite. The hyperfine splitting will then be calculated perturbatively, using the heavy quark mass obtained in our Schrödinger fit. The radial Dirac eq. in matrix form for the small and large compoenents is

$$\begin{pmatrix} V_v + m + V_s & -\frac{d}{dr} - \frac{l+1}{r} \\ \frac{d}{dr} - \frac{l}{r} & V_v - m - V_s \end{pmatrix} \begin{pmatrix} g \\ f \end{pmatrix} = E \begin{pmatrix} g \\ f \end{pmatrix} \tag{24}$$

where V_s is the scalar potential and V_v is the 4th component of a 4-vector. The conventional view is that the confining potential is a scalar and that the Coulomb term is a vector. For our potentials the fact that the string tension appears in V_1' tends to support this division, although in this application we really have no choice because there is no solution to the Dirac eq. for a linear potential in V_v.

Each of the four matrices on the left of Eq.(24) is expressed as an NxN matrix using the same basis function as in the solution of the SP equation. The resulting eigenvalues and eigenvectors are then used to calculate the hyperfine splitting. For a pure Coulomb potential (α = .5 and m=1) using a 20x20 representation for the individual matrices, we obtained .866028 for the groundstate energy to be compared to .866025 for the exact result. For the linear potential we compared our results with those of Mur et al[9] and found excellent agreement.

In the calculation of the masses of the heavy-quark light-quark systems, we first adjusted the light-quark mass (the only free parameter in this calculation) to give the correct center of mass of the s-wave states and then calculated the splitting using this mass and the heavy-quark mass.

The our results are given in Table 4.

Table 4. Predicted masses of B, B^*, D and D^* mesons[5]. m_u is the light quark mass.

m_u=84MeV			m_u=251MeV		
state	**Theory**	**Expt.**	**state**	**Theory**	**Expt.**
B	5.277	5.279	D	1.814	1.869
B^*	5.325	5.325	D^*	2.029	2.010

The results are very good except for the fact that one would like the same light quark mass for both the $D - D^*$ system and the $B - B^*$ system. Including a nonrelativistic kenetic energy term for the heavy quark is a simple process and might both improve the fit to the D and D^* masses and change the required light-quark mass.

6. References

1. W. Lucha, F. Schöbert and D. Gromes,*Phys. Rep.* **200** (1990) 127; E. Eichten and F. Feinberg,*Phys. Rev.***23** (1981) 2724.
2. M. Baker, J.S. Ball and F. Zachariasen,*Phys. Rev.* **D44** (1991) 3949. M. Baker, J.S. Ball and F. Zachariasen,*Phys. Lett.* **B283** (1992) 360. M. Baker, J.S. Ball and F. Zachariasen,*Phys. Rev.* **D47** (1993) 3021.
3. D. Gromes, *Z. Phys.* **C26** (1984) 401.
4. A. Barchielli,N. Brambilla and G. M. Prosperi, *Nuovo Cimento***103** (1990) 59
5. Particle Data Group, *Phys. Rev.***D50** (1994) 1173
6. L. P. Fulcher,Z. Chen and K. C. Yeong, *Phys. Rev.* **D47** (1993) 4122.
7. L. J. Nickisch, L. Durand and B. Durand, *Phys. Rev.* **D30** (1984) 660.
8. L. Durand, *Phys. Rev.* **D32** (1985) 1257.
9. V. D. Mur, V. S. Popov, Yu. A. Simonov and V. P. Yurov,*J. Expt. Theor. Phys.* **78** (1994) 1. (HEP-PH-9401203)

Effective Lagrangians and QCD

Linear response theory : The Roper resonance*

Abdellatif Abada †

Laboratoire de Physique Nucléaire, CNRS/IN2P3 Université de Nantes,
2, Rue de la Houssinière, 44072 Nantes Cedex 03, France

Abstract

After discussing briefly the necessity to improve the standard Skyrme model for the description of low energy hadron physics, I consider two extended Skyrme models. The first one includes fourth and sixth-order terms while the second one incorporates the ρ and the ω mesons explicitly. In the framework of the linear response theory, I investigate the soliton breathing mode within these two effective Lagrangians of QCD. In the two cases I find that the monopole response function has a pronounced peak which I identify to the Roper resonance $N(1440)$. I compare my result to the one obtained with phase shifts method.

1. There exists a well known and a powerful method extensively used in nuclear physics to describe giant resonances of nuclei. Namely, the Linear Response Theory [1]. The aim of this contribution is to show how this method can be applied in hadronic physics. My purpose is to describe the simplest vibrational excitation of the nucleon : the breathing mode. This mode, called Roper resonance, is an excited state of the nucleon which has the same quantum numbers ($I(J^P) = \frac{1}{2}(\frac{1}{2}^+)$). Its experimental excitation energy is 500 MeV while its width stands between 250 and 450 MeV [2].

In order to describe hadrons one should use in principle the Quantum Chromo Dynamics (QCD) since this theory is believed to be the fundamental theory of hadronic physics. However, in QCD, one is confronted to the problem of the running coupling constant α_s. For processes with high momentum transfer α_s is small, because of the asymptotic freedom of QCD, and one can use perturbation theory. But at low energy, the running coupling constant becomes large which renders the standard perturbation theory with respect to α_s inapplicable. Therefore in the domain of energy of interest in nuclear physics, non perturbative methods are indispensable. In this work I have considered *effective* models of QCD whose degrees of freedom are the hadronic observable instead of the fundamental constituents, quarks and gluons, which are confined. This point of view has been supported by the works of t'Hooft and Witten. Namely, Witten [3], in analysis of QCD diagrams in the $1/N_c$ expansion (proposed by t'Hooft [4]), was led to the conjecture that in the large N_c limit QCD is equivalent to an effective field theory of mesons. Soon after Witten [5] recognized

*Talk presented at the *American University of Paris* for the workshop on Quantum Infrared Physics, June (1994).

†email : "ABADA@nanvs2.in2p3.fr"

that baryons can emerge as topological solitons of this meson theory. Therefore one arrives to an unified effective theory describing both mesons and baryons. This was the aim of Skyrme [6] in 1961 well before the advent of QCD. He constructed a model in which the fundamental field is the pion field and where baryons are formed from meson solitons with non-zero winding number. After several years of neglect, the Skyrme model made a comeback in 1983 with the famous paper of Adkins, Nappi and Witten [7]. Since then it has been used extensively to describe static and dynamical properties of baryons. The Skyrme model is now recognized as the simplest chiral realization of QCD at low energy and large N_c [5].

2. Expressed in terms of an SU(2) matrix field U which characterizes the pion field

$$U = \exp[\ i\vec{\phi}.\vec{\tau}/f_\pi\ ]\ , \tag{1}$$

the τ_a's being the usual Pauli matrices, the chiral effective Lagrangian density proposed by Skyrme [6] reads:

$$\mathcal{L}_{SK} = \frac{f_\pi^2}{4}\ \mathrm{Tr}\ (\partial_\mu U \partial^\mu U^+) + \frac{1}{32e^2}\ \mathrm{Tr}\ \{\ [(\partial_\mu U)U^+,(\partial_\nu U)U^+]^2\ \}\ . \tag{2}$$

The first term in the Lagrangian (2) corresponds to the nonlinear σ model in which $f_\pi = 93$ MeV is the pion decay constant. The second term, which contains the dimensionless parameter e, is of fourth-order in powers of the derivatives of the pion field and was introduced by Skyrme to stabilize energetically the soliton. The Skyrme model (2) is usually investigated by assuming the hedgehog ansatz, i.e., the pion field is directed radially in the configuration space :

$$\vec{\phi}(\mathbf{r},t) = f_\pi F(r,t)\hat{\mathbf{r}}\ . \tag{3}$$

Despite of its success, the Skyrme model presents, however, a lot of problems. Concerning the static properties of baryons, the predictions of the model are within 30% of the experimental values when the parameters are adjusted to fit the nucleon and the Δ masses [7, 8]. I would like to stress at this point that this manner to choose the parameters is not correct. Since it is a meson theory, the parameters of the model have to be fixed by fitting to the low-energy meson observables and not to baryon data. The lack of attraction in the central potential [8] as well as in the isospin independent spin-orbit force [9] are other drawbacks of the Skyrme model. Moreover, in taking into account the Casimir energy of the soliton, the author of Ref. [10] has found that the nucleon mass is lowered to a value between 0 and 400 MeV which is much too small ! He also proved that the $O(N_c^1)$ and $O(N_c^0)$ mass contributions are of same order. Thus the large N_c expansion is a poor approximation within the Skyrme model. An other weakness of this model (2) concerns the breathing mode (the main subject of this contribution). One problem regarding this mode in the Skyrme model is that it does not show up in analysis based on phase shifts, whereas when other calculations found it, its excitation energy stands between 200 and 300 MeV.

Thus, in order to describe low energy hadron physics one should improve the Skyrme model. One possible way to extend the model consists to add higher order terms in derivatives in the corresponding chiral Lagrangian as the sixth-order term generated by ω-meson exchange [11]. A second way to improve the model is to consider an effective Lagrangian which incorporates the low mesons explicitly, since the 4-th and 6-th order terms can be derived from a local approximation of a model with ρ [12] and ω mesons [11] respectively. In this paper I will explore the soliton breathing mode in the framework of the linear response theory within these two extensions of the Skyrme model. Let me start with the first extension.

3. The term of order six generated from ω-meson exchange reads:

$$\mathcal{L}_6 = -\frac{1}{2}\frac{\beta_\omega^2}{m_\omega^2} B_\mu B^\mu \quad \text{with} \quad B^\mu = \frac{1}{24\pi^2}\epsilon^{\mu\nu\alpha\beta}\mathrm{Tr}\,\{\,(\partial_\nu U)U^+(\partial_\alpha U)U^+(\partial_\beta U)U^+\,\}\,. \tag{4}$$

B^μ has been constructed and identified to the baryon current by Skyrme [6]. The two new constants appearing in Eq. (4) are the ω-meson mass m_ω and the parameter β_ω which is related to the $\omega \to \pi\gamma$ width [13]. Let me consider the following Lagrangian [14]

$$\mathcal{L} = \mathcal{L}_{SK} + \mathcal{L}_6 + \frac{1}{4} f_\pi^2 m_\pi^2 \mathrm{Tr}\,(\,U + U^+ - 2\,)\ . \tag{5}$$

where the last term implements a small explicit breaking of chiral symmetry, m_π being the pion mass. Assuming the hedgehog ansatz (3) the classical Euler-Lagrange equation for the chiral function F reads:

$$g^2\ddot{F} + \alpha\dot{F}^2 = \left(g^2F'\right)' - \left(\theta + \alpha F'^2\right)\ , \tag{6}$$

where the space and time dependent functions g, α and θ are, respectively,

$$\begin{aligned} g(r,t) = &\ \left((ef_\pi r)^2 + 2s^2 + \frac{\beta_\omega^2}{4m_\omega^2}\frac{e^2}{\pi^4}\frac{s^4}{r^2}\right)^{\frac{1}{2}}\ , \quad \alpha(r,t) = 2sc\left(1 + \frac{\beta_\omega^2}{4m_\omega^2}\frac{e^2}{\pi^4}\frac{s^2}{r^2}\right)\ , \\ \theta(r,t) = &\ 2sc\left((ef_\pi)^2 + \frac{s^2}{r^2}\right) + (ef_\pi m_\pi\, r)^2 s\ , \quad \text{with}\ \ s = \sin F\ ,\ c = \cos F \end{aligned} \tag{7}$$

Primes and dots in Eq. (6) indicate radial coordinate and time differentiations respectively. The equation of motion (6) has to be solved with the usual conditions for winding number one: $F(0,t) = \pi$ and $F(\infty,t) = 0$.

In order to describe the soliton breathing mode, I explore the response of the Skyrmion to an external infinitesimal monopole field. The response function is determined from the evolution of the isoscalar mean square radius of the Skyrmion with respect to the frequency Ω of the excitation field. An external time-dependent monopole field corresponds to the addition of the following term to the Lagrangian density (5)

$$\mathcal{L}_{\text{int}} = ef_\pi^3 r^2 B^0(r,t)\ \epsilon\sin(\Omega t)\exp(\eta t) \tag{8}$$

where $B^0(r,t)$ is the time-component of the baryon current, $B^0(r,t) = -\frac{1}{2\pi^2}\frac{s^2}{r^2}\frac{\partial F}{\partial r}$, and η a vanishingly small positive number. With this interaction term (8) the new corresponding Euler-Lagrange equation reads:

$$g^2\ddot{F} + \alpha\dot{F}^2 = \left(g^2F'\right)' - \left(\theta + \alpha F'^2\right) + \epsilon\frac{(ef_\pi)^3 r}{\pi^2} s^2\sin(\Omega t)\exp(\eta t)\ . \tag{9}$$

Because the term (8) is weak (in the domain $t \in [-\infty, 0]$), it introduces only small changes of the classical solution. One thus can treats this solution in a linear approximation. To first order in ϵ, one has

$$F(r,t) = F_s(r) + \delta F(r,t)\ + \delta F^*(r,t) \quad \text{with} \quad \delta F(r,t) = -i\frac{\epsilon}{2}R(r)\exp(i(\Omega - i\eta)t)$$

where $F_s(r)$ is the static solution of Eq. (6). Thus, linearizing equation (9), one obtains

$$((\Omega - i\eta)^2 - \mathcal{A}_s)(g_s R) = -\frac{(ef_\pi)^3}{\pi^2}\frac{r \sin^2 F_s}{g_s} \tag{10}$$

where $g_s(r)$ is the function g (c.f. (7)) in the static case, and the operator $\mathcal{A}_s$ depends only on F_s :

$$\begin{aligned} \mathcal{A}_s = -\frac{\mathrm{d}^2}{\mathrm{d}r^2} + \frac{g_s''}{g_s} - \frac{2}{g_s^2}\Big(2sc(F_s'' + \frac{\beta_\omega^2\, e^2}{4m_\omega^2\, \pi^4}(\frac{s^2}{r^2}F_s')' \,) + (c^2 - s^2)(F_s'^2 - (ef_\pi)^2 - 2\frac{s^2}{r^2}\,) \\ -\frac{s^2}{r^2}(1 + \frac{\beta_\omega^2\, e^2}{4m_\omega^2\, \pi^4}\, F_s'^2\,) - \frac{1}{2}(ef_\pi m_\pi r)^2 c \,\Big) \end{aligned} \tag{11}$$

Up to first order in ϵ the isoscalar mean square radius [7], $\langle r^2 \rangle = \int \mathrm{d}^3r\, r^2 B^0(r,t)$, reads:

$$\langle r^2 \rangle = \langle r^2 \rangle_s - i\frac{\epsilon}{2}(\, f(\Omega)\exp(i(\Omega - i\eta)t) - f^*(\Omega)\exp(-i(\Omega + i\eta)t)\,)$$

where $\langle r^2 \rangle_s$ is the static mean square radius and f the response function

$$f(\Omega) = \frac{4}{\pi}\int_0^\infty r\frac{\sin^2 F_s}{g_s}(g_s R)\, \mathrm{d}r. \tag{12}$$

By using equation (10), one can extract the following spectral representation of the response function

$$f(\Omega) = -\frac{1}{\pi}\sum_n \frac{|\langle\phi|\phi_n\rangle|^2}{(\Omega - i\eta)^2 - \omega_n^2}\,, \tag{13}$$

where the state ϕ is defined by

$$\langle r|\phi\rangle = \frac{2ef_\pi}{\pi}\frac{r \sin^2 F_s}{g_s}, \tag{14}$$

and the ϕ_n are the eigenstates of the operator $\mathcal{A}_s$ (see Eq. (11)), with the eigenvalues ω_n^2, normalized according to $\int_0^\infty \phi_n(r)\phi_m(r)ef_\pi \mathrm{d}r = \delta_{nm}$. In Eq. (13) the limit $\eta \to 0^+$ is implicit and corresponds to the boundary condition specified above. The quantity of interest here is the imaginary part of the response function, which is directly related to the distribution of collective strength (see e.g. [1]). The energy at which Im (f) exhibits an unbound peak is identified with the excitation energy of the Roper resonance.

Let me now present the results for some static properties and for the energy of the Roper resonance. First of all, because it is a meson theory, the parameters of the model have to be fixed by fitting to the low energy meson observables. This gives: $f_\pi = 93$ MeV, $m_\pi = 139.5$ MeV, $m_\omega = 782$ MeV [2], $e = 7.1$ [10] and $\beta_\omega = 9.3$ [13] (and $= 0$ in the case of the minimal Skyrme model of course). In Tab.1, I report the soliton mass, the isoscalar root mean square radius, the axial vector coupling constant and the Δ-nucleon mass splitting for the different sets of parameters. In the third row I report the experimental values [2]. Concerning the soliton mass I find 972 MeV and 1563 MeV in the case [i] and [ii], respectively. one has to subtract ~ 1 GeV (Casimir energy) in the case of the minimal model [i] which leads to a value of ~ 0 MeV [10] ! In the extended Skyrme model (case [ii]) I must subtract ~ 500 MeV [1][10] which leads to a reasonable value of ~ 1 GeV. Regarding the Δ-nucleon mass splitting, which is not affected by the Casimir effect, I find

[1]Indeed, it has been shown in Ref. [10] that the $O(N_c^0)$ mass contribution is $\sim O(N_c^0)/3$ with a negative sign.

1192 MeV in the case [i] which is not a realistic value compared to the experimental one 290 MeV. However, in the case [ii] I obtain 227 MeV which is more acceptable. Furthermore, the axial vector coupling constant g_A is found to be 0.34, in the case [i], instead of the experimental value 1.23 which is very close to the value predicted by the improved Skyrme model.

	e	β_ω	M_S MeV	r_0 fm	g_A	$M_\Delta - M_N$ MeV	$\hbar\Omega_{Roper}$ MeV
[i]	$7.1^{(a)}$	0	972	0.30	0.34	1192	500
[ii]	$7.1^{(a)}$	$9.3^{(b)}$	1563	0.61	1.24	227	395
			939	0.72	1.23	290	500

Table 1: Are reported the soliton mass M_S, the isoscalar root-mean-square radius r_0, the axial vector coupling constant g_A, the Δ-nucleon mass splitting and the Roper-nucleon mass splitting $\hbar\Omega_{Roper}$, for the Skyrme and extended models. The last line is derived from the data [2].

For the breathing mode of the Skyrmion, I display in Fig. 1 the imaginary part of the response function (13) for the two cases $\beta_\omega = 0$ and $\beta_\omega = 9.3$. In both cases I find an unbound peak. In case [i] I find this peak to be located at 500 MeV. Nevertheless, this result cannot be reliable since the other predicted properties are unrealistic (see Tab.1). Contrary, in the case of the extended Skyrme model, the response function exhibits a pronounced sharp peak located at $\sim$ 400 MeV which I identify to the Roper resonance. This value corresponds to the Roper-nucleon mass splitting and, consequently, is not sensitive to the Casimir effect.

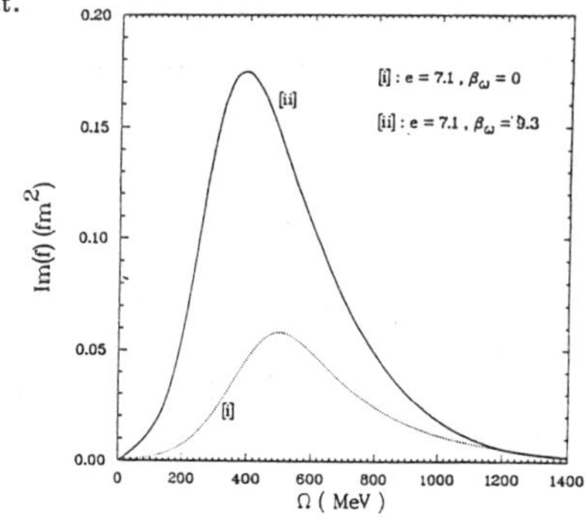

Figure 1: Imaginary part of the response function f in fm^2 (c.f. Eq. (13)) versus the energy Ω in MeV.

Before continuing I want to stress that, by using phase shifts analysis, the authors of Ref. [15] found no trace of the Roper resonance within the same extended Skyrme model. So it is natural to check if this result is due to the parameters they chose. When I take their parameters ($f_\pi = 71.2$ MeV, $e = 9.92$ and $\beta_\omega = 13.6$) to investigate the Roper resonance with my approach, I find a sharp peak located at 320 MeV. This leads me to think that the difficulty in finding the breathing mode resonance in Ref. [15] is due to the *implementation* of the phase shifts method.

4. Since the fourth and sixth-order terms in Eq. (5) can be derived from a local approximation of an effective model with ρ [12] and ω mesons [11] respectively, it seems therefore interesting to investigate the soliton breathing mode in the framework of linear response theory within an effective Lagrangian which incorporates these two mesons *explicitly* with finite masses. Let me consider the following Skyrme Lagrangian vector meson stabilized [16]:

$$\mathcal{L} = \frac{f_\pi^2}{4}\,\mathrm{Tr}\,(\partial_\mu U \partial^\mu U^\dagger) + \mathcal{L}_{\pi\rho} + \mathcal{L}_{\pi\omega} + \frac{1}{4} f_\pi^2 m_\pi^2 \mathrm{Tr}(U + U^\dagger - 2)\ . \tag{15}$$

The first and the last term in this Lagrangian have been discussed above. The second one corresponds to the coupling between the isospin one ρ-meson and the pion field. Here I will follow the standard prescription of nonlinear chiral symmetry for massive particles [17] to introduce the ρ vector meson in the chiral Lagrangian. In a recent paper [18] it was proven that in this case the soliton is stable with respect to the breathing fluctuations contrary to the usual way to introduce iso-vector mesons (massive Yang Mills particles or hidden gauge particles). $\mathcal{L}_{\pi\rho}$ is given by

$$\mathcal{L}_{\pi\rho} = \frac{M_\rho^2}{2}\mathrm{Tr}(V_\mu V^\mu) - \frac{1}{4}\mathrm{Tr}\left\{ \left(\nabla_\mu V_\nu - \nabla_\nu V_\mu + i\frac{g_V}{\sqrt{2}}[u_\mu, u_\nu] \right)^2 \right\} \tag{16}$$

where V_μ is the ρ-meson field and

$$\nabla_\mu = \partial_\mu + [\Gamma_\mu,\ ],\ \Gamma_\mu = \frac{1}{2}[u^\dagger, \partial_\mu u],\ u = U^{\frac{1}{2}},\ u_\mu = i u^\dagger \partial_\mu U u^\dagger\ . \tag{17}$$

The two constants appearing in Eq. (16) are the ρ-meson mass M_ρ and the dimensionless parameter g_V which can be related to the $\rho \to \pi\pi$ decay width. The third term in Eq. (15) corresponds to ω-meson exchange [19]. It reads:

$$\mathcal{L}_{\pi\omega} = -\frac{1}{4}\left(\partial_\mu\omega_\nu - \partial_\nu\omega_\mu \right)^2 + \frac{1}{2}m_\omega^2 \omega_\mu \omega^\mu + \beta_\omega \omega_\mu B^\mu \tag{18}$$

where ω_μ is the omega field. It is easy to show that if one increases the masses M_ρ and m_ω and the coupling constant β_ω to infinity, keeping g_V and the ration β_ω/m_ω finites, the Lagrangian (15) becomes the extended Lagrangian (5) with the Skyrme parameter $e = 1/2g_V$. This is called the local approximation of (15).

In Ref. [16] I have investigated in details the model (15). More precisely I worked with the Hamilton density of the theory (15). By making the most general most general spherical ansatz for the ρ and the ω fields:

$$\vec{\pi} = f_\pi F\ \hat{\mathbf{r}},\ \ V_i = f_\pi \left\{ v_1(\tau_i - (\vec{\tau}.\hat{\mathbf{r}})\hat{r}_i) + v_2(\vec{\tau}.\hat{\mathbf{r}})\hat{r}_i - v_3(\vec{\tau}\times\hat{\mathbf{r}})_i \right\},\ \ \omega_i = f_\pi\omega\ \hat{r}_i, \tag{19}$$

I first solved the static Hamilton equations to build the soliton, with the following set of physical parameters (fixed by fitting to the low energy meson observables) :

$$f_\pi = 93,\ M_\rho = 769,\ m_\omega = 782,\ m_\pi = 139.5\ \ (\mathrm{MeV})\ \ ;\ \ \beta_\omega = 9.3, g_V = 0.09\ . \tag{20}$$

After that I investigated the soliton breathing mode with the method described above. Namely, I considered the linearized time-dependent Hamilton equations in presence of an external infinitesimal monopole field by adding the following perturbation term

$$\delta E_{\text{int}} = \epsilon f_\pi^3 \langle r^2 \rangle \sin(\Omega t) \exp(\eta t) \tag{21}$$

to the energy functional. The stages of calculations are the same as in Sec. 3 but rather more technical because in this case there are several fields to treat (c.f. Eq. (19)). All the details of calculations can be found in Ref. [16]. The fact remains that I was able to extract the monopole response function from the evolution of the mean square radius. I display in Fig. 2 the imaginary part of such a function with respect to the frequency Ω of the external field (21) for the parameters given in Eq. (20). I find that it exhibits a pronounced peak at 400 MeV which I identify with the excitation energy of the Roper resonance. Within the model (5), the local approximation to the Lagrangian (15), I have found nearly the same value except for the detail that the dimensionless Skyrme term parameter e was equal to 7.1 while the value of e which corresponds to the parameters (20) I use in this work is $e = 1/2g_V = 5.57$. I show also in Fig. 2 the response function which corresponds to the local approximation taking $e = 5.57$. One sees from this figure that there is an improvement over the local case, although a slight one ($\sim$ 40 MeV) but in the good direction.

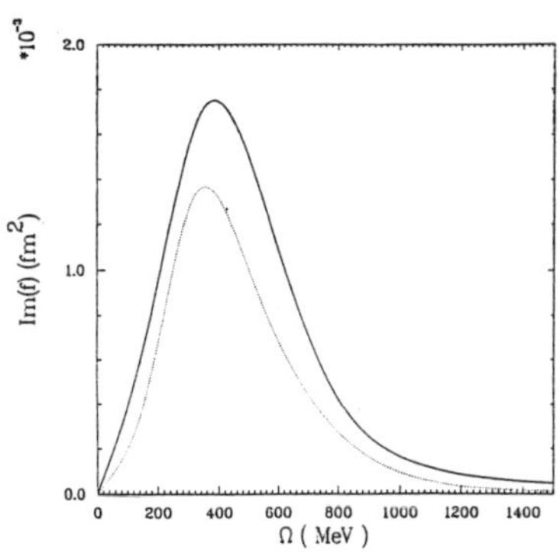

Figure 2: Imaginary part of the response function f (fm^2) versus the energy Ω (MeV). The full line corresponds to the Lagrangian (15) with the parameters given in Eq. (20) while the dotted line corresponds to the local approximation (5).

5. To summarize, the message that I want to transmit through this work is that one should not restrict oneself to the standard Skyrme model (2) for the description of low energy hadron physics, but consider extensions of this model including higher order terms in powers of the derivatives of the pion field. This is obvious from Tab. 1. The model (5) can be considered as a minimal extension of the Skyrme Lagrangian since there are several other higher order terms. An other kind of extension consists of considering effective Lagrangians which incorporate low mass mesons with finite mass like the model (15. However, to get a more realistic improvement, it is suitable to include the axial vector meson-A_1 and the scalar meson-ϵ.

Finally, the difficulty to find the $P11$ resonance $N(1440)$ in analysis based on phase shifts method [15] and the encouraging results obtained in this work lead me to think that the linear response theory is the most adapted method to describe the Roper resonance within effective Lagrangians of QCD.

References

[1] P. Ring and P. Schuck, *The nuclear many body problem*, Springer (1980).

[2] Particle data booklet, *review of particle properties* Phys. Rev. D **45**, Part 2 (june 1992).

[3] E. Witten, Nucl. Phys. **B160**, 57 (1979).

[4] G. t'Hooft, Nucl. Phys. **B72**, 461 (1974).

[5] E. Witten, Nucl. Phys. **B223**, 422 and 433 (1983).

[6] T. H. R. Skyrme, Proc. Roy. Soc. **A260**, 127 (1961) ; Nucl. Phys. **31**, 550 (1962).

[7] G. S. Adkins, C. R. Nappi and E. Witten, Nucl. Phys. **B228**, 552 (1983).

[8] R. Vinh Mau, Prog. Part. Nucl. Phys. **20**, 221 (1988) and references therein.

[9] A. Abada, to be published in Z. Physik A, and references therein.

[10] B. Moussallam, Ann. Phys. (NY) **225**, 264 (1993).

[11] A. Jackson, A. D. Jackson, A. S. Goldhaber, G. E. Brown and L. C. Castillejo, Phys. Lett. **154B**, 101 (1985).

[12] K. Iketani, Kyushu Univ. preprint, 84-HE-2 (1984).

[13] M. Lacombe, B. Loiseau, R. Vinh Mau and W. N. Cottingham, Phys Rev. D**38**, 1491 (1988).

[14] A. Abada and H. Merabet, Phys. Rev. D **48**, 2337 (1993).

[15] B. Schwesinger and H. Weigel, Nucl. Phys. **A465**, 733 (1987).

[16] A. Abada, Phys. Lett. B 322, 103 (1994).

[17] S. Coleman, J. Wess and B. Zumino, Phys. Rev. **177**, 2239 (1969) ; C. G. Callan, S. Coleman, J. Wess and B. Zumino, *ibid* p. 2247.

[18] A. Abada, D. Kalafatis and B. Moussallam, Phys. Lett. B **300**, 256 (1993).

[19] G. S. Adkins and C. R. Nappi, Phys. Lett. **137B**, 251 (1984).

NONPERTURBATIVE EFFECTS IN LOW ENERGY EFFECTIVE THEORIES OF QCD

DIMITRI KALAFATIS
Theoretical Physics Group
Department of Physics and Astronomy
University of Manchester
MANCHESTER M13 9PL, UK

ABSTRACT

In this talk I concentrate on the role of chiral symmetry realisation by spin-1 fields in the low energy QCD effective lagrangian. I assume that chiral symmetry is nonlinearly realised and that spin-1 fields transform homogeneously under chiral rotations, which is in sharp contrast to previous works where the ρ and the a_1 mesons were treated as approximate gauge bosons of some chiral group. I emphasise the role played by four-meson couplings which in our scheme are essential for the theory to make sense. By requiring the kinetic energy density of the theory to be bounded from below we find inequalities relating three- and four-point meson couplings. It is finally shown how the combination of our analysis and of unitarity requirements naturally leads to a low-energy phenomenological lagrangian for the nonanomalous sector of $\pi\rho a_1$ strong interactions.

1. Introduction

Strong interactions at low energies are quite well understood in terms of an effective meson lagrangian[1]. The starting point for such an effective lagrangian is the nonlinear sigma model of pseudoscalar pions with spontaneous breaking of chiral symmetry, a central feature of low energy QCD. The experimental discovery of meson resonances as well as some theoretical notions such as the large N_c expansion of QCD[2], strongly support the idea of introducing mesons other than the pion into this model. There is a considerable amount of work in the literature treating the role played by massive spin-1 mesons (the ρ- and the a_1-mesons) in low-energy lagrangians. In most of these works isovector resonances are introduced as massive Yang-Mills particles[3] or as gauge bosons of local chiral symmetry[4], and low energy phenomena such as ρ-coupling universality, the KSRF relation are then nicely described.

It should be made clear however that there is neither experimental evidence nor theoretical prejudice from QCD to support an even approximate dynamical gauge boson character of the spin-1 resonances and therefore justify among other things the conspicious emergence of Yang-Mills self-couplings of the ρ and the a_1

fields. As we stated above, the nice feature of the "gauge" models is their natural compliance with the phenomenologically successful notion of vector meson dominance, but as other authors have shown[5], this feature is not unique to the models of Refs. 3-4. It can also be obtained in cases where chiral symmetry is realised in a less exotic manner.

In our approach we construct a lagrangian consistent with general principles of quantum field theory and chiral symmetry. Vector meson dominance can be implemented later, if so desired. We assume therefore a homogeneous transformation law for isovector spin-1 fields. One of the purposes of my talk is to convince you that even in that case constraints relating three- and four-point coupling strengths do exist. These derive from demanding the hamiltonian to be bounded from below. I will extend our previous analysis of the $\pi\rho$ system[6] to the description of interacting pions, ρ- and a_1-mesons. In section 2 I introduce you to the symmetry structure of our scheme and build the basic interaction lagrangian. Section 3 is devoted to the investigation of the energies of some nonperturbative field configurations. It is shown that these energies are unbounded from below at the three-meson coupling level. In section 4 I demonstrate how the inclusion of four-point effective couplings counterbalances the dangerous contributions of the three-point terms to these energies and derive inequalities between three- and four- meson couplings for the theory to make sense. We finally discuss in section 5 how unitarity arguments based on vector dominance could lead to saturation of these inequalities and suggest a novel low-energy $\pi\rho a_1$ effective lagrangian consistent with chiral symmetry and general field theoretical principles.

2. Nonlinear realisations of chiral symmetry

Let us start with the lagrangian of the nonlinear sigma model defined in terms of the $SU(2)$ field U as:

$$\mathcal{L}_{NL\sigma} = \frac{f^2}{4} < \partial^\mu U \partial_\mu U^\dagger >, \tag{1}$$

f being the pion decay constant. We define U as $U = \exp(i\vec{\tau}.\vec{F}(x))$ with the pion field given by $\vec{F} = F\hat{F}$. Other parametrisations are perhaps more suitable for perturbative evaluations of Green' s functions, but are not as convenient for investigations of the large field region, which is of interest for our purposes.

The lagrangian (1) is invariant under the linear $SU(2)_L \otimes SU(2)_R$ group rotation $U \to g_L U g_R^\dagger$. It is also invariant under the following nonlinear rotation[7]:

$$u(\vec{F}) \;\to\; g_L u(\vec{F}) h^\dagger(\vec{F}) \;=\; h(\vec{F}) u(\vec{F}) g_R^\dagger, \tag{2}$$

where u is the square root of U and $h(\vec{F})$ is a compensating transformation ensuring that U transforms linearly. Consider now the following axial-vector and vector respectively field gradients:

$$\begin{aligned} u_\mu &= i(u^\dagger \partial_\mu u - u \partial_\mu u^\dagger) \\ \Gamma_\mu &= \frac{1}{2}(u^\dagger \partial_\mu u + u \partial_\mu u^\dagger). \end{aligned} \tag{3}$$

These gradients transform as

$$\begin{aligned} u_\mu &\rightarrow h(\vec{F}) u_\mu h^\dagger(\vec{F}) \\ \Gamma_\mu &\rightarrow h(\vec{F}) \Gamma_\mu h^\dagger(\vec{F}) + h(\vec{F}) \partial_\mu h^\dagger(\vec{F}). \end{aligned} \tag{4}$$

It is seen that while u_μ transforms homogeneously the transformation of Γ_μ contains an inhomogeneous part as a result of the field dependence of $h(\vec{F})$. How should spin-1 fields transform in this framework? Basically there are two possibilities forming a group:homogeneous or inhomogeneous. The inhomogeneous group was chosen by Weinberg[1] in its study of the $\pi\rho$ system. This approach later acquired the name of "hidden gauge symmetry"[4] approach, because the associated lowest order invariant lagrangian preserves not only chiral symmetry but also a certain sort of a gauge symmetry. It is one of my purposes to point out that this symmetry was never revealed by low energy $\pi\pi$ scattering experiments. Furthermore in the inhomogeneous approach it is impossible to define similar transformations for both the ρ and the a_1 fields, simply because the associated particles have opposite parity.

In contrast the homogeneous transformation is the simplest one consistent with chiral symmetry and it can be applied to both the ρ and the a_1 fields. We adopt this point of view and assume that the ρ- and the a_1-mesons transform as

$$\begin{aligned} V_\mu &\rightarrow h(\vec{F}) V_\mu h^\dagger(\vec{F}) \\ A_\mu &\rightarrow h(\vec{F}) A_\mu h^\dagger(\vec{F}) \end{aligned} \tag{5}$$

under the nonlinear group. In this expression $V_\mu = \vec{\tau}.\vec{V}_\mu$ and $A_\mu = \vec{\tau}.\vec{A}_\mu$. To build invariant couplings we define covariant derivatives of spin-1 fields transforming as in eq. (5)

$$\nabla_\mu = \partial_\mu + [\Gamma_\mu, \ ], \tag{6}$$

in such a way that $\nabla_\mu V_\nu$ and $\nabla_\mu A_\nu$ also transform homogeneously: $\nabla_\mu V_\nu \rightarrow h \nabla_\mu V_\nu \, h^\dagger$ and similarly $\nabla_\mu A_\nu \rightarrow h \nabla_\mu A_\nu \, h^\dagger$.

The invariant lagrangian at quadratic order in the fields is given by

$$\begin{aligned} \mathcal{L}^{(2)}_{\pi\rho a_1} = &\frac{f^2}{4} < u_\mu u^\mu > - \frac{1}{4} < V_{\mu\nu} V^{\mu\nu} > - \frac{1}{4} < A_{\mu\nu} A^{\mu\nu} > \\ &+ \frac{M_\rho^2}{2} < V_\mu V^\mu > + \frac{M_a^2}{2} < A_\mu A^\mu >, \end{aligned} \tag{7}$$

where $V_{\mu\nu} = \nabla_\mu V_\nu - \nabla_\nu V_\mu$ and $A_{\mu\nu} = \nabla_\mu A_\nu - \nabla_\nu A_\mu$ are the covariant field strengths of the spin-1 resonances. We introduce chirally invariant mass terms for the ρ- and the a_1-mesons and we assume that the coupling $c < A_\mu u^\mu >$ is not present. Actually with the choice $c = 0$ no diagonalisation of $\pi\rho a_1$ interactions is needed - obviously not a disadvantage of our framework. At the three-point level we shall consider some chirally invariant terms consistent with charge conjugation and parity invariance, containing at least one pion field gradient:

$$\mathcal{L}^{(3)}_{\pi\rho a_1} = -\frac{i}{2\sqrt{2}}\Big\{ g_1 < V_{\mu\nu}[u^\mu, u^\nu] > \; + g_2 < A_{\mu\nu}([V^\mu, u^\nu] - [V^\nu, u^\mu]) > \\ + g_3 < V_{\mu\nu}([A^\mu, u^\nu] - [A^\nu, u^\mu]) > \Big\}. \tag{8}$$

The lagrangian $\mathcal{L}^{(2)}_{\pi\rho a_1} + \mathcal{L}^{(3)}_{\pi\rho a_1}$ has six free parameters that one would ideally like to determine from QCD. But this problem seems to be elusive since a sensible method to perform such an extraction from QCD has not yet been discovered.

3. Nonperturbative pathologies

The issue I address here is rather different: assuming that g_1, g_2, g_3 are given by the underlying QCD dynamics, are there any relations between these parameters and higher order ones? Previous investigations[6] suggest that this question should be addressed in a nonperturbative framework. In particular does the theory (7, 8) yield a hamiltonian that is bounded from below? To find an answer let us investigate the hamiltonian associated with the lagrangian $\mathcal{L}^{(2)}_{\pi\rho a_1} + \mathcal{L}^{(3)}_{\pi\rho a_1}$ in terms of the canonical degrees of freedom: the fields $\vec{F}$, $\vec{V}_i$, $\vec{A}_i$ and their conjugate momenta, respectively $\vec{\phi}$, $\vec{\pi}_i, \vec{\chi}_i$. The energy can be written as a sum of two terms $H = H_T + H_V$, where the kinetic energy is H_T and the potential energy is H_V. The potential part contains only space components and in the three-point case is given by

$$H_V = \int d^3x \Big\{ \frac{f^2}{2}(u_i)_k^2 + M_a^2(A_i)_k^2 + \frac{1}{2}(A_{ij})_k \big[\, A_{ij} + i\sqrt{2}g_2([V_i, u_j] - [V_j, u_i]) \,\big]_k \\ + M_\rho^2 (V_i)_k^2 + \frac{1}{2}(V_{ij})_k \big[\, V_{ij} + i\sqrt{2}g_1[u_i, u_j] + i\sqrt{2}g_3([A_i, u_j] - [A_j, u_i]) \,\big]_k \Big\}. \tag{9}$$

The kinetic piece needs some work in order to eliminate the dependent variables $\vec{V}_0$, $\vec{A}_0$. It turns out that the generic expression for H_T in the case of a quadratic in momenta theory is:

$$H_T = \int d^3x \Big\{ \frac{1}{2}\vec{\Phi}\mathcal{A}^{-1}\vec{\Phi} + \frac{\vec{\pi}_i^2}{4} + \frac{\vec{\chi}_i^2}{4} + \frac{1}{2}\vec{\Gamma}\mathcal{P}^{-1}\vec{\Gamma} \Big\}, \tag{10}$$

where $\vec{\Phi}$, $\vec{\Gamma}$ are linearly related to the momenta $\vec{\phi}$, $\vec{\pi}_i$, $\vec{\chi}_i$ and $\mathcal{A}$, $\mathcal{P}$ are isospin tensor functions of $\vec{F}$, $\vec{V}_i$, $\vec{A}_i$, which depend on the detailed structure of the dynamics.

Let us now investigate the energy of a classical meson mapping. The simplest such object one can imagine has an isospin content specified by a constant unit vector $\hat{F}$:

$$\vec{F}_0(\vec{x}) = F(\vec{x})\ \hat{F}, \tag{11}$$

where $F(\vec{x})$ is a regular function of space. Such a configuration is topologically trivial. For the vector and the axial vector fields I will assume that they are parallel to the pion field:

$$\begin{aligned} \vec{V}_i(\vec{x}) =& V_i(\vec{x})\ \hat{F} \\ \vec{A}_i(\vec{x}) =& A_i(\vec{x})\ \hat{F}. \end{aligned} \tag{12}$$

The special form of our ansatz result in a potential energy which is simply given by:

$$H_V = \int d^3x \left\{ \frac{f^2}{2}(\partial_i F)^2 + M_a^2 A_i^2 + \frac{1}{2}(\partial_i A_j - \partial_j A_i)^2 + M_\rho^2 V_i^2 + \frac{1}{2}(\partial_i V_j - \partial_j V_i)^2 \right\}, \tag{13}$$

the potential energy of a free theory, an obviously positive quantity. We therefore concentrate in the kinetic energy of the theory as given by H_T. Because of the particular isospin structure we consider here only momenta that point in a direction perpendicular to that of the pion actually "see" the couplings to the vector mesons. We assume:

$$\begin{aligned} \vec{\phi} =& \phi(\vec{x})\ \hat{\phi} \\ \vec{\pi}_i =& \pi_i(\vec{x})\ \hat{F} \\ \vec{\chi}_i =& \chi_i(\vec{x})\ \hat{F}, \end{aligned} \tag{14}$$

with $\hat{\phi} \cdot \hat{F} = 0$. The kinetic energy of this field configuration reads

$$H_T = \int d^3x \left\{ \frac{\phi^2}{2f^2 s^2 \mathcal{I}} + \frac{1}{4}\left[\pi_i^2 + \chi_i^2 + \frac{(\partial_i \pi_i)^2}{M_\rho^2} + \frac{(\partial_i \chi_i)^2}{M_a^2} \right] \right\}, \tag{15}$$

where s is a shorthand notation for $\sin F/F$. Observe now the structure of the dimensionless "inertial" parameter $\mathcal{I}$ which contains all the nontrivial effects due to the spin-1 fields:

$$\begin{aligned} \mathcal{I} = \frac{1}{f^4 M_1 M_2} \Big(& f^4 M_1 M_2 - 8(g_2^2 V_i^2 M_2 M_\rho^2 + (g_1 \partial_i F - g_3 A_i)^2 M_1 M_a^2) \\ & + 16\left[g_2^4 M_2 (V_i^2 (\partial_i F)^2 - (V_i \partial_i F)^2) + g_3^4 M_1 (A_i^2 (\partial_i F)^2 - (A_i \partial_i F)^2) \right] \Big), \end{aligned} \tag{16}$$

with $M_1 = (1/f^2)\left[2M_\rho^2 - 4g_2^2(\partial_i F)^2\right]$ and $M_2 = (1/f^2)\left[2M_a^2 - 4g_3^2(\partial_i F)^2\right]$. While in the case of the nonlinear sigma model with vanishing couplings $\mathcal{I}$ is simply equal to

1, here it acquires negative contributions. For small fields $F,\ V_i,\ A_i \approx 0$ appropriate to perturbation theory one has $\mathcal{I} \approx 1$ so no problem seems to appear in perturbative expansions of scattering amplitudes. For nonperturbative configurations however the situation changes dramatically since then negative contributions proportional to quadratic powers of the couplings can drive $\mathcal{I}$ to zero or negative values. The hamiltonian density acquires poles and the energy is not bounded from below. Such troubles can arise at fairly low energy scales. Consider a localised meson wave carrying momentum k_i and of amplitude $F \approx 1$. Assume to simplify further that all classical fields vanish except F and ϕ. The gradient $\partial_i F$ is roughly approximated by k_i and $\mathcal{I}$ inside the meson wave looks like:

$$\mathcal{I} \approx \frac{1 - 2\left(\dfrac{g_3^2}{M_a^2} + 2\dfrac{g_1^2}{f^2}\right)k^2}{1 - 2\dfrac{g_3^2}{M_a^2}k^2}. \tag{17}$$

At small or very large momenta k the inertial parameter is positive but for k^2 in the intermediate range

$$\left(2\frac{g_3^2}{M_a^2} + 4\frac{g_1^2}{f^2}\right)^{-1} < k^2 < \frac{M_a^2}{2g_3^2} \tag{18}$$

$\mathcal{I}$ becomes negative and as a consequence the kinetic energy density is negative, making the theory ill defined in these regions. Taking reasonable numerical values for the coupling constants[8], the region of dangerous momenta is found to be 0.4 GeV $< k < 2.0$ GeV, which includes the range of masses of the ρ and the a_1 resonances. And this is precisely the range that one would like to describe by extending the low-energy effective theories to include spin-1 mesons !

We conclude that the hamiltonian associated with the simplest three-point $\pi\rho a_1$ interactions *is not bounded from below* which is of course unacceptable.

4. Constraints on four-meson coupling strengths

Let us now consider the effect of some four-meson couplings that are relevant to our analysis:

$$\mathcal{L}^{(4)}_{\pi\rho a_1} = \frac{1}{8}\Big\{ g_4 < [u_\mu, u_\nu]^2 > +2g_5 < [u_\mu, u_\nu][A^\mu, u^\nu] > +2g_6\big(< [V_\mu, u_\nu]^2 > \\ - < [V_\mu, u_\nu][V^\nu, u^\mu] > \big) + 2g_7\big(< [A_\mu, u_\nu]^2 > - < [A_\mu, u_\nu][A^\nu, u^\mu] > \big)\Big\}. \tag{19}$$

We introduce four new coupling constants $g_4,\ g_5,\ g_6,\ g_7$. Amongst these terms one can recognise a local four-point pion vertex, the so-called "Skyrme term", as well as a term contributing to the decay $a_1 \to \pi\pi\pi$.

Concerning now the energy of the charge-zero meson configuration defined in the previous section, we note first that its potential energy is unaffected by the new couplings and is still given by eq. (13). The kinetic piece has the same form as in eq. (15) but with a new inertial function, $\tilde{\mathcal{I}}$. After a tedious but straightforward calculation one finds:

$$\begin{aligned}\tilde{\mathcal{I}} = \frac{1}{f^4\tilde{M}_1\tilde{M}_2}\Big\{ & f^4\tilde{M}_1\tilde{M}_2 - 8(g_2^2-g_6)V_i^2M_\rho^2\tilde{M}_2 \\ & + 16\Big[(g_2^2-g_6)^2\big(V_i^2(\partial_iF)^2-(\partial_iFV_i)^2\big)\Big]\tilde{M}_2 \\ & - 8\Big[(g_1^2-g_4)(\partial_iF)^2+(g_3^2-g_7)A_i^2-2\left(g_3g_1-\frac{g_5}{2}\right)(\partial_iFA_i)\Big]M_a^2\tilde{M}_1 \\ & + 16\Big[\Big((g_1^2-g_4)(g_3^2-g_7)-\left(g_3g_1-\frac{g_5}{2}\right)^2\Big)(\partial_iF)^4 \\ & \quad + (g_3^2-g_7)^2\big(A_i^2(\partial_iF)^2-(A_i\partial_iF)^2\big)\Big]\tilde{M}_1\Big\}, \end{aligned} \tag{20}$$

with $\tilde{M}_1=(1/f^2)\left[2M_\rho^2-4(g_2^2-g_6)(\partial_iF)^2\right]$ and $\tilde{M}_2=(1/f^2)\left[2M_a^2-4(g_3^2-g_7)(\partial_iF)^2\right]$. By requiring that for any value of the classical profiles ∂_iF, V_i, A_i the function $\tilde{\mathcal{I}}$ is non-negative, we find constraints *on the couplings*. Consider the following simplifying cases:

$$\begin{aligned} &a)\quad \partial_iF=A_i=0 \quad\Rightarrow\quad \tilde{\mathcal{I}}_a = 1-\frac{4}{f^2}(g_2^2-g_6)V_i^2 \\ &b)\quad \partial_iF=V_i=0 \quad\Rightarrow\quad \tilde{\mathcal{I}}_b = 1-\frac{4}{f^2}(g_3^2-g_7)A_i^2 \\ &c)\quad V_i=A_i=0 \quad\Rightarrow\quad \tilde{\mathcal{I}}_c = \frac{1}{f^2\tilde{M}_2}\Big[2M_a^2-\big[4(g_3^2-g_7)+8(g_1^2-g_4)\frac{M_a^2}{f^2}\big](\partial_iF)^2 \\ &\qquad\qquad + \frac{16}{f^2}\big[(g_1^2-g_4)(g_3^2-g_7)-(g_3g_1-\frac{g_5}{2})^2\big](\partial_iF)^4\Big]. \end{aligned} \tag{21}$$

Requiring positiveness of $\tilde{\mathcal{I}}_{a,b,c}$ for all possible values of the fields leads to:

$$\begin{aligned} g_4 &\geq g_1^2 \\ g_6 &\geq g_2^2 \\ g_7 &\geq g_3^2 \\ (g_1^2-g_4)(g_3^2-g_7) &\geq (g_3g_1-\frac{g_5}{2})^2. \end{aligned} \tag{22}$$

These conditions show that the Skyrme term and other four-point interactions are essential if the hamiltonian is to be bounded from below. In an effective theory where the spin-1 fields transform homogeneously they arise as counterterms for the bad behaviour of the vector-meson contributions. This is in sharp contrast to other approaches[3,4], where the same Skyrme term emerges from the exchange of a very heavy ρ-meson.

5. Discussion

Our investigation of classical nonperturbative effects in low-energy chiral theories shows that the constraints (22), relating three- and four-point couplings, must be satisfied for a consistent description of the interactions between pions and spin-1 isovector mesons. We stress that chiral symmetry is implemented nonlinearly in this approach and the vector mesons are naturally assumed to transform homogeneously under chiral rotations. The constraints arise from demanding the hamiltonian to be bounded from below. They do not depend on phenomenological ideas such as vector dominance.

One might ask now whether there are any other constraints on the couplings from first principles. For instance another nonperturbative notion that one could invoke in this context is the unitarity of the scattering matrix. This was previously studied[9] in the special case of the lagrangian (7, 8) without the a_1 ($g_2 = g_3 = 0$). Working at tree-level it was found that further local pion interactions must be added by hand if the forward elastic $\pi\pi$ scattering amplitude is to obey the Froissart bound[10]. These local interactions compensate for the most divergent contribution produced by ρ-exchange. The result, in the SU(2) sector is:

$$\mathcal{L}_{local}^{SU(2)} = \frac{g_1^2}{8} < [u_\mu, u_\nu]^2 > . \tag{23}$$

This is just the Skyrme term, with a coefficient that is fixed by the three-point coupling g_1. If one works at tree level, the a_1 does not contribute to $\pi\pi$ scattering. Imposing unitarity therefore leads to saturation of the lower bound on g_4 in (22):

$$g_4 - g_1^2 = 0. \tag{24}$$

Combining this with the final constraint in (22), we obtain a relation expressing the implications of unitarity for the couplings of the *axial* meson:

$$g_5 = 2g_1 g_3. \tag{25}$$

This is nontrivially relating the strength of the $a_1 \to \pi\pi\pi$ decay to those of the processes $\rho \to \pi\pi$ and $a_1 \to \rho\pi$.

The saturation of two of our constraints follows from the assumption that an extreme version of vector dominance holds for strong interactions. In fact the value for g_4 determined assuming ρ-meson dominance agrees well with that from chiral perturbation theory[11], suggesting that at least at low energies, vector dominance is really making sense. We speculate that dominance of a single resonance may also hold in the axial-vector channel, leading to saturation of the remaining constraints in (24). In this case our lagrangian would simplify to:

$$\begin{aligned} \mathcal{L}_{\pi\rho a_1} =& \frac{f^2}{4} < u_\mu u^\mu > + \frac{M_\rho^2}{2} < V_\mu V^\mu > + \frac{M_a^2}{2} < A_\mu A^\mu > \\ & - \frac{1}{4} < \left(V_{\mu\nu} + \frac{i}{\sqrt{2}} (g_1 [u_\mu, u_\nu] + g_3 ([A_\mu, u_\nu] - [A_\nu, u_\mu])) \right)^2 > \\ & - \frac{1}{4} < \left(A_{\mu\nu} + \frac{i}{\sqrt{2}} g_2 ([V_\mu, u_\nu] - [V_\nu, u_\mu]) \right)^2 > . \end{aligned} \tag{26}$$

This constitutes an effective lagrangian describing the strong interactions of $\pi\rho a_1$ mesons with a minimal number of free coupling constants. It is the simplest one compatible with chiral symmetry and leading to a hamiltonian which is free of pathologies.

To summarise: in our framework theory chiral symmetry is implemented in the simplest possible way and no speculative gauge symmetry assumption is made. Constraints between couplings are there to ensure that the hamiltonian is bounded from below, and vector meson dominance can be implemented by specific choices of parameters. It is therefore most natural to regard our lagrangian (26) as the starting point for any extension of chiral perturbation theory of pseudoscalar pions to the resonance region.

6. Acknowledgements

This work was done in collaboration with Mike Birse. I would like to thank him for his contribution to my understanding of low energy phenomena. I gratefully acknowledge financial support from the SERC.

7. References

1. S. Weinberg, *Phys. Rev.* **166** (1968) 1568; I. Gerstein, R. Jackiw, B. Lee and S. Weinberg, *Phys. Rev.* **D3** (1971) 2486; A. Salam and J. Strathdee, *Phys. Rev.* **D2** (1970) 2869
2. G. 't Hooft, *Nucl. Phys.* **B79** (1974) 276
3. J. J. Sakurai, *Currents and Mesons*, University of Chicago Press, Chicago 1969; J. Schwinger, *Particles and Sources*, Clarendon Press, Oxford, UK, 1969
4. M. Bando, T. Kugo and K. Yamawaki, *Phys. Rep.* **164** (1988) 217

5. G. Ecker, J. Gasser, A. Pich and E. de Rafael, *Nucl. Phys.* **B321** (1989) 311
6. D. Kalafatis, *Phys. Lett.* **B313** (1993) 115
7. S. Coleman, J. Wess and B. Zumino, *Phys. Rev.* **177** (1969) 2239; C. G. Callan, S. Coleman, J. Wess and B. Zumino, *ibid* 2247
8. B. Moussallam, private communication
9. G. Ecker, J. Gasser, H. Leutwyler, A. Pich and E. de Rafael, *Phys. Lett.* **B223** (1989) 425
10. M. Froissart, *Phys. Rev.* **123** (1961) 1053; A. Martin, *Phys. Rev.* **129** (1963) 1432
11. J. Gasser and H. Leutwyler, *Ann. Phys.* (NY) **158** (1984) 142

GENERALIZED CHIRAL PERTURBATION THEORY

M. KNECHT
Division de Physique Théorique [1], Institut de Physique Nucléaire
F-91406 Orsay Cedex, France

ABSTRACT

A generalization of the standard χPT expansion scheme is presented, where the quark mass ratio $m_s/\hat{m}$ is not fixed *a priori*, but remains a free quantity, to be determined experimentally, along with the other low energy constants of χPT. Theoretical and phenomenological arguments supporting the necessity to consider such a generalized χPT are put forward, and the possibilities to extract the value of $m_s/\hat{m}$ from future high precision low energy experiments at Daϕne and at CERN are discussed.

1. Introduction

Up to now, very little is known about the chiral structure of the QCD ground state. The fact that in the limit of vanishing quark masses $m_u = m_d = m_s = 0$, the chiral symmetry of QCD is spontaneously broken down to $SU_V(3)$ just ensures the existence of 8 Goldstone bosons coupled to 8 conserved axial currents. The strength of this coupling defines a mass scale $F_0 = F_\pi|_{m_u=m_d=m_s=0} \simeq 90$ MeV, which is characteristic of spontaneous chiral symmetry breaking (χSB). Indeed, F_0 is a long range order parameter which plays a fundamental role : a non-zero value of F_0 not only signals a broken symmetry phase but, moreover, it is necessary for the spontaneous breakdown to occur. The fact that F_0 is about ten times smaller than the typical mass-scale $\Lambda_H \sim 1$ GeV of lightest massive bound states $(\rho, N, \cdots)$ remains unexplained and it raises a question: What is the natural size of other order parameters, such as the quark condensate B_0, with $-F_0^2 B_0 =< \bar{u}u >_0 =< \bar{d}d >_0 =< \bar{s}s >_0$, which describe the chiral structure of the massless QCD vacuum ? Generalized Chiral Perturbation Theory (GχPT) [1] is motivated by the observation that nothing from our present theoretical or experimental knowledge allows us to answer the above question in one way or the other : B_0 could be as large as the bound state scale $\Lambda_H \sim 1$ GeV or as small as the fundamental order parameter $F_0 \sim 90$ MeV. GχPT provides a systematic extension of the standard χPT [4] and does not assume *a priori* that $B_0 \sim \Lambda_H$.

[1] Unité de Recherche des Universités Paris XI et Paris VI associée au CNRS.

2. Expansion of Goldstone Boson Masses

If quark masses are switched on, Goldstone bosons acquire a mass. Since in the real world m_u, m_d, $m_s \ll \Lambda_H$, the Goldstone boson masses can be expanded in powers of quark masses.

If, for a given value of $B_0 \neq 0$, quark masses are sufficiently small, the linear term in this expansion dominates. Otherwise, the prominent role is played by a different order parameter A_0, defined by a two point correlator of the scalar and pseudoscalar densities, $S^a(=\bar{\psi}(\lambda^a/2)\psi)$ and $P^a(=\bar{\psi}(\lambda^a/2)i\gamma_5\psi)$:

$$\frac{i}{F_0^2}\int dx\, e^{iqx} < 0\,|T\left\{S^a(x)S^b(0) - P^a(x)P^b(0)\right\}|\,0 >= \delta^{ab}\Pi_{SP}(q^2) \tag{1}$$

where $a, b = 1\cdots 8$, and $|\,0>$ denotes the ground state of the massless QCD. The small q^2 behaviour of the two point function (1) is dictated by a chiral Ward identity (WI),

$$\Pi_{SP}(q^2) = \frac{B_0^2}{q^2} + A_0(\mu) + 5B_0^2/96\pi^2F_0^2\ln\left(\mu^2/-q^2\right) + O(q^2)\ , \tag{2}$$

where the μ dependence of A_0 compensates the scale dependence of the logarithm. A_0 is also an order parameter of χSB: The operator in (1) does not contain the singlet representation of the chiral symmetry group.

In order to obtain the expansion of Goldstone boson masses explicitly, it is sufficient to consider for $m_q \neq 0$ the two-point function of the divergence of the axial current at zero momentum transfer, where it is related by a well known WI to the quark condensate. Separating in this WI the pseudoscalar meson contribution and expanding the remainder in powers of m_q, one obtains ($m_u = m_d = \widehat{m}$ for simplicity):

$$\begin{aligned} M_\pi^2 &= 2\widehat{m}B(\mu) + 4\widehat{m}^2A_0(\mu) + \delta M_\pi^2(\mu) \\ M_K^2 &= (m_s+\widehat{m})B(\mu) + (m_s+\widehat{m})^2A_0(\mu) + F_K^2\delta M_\pi^2(\mu)\ . \end{aligned} \tag{3}$$

$B(\mu)$ is related to the vacuum condensate B_0 by $B(\mu) = B_0 + 2(m_s + 2\widehat{m})Z_0^S(\mu)$ and Z_0^S is defined in terms of the low energy behaviour of the two point function $< S^0S^0 - S^8S^8 >$. Since it violates the Zweig rule, one expects $\zeta \equiv Z_0^S/A_0 \ll 1$. Like B_0 and A_0, Z_0^S is an order parameter of χSB. Finally, δM_π^2 and δM_K^2 contain chiral logarithms, which compensate for the scale dependences of A_0 and Z_0^S, and higher powers of quark masses. A few comments are in order:

i) The overall convergence of the expansions (3) is controlled by the small parameter

$$m_q/\Lambda_H\ ,\ \Lambda_H \sim 1\text{GeV}\ . \tag{4}$$

This statement becomes particularly transparent in the large N_c limit, in which the chiral logarithms drop out. In general, the coefficient of the n-th power of quark masses is defined by means of the low momentum behaviour of a n-point function of S^a and P^a, cf. A_0 in (2). These n-point functions satisfy superconvergent dispersion relations (they are order parameters) which can be saturated by the lowest massive bound states of masses $\sim \Lambda_H \sim 1$ GeV. This leads to the estimate that the coefficient of m_q^n $(n \geq 2)$ in (3) should indeed be of the order Λ_H^{2-n} times a factor of order unity. Notice that quark confinement makes it difficult to extend the above estimate to the quark condensate B_0 itself.

ii) The relative importance of the linear term in (3) is controled by another parameter,

$$m_q/m_0 \ , \quad m_0 = B_0/2A_0 \ , \tag{5}$$

which indicates how small the quark masses should actually be such as to ensure the validity of the Gell-Mann–Oakes–Renner formula [5]. If $m_s \sim m_0$, the contributions of the first and second order terms in the expansion of M_K^2 are comparable, and if even m_u, $m_d \sim m_0$, the same would happen for the pion mass. There is no compelling reason why m_0 should be as large as Λ_H and there is nothing unnatural in having $m_q \sim m_0 \ll \Lambda_H$. As already pointed out, B_0 might be as small as 100 MeV $\sim F_0$, and an estimate of A_0 along the lines described above gives $A_0|_{1 \text{ GeV}} \simeq 1 \div 5$. Hence, the $B_0 \sim F_0$ alternative for the QCD vacuum is perfectly consistent with a low value of $m_0 = (10 \div 50)$ MeV, indicating the possibility of a strong violation of the Gell-Mann–Oakes–Renner formula.

iii) The parameter which actually controls the relative size of the first and second order terms in (3) is **the quark mass ratio** $r = 2m_s/(m_u + m_d) = m_s/\widehat{m}$. It is well known that provided the first term dominates, r is given by the ratio of M_K^2 and M_π^2, $r \simeq 26$ [5, 6, 7]. To the extent that B_0 decreases, the ratio r becomes smaller, until it reaches a critical value r_1^* corresponding to vanishing B_0. For r below this critical value, the vacuum would become unstable with respect to small perturbations produced by quark masses. Similarly, as A_0 decreases, r becomes larger until it reaches a critical value r_2^* where A_0 vanishes. The QCD vacuum is likely to be in the phase in which the order parameters $m_q B_0$, A_0 and Z_0^S are all non-negative. This phase is characterized in particular by $r_1^* \leq r \leq r_2^*$, $\zeta \geq 0$. The values of r_1^* and r_2^* may be obtained from (3), $r_1^* = r_1 + O(m_q)$, $r_2^* = r_2 + O(m_q)$, where the leading order values r_1 and r_2 are known [8]: $r_1 = 2\,M_K/M_\pi - 1 \simeq 6.3$, $r_2 = 2\,M_K^2/M_\pi^2 - 1 = 25.9$. The corrections push both r_1^* and r_2^* upwards: r_1^* can reach the value $8 \div 9$, whereas r_2^* can actually be as large as 38.

In GχPT, $r = m_s/\widehat{m}$ is a free parameter, to be determined experimentally. The evaluation of the Dashen-Weinstein sum rule for the deviations from the Goldberger-Treiman relations, yields a first order measurement of $m_s/\widehat{m}$ [8]. The result is very sensitive to the precise values of the strong coupling constants $g_{\pi NN}$, $g_{K\Lambda N}$ and $g_{K\Sigma N}$. Present values lead to the bound $r \leq 10.6 \pm 4.2$. A new and more precise determination of the hyperon coupling constants at, e.g., DaΦne, would be welcome.

3. Expansion of the Effective Lagrangian

The above remarks do not affect the construction of the low energy effective Lagrangian [9]. Its form is merely dictated by the chiral symmetry [10] and by the transformation properties of the symmetry breaking quark mass term. $\mathcal{L}^{eff}$ is a function of 8 Goldstone boson fields and of external sources. The formalism and notation used here are standard [4], unless otherwise stated. $\mathcal{L}^{eff}$ consists of an infinite tower of invariants, $\mathcal{L}^{eff} = \sum_{(k,l)} \mathcal{L}_{(k,l)}$, where $\mathcal{L}_{(k,l)}$ contains k powers of covariant derivatives and l powers of scalar or pseudoscalar sources. In the low energy limit, $\mathcal{L}_{(k,l)} \sim \mathrm{p}^k m_q^l$, where p represents external momenta. Chiral perturbation theory is an expansion of $\mathcal{L}^{eff}$ in powers of the pion (kaon) mass assuming that all external momenta are of that size. For sufficiently small quark masses, such that both $m_q \ll \Lambda_H$ **and** $m_q \ll m_0$ hold, one has $\mathcal{L}_{(k,l)} = O(\mathrm{p}^{k+2l})$. In this case, one can write

$$\mathcal{L}^{eff} = \mathcal{L}^{(2)} + \mathcal{L}^{(4)} + \mathcal{L}^{(6)} + \cdots \tag{6}$$

where [4] $\mathcal{L}^{(d)} = \sum_{k+2l=d} \mathcal{L}_{(k,l)}$. This expansion defines the standard χPT. If for actual values of quark masses one has instead $m_q \sim m_0 \ll \Lambda_H$, both m_q and B_0 should count as parameters of the size of the pion mass and, consequently, $\mathcal{L}_{(k,l)} = O(\mathrm{p}^{k+l})$. This new counting yields a different expansion of $\mathcal{L}^{eff}$,

$$\mathcal{L}^{eff} = \tilde{\mathcal{L}}^{(2)} + \tilde{\mathcal{L}}^{(3)} + \tilde{\mathcal{L}}^{(4)} + \tilde{\mathcal{L}}^{(5)} + \tilde{\mathcal{L}}^{(6)} + \cdots \tag{7}$$

where [1] $\tilde{\mathcal{L}}^{(d)} = \sum_{k+l+n=d} B_0^n \mathcal{L}_{(k,l)}$. It should be stressed that Eqs. (6) and (7) represent two different expansions of the *same* effective Lagrangian. To all orders they are identical, at a given finite order they may differ.

It is straightforward to write down the most general expression [1] of $\tilde{\mathcal{L}}^{(2)}$ which defines the leading $O(\mathrm{p}^2)$ order of GχPT :

$$\begin{aligned} \tilde{\mathcal{L}}^{(2)} &= \frac{1}{4} F_0^2 \left\{ \langle D_\mu U^+ D^\mu U \rangle + 2B_0 \langle U^+ \chi + \chi^+ U \rangle + A_0 \langle (U^+ \chi)^2 + (\chi^+ U)^2 \rangle \right. \\ &\quad \left. + Z_0^S \langle U^+ \chi + \chi^+ U \rangle^2 + Z_0^P \langle U^+ \chi - \chi^+ U \rangle^2 + H_0 \langle \chi^+ \chi \rangle \right\} . \end{aligned} \tag{8}$$

χ collects the scalar and pseudoscalar sources, $\chi = s + ip = \mathcal{M} + \cdots$, $\mathcal{M} = \mathrm{diag}(m_u, m_d, m_s)$. Notice the absence of the factor $2B_0$, which appears in the standard definition of χ [4]. Compared to $\mathcal{L}^{(2)}$, Eq. (8) contains additional terms. The constants A_0 and Z_0^S are the same as introduced before (Z_0^P appears only in the expansion of the M_η): Eq. (3) is indeed a straightforward consequence of the Lagrangian (8). The fact that the A_0, Z_0^S and Z_0^P terms now appear at the same order $O(\mathrm{p}^2)$ as the B_0 term reflects the possibility that in (3) the first and second order terms are of comparable size. To order $O(\mathrm{p}^2)$, the low energy constants in (8) can be expressed in terms of physical masses M_π, M_K, M_η, of the quark mass ratio $r = m_s/\widehat{m}$ and of the

Zweig rule violating parameter ζ

$$2\widehat{m}B_0 = \{1 - \epsilon[1 + \zeta(2 + r)]\} M_\pi^2 \quad , \quad 4\widehat{m}^2 A_0 = \epsilon M_\pi^2 \quad ,$$
$$8\widehat{m}^2 Z_0^P = \Delta_{GMO}/(r-1)^2 - \epsilon M_\pi^2 \quad , \quad 4\widehat{m}^2 Z_0^S = \epsilon\zeta M_\pi^2 \quad , \tag{9}$$

where $\Delta_{GMO} \equiv (3M_\eta^2 - 4M_K^2 + M_\pi^2)$ and $\epsilon = 2(r_2 - r)/(r^2 - 1)$. For particular values of the parameters, $r = r_2$, $\Delta_{GMO} = 0$, Eqs. (9) imply $A_0 = Z_0^S = Z_0^P = 0$ and $2\widehat{m}B_0 = M_\pi^2$, i.e. one recovers the standard $O(\mathrm{p}^2)$ Lagrangian $\mathcal{L}^{(2)}$. Order by order, GχPT contains the standard χPT as a special case.

The next to the leading term of the expansion (7) is of odd order $O(\mathrm{p}^3)$, which is a new feature, absent in the standard χPT. One has [1]

$$\begin{aligned}\tilde{\mathcal{L}}^{(3)} &= \frac{1}{4}F_0^2\left\{\xi\langle D_\mu U^+ D^\mu U(\chi^+ U + U^+\chi)\rangle + \tilde{\xi}\langle D_\mu U^+ D^\mu U\rangle\langle\chi^+ U + U^+\chi\rangle\right. \\ &\quad \left. +\rho_1\langle(\chi^+ U)^3 + (U^+\chi)^3\rangle + \rho_2\langle(\chi^+ U + U^+\chi)\chi^+\chi\rangle + \cdots\right\} \quad , \end{aligned}\tag{10}$$

where the ellipsis denotes Zweig rule violating terms of $\mathcal{L}_{(0,3)}$. The main physical effect described by the Lagrangian $\tilde{\mathcal{L}}^{(3)}$ is the splitting of the decay constants F_π, F_K, and F_η, which gives $\widehat{m}\xi = (F_K^2/F_\pi^2 - 1)/(r-1)$. The $\mathcal{L}_{(0,3)}$ part of $\tilde{\mathcal{L}}^{(3)}$, described by the constants ρ_1, ρ_2, $\cdots$, generates an $O(m_q^3/\Lambda_H)$ contribution to the pseudoscalar masses. In the standard χPT these terms would count as $O(\mathrm{p}^6)$.

$\tilde{\mathcal{L}}^{(3)}$ provides the simplest example of odd chiral orders characteristic of GχPT. They do not correspond to an increase in the number of loops, but to additional corrections in powers of the quark masses. In the standard χPT [4], the splitting in the decay constants is a $O(M_\pi^2)$ effect arising from loops and from the corresponding counterterms in $\mathcal{L}^{(4)}$. Here, the leading contribution comes from $\widehat{m}\xi$, and the loop effects only show up at order $O(\mathrm{p}^4)$.

The $O(\mathrm{p}^4)$ Lagrangian $\tilde{\mathcal{L}}^{(4)}$ consists of several components : $\tilde{\mathcal{L}}^{(4)} = \mathcal{L}_{(4,0)} + \mathcal{L}_{(2,2)} + \mathcal{L}_{(0,4)} + B_0^2\mathcal{L}'_{(0,2)} + B_0\mathcal{L}'_{(2,1)} + B_0\mathcal{L}'_{(0,3)}$. $\mathcal{L}_{(4,0)}$ consists of terms with four derivatives and no quark masses: it is given by the five terms described by the low energy constants L_1, L_2, L_3, L_9 and L_{10} of the standard $\mathcal{L}^{(4)}$ [4].

$\mathcal{L}_{(2,2)}$ and $\mathcal{L}_{(0,4)}$ are new terms, which in the standard χPT would count as $O(\mathrm{p}^6)$ and $O(\mathrm{p}^8)$, respectively. They involve about 20 independent terms each. An experimental determination of all the corresponding low energy constants is obviously hard to imagine. However, a few particular combinations of these constants contribute, for instance, to K_{l3} and K_{l4} decays, or to the $\pi\pi$ scattering amplitude at the one loop level. The last three terms in $\tilde{\mathcal{L}}^{(4)}$ represent B_0-dependent counterterms of order $O(\mathrm{p}^4)$ which are needed to renormalize the divergences of the constants of $\tilde{\mathcal{L}}^2 + \tilde{\mathcal{L}}^3$ that arise from the $O(\mathrm{p}^4)$ loops: In generalized χPT, renormalization proceeds order by order in the expansion in powers of B_0.

4. Differences between Standard and Generalized χPT and Experimental Tests

The difference between standard and generalized χPT merely concerns the explicit symmetry breaking sector of the theory. In the chiral limit $m_q = 0$, observables described by the purely derivative terms $\mathcal{L}_{(k,0)}$ of $\mathcal{L}^{eff}$ are essentially given by soft pion theorems which are identical in both schemes. This concerns, in particular, the electromagnetic radius of the pion, the decay $\pi \to e\nu\gamma$, the $\pi\pi$ scattering in the P-wave, the leading order K_{l3} and K_{e4} form-factors, etc... The difficulty of disentangling both schemes resides in the fact that symmetry breaking effects are small and not well known experimentally.

As an example among many (see Ref. [11]), let us consider the χPT prediction for the low Q^2 behavior of the spectral function associated with the divergence of the axial current,

$$\rho(Q^2) = \frac{1}{2\pi}\sum_n (2\pi)^4\delta^4(Q-P_n)\left|< n|\partial^\mu(\bar{d}\gamma_\mu\gamma_5 u)|0>\right|^2 . \tag{11}$$

At low Q^2, the continuum part of $\rho(Q^2)$ is dominated by the contribution of 3π intermediate states. Using for the latter the soft pion theorem (and neglecting in the phase space integral (11) the pion mass) one gets the standard result [12]

$$\rho_{3\pi}(Q^2) = \frac{1}{768\pi^4}\frac{M_\pi^4}{F_\pi^2}\,Q^2 + \cdots \tag{12}$$

Within the $O(p^2)$ GχPT, this result is considerably modified. Still neglecting the pion mass in the phase space integral (in order to facilitate a comparison with the standard result) one gets [13]

$$\rho_{3\pi}(Q^2) = \frac{1}{768\pi^4}\frac{M_\pi^4}{F_\pi^2}\,Q^2\left\{1 + 10\,\frac{r_2 - r}{r^2-1} + 30\left(\frac{r_2-r}{r^2-1}\right)^2\right\} + \cdots \tag{13}$$

As r decreases from r_2 down to r_1, the enhancement factor in Eq. (13) increases from 1 to 13.5. This enhancement would considerably affect the existing estimates of $\widehat{m}$ using the QCD sum rules [12]. In GχPT, the decrease of the quark mass ratio r from $r \sim 25$ to $r \sim 10$ is likely to be interpreted as an increase in the value of the running quark mass $\widehat{m}(\mu)$ by a factor of $2 \div 3$. The order of magnitude of the mass differences $m_s - \widehat{m}$ and $m_d - m_u$ should remain essentially unchanged. In this connection, it should be stressed that all existing estimates of $\widehat{m}(\mu)$ use in one way or the other the assumption $r \sim 25$. This is obvious for those approaches which deduce the value of $\widehat{m}(\mu)$ from the estimates of $m_s - \widehat{m}$ [6, 7]. It is however even true in the case of direct quantitative determinations of $\widehat{m}(\mu)$ from QCD sum rules [12]. The latter express the square of $\widehat{m}(\mu)$ as a weighted integral of the spectral function (43). Nothing is known experimentally about $\rho(Q^2)$ beyond the one pion intermediate state contribution. The only existing attempt to fill this gap makes use of the low Q^2 behaviour (44) of $\rho_{3\pi}(Q^2)$ as given by the standard χPT in order to normalize the whole $\rho_{3\pi}(Q^2)$ contribution represented by a broad π' Breit-Wigner peak. In this way, the result $\widehat{m}(1\text{GeV}) = (7 \pm 1)\text{MeV}$ is obtained [12]. If, instead, the 3π contribution is normalized using the GχPT result (45), the above value of $\widehat{m}(\mu)$ is increased by a factor $2 \div 3$. (Let us note in passing that within GχPT, the possibility of having $m_u = 0$ appears even less likely than in the standard case [14].)

In general, GχPT admits and expects a larger absolute strength of the divergence of the $\Delta S = 0$ axial current away from the pion pole than the standard scheme could possibly tolerate. The physical origin of this increase would be ascribed to the importance of the "π' contribution", which can and should be checked experimentally: The $\rho_{3\pi}(Q^2)$ component of the spectral function (43) can be measured in high statistics $\tau \to 3\pi\nu_\tau$ decay experiments [13] at a future Tau-Charm Factory. In this way, the determination of $\widehat{m}(\mu)$ could be put on a solid experimental basis.

Another important difference, also of direct experimental relevance, between standard χPT and GχPT concerns the low energy $\pi\pi$ [1] and πK [2] scattering. The $\pi\pi$ amplitude predicted by the GχPT lagrangian $\tilde{\mathcal{L}}^{(2)}$ reads $A_{lead}(s|t,u) = (s - 2\widehat{m}\tilde{B})/F_0^2$ where $\tilde{B} = B_0 + 2m_s Z_0^S$ is the non strange quark condensate in the SU(2)$\times$SU(2) chiral limit. This expression for $A_{lead}(s|t,u)$ holds both in the standard case and in GχPT. In the standard case, however, $2\widehat{m}\tilde{B} \sim M_\pi^2$, and one recovers the well known Weinberg formula [15]. In GχPT, at order $O(\text{p}^2)$, $2\widehat{m}\tilde{B}$ can be considerably smaller than M_π^2 by an amount which depends on r. For r decreasing from r_2=25.9 to r=6.3, $2\widehat{m}\tilde{B}$ decreases from M_π^2 to zero. The low energy $\pi\pi$ scattering thus provides us with a quasi-unique experimental access to the order parameter $\widehat{m}B_0$ (or to r). The corresponding GχPT predictions, endowed with the necessary loop corrections, are presented in detail in Ref. [16].

DaΦne will give us an accurate measurement of the low energy phase shifts $\delta_0^0 - \delta_1^1$ as a function of energy. This information, taken together with the direct determination of $a_0^0 - a_0^2$ to 5% accuracy from the $\pi^+\pi^-$ atom lifetime experiment which is planed at CERN [17], could well provide a decisive measurement of $m_s/\widehat{m}$ and a crucial test of χPT.

References

[1] N. H. Fuchs, H. Sazdjian and J. Stern, Phys. Lett. **B269** (1991) 183.
J. Stern et al., Phys. Rev. **D47** (1993) 3814.

[2] M. Knecht, H. Sazdjian, J. Stern and N. H. Fuchs, Phys. Lett. **B313** (1993), 229.

[3] M. Knecht, B. Moussallam and J. Stern, IPNO-TH 94-08, Nucl. Phys. **B** (to appear).

[4] J. Gasser and H. Leutwyler, Ann. Phys. **158** (1984) 142; Nucl. Phys. **B250** (1985) 465.

[5] M. Gell-Mann, R. J. Oakes and B. Renner, Phys. Rev. **175** (1968) 2195.

[6] S. Weinberg, in *A Festschrift for I. I. Rabi*, edited by L. Motz (New York Academy of Sciences, New York, 1977), p. 185.

[7] J. Gasser and H. Leutwyler, Phys. Rep. **87** (1982) 77.

[8] N. H. Fuchs, H. Sazdjian and J. Stern, Phys. Lett. **B238** (1990) 380.

[9] S. Weinberg, Physica **96A** (1979) 327.

[10] H. Leutwyler, preprint BUTP-93/24, Ann. Phys. (in press).

[11] M. Knecht and J. Stern, preprint IPNO-TH 94-53, contribution to the second edition of the DaΦne Handbook, L. Maiani, G. Pancheri and N. Paver, eds.

[12] C. A. Dominguez and E. de Rafael, Ann. Phys. **174** (1987) 372.

[13] J. Stern, N. H. Fuchs and M. Knecht, "Light quark masses from exclusive tau decays: An experimental proposal", in "Proceedings of the Third Workshop on the Tau-Charm Factory, Marbella, Spain, 1-5 June 1993", Eds. J. Kirkby and R. Kirkby (Editions Frontieres, 1994).

[14] H. Leutwyler, Nucl. Phys. **B337** (1990) 108; "Masses of the light quarks", preprint BUTP-94/8, May 1994.

[15] S. Weinberg, Phys. Rev. Lett. **17** (1966) 616.

[16] M. Knecht, B. Moussallam and J. Stern, preprint IPNO-TH 94-54, contribution to the second edition of the DaΦne Handbook, L. Maiani, G. Pancheri and N. Paver, eds.

[17] G. Czapek et al., Letter of intent: Lifetime measurements of $\pi^+\pi^-$ atoms to test low energy QCD predictions, preprint CERN/SPSLC 02-44.

L. Montanet, private communication.

Gauged Linear σ-Model and Nucleon Structure

M. M. Islam
Department of Physics, University of Connecticut, Storrs, CT 06269, U.S.A.

Abstract

The gauged linear σ-model brings an explicit quark sector to the gauged nonlinear σ-model. Interaction in this sector can lead to a condensed ground state of chiral quarks. The nucleon emerges as a topological soliton surrounded by a chiral condensate. High energy elastic pp and $\bar{p}p$ scattering data from CERN ISR and SPS Collider appear to provide strong evidence in favor of this nucleon structure.

I. INTRODUCTION

Last three decades of intensive research on the structure of the nucleon following the birth of the quark model has not led to a unique description of the nucleon. Instead, it has generated a wide array of models:[1–3] the constituent quark model, the MIT bag model, the chiral bag model, the topological soliton model, the non-topological soliton model, the color dielectric model etc. Experimentally established low energy properties of the nucleon and low energy nucleon-nucleon interaction have not been able to single out one model as incorporating the key features. Surprisingly, from high energy elastic proton-proton and antiproton-proton scattering, there seems to be strong evidence in favor of one of the models: namely, the gauged nonlinear σ- model that describes the nucleon as a topological soliton and introduces the vector mesons ω, ρ, A_1 as gauge bosons.[4] This model, however, needs to be extended (i.e., made linear) to include an explicit quark sector, where left and right quarks can interact via a scalar field and can form a condensed ground state (a chiral condensate).[5] The nucleon appears as a topological soliton surrounded by a chiral condensate.[6,7]

II. GAUGED NONLINEAR σ-MODEL

Our first step is to understand how the gauged nonlinear σ-model is built. To this end, we begin with a much simpler model—the Gell-Mann–Levy σ-model with $SU(2)_L \times SU(2)_R$ global symmetry and Higgs potential. The model is described by the Lagrangian

$$\mathcal{L} = \bar{\Psi} i\gamma^\mu \partial_\mu \Psi + \frac{1}{2}(\partial_\mu \sigma \partial^\mu \sigma + \partial_\mu \boldsymbol{\pi} \cdot \partial^\mu \boldsymbol{\pi}) \\ -G\bar{\Psi}(\sigma + i\boldsymbol{\tau} \cdot \boldsymbol{\pi}\gamma^5)\Psi - \lambda(\sigma^2 + \boldsymbol{\pi}^2 - f_\pi^2)^2 , \quad (1)$$

where Ψ is the fermion or quark field and f_π is the pion decay constant. We next introduce the unitary field $U(x) = \exp(i\tau^a \phi^a(x)/f_\pi)$, and write $\sigma(x) + i\boldsymbol{\tau} \cdot \boldsymbol{\pi}(x) = \zeta(x)U(x)$. Here, $\phi^a(x)$'s (a= 1,2,3) are the pions or the Goldstone bosons, and $\zeta(x)$ is a scalar field ($\sigma^2 + \boldsymbol{\pi}^2 = \zeta^2$). Eq.(1) can be written as

$$\mathcal{L} = \bar{\Psi}_L i\gamma^\mu \partial_\mu \Psi_L + \bar{\Psi}_R i\gamma^\mu \partial_\mu \Psi_R + \frac{1}{4}\zeta^2 \mathrm{tr}[\partial_\mu U \partial^\mu U^\dagger] + \frac{1}{2}\partial_\mu \zeta \partial^\mu \zeta \\ -G\zeta(\bar{\Psi}_L U \Psi_R + \bar{\Psi}_R U^\dagger \Psi_L) - \lambda(\zeta^2 - f_\pi^2)^2 , \quad (2)$$

where $\Psi_L = \frac{1}{2}(1-\gamma^5)\Psi$ and $\Psi_R = \frac{1}{2}(1+\gamma^5)\Psi$ are the left and right chiral fields.

The model given by Eq.(2) can be easily generalized to $SU(3)_L \times SU(3)_R$ symmetry with local gauge invariance by considering the unitary field $U(x) = \exp(i\lambda^a \phi^a(x)/f_\pi)$ and by introducing the left gauge field $\mathcal{A}_\mu$ and the right gauge field $\mathcal{B}_\mu$ (λ^a's are the Gell-Mann matrices; a=1–8). We now write a path-integral representation for the partition function of this generalized model:

$$Tr\left[e^{-iHT}\right] = \frac{1}{\mathcal{N}} \int DU D\zeta D\psi D\bar{\psi} \exp\Big[i \int d^4x \Big(\frac{1}{4}\zeta^2 tr[\mathcal{D}_\mu U (\mathcal{D}^\mu U)^\dagger] \\ + \frac{1}{2}\partial_\mu \zeta \partial^\mu \zeta - V(\zeta) + \bar{\psi}^A i \not{\mathcal{D}}(\mathcal{A}) a_- \psi^A + \bar{\psi}^B i \not{\mathcal{D}}(\mathcal{B}) a_+ \psi^B \\ -G\zeta[\bar{\psi}^A U a_+ \psi^B + \bar{\psi}^B U^\dagger a_- \psi^A]\Big)\Big] ; \quad (3)$$

here, $\mathcal{D}_\mu U$ is the covariant derivative of U: $\mathcal{D}_\mu U = \partial_\mu U + \mathcal{A}_\mu U - U\mathcal{B}_\mu$; $V(\zeta) = \lambda(\zeta^2 - f_\pi^2)^2$, $\not{\mathcal{D}}(\mathcal{A}) = \gamma^\mu(\partial_\mu + \mathcal{A}_\mu)$, $\not{\mathcal{D}}(\mathcal{B}) = \gamma^\mu(\partial_\mu + \mathcal{B}_\mu)$, $a_\pm = \frac{1}{2}(1 \pm \gamma^5)$.

The path-integral representation shows how quantum effects come into the picture. The fermion measure in Eq.(3) is gauge dependent and can be written as the product of a Jacobian and an invariant measure:[5]

$$D\psi D\bar{\psi} = \mathcal{J}[\mathcal{A}, \mathcal{B}, U] D\psi^o D\bar{\psi}^o . \quad (4)$$

The Jacobian can be identified as $\exp[i\Gamma_{WZ}]$, where $\Gamma_{WZ} = \Gamma_{WZ}[\mathcal{A}, \mathcal{B}, U]$ is the Wess-Zumino action. The net result, therefore, is the appearence of an additional piece of action in Eq.(3). It is not surprising that currents which were conserved before are no longer conserved, because of the contribution of Γ_{WZ}. This is known as the anomaly; namely, a current that is conserved at the classical level is no longer conserved when quantum effects are taken into account.[8]

A simple model can now be built if we: (i) replace the scalar field $\zeta(x)$ by its vacuum value f_π from the very beginning, (ii) keep the WZ action arising from the fermion measure, and (iii) disregard the interaction between the left and right quarks. This is the nonlinear σ-model described by the partition function:

$$Tr\left[e^{-iHT}\right] \simeq \frac{1}{\mathcal{N}'}\int DU \exp\left[i\int d^4x \frac{1}{4}f_\pi^2 tr[\mathcal{D}_\mu U(\mathcal{D}^\mu U)^\dagger] + i\Gamma_{WZ}[\mathcal{A},\mathcal{B},U]\right] . \quad (5)$$

To describe the nucleon using this model, one further includes an additional $U(1)_V$ gauge invariance connected with the baryonic charge conservation, and one introduces the vector meson ω as the gauge boson coupled to the baryonic charge. Limiting oneself to the $SU(2)_L \times SU(2)_R$ sector, one now has ω, ρ, and A_1 introduced as gauge bosons and a WZ action that involves these bosons.[4,9]

The WZ action shows that the ω couples to the pseudoscalar mesons via the interaction $g_\omega \omega_\mu j^\mu$, where

$$j^\mu = \frac{1}{24\pi^2}\,\epsilon^{\mu\nu\rho\sigma}\,\mathrm{tr}\left[U^{-1}\partial_\nu U U^{-1}\partial_\rho U U^{-1}\partial_\sigma U\right] \quad (6)$$

is the baryonic current. This current is conserved algebraically and can be identified as the topological current connected with the configuration of the unitary field $U(x)$. The total baryonic charge $\int j_0(\mathbf{x},t)d^3x$ then becomes the topological charge, or, the winding number of the unitary field. This model has been extensively studied by three groups in order to see whether the nucleon can be described as a topological soliton.[10,11,12] To predict the low energy properties of the soliton, the following ansatz, known as the hedgehog ansatz, is made : $U(\mathbf{r}) = \exp(i\boldsymbol{\tau}\cdot\hat{\mathbf{r}}\theta(r))$, i.e., $\boldsymbol{\phi}(\mathbf{r}) = \hat{\mathbf{r}} f_\pi \theta(r)$.We refer to $\theta(r)$ as the pion profile function, or, the pion field configuration. What these groups have accomplished is quite significant. They fix all the parameters of the model by studying the meson sector. Then, they proceed to predict the properties of the nucleon as a soliton.[9] All three groups have found that the model describes the low energy properties of the nucleon quite satisfactorily except in one respect. The mass of the soliton comes out consistently too large ($\sim$ 1500 MeV) compared to the actual mass of the nucleon (939 MeV).

Before addressing the problem of large soliton mass, we examine how the topological soliton description of the nucleon gets linked in our work with the high energy elastic scattering.

III. HIGH ENERGY ELASTIC PP AND $\bar{P}P$ SCATTERING AND NUCLEON STRUCTURE

High energy elastic pp and $\bar{p}p$ scattering at the CERN ISR and SPS Collider[13–15] have been analyzed by my collaborators and me over a number of years.[16–18] From our analyses, we have arrived at the following phenomenological description. The nucleon has an inner core and an outer cloud (Fig.1). Elastic scattering at high energy is primarily due to two processes: (1) a glancing collision in which the outer cloud of one nucleon interacts with that of the other giving rise to diffraction scattering; (2) a hard collison in which one nucleon core scatters off the other core via vector meson ω exchange, while their outer clouds overlap and interact independently (Fig. 2). In the small momentum transfer region, diffrac-

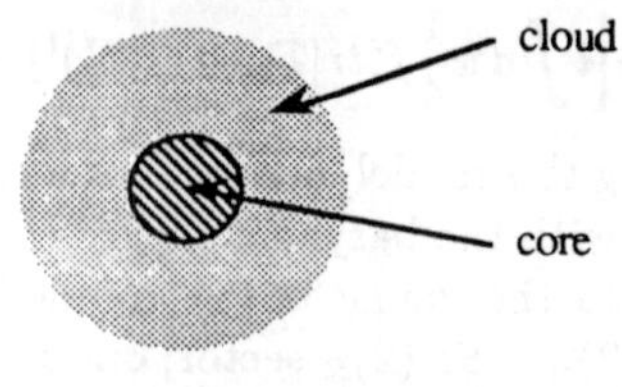

Fig.1. Physical picture of the nucleon from high energy elastic scattering.

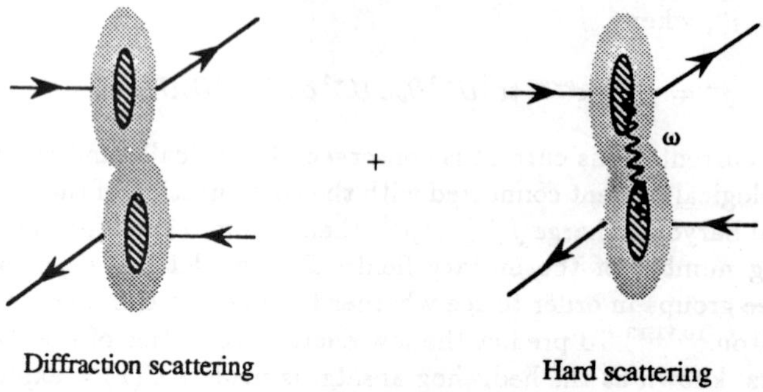

Fig.2. Primary processes responsible for high energy elastic scattering in our description.

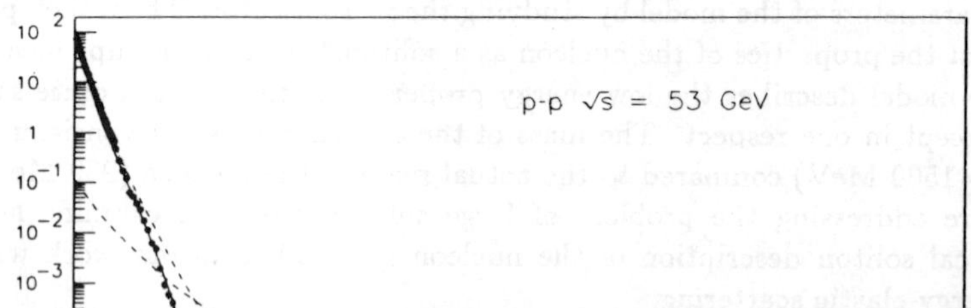

Fig.3. Solid curve represents calculated pp elastic differential cross section at $\sqrt{s} = 53$ GeV [Ref.18]. Differential cross section due to diffraction alone (dot-dashed line) and due to ω-exchange alone (dashed line) are also shown. ISR data from Nagy et al. [Ref.13].

tion dominates. As the momentum transfer increases, the hard scattering takes over. Fig.3 shows our fit to the ISR pp elastic data at $\sqrt{s} = 53$ GeV.[18] The solid curve represents the pp differential cross section calculated by us. The dot-dashed curve represents the differential cross section due to diffraction alone, while the dashed curve represents that due to hard scattering alone, i.e., due to ω exchange. As can be seen, diffraction dominates in the forward direction, but the hard scatterring takes over as the momentum transfer increases.

In our analyses, ω behaves as an elementary spin-1 boson. This can be seen by noticing that the ω exchange amplitude is dominantly of the form:[18]

$$T_H(s,t) \sim e^{i\chi_D(s,b=0)} s \,\frac{F^2(t)}{m_\omega^2 - t} \,.$$

The factor of s originates from the spin 1 of ω. The t-dependence (product of two form factors and the ω propagator) shows that ω exchange corresponds to an elementary meson exchange between two sources. (The factor $e^{i\chi_D(s,b=0)}$ reflects the absorption effect due to diffraction scattering.) From the viewpoint of conventional high energy scattering models, this behavior of ω is hard to understand. In such models, ω is a composite object, and therefore, behaves as a Regge pole. Its contribution away from the forward direction is then negligible. In contrast, the behavior of ω as a spin-1 boson even at high energy is easy to understand in the gauged nonlinear σ-model. This is an effective QCD model in which baryonic charge is topological, and ω is the gauge boson coupled to this charge. As long as the model remains valid, baryonic charge continues to behave as topological charge and ω continues to act as a gauge boson, i.e., as an elementary spin-1 boson. Apparently, this is the behavior we seem to have observed in our phenomenological analyses of elastic scattering. This conclusion, interestingly enough, can be subjected to an important test.[6]

Let us recall the hedgehog ansatz for the soliton model. This ansatz leads to the following expression for the baryonic charge density:

$$\tilde{\rho}(r) = -\frac{1}{2\pi^2}\,\frac{\sin^2\theta(r)}{r^2}\,\frac{d\theta(r)}{dr}\,, \tag{7}$$

where $\theta(r)$ is the pion field configuration introduced earlier. If we identify ω as the gauge boson probing the baryonic charge, then from the ωNN form factor found in elastic scattering, we can obtain the baryonic charge density $\tilde{\rho}(r)$, and from $\tilde{\rho}(r)$ we can determine via Eq.(7) the pion field configuration in a nucleon as seen in elastic scattering. This can then be compared with the pion field configuration obtained from the low energy soliton model calculation. The result of such a comparison is shown in Fig.4. The solid curve shows $\theta(r)$ obtained from elastic scattering.The dotted and dashed curves represent the calculations of Meissner et al.[10] in the minimum soliton model (only ω present) and in the complete soliton model (equivalent to having ω, ρ, A_1—all three present). As can be seen, the pion field configuration from two totally different regimes of physics are not that far apart.

A question that naturally arises in our description of elastic scattering is the following: ω, in our analyses, plays a dominant role, but, ρ and A_1, which are similar low energy vector mesons, play no role. Why does this happen? The nonlinear σ-model again provides an answer. In this model, one introduces ω, ρ and A_1— all on the same footing as gauge bosons. What one finds is that ρ and A_1 couple via isospin vector and axial vector currents, and these currents are dominantly Noether currents.[19] In a path-integral representation, they occur in the usual action. ω, on the other hand, couples via a topological current that occurs in the WZ action. Thus ρ and A_1 couple in a conventional way, and we expect them to behave in a perturbative field theory manner and to Reggeize. Their contributions away from the forward direction would then be negligible. In contrast, ω couples in a nonperturbative way and can continue to act as a gauge boson coupled to the baryonic charge, so that its contribution away from the forward direction can remain significant.

IV. LARGE SOLITON MASS PROBLEM AND GAUGED LINEAR σ-MODEL

We return to the problem of large soliton mass predicted by the gauged nonlinear σ-model. The problem now takes on added importance, since we have found evidence in favor of the model from high energy elastic scattering. To tackle the problem, we go back to the path-integral representation for the gauged linear σ model, extract the WZ action from the fermion measure, and make the approximation that the scalar field $\zeta(x)$ can be replaced by its vacuum value f_π in the meson sector, but not in the quark sector. The partition function then takes the form:

$$\begin{aligned} Tr\left[e^{-iHT}\right] \simeq \frac{1}{\mathcal{N}}\int DU \exp\left[i\int d^4x\, \frac{1}{4}f_\pi^2 tr[\mathcal{D}_\mu U(\mathcal{D}^\mu U)^\dagger] + i\Gamma_{WZ}[\mathcal{A},\mathcal{B},U]\right] \\ \times \int D\zeta D\psi^o D\bar{\psi}^o \exp\left[i\int d^4x \left(\bar{\psi}_L^o i\gamma^\mu(\partial_\mu + \mathcal{A}_\mu^o)\psi_L^o + \bar{\psi}_R^o i\gamma^\mu(\partial_\mu + \mathcal{B}_\mu^o)\psi_R^o \right.\right. \\ \left.\left. + \frac{1}{2}\partial_\mu\zeta\partial^\mu\zeta - \lambda(\zeta^2 - f_\pi^2)^2 - G\zeta(\bar{\psi}_L^o\psi_R^o + \bar{\psi}_R^o\psi_L^o)\right)\right]\ ; \qquad (8) \end{aligned}$$

it becomes the product of two functional integrals.The first integral is exactly the same as before (Eq.(5)). It represents the gauged nonlinear σ-model. The second integral shows that, we now have an explicit quark sector where left and right quarks interact via a scalar field, and that this sector can contribute to the total energy of the system. The superscript zero indicates that the fields in the quark sector are all in the unitary gauge (U=1). Further investigation reveals the following.[5]

The nonlinear σ-model, by ignoring interactions in the quark sector, is stating that all the important interactions are in the pseudoscalar meson sector. Quarks are present, but they can be regarded as noninteracting. Therefore, in this model, one has a soliton surrounded by a noninteracting Dirac sea of quarks.On the other hand, if we adopt the linear σ-model and keep the scalar field, then the scalar field mediates an interaction between the left and right quarks, and we have a soliton surrounded by an interacting Dirac sea. What one finds is that if the scalar field has a critical behavior (i.e., zero at small distances, but rises sharply to its vacuum value f_π at some distance $r = R$), then the energy of the interacting Dirac sea can be significantly less than the energy of the noninteracting Dirac sea. The system, consequently, can make a phase transition to the interacting ground state and reduce its total energy or mass by the condensation energy. This mechanism can solve the large soliton mass problem. Furthermore, since it does not affect the pseudoscalar meson sector, all the good predictions of the gauged nonlinear σ-model remain valid. The condensed ground state originating from the interaction between left and right quarks is analogous to a superconducting ground state and can be explicitly written as:

$$|\Psi_o\rangle = \frac{1}{\sqrt{N}} \prod_k \xi_{pk} \xi_{\bar{p}k} \, |\Phi_o\rangle \ ; \tag{9}$$

here, ξ_{pk} and $\xi_{\bar{p}k}$ are the quasi-quark and quasi-antiquark annihilation operators and are related to the usual quark and antiquark annihilation operators by a Bogoliubov transformation. $|\Psi_o\rangle$ is the quasiparticle vacuum, and $|\Phi_o\rangle$ is the particle vacuum. A number of consequences now follow, and let us briefly examine them.

V. CONSEQUENCES OF A CONDENSED GROUND STATE OF CHIRAL QUARKS

1. In our phenomenological description of high energy elastic scattering, the nucleon has an outer cloud. Such a cloud does not occur in the usual low energy soliton model calculations, where the soliton is surrounded by an ordinary vacuum.We now notice that a condensed ground state of chiral quarks (the chiral vacuum) can, in fact, provide the soliton an outer cloud making the low energy picture of the nucleon coincide with its high energy description.

2. Confinement of quarks now becomes automatic.The single-particle excitations of the condensed chiral ground state are quasi-quarks. They exist in the Hilbert space based on the chiral vacuum $|\Psi_o\rangle$, but not in the Hilbert space based on the ordinary vacuum $|\Phi_o\rangle$. Nambu and Jona-Lasinio[20] pointed out a long time ago that the two vacua $|\Psi_o\rangle$ and $|\Phi_o\rangle$ do not overlap, and that there cannot be any transition between the two Hilbert spaces. Therefore, we are not going to see

the quasi-quarks as free particles in the Hilbert space based on $|\Phi_o\rangle$, the ordinary vacuum.

3. If we write the scalar field $\zeta(x) = f_\pi + h(x)$, then the fluctuation field $h(x)$ is the Higgs field in our present context. Its quanta are going to be isoscalar scalar particles, and when two nucleons come close to each other, they will exchange such a Higgs scalar. This Higgs scalar, however, will not be observed as a physical particle or resonance, since it exists only in the condensed chiral vacuum. We can identify it with the σ particle of low energy *NN* interaction. It has long been known that the attractive intermediate range *NN* interaction can be attributed to the exchange of an isoscalar scalar particle of mass ~ 500 MeV, which, however, has never been observed experimentally.[21]

4. The condensed ground state $|\Psi_o\rangle$ exhibits spontaneous breakdown of chiral symmetry. If $\hat{Q}_A$ is the conserved axial or chiral charge operator, then it transforms the left and right chiral fields $\hat{\psi}_L$ and $\hat{\psi}_R$ in opposite ways. Consequently, the operator $e^{-i\alpha\hat{Q}_A}$ (α a constant) acting on the vacuum $|\Psi_o\rangle$ transforms it to a new vacuum $|\Psi'_o\rangle$:

$$\begin{aligned} |\Psi'_o\rangle &= e^{-i\alpha\hat{Q}_A}\,|\Psi_o\rangle \\ &= e^{-i\alpha\hat{Q}_A}\frac{1}{\sqrt{N}}\prod_k \xi_{pk}\xi_{\bar{p}k}\,|\Phi_o\rangle \\ &= \frac{1}{\sqrt{N}}\prod_k \xi'_{pk}\xi'_{\bar{p}k}\,|\Phi_o\rangle \quad , \end{aligned} \tag{10}$$

where ξ'_{pk} and $\xi'_{\bar{p}k}$ are the new quasiparticle operators. Since $|\Psi'_o\rangle \neq |\Psi_o\rangle$, we have spontaneous breakdown of chiral symmetry. This is analogous to the spontaneous breakdown of the total charge operator in the BCS superconducting ground state.

VI. CONCLUDING REMARKS

In summary, we see that the gauged linear σ-model brings an explicit quark sector to the gauged nonlinear σ-model. In this sector, left and right quarks interact via a scalar field. A critical behavior of the scalar field can lead to a condensed ground state analogous to a superconducting ground state, and the system can significantly reduce its total energy or mass by the condensation energy. This can solve the large soliton mass problem of the gauged nonlinear σ-model. The nucleon emerges as a topological soliton surrounded by a chiral condensate (Fig.5).[22] The low energy soliton model and the high energy phenomenological description combine to provide this physical picture of the nucleon. If future quantitative calculations bear out the predictions of this combined model, then we would conclude that the elastic data from the CERN ISR and SPS Collider had been pivotal in determining the structure of the nucleon.

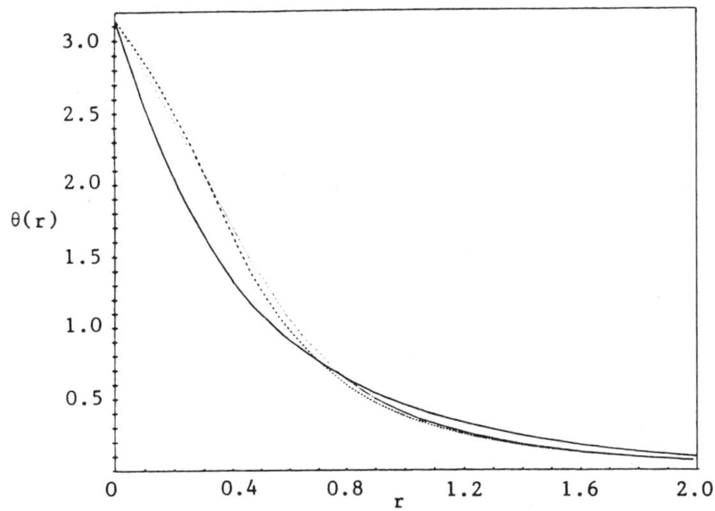

Fig.4. The pion field configuration $\theta(r)$ as a function of r in fermi. The continuous curve represents $\theta(r)$ obtained from high energy elastic scattering. The dotted and the dashed curves represent $\theta(r)$ calculated from low energy in the minimal and in the complete soliton model by Meissner et. al. [Ref.10].

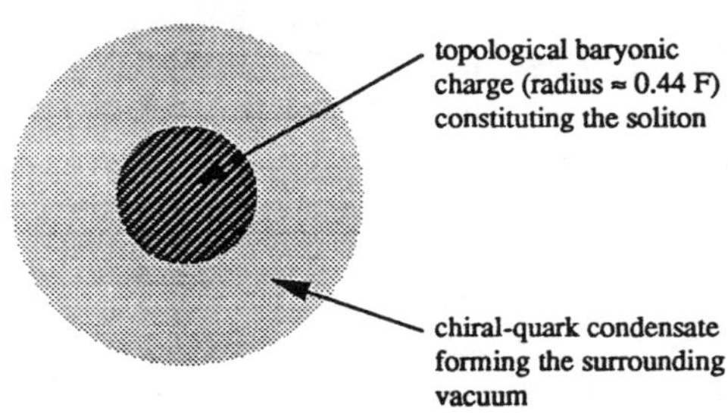

Fig.5 Structure of the nucleon as a topological soliton surrounded by a chiral condensate.

Acknowledgment

This work has been supported in part by the U.S. Department of Energy. The author wishes to thank Timothy Leek for assistance in computation.

References

1. G. Ripka, "Can We control the proliferation of nucleon models?", in *Windsurfing The Fermi Sea*, vol.2 (North-Holland, Amsterdam, 1987).
2. M.C. Birse, *Prog. Part. Nucl. Phys.* **25** (1990) 1.
3. R.D. Ball, *Int. J. Mod. Phys.* **A 5** (1990) 4391.
4. I. Zahed and G.E. Brown, *Phys. Reports* **142** (1986) 1.
5. M.M. Islam, *Z. Phys.* **C 53** (1992) 253.
6. M.M. Islam, in *Proceedings of the Vth Blois Workshop–Int. Conf. on Elastic and Diffractive Scattering*, ed. H.M. Fried, K.Kang, and C-I Tan (World Scientific, Singapore, 1994) p.230.
7. M.M. Islam, *Foundations of Physics*, **24** (1994) 419.
8. That anomalies originate from the gauge dependence of the fermion measure was pointed out by K. Fujikawa, *Phys. Rev.* **D 21** (1980) 2848; **29** (1984) 285.
9. U.-G. Meissner, *Phys. Rep.* **161** (1988) 213.
10. U.-G. Meissner, N. Kaiser, and W. Weise, *Nucl. Phys.* **A 466** (1987) 685.
11. M. Lacombe, B. Loiseau, R. Vinh Mau, and W.N. Cottingham, *Phys. Rev.* **D 38** (1988) 1491.
12. P. Jain, R. Johnson, U.-G. Meissner, N.W. Park, and J. Schechter, *Phys. Rev.* **D 37** (1988) 3252.
13. E. Nagy et al., *Nucl. Phys.* **B 150** (1979) 221.
14. M. Bozzo et al., *Phys. Lett.* **B 155** (1985) 197.
15. D. Bernard et al., *Phys. Lett.* **B 171** (1986) 142.
16. G.W. Heines and M.M. Islam, *Nuovo Cimento* **A 61** (1981) 149.
17. M.M. Islam, T. Fearnley, and J.P. Guillaud, *Nuovo Cimento* **A 81** (1984) 737.
18. M.M. Islam, V. Innocente, T. Fearnley, and G. Sanguinetti, *Europhys. Lett.* **4** (1987) 189.
19. Explicit expressions for these currents are given by Lacombe et al., loc. cit.
20. Y. Nambu and G. Jona-Lasinio, *Phys. Rev.* **122** (1961) 345.
21. R. Machleidt, K. Holinde, and Ch. Elster, *Phys. Rep.* **149** (1987) 1.
22. Recently a model of high energy elastic hadron scattering has been proposed based on our description of the nucleon having an outer chiral condensate, but with a core of valence quarks: S.M. Troshin and N.E. Tyurin,*Phys. Rev.* **D 49** (1994) 4427.

Microscopic description of chiral solitons[1]

H. REINHARDT

Institut für Theoretische Physik
Universität Tübingen
D-72076 Tübingen, FRG

Abstract

The quark-flavour dynamics is approximated by the Nambu-Jona-Lasinio (NJL) model. Within the bosonized version of this model the emergence of baryons as chiral solitons of the meson fields is explained. Recent results obtained in the description of hyperons as solitons of the bosonized NJL model are reviewed.

1. Introduction

In the limit of a large number of colors $N_c \to \infty$ QCD is known to reduce to an effective theory of weakly interacting mesons and glueballs [1]. Witten conjectured that baryons emerge as soliton solutions of this effective theory [2]. While in two space-time dimensions $D = 2$ QCD may be bosonized exactly [3] this is impossible in our real world $D = 4$. At low energies the pseudoscalar mesons, the would-be Goldstone bosons of spontaneous breaking of chiral symmetry, are most relevant. Witten argued that in leading order low energy (gradient) expansion the effective meson theory of QCD should be given by the non-linear σ-model supplemented by the Wess-Zumino action. Introducing external gauge fileds and requiring gauge covariance the Wess-Zumino action has to be augmented with additional terms [4]. They allow to extract the low-energy bosonized version of the relevant currents. In particular, the baryon current is found to coincide with the topological current $B_\mu(x)$ of the Skyrme model [5]. The identification of the topological current with the baryon current - which we shall refer to as "Wittens's conjecture" in the following - is the crucial assumption of all purely mesonic topological chiral soliton models of baryons, like the Skyrme model [5] or extensions of it [6].

[1]Supported by Deutsche Forschungsgemeinschaft under contract Re 856/2-2

All the above discussed large N_C-QCD considerations, in particular the discussion of the topological properties of the Wess-Zumino action, are based on Euclidean field theroy ,which yields (for large Euclidean times) the vacuum part of the observables only. Hence the non-linear-σ-model (supplemented with the Wess-Zumino action) and the topological current $B_\mu(x)$ should be considered as the vacuum part of the effective meson action and of the baryon current, respectively. Therefore Witten's conjecture implies that valence quarks in the baryons have joined the Dirac sea such that the vacuum distorted by the mesonic soliton fields carries the baryonic charge. In a direct bosonization of QCD all these assumptions underlying the chiral soliton picture of the baryons could be explicitly tested.

Since an exact bosonization of QCD is out of reach it is useful to investigate all these assumptions and expected results within simpler models of the quark flavour dynamics of QCD. A convenient model for this purpose is the one of Nambu-Jona-Lasinio (NJL) [7] which shares the important property of chiral symmetry and spontaneous breaking of the latter with QCD. In fact, the NJL model can be strictly bosonized yielding an effective meson theory which reduces in leading order gradient expansion precisely to the gauged linear σ-model supplemented with the Wess-Zumino term [8]. This effective meson theory is in good agreement with the low energy meson data. Furthermore, the bosonization of the NJL model also shows that the topological current of the Skyrme model in fact arises in leading order gradient expansion of the vacuum part of the baryon current [8] as can be also infered from the quark loop expansion of Goldstone and Wilczek [9].

In this talk I will explain, in the context of the NJL model, how baryons emerge as chiral solitons of the meson fields. I will also discuss recent results obtained in the description of baryons as solitons of the bosonized NJL mode..

2. The bosonized NJL model

The NJL model is defined by the following lagrangian density

$$L = \bar{q}(i - \hat{m}_o)q + \frac{\kappa}{2}(\bar{q}\gamma_\mu \frac{\lambda^a}{2} q)^2 \tag{1}$$

Here $q = (u, d, s)$ denotes the light flavour quark spinors and m_0 denotes the current quark mass. Furthermore λ^a denotes the generator of the colour group and κ is a dimension full coupling constant. The first term in the lagrangian density describes the free quarks. This term is inherited from the QCD lagrangian. The second term describes the interaction of the quarks and replaces the interaction of the quarks with the gluon and the gluon interactions of QCD. It is this part where the NJL model greatly simplifies the QCD dynamics. In particular local colour gauge invariants is lost but global colour symmetry is still preserved. More important is that, like QCD, the NJL model has for $\hat{m}_0 = 0$ global chiral symmetry.

While QCD has been proved renormalizable the NJL model cannot be renormalizable due to the presence of the dimension full coupling constant. Therefore the model lagrangian has to be suplemented by an ultraviolet cutoff Λ which indicates the range of validity of the model. Since in the NJL model the dynamics of the gluons is left out, we expect that the cutoff Λ should not exceed the first glue ball mass which occures around 1.5 GeV. We will later see that in the application of the NJL model the cutoff takes in fact these values. Finally the NJL model does not confine quarks. This is considered to be a minor short comings of the model in the description of low energy hadron dynamics. This is because inside hadrons the quarks stay so closely together that they will presumably not feel the long range confinement forces.

By using functional integral technics one can exactly transform the NJL model into an effective meson theory [8]. Thereby the meson field ϕ emerges as a composite quark anti-quark colour singlet field. The obtained effective meson action has the generic form [8]

$$S[\Phi] = \frac{1}{2\kappa} Tr(\phi - \hat{m}_0)^2 + Trlog(i\partial\!\!\!/ - \phi) \tag{2}$$

where Tr includes also the functional trace in addition to the ordinary matrix trace. The first term of the action which is quadratic in the meson field will eventually give rise to the meson masses. The dynamics of the mesons is contained in the second part the so-called quark loop. Due to the fact that the momentum operator does not commute with the local meson field this term is highly nonlocal. In the meson sector we do not need the full nonlocal structure of the quark loop. At low energies a gradient expansion in powers of derivatives of the meson field seems to be appopriate. In leading order of the gradient expansion the effective meson theory reduces to the gauged linear σ model supplemented by the Wess-Zumino action [8]

$$S[\phi] = \int d^4x (\frac{1}{2} tr \partial_\mu \phi \partial_\mu \phi + w[\phi] + ...) + S_{WZ}. \tag{3}$$

For simplicity I have written down here only the ungauged linear σ model. The first term here represents the ordinary kinetic energy of the free mesons while the second term represents the effective meson potential. The minimum of the effective potential occurs at a non-zero expectation value of the meson field $< \phi >= m$, which represents a mass for the quarks. Since a mass term violates chiral symmetry the appearance of the vacuum expectation value of the meson field ,m, signals the spontaneous breakdown of chiral symmetry. The corresponding Goldstone bosons are pseudoscalar mesons. By giving the quarks a small but finite current mass $m_0 \neq 0$ a finite mass is induced for the pseudoscalar mesons. In this way the bosonized NJL model describes the dynamical generation of the constituent quark mass m by spontaneous breaking of chiral symmetry.

The physical mesons represent quantum fluctuations in the field ϕ around its vacuum expectation value $\phi' = \phi - < \phi >$. By studying the effective meson theory (3) for small fluctuations in the meson field ϕ one can describe the low energy meson dynamics. This program has been very successfully performed in the past (see ref. 10).

3. Baryons as chiral solitons

The bosonized NJL model shows in a very transparent way the emergence of baryons as chiral solitons of the meson fields. At first side it seems strange that baryons carrying have integer spin should be describable by meson fields which have integer spin. This is because by ordinary angular momentum coupling half- integer spin cannot be generated by coupling of various integer spins. The soliton picture of baryons emerges due to topological properties of the solitonic meson field configuration which give rise to a transmutation of isospin into spin.

To understand the emergence of baryons as solitons let us consider the spectrum of quarks in a chiral field having the celebrated hedgehog form

$$U(x) = exp(i\vec{\tau}\hat{r}\Theta(r)). \tag{4}$$

Here $\Theta(r)$ is a scalar function satisfying the boundering conditions $\Theta(0) = -n\pi, \Theta(\infty) = 0$ where n denotes so called winding number. Figure 1 shows the eigenvalues of the Dirac Hamiltonian of the quarks in a chiral hedgehog field with winding number $n = 1$

$$h = \vec{\alpha}\vec{p} + \beta m(U)^{\gamma_5} \tag{5}$$

The energy eigenvalues of the quarks are shown as function of the strength of the hedgehog field which is given by the product of the spatial extension R of the hedgehog times the constituent quark mass m. As can be seen from the figure the positive and negative energy contiuum states are in general little influenced by the presence of the hedgehog field, except for the lowest positive energy state which gets strongly bound as the strength of the chiral field is increased. Eventually this state acquires a negative energy and joins the Dirac sea. Since in the vacuum by definition all negative energy states are occupied we obtain for strong hedgehog fields a vacuum which carries baryonic charge equal to the winding number.

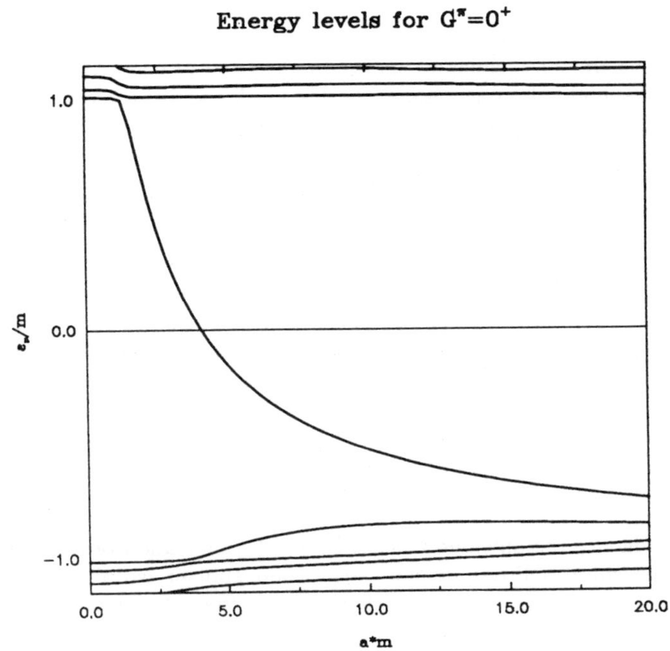

Fig. 1

The spectrum of the quarks with constituent mass in the background of a chiral hedgehog field with spatial extansion R as functionn of the strength of the hedgehog mR taken from ref. 20.

Let me emphasize that the baryonic charge is carried by the polarized Dirac sea and not by the meson cloud. However, the topologically non-trivial meson cloud with winding number n gives rise to the binding of n valence quarks states in the negative energy Dirac sea. So it is the meson cloud which generates the baryonically charged vacuum. In this sense we speak about baryons as chiral solitons of the meson field. Let me stress, however, that it is not the meson cloud itself which carries the baryonic properties but rather the vacuum generated by the mesonic cloud. In the pure mesonic soliton models like the Skryme model [5] or the nonlinear σ model with vector mesons one considers only the mesonic cloud but not the vacuum of the quarks. In these models one then interpretes the mesonic cloud itself as the baryon. The upshot of the above discussion is that the soliton picture of baryon assumes that the valence quarks are deeply bound in the Dirac sea.

Using the NJL model we can perform self-consistent Hartree–Fock type calculations for the baryons. In these calculations the mesons and in particular the topologically non-trivial meson cloud are generated by the quarks, and this cloud acts back on the quarks giving arise to a distortion of the quark spectrum. From the conceptual point of view the interesting question is whether in the self-consistent field configurations of the baryons the valence quarks have a positive energy or are deeply bound in the Dirac sea. In the first case we speak about a valence quark dominated baryon while in the later case the baryons appear as solitons of the meson fields.

The first self-consistent calculations were performed in ref. 11. In these calculations only the chiral hedgehog field was included. It turned out that the valence quarks got a positive energy. The same calculations were subsequently repeated in ref. 12, which confirmed this result. More recently we have performed self-consistent calculations, where all the non-strange meson fields up to 1.2 GeV were included, i.e. besides the pion field the ρ, the ω and the a_1 were included [13]. Then one finds in fact that the valence quarks are deeply bound in the Dirac sea and the baryons appear as chiral solitons. This was the first direct support of Witten's conjecture.

The chiral hedgehog meson cloud violates both rotational and isorotational invariance. Therefore the chiral solitons do not carry good spin or isospin. To compare the predictions of the soliton picture with the properties of the physical baryons one has first to perform a quantization of the chiral soliton. This is performed in a semiclassical way using the familiar cranking approach from nuclear physics [14], i.e. one rotates the hedgehog meson cloud and observes the increase in energy due to rotation. For small rotational frequency ω one can perform a Taylor expansion of the energy

$$E(\omega) = E(0) + \frac{1}{2}\vec{\omega}\lambda\vec{\omega} + ... \tag{6}$$

where the term linear in the rotational frequency ω is missing. From the quadratic term one extracts the moment of inertia λ. Performing a canonical quantization

$$\lambda\vec{\omega} \rightarrow \hat{I} = \hat{J}, \tag{7}$$

where $\hat{I}$ and $\hat{J}$ are the operators of the isospin and orbital angular momentum, one finds than the (iso-)rotational energy

$$E_I = E(0) + \frac{I(I+1)}{2\lambda} \tag{8}$$

Due to the special form of a hedgehog, rotation in space can be undone by a reversed rotation in isospace. As a result the quantized hedgehog carries the same quantum numbers for spin and isospin.

The nucleon and the Δ appear as the lowest energy states of a isorotational band with the nucleon as ground state and the Δ as first rotationally excited state.

The aboved described familiar cranking concept for isorotations can be extended to rotations in $SU(2)$ flavour space. In the frame work of the Skryme model [6] this extension was performed by Yabu and Ando [15]. In this approach one embeds the $SU(2)$ hedgehog in an $SU(3)$ flavour matrix and rotates it in flavour space. Thereby one includes explicitely the flavour symmetry breaking induced by the different current quark masses for the up, down and strange quarks. Due to the symmetry breaking, baryons with different flavour content acquire different masses. Table 1 shows the results obtained for the mass splittings of the hyperons within the soliton of the NJL model quantized according to the Yabu-Ando approach [16]. In these calculations the experimental splitting between the decay constants F_π and F_K were used as input.

Table 1: The mass splittings of the low-lying $\frac{1}{2}^+$ and $\frac{3}{2}^+$ baryon states with respect to the nucleon. All numbers are in MeV. For more details see ref. 16.

	Λ	Σ	Ξ	Δ	Σ^*	Ξ^*	Ω
Calc.	175	248	396	291	449	608	765
Expt.	177	254	379	293	446	591	733

Though the solition picture describes many low energy spectroscopic properties of baryons quite reasonable, there are still a few open problems. The absolute mass of the baryons come out much too large. For example the mass of the nucleon is obtained in the order of 1.4 ... 1.7 GeV. Furthermore the axial coupling constant G_A is obtained in the range between 0.6 and 0.8 while the experimental value is 1.25.

In the context of the Skryme model is has been shown that the large baryon mass is considerably lowered by including quantum fluctuations in the meson fields (one-meson loops) [17]. The importance of meson fluctuations is also seen in the $1/N$ expansion of the baryon mass

$$E_I = E_{sol} + E_{vib} + E_{rot} + \ldots = \sim N + \sim N^0 + \sim N^{-1}. \tag{9}$$

There the first term represents the energy of the soliton which is proportional to the number of colours N. The third term represents the rotational energy originating from the canonical quantization. This term is of order $1/N$, since the moment of inertia is of order N. These two terms have been included in the semiclassical quantization of the soliton. What has been left out, however, are quantum fluctuations of the vibrational type which do not correspond to the zero modes included in the semiclassical quantization. These vibrational fluctuations give contributions of the order 1 and are thus suspected to be larger than the rotational corrections included so far.

Recently the vibrational quantum fluctuation around the chiral soliton of the NJL model have been calculated in ref. 18. For a constituent quark mass of 400 MeV and a pion mass of 138 MeV one finds a soliton mass of 1250 MeV and a correction from the vibrational quantum fluctuations of - 338 MeV. This gives rise to a nucleon mass of 976 MeV and the Δ mass of 1236 MeV. The Δ mass compares well to the experimental value of 1232 MeV while the nucleon mass is still somewhat larger than the experimental value. However the vibrational quantum fluctuations have given a substantial improvement for the absolute baryon mass as compared to the result obtained in the naive semiclassical quantization. One might hope that the inclusion of the vibrational quantum fluctuations leed to a similar improvement for the axial coupling constant G_A towards the experimental value.

4. Conclusions

In my talk I have shown that the NJL model successfully describes the chiral quark flavour dynamics at low energy. In particular this model gives rise to a vacuum which shows spontaneous breaking of chiral symmetry and the pseudoscalar mesons arise as Goldstone bosons of the spontaneous breaking of chiral symmetry. The bosonized version of this model explains how baryons emerge as solitons of the meson fields and the actual numerical calculations give strong support for Witten's conjecture. Quantizing the NJL soliton following the method of Yabu and Ando gives excellent results for the mass splittings of the hyperons. Finally, the model can be easily extended to include heavy quark flavours as well [19]. In this case the NJL model contains also the heavy quark symmetries and its bosonized version adequatly describes the interplay between chiral symmetry and heavy quark symmetries.

References

1. G. 't Hooft,*Nucl. Phys.* **B72** (1974) 461

2. E. Witten, *Nucl. Phys.* **B160** (1979) 57

3. E. Witten, *Comm. Math. Phys.* **92** (1984) 455 D. Gepner, *Nucl. Phys.* **B252** (1985) 481

4. E. Witten, *Nucl. Phys.* **B223** (1983) 422; 433 see also: Ö. Kaymakcalan, S. Rajeev and J. Schechter, *Phys. Rev.* **D30** (1984) 594

5. T.H.R. Skyrme, *Proc. Roy. Soc.* **A260** (1961) 127

6. For a review see e.g.: U.-G. Meissner, *Phys. Rep.* **161** (1988) 213

7. Y. Nambu and G. Jona-Lasinio, *Phys. Rev.* **122** (1961) 345, **124** (1961) 246

8. D. Ebert und H. Reinhardt, *Nucl. Phys.* **B271** (1986) 188

9. J. Goldstone and F. Wilczek, *Phys. Rev. Lett.* **47** (1981) 986

10. D. Ebert, H. Reinhardt and M.K. Volkov, Effective hadron theory of QCD, *Progr. Part. Nucl. Phys.*, **33**, (1994) 1

11. H. Reinhardt and H. Wünsch, *Phys. Lett* **B215** (1988) 577

12. T. Meissner, F. Grümmer and K. Goeke, *Phys. Lett.* **B227** (1989) 296 R. Alkofer, *Phys. Lett.* **B236** (1990) 310

13. R. Alkofer, H. Reinhardt, H. Weigel and U. Zückert, *Phys. Rev. Lett.* **69** (1992) 1874

14. H. Reinhardt, *Nucl. Phys.* **A 503** (1989) 825

15. H. Yabu and K. Ando, *Nucl. Phys.* **B 301** (1988) 601

16. H. Weigel, R. Alkofer and H. Reinhardt, *Nucl. Phys.* **B 387** (1992) 638

17. B. Moussallam and D. Kalafatis, *Phys.* **B 272** (1991) 196, G. Holzwarth, *Phys. Lett.* **B 291** (1992) 196, *Nucl. Phys.* **A 572** (1994) 69

18. H. Weigel, R. Alkofer and H. Reinhardt, **UNITUE- THEP-16/1994** *Nucl. Phys. B*, in press

19. D. Ebert, T. Feldmann, R. Friedrich, H. Reinhardt, **UNITUE-THEP-11/94**

20. R. Alkofer and H. Reinhardt, **Chiral quark flavour dynamics**, to be published

ABELIAN ACTION FOR QUARK CONFINEMENT

KEN YEE
Physics and Astronomy, LSU, Baton Rouge, LA 70803-4001, USA
kyee@rouge.phys.lsu.edu

ABSTRACT

We evaluate S_{APQCD}, the Abelian projection QCD(APQCD) action, using the microcanonical demon method. For $SU(2)$, S_{APQCD} at strong coupling is essentially the compact QED(CQED) action with $\beta_{CQED} = \frac{1}{2}\beta_{SU(2)}$. Since CQED deconfines when $\beta_{CQED} > 1$, this relation must break down as $\beta_{SU(2)} \to 2$. Indeed S_{APQCD} mutates: near $\beta_{SU(2)} \sim 2$ it gains additional operators, including an exogenous *negative* magnetic monopole mass shift. Since monopoles are condensed in CQED when $\beta_{CQED} < 1$, a vicarious corollary of our results is that $SU(2)$ monopoles are condensed when $\beta_{SU(2)} < 2$.

A clear demonstration that monopole condensation is the origin of quark confinement would be a notable achievement. To this end, 't Hooft[1-3] proposed that QCD monopoles are magnetic with respect to the $[U(1)]^{N-1}$ Cartan subgroup of color $SU(N)$. Full $SU(N)$ gauge symmetry obscures these charges and it is necessary to gauge fix at least the $SU(N)/[U(1)]^{N-1}$ symmetry to expose them. In this scenario monopoles are fixed-gauge manifestations of gauge field features responsible for quark confinement. Only in special gauges does one have a picture of confinement caused by monopole condensation. In other gauges the gauge field features causing confinement are still present but they do not look like magnetic monopoles.[4]

Numerical studies have found that maximal Abelian(MA) gauge[5] is compelling for 't Hooft's hypothesis. Upon decomposing gauge field A into purely diagonal(n) and purely off-diagonal(ch) parts

$$A = A^n + A^{ch}, \tag{1}$$

the MA gauge condition $D^n_\mu A^{ch}_\mu \equiv \partial_\mu A^{ch}_\mu - ig[A^n_\mu, A^{ch}_\mu] = 0$ leaves a residual $[U(1)]^{N-1}$ gauge invariance under $\Omega_{\text{residual}} = \text{diag}(\exp^{-i\omega_1}, \cdots, \exp^{-i\omega_N})$ where $\sum_{i=1}^N \omega_i = 0$. Under Ω_{residual} the N diagonal elements $(A^n)_{ii}$ transform as neutral photon fields whereas the $N(N-1)$ offdiagonal elements $(A^{ch})_{ij}$ transform as charged matter fields: $(A^n_\mu)_{ii} \to (A^n_\mu)_{ii} - \frac{1}{g}\partial_\mu\omega_i$ and, for $i \neq j$, $(A^{ch}_\mu)_{ij} \to (A^{ch}_\mu)_{ij}\exp^{-i(\omega_i-\omega_j)}$. Since $(A^{ch})_{ij}$ carries two different $U(1)$ charges, the A^{ch} fields induce "interspecies" interactions between the N photons. On the lattice the monopole currents are identified according to discretized versions[6] of $k_\mu \equiv \frac{1}{2\pi}\epsilon_{\mu\nu\lambda\delta}\partial_\nu f_{\lambda\delta}$ and $f_{\mu\nu} \equiv \partial_\mu A^n_\nu - \partial_\nu A^n_\mu$.

This procedure where only the diagonal A^n components of the $SU(N)$ gauge fields are used for measuring k_μ and $f_{\mu\nu}$ is called *Abelian projection*. Since $\sum_{i=1}^N (A^n_\mu)_{ii}$ is

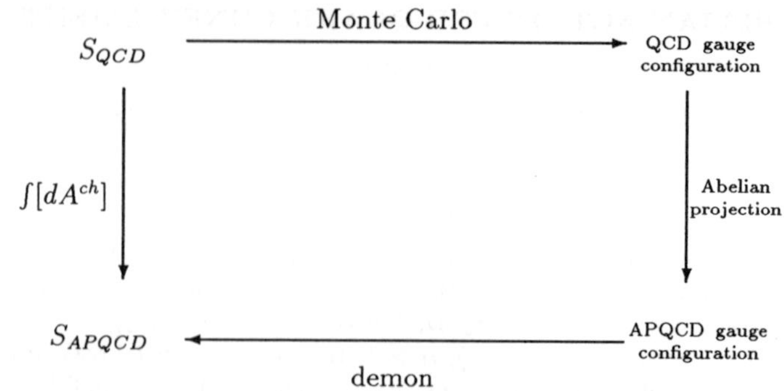

Fig. 1. To integrate out A^{ch} we: (i)generate an importance sampling $SU(N)$ gauge configuration; (ii)project this configuration to a $[U(1)]^{N-1}$ configuration; and (iii)compute the S_{APQCD} couplings using the microcanonical demon.

invariant under Ω_{residual}, an irreducible representation of $[U(1)]^{N-1}$ is $\theta^i_\mu \equiv (A^n_\mu)_{ii} - \Lambda_\mu$ where $\Lambda_\mu \equiv \frac{1}{N}\sum_{j=1}^N (A^n_\mu)_{jj}$. While vector field Λ is $[U(1)]^{N-1}$ invariant, the θ^i transform as $\theta^i_\mu \to \theta^i_\mu - \frac{1}{g}\partial_\mu\omega_i$ and obey constraint $\sum_{i=1}^N \theta^i_\mu = 0$. We shall refer to the quantum dynamics of the N angles θ^i as Abelian projected QCD or APQCD. Equivalently, APQCD is the field theory obtained by integrating out A^{ch} and Λ from QCD in MA gauge.[7] Its action S_{APQCD} is formally defined as

$$-S_{APQCD}[\theta^1, \cdots, \theta^N] \equiv \log\Big\{\int [dA^{ch} d\Lambda]\ \exp(-S_{QCD})\ \Delta_{FP}\delta[D^n_\mu A^{ch}_\mu]\Big\} \tag{2}$$

where Δ_{FP} is the Faddeev-Popov determinant. Monopoles arise in APQCD due to topological quantum fluctuations in the compact fields θ^i.

While there is no guarantee that S_{APQCD} has a simple form or is otherwise well-behaved, it is of central import due to *Abelian dominance*,[8] the fact that θ^i Wilson loops in APQCD have predominantly the same string tension as $SU(N)$ Wilson loops in QCD. Abelian dominance has the following formal implication. If $\mathrm{tr}\mathbf{W}$ is an $SU(N)$ Wilson loop and if $\langle\cdot\rangle_{QCD}$ and $\langle\cdot\rangle_{APQCD}$ refer respectively to S_{QCD} and S_{APQCD} expectation values, the APQCD operator $\mathcal{W}$ which obeys

$$\langle \mathcal{W}\rangle_{APQCD} = \mathrm{tr}\langle \mathbf{W}\rangle_{QCD} \tag{3}$$

is

$$\mathcal{W} = \exp(+S_{APQCD}) \int [dA^{ch} d\Lambda]\ \exp(-S_{QCD})\ \Delta_{FP}\delta[D^n_\mu A^{ch}_\mu]\ \mathrm{tr}\mathbf{W}. \tag{4}$$

Abelian dominance means that the complicated operator $\mathcal{W}$, which in other gauges would be a superposition of assorted $[U(1)]^{N-1}$-invariant operators of various sizes and shapes, is (for string tension) well-approximated in MA gauge by a θ^i loop of the *same*

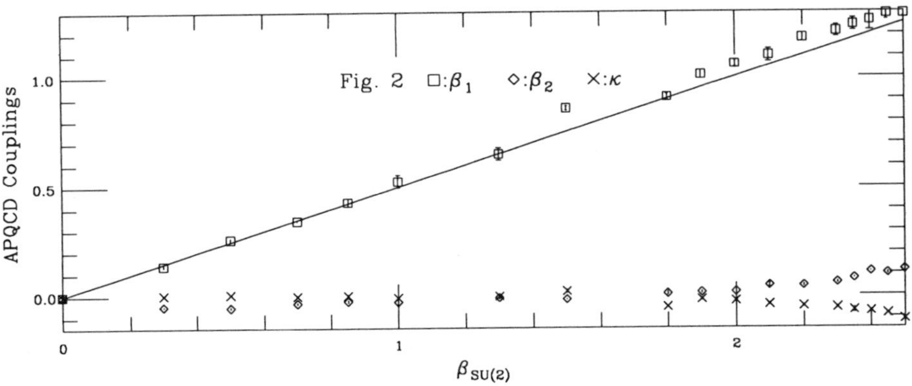

Fig. 2. Figure 2 depicts S^{ansatz} coefficients β_1, β_2 and κ as a function of $\beta_{SU(2)}$. $|\beta_3|$, not depicted, is always smaller than $|\beta_2|$, typically by a factor of $3-5$. Our $20^3 \times 16$ lattices are all well inside the zero temperature phase for the $\beta_{SU(2)}$ range depicted. The bold $\beta_1 = \frac{1}{2}\beta_{SU(2)}$ line is a guide-to-eye.

size and shape as tr$\mathbf{W}$. In other gauges the $\langle \mathcal{W} \rangle_{APQCD}$ string tension would be due to a *combination* of S_{APQCD} effects *and* properties of $\mathcal{W}$. An extreme example in which S_{APQCD} is immaterial is in a (hypothetical) gauge where $\mathcal{W} = \exp\{-\lambda RT\} \cdot 1$—the constant APQCD operator with area law coefficient. In such a gauge, the coefficient in $\mathcal{W}$ hordes the area law and S_{APQCD} is immaterial because $\mathcal{W}$ is simply 1. In stark contrast, S_{APQCD} alone determines string tension in MA gauge: given S_{APQCD} one can reconstruct the QCD string tension using APQCD Wilson loops without reference to the RHS of (4). In MA gauge S_{APQCD} apparently knows all there is to tell about quark confinement.

Our numerical procedure for evaluating S_{APQCD} is depicted in Figure 1. Let us temporarily focus on $SU(2)$. First, we make a representative importance sampling APQCD gauge configuration by applying the Abelian projection to a Monte Carlo lattice $SU(2)$ gauge configuration at some chosen coupling $\beta_{SU(2)}$. Seeking the action S_{APQCD} which would reproduce this APQCD configuration in a Monte Carlo simulation, we state an ansatz for S_{APQCD} and apply the microcanonical demon technique[9] to compute the parameters of this ansatz. This "inverse Monte Carlo" procedure is repeated to determine how S_{APQCD} fluctuates between different importance sampling $SU(2)$ configurations.

The general $U(1)$-invariant action consistent with APQCD involves an infinity of operators. Fortunately, previous studies[7,10] and, independently, the demon technique indicate that neither extended nor highly charged Wilson loops contribute substantially to S_{APQCD}. In particular, we have applied the demon to an ansatz consisting

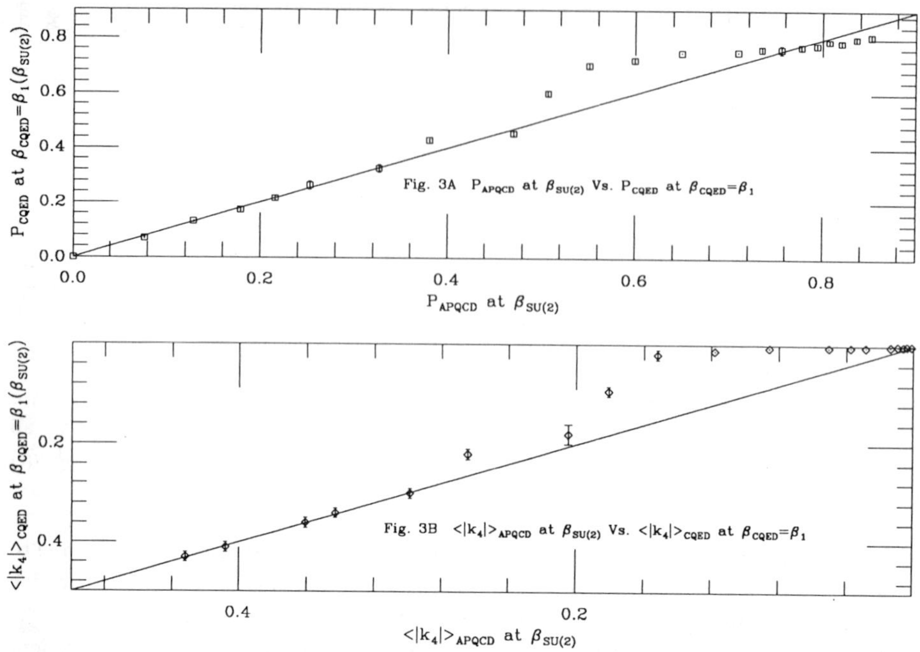

Fig. 3. 3A compares APQCD plaquettes P_{APQCD} at $\beta_{SU(2)}$ to CQED plaquettes P_{CQED} at $\beta_{CQED} = \beta_1(\beta_{SU(2)})$ for a range of $\beta_{SU(2)}$ values. When $\beta_{SU(2)} < 2$ the data points lie on the bold $P_{CQED} = P_{APQCD}$ line showing that S_{CQED} is a good model of S_{APQCD}. The set of points wandering off of the $P_{CQED} = P_{APQCD}$ line corresponds to $\beta_{SU(2)} > 2$, when S_{CQED} is not a good model of S_{APQCD}. 3B is an analogous plot using monopole densities.

of 1×1, 2×2, and 3×3 plaquettes. Over a wide range of $\beta_{SU(2)}$, we are unable to resolve a nonzero signal for any of the 2×2 or 3×3 plaquette couplings. Therefore, we focus now solely on a 1×1 ansatz

$$-S^{\text{ansatz}} = \sum_{q=1}^{3} \sum_{x,\mu<\nu} \beta_q \cos q\Theta_{\mu\nu} - \kappa \sum_{x,\mu} k_\mu(x) k_\mu(x). \tag{5}$$

$\cos q\Theta_{\mu\nu}$ is a 1×1 plaquette in $U(1)$ representation q given in terms of link angles θ^1_μ. κ shifts the $q = 1$, 1^3 monopole mass[11] implicit in $\beta_1 \cos \Theta_{\mu\nu}$, allowing the APQCD monopole mass to be independent of β_1. Of course, β_i and κ vary with $\beta_{SU(2)}$.

In our version of the demon technique, imagine a battalion of demons each carrying M *coupled* thermometers corresponding to the M undetermined coefficients in the ansatz action. ($M = 4$ for S^{ansatz}.) The demons thermalize with the APQCD configuration by hopping from link to link and exchanging energy(≡action) with the configuration.* The thermometers are coupled by requiring all of their energies to remain within a given range $[-E^0, E^0]$; if any proposed energy exchange violates this range it is rejected. Upon thermalization the couplings are read off from the battalion of energies, which has a Boltzmann distribution. Statistical errors are computed by jackknifing the demons. (The independent errors from jackknifing $SU(2)$ configurations are comparable in size.) In principle, if S^{ansatz} contains all the operators of S_{APQCD} the demon technique yields all the coupling constants exactly (modulo statistics). In practice, S^{ansatz} is a truncated action which is unlikely to contain all S_{APQCD} operators. Extensive numerical experiments with idealized configurations reveal that if operators are missing the method yields effective values for the couplings. These effective values would not be the same as the "true" values when all operators are present.

Figure 2 shows S^{ansatz} coefficients β_1, β_2 and κ, computed by the demon, as a function of $\beta_{SU(2)}$. $|\beta_3|$, not depicted, is always smaller than $|\beta_2|$, typically by a factor of $3-5$. Each $\beta_{SU(2)}$ configuration is generated fresh from a cold start so our data points do not contain any spurious correlations. Our $N_S^3 \times N_T = 20^3 \times 16$ lattices are all well inside the zero temperature phase; for the range of $\beta_{SU(2)}$ shown the APQCD Polyakov loop vanishes. As depicted, at strong coupling($\beta_{SU(2)} < 2$)

$$\beta_1 \sim \frac{1}{2}\beta_{SU(2)}, \quad \beta_{2,3} \sim 0, \quad \kappa \sim 0, \tag{6}$$

that is, S_{APQCD} reduces to the compact QED(CQED) action at strong coupling. At weaker coupling($\beta_{SU(2)} > 2$) β_2 and κ grow in magnitude but β_1 always remains the largest coupling.

Since monopoles are condensed when $\beta_{CQED} < 1$ in[12] CQED, Figure 2 or Eq. (6) proves (albeit vicariously) that $SU(2)$ monopoles are condensed when $\beta_{SU(2)} < 2$.

When $\beta_{SU(2)} > 2$, the situation is not so clear. In fact, Figure 2 suggests a paradox in the $\beta_{SU(2)} > 2$ region: how can APQCD maintain confinement in the continuum limit if CQED deconfines when $\beta_{CQED} > 1$? Clearly, either the meaning or validity of relation (6) must break down when $\beta_{SU(2)}$ is sufficiently large. Either (I)Abelian dominance does not survive the $\beta_{SU(2)} \sim 2$ crossover making S_{APQCD} less pertinent at weaker coupling—see discussion pertaining to Eq. (4); or (II)S_{APQCD} gains additional operators near $\beta_{SU(2)} \sim 2$; or a combination of (I) and (II). We do not have anything to say about (I) in this talk except to observe that there has been no definitive study

*Contrary to Ref. 9 we do not update the APQCD configuration so that energy is figuratively rather than literally "exchanged." This shortens the numerical algorithm and avoids the possibility of damaging the APQCD configuration if the battalion of demons absorbs too much energy.

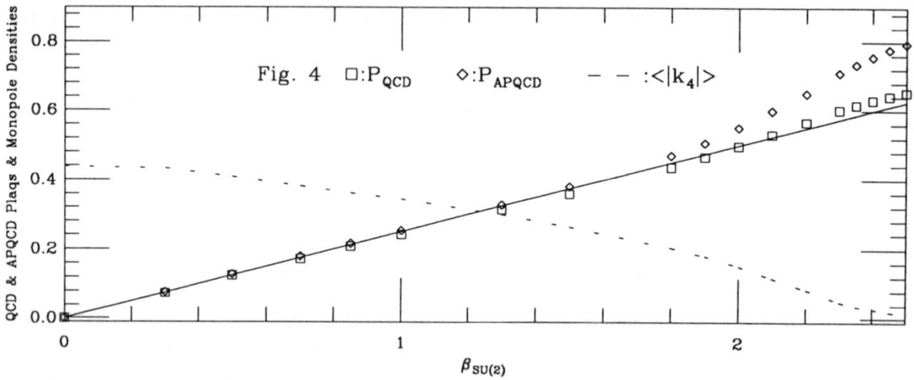

Fig. 4. Figure 4 depicts the APQCD and $SU(2)$ plaquettes as a function of $\beta_{SU(2)}$. In the strong coupling region($\beta_{SU(2)} < 2$), both the APQCD and $SU(2)$ plaquettes grow like $\frac{1}{4}\beta_{SU(2)}$, the guide-to-eye line's slope. At weaker coupling($\beta_{SU(2)} > 2$) the APQCD plaquettes deviate substantially from the $SU(2)$ plaquettes. Correspondingly, the monopole density decelerates noticeably near $\beta_{SU(2)} \sim 2$.

of Abelian dominance at larger $\beta_{SU(2)}$ values.†

(II) requires that S_{APQCD} is not well described by S_{CQED} when $\beta_{SU(2)} > 2$. Indeed, we can demonstrate this by simulating

$$- S_{CQED} = \sum_{x,\mu<\nu} \beta_{CQED} \cos \Theta_{\mu\nu} \Big|_{\beta_{CQED}=\beta_1(\beta_{SU(2)})} \tag{7}$$

(also on a $20^3 \times 16$ lattice) to see if it reproduces corresponding APQCD expectation values. As depicted in Figure 3, S_{CQED} reproduces APQCD plaquette averages and monopole densities only in the $SU(2)$ strong coupling region. At weaker coupling the CQED simulations start to disagree dramatically with APQCD. This implies that at weaker coupling either other terms of S^{ansatz} have become important or S^{ansatz} itself is inadequate. In any case, this means S_{APQCD} is not form-invariant between the strong and weak coupling regimes: at strong coupling S_{APQCD} is well approximated by S_{CQED}; at crossover region $\beta_{SU(2)} \sim 2$ S_{APQCD} mutates and develops substantial deviations from S_{CQED}. Inspection of Figure 2 reveals that a possible scenario might be that κ, the exogenous magnetic monopole mass shift, becomes more and more *negative* at larger $\beta_{SU(2)}$. As negative monopole mass favors monopole condensation (compensating for a large β_1), the occurrence of a sufficiently negative κ in S_{APQCD} at $\beta_{SU(2)} > 2$ could maintain APQCD confinement.

†In typical string tension measurements, the static quark potential is determined by superimposing data points from a range of different $\beta_{SU(2)}$ values.

We note that Eq. (6) "explains" Abelian dominance in the strong coupling regime. The $SU(2)$ plaquette in the strong coupling expansion behaves like $P_{QCD} \sim \frac{1}{4}\beta_{SU(2)}$ and the CQED plaquette like $P_{CQED} \sim \frac{1}{2}\beta_{CQED}$. Therefore, identifying $P_{CQED}(\beta_{CQED} = \beta_1)$ with P_{APQCD} and applying Eq. (6) yields

$$P_{APQCD} \sim \frac{1}{4}\beta_{SU(2)} \sim P_{QCD} \implies P_{APQCD} \sim P_{QCD}. \tag{8}$$

Carrying this argument over to larger Wilson loops leads to a (strong) statement of Abelian dominance: at sufficiently strong coupling APQCD and QCD Wilson loop averages and, hence, string tensions are equal. Figure 4 confirms (8) and shows how it breaks down at weaker coupling. Eq. (8) contradicts the naive expectation, based on P_{QCD} containing a trace over a 2×2 matrix and P_{APQCD} involving no trace, that $P_{APQCD} = \frac{1}{2}P_{QCD}$.

We have obtained similar results for the $SU(3)$ Abelian projection. For $SU(3)$, S_{APQCD} is more complicated due to interspecies dynamics.[10,13,14] Nonetheless the same crossover behavior occurs in $SU(3)$: at strong coupling, S_{APQCD} is simple; at weaker coupling, it crosses over to a more complicated action. A preliminary indication of this can be seen in the data reported in Ref. 3.

I thank M.I. Polikarpov for discussions and ITEP(Moscow) for its hospitality.

1. G. 't Hooft, Nucl. Phys. B190 (1981) 455.
2. A. Kronfeld, G. Schierholz, U. Wiese, Nucl. Phys. B293 (1987) 461.
3. A review with more references is K. Yee, Proceedings of Lake Louise Winter Institute (February, 1994) to be published by World Scientific (hep-ph/9404363).
4. M.N. Chernodub, M.I. Polikarpov, M.A. Zubkov, Nucl. Phys. B(Proc. Suppl.) 34 (1994) 256; M.I. Polikarpov and K. Yee, Phys. Lett. B333 (1994) 452.
5. A. Kronfeld, M. Laursen, G. Schierholz, U. Wiese, Phys. Lett. B198 (1987) 516.
6. D. Toussaint and T. DeGrand, Phys. Rev. D22 (1980) 2478.
7. K. Yee, Phys. Rev. D49 (1994) 2574.
8. T. Suzuki *et. al.*, Phys. Rev. D42 (1990) 4257; Phys. Lett. B272 (1991) 326.
9. M. Creutz, A. Gocksch, M. Ogilvie, M. Okawa, Phys. Rev. Lett. 53 (1984) 875.
10. K. Yee, Phys. Rev. D50 (1994) 2309.
11. J. Smit and A. van der Sijs, Nucl. Phys. B355 (1991) 603.
12. M.I. Polikarpov, L. Polley, U.J. Wiese, Phys. Lett. B253 (1991) 212.
13. M. I. Polikarpov and K. Yee, Phys. Lett. B316 (1993) 333.
14. K. Yee, preprint LSUHEP-1010-93 accepted by Mod. Phys. Lett. A.

FINITE TEMPERATURE EFFECTIVE POTENTIAL FOR SPONTANEOUSLY BROKEN $\lambda\Phi^4$ THEORY

HERBERT NACHBAGAUER *
Laboratoire de Physique Théorique ENSLAPP
Chemin de Bellevue, BP 110, F-74941 Annecy-le-Vieux Cedex, France

ABSTRACT

We present a self-consistent calculation of the finite temperature effective potential for $\lambda\Phi^4$ theory in four dimensions using a composite operator effective action. We find that in a spontaneously broken theory not only the so-called daisy and superdaisy graphs contribute to the next-to-leading order thermal mass, but also resummed non-local diagrams are of the same order, thus altering the effective potential at small effective mass.

1. Introduction

The study of $\lambda\Phi^4$-theory at finite temperature is of great interest for a wide field of applications. A self-interacting scalar field serving as a simple model for the Higgs particle in the standard model of electroweak interactions may allow the study of symmetry changing phase transitions. In fact, the order of the electroweak phase transition plays a crucial rôle in the framework of cosmological scenarios as well as for the badly understood process of baryogenesis[1]. Despite the simplicity of the model, one may at least hope to gain some insight in the mechanism of the phase transition. Moreover, the theory is a suitable test ground for analytic non-perturbative methods, e.g. variational methods[2] as well as for lattice simulations.

High temperature symmetry restoration in a spontaneously broken theory was already noted by Kirzhnits and Linde[3] and worked out quantitatively subsequently[4]. The convenient tool for studying the behavior of the theory turns out to be the effective potential. Whereas the approximate critical temperature is already determined by the one loop potential, the order of the phase transition depends on the detailed shape of the potential requiring also the analysis of higher loop contributions, even for small coupling constant.

In particular, one finds quantizing the theory around the classical non-trivial minimum that the one loop self-energy at high temperature behaves like $m_T \sim \sqrt{\lambda}T$, independent of external momenta and the value of the classical minimum. Thus, for large T the thermal mass dominates over the tree-level mass and the minimum of the effective action becomes the trivial one. The theory exhibits two important features: non-temperature stable vacuum and effective temperature dependent mass. It was consequently proposed[2] to move the tree level mass into the interaction part of the Lagrangian and to start perturbation theory with the free Lagrangian

*email: herby @ lapphp8.in2p3.fr

$\frac{1}{2}\Phi(\Box+\Omega^2)\Phi$ including a yet undetermined mass parameter Ω. Then one calculates the effective potential and fixes the parameter by the 'principle of minimal sensitivity' $\partial V(\Phi_{min})/\partial\Omega = 0$.

A more systematic approach of a self-consistent loop expansion was developed already some time ago by Cornwall, Jackiw and Tomboulis (CJT) in their effective action formalism for composite operators[5]. The basic idea is to introduce a bilocal mass operator instead of the local mass Ω in the generating functional. Then one defines a generalized effective action as the double Legendre transform of the generating functional, which is now not only a functional of the expectation value of the field, $\phi = \langle\Phi\rangle$ but also depends on the expectation value of the time-ordered product $\mathrm{T}\Phi(x)\Phi(y)$. The principle of minimal sensitivity gets replaced by the functional condition $\delta\Gamma(\phi,G)/\delta G(x,y) = 0$ which is employed to solve for the unknown exact two-point Green function $G_0(\phi;x,y)$, and the conventional effective action can be reconstructed as $\Gamma(\phi, G_0(\phi;x,y))$. For a given action $I(\Phi)$ one finds[5]

$$\Gamma(\phi,G) = I(\phi) + \frac{1}{2}\mathrm{Tr}\ln(D_0G^{-1}) + \frac{1}{2}\mathrm{Tr}(D^{-1}G-1) + \Gamma^{(2)}(\phi,G),$$

$$D^{-1}(\phi;x,y) = \frac{\delta^2 I(\phi)}{\delta\phi(x)\delta\phi(y)},$$

where $D_0(x,y)$ is the free propagator derived from the part of the action which is quadratic in the fields. The quantity $\Gamma^{(2)}(\phi,G)$ contains all two–loop contributions and higher and has to be calculated as follows: Shift the field Φ in the classical action by ϕ. Then $I(\Phi+\phi)$ contains terms cubic and higher in Φ which define the vertices. $\Gamma^{(2)}(\phi,G)$ contains all two-particle-irreducible vacuum graphs with the given set of vertices and the propagators replaced by $G(x,y)$. The gap equation now reads

$$G^{-1}(x,y) = D^{-1}(x,y) + 2\frac{\delta\Gamma^{(2)}(\phi,G)}{\delta G(x,y)}.$$

2. $\lambda\Phi^4$ - theory

The graphs naively contributing up to order λ^2 are shown in Fig. 1. Although the non-local contribution Fig. 1b is formally of order λ^2 it contributes to order λ in the false vacuum since already the tree-level value of the expectation value of the field is $\phi \sim m/\sqrt{\lambda}$. Consequently, the 3-vertex given by $\lambda\phi$ is of order $\sqrt{\lambda}$ which means that in the case of a spontaneously broken theory the non-local contribution (b) is as important as the local one (a). We will show that this is also true in the quantized theory. Thus, the Hartree–Fock approximation involving only the local graph which corresponds to the superdaisy or hard thermal loop resummation at finite temperature fails to be consistent. We therefore include the non-local graph Fig. 1b in the self energy and the gap equation in Fourier-space reads

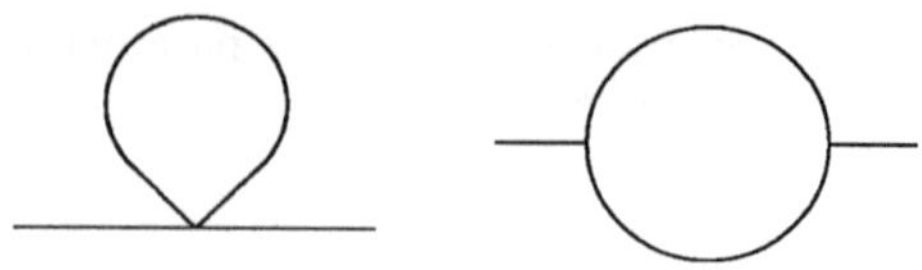

Fig. 1. Feynman graphs contributing to lowest order to the self energy. Lines denote the function $G(x, y)$, the three-vertex is given by $-\lambda\phi/3!$ and the four-vertex by $-\lambda/4!$.

$$G^{-1}(\vec{k}, \omega) = \vec{k}^2 + \omega^2 + (-m^2 + \frac{\lambda}{2}\phi^2) + \frac{\lambda}{2} T \sum_{\omega'} \int \frac{d^3p}{(2\pi)^3} G(\omega', \vec{p}) + \\ + \frac{1}{2}(\lambda\phi)^2 T \sum_{\omega'} \int \frac{d^3p}{(2\pi)^3} G(\omega, \vec{p}) G(\omega + \omega', \vec{p} + \vec{k}) \qquad (1)$$

where we have written down the expression in imaginary time formalism. The first term corresponding to graph (a) does not depend on external momenta and energies and thus can be treated as a mass term. A solution for the gap equation involving only the constant term can easily be found[6] and corresponds to superdaisy resummation. However, we are interested in the behavior of a broken theory, where the last term in Eq. (1) is of the same order. We employ the Ansatz

$$G^{-1}(\omega, \vec{k}) = \vec{k}^2 + \omega^2 + \mu^2 + \Pi(\omega, \vec{k})$$

and subsequently absorb all local contributions in a mass parameter μ which will be determined self-consistently afterwards. Plugging the Ansatz into Eq. (1) recasts the gap equation into an equation for $\Pi(\omega, \vec{k})$. Simple power counting now reveals that the second term in Eq. (1) diverges for large momenta $\vec{p}$. However, one may absorb the divergent part in a mass renormalization rendering the self energy $\Pi(\omega, \vec{k})$ a finite quantity.

Since we are only interested in the high temperature ($T \gg \mu$) regime of the theory one may further simplify Eq. (1) by taking only the $\omega' = 0$ mode in the second sum, since all other terms are at least suppressed by a factor $1/T$. This in turn means that it is also only the zero mode of the self energy, that contributes most. Putting $\Pi(k) = \Pi(\omega = 0, |\vec{k}|)$ we end up with the integral equation

$$\Pi(k) = \\ \frac{T}{2}(\lambda\phi)^2 \int_0^\infty \frac{dp p^2}{(2\pi)^2} \int_{-1}^{1} \frac{dz}{(p^2 + \mu^2 + \Pi(p))(p^2 + k^2 + 2pkz + \mu^2 + \Pi(\sqrt{p^2 + k^2 + 2pkz}))} \qquad (2)$$

which we are now going to solve.

3. Solution of the gap equation and effective potential

First, from Eq. (2) one immediately reads off the high momentum behavior $\Pi(k \to \infty) = 0$. Secondly, we note that due to the non-local character of the integral equation we have to know the function $\Pi(k)$ for all momenta, even if we were only interested in the low momentum behavior of the self energy. However, numerical study of this equation reveals that $\Pi(k)$ can be approximated by

$$\Pi(k) = \mu^2 \frac{A(\kappa)^2}{a(\kappa)^2 + (k/\mu)^2}, \qquad \kappa = \frac{T(\lambda\phi)^2}{8\pi^2\mu^3} \tag{3}$$

where we introduced the dimensionless quantity κ. Since $\Pi(k)$ is of relevance in the denominator of Eq. (2) only in the infrared region $k \ll \mu$, we choose $A(\kappa)$ and $a(\kappa)$ such that Eq. (2) is fulfilled exactly for $\Pi(0)$ and $\Pi''(0)$. $A(\kappa)$ and $a(\kappa)$ are then determined by two coupled complicated transcendental equations[7]. We note that the Ansatz (3) may be interpreted as Yukawa-like potential with screening mass μa and strength $\sim A^2$

Putting all together one finally gets the gap equation for the mass parameter μ,

$$-m^2 + \frac{\lambda}{2}\phi^2 + \frac{\lambda}{2}(\frac{T^2}{12} + \frac{T\mu}{8\pi} f(\kappa)) = \mu^2 \tag{4}$$

where we have absorbed the divergencies in redefined quantities m, λ and $f = f(A(\kappa), a(\kappa))$. We find $f(0) = -2$ which corresponds to the value if one only had included the local contribution of graph (a). On the other hand for large values of the parameter κ, which in the non-trivial vacuum $\phi \neq 0$ means small mass parameter μ, one finds the asymptotic behavior $f(\kappa) \sim C\kappa^{\frac{1}{3}}$ with $C \sim -0.0176\ldots$. Remarkably enough, the mass parameter cancels out in the next to leading order contribution in Eq. (4) and we are left with a term $T\mu f(\kappa) \sim T^{\frac{4}{3}}(\lambda\phi)^{\frac{2}{3}}$. In view of the result, this quite surprising behavior can already estimated from Eq. (2). Rescaling by μ such that the self-energy becomes a dimensionless, the equation may roughly be read as $\Pi^3 \sim \kappa$.

We thus find that the non-local term included in this investigation significantly alters the gap equation (4) at vanishing mass μ, that means when perturbation theory becomes problematic and infrared effects enter the game.

Since, to lowest order, the effective mass is just the second derivative of the effective potential with respect to ϕ, we are now able to reconstruct the lowest order effective potential just by twice integrating the gap equation over to ϕ. Keeping only the asymptotic form of f in the gap equation, different temperature regions may distinguished by their limiting values $T_i = m\sqrt{\frac{24}{\lambda}}(1 - F_i\lambda)^{-\frac{1}{2}}$, where $F_i = (0, \frac{216}{20\sqrt{10}}|C|^{3/2}, \frac{24\sqrt{3}}{5\sqrt{5}}|C|^{3/2}, 24 \cdot 3^{-3/2}|C|^{3/2})$. The potential has one non-trivial minimum for $T < T_0$ and develops another for $T_0 < T < T_1$. At $T = T_1$ the trivial and the non-trivial minimum have equal depth. As temperature is further raised until $T = T_2$, the non-trivial minimum vanishes completely. There exists however still a value $\phi \neq 0$, where the bending of the potential changes sign as long as the

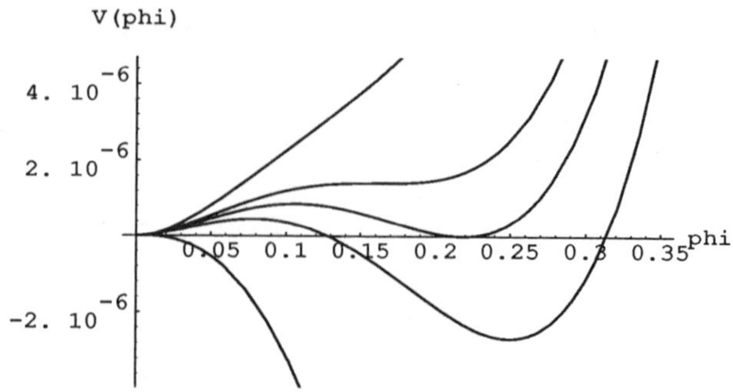

Fig. 2. The finite temperature effective potential for $\lambda = 0.1$ in units of the redefined mass m. The upper three curves correspond to the temperatures T_3, T_2, T_1 respectively, the lowest curve to the temperature T_0.

temperature is lower than T_3. The potential for $\lambda = 0.1$ is drawn in Fig. 2 for the corresponding values of the limiting temperatures where in addition, we added a curve for a temperature between $T_0 < T < T_1$.

4. Conclusion

We have shown, to lowest order however, that the naively expected resummed effective potential in a spontaneously broken theory fails to be consistent. Instead we encountered an additional contribution to the effective mass in the gap equation behaving like $\phi^{2/3}$ which cannot be found in a perturbative calculation. This may point towards a breakdown of the resummed perturbative calculation thus encouraging the extensive study of alternative methods. Nevertheless, we find that the phase transition is still of first order. A refinement of the present approach is currently under investigation.

1. E.W. Kolb and M. S. Turner, *The Early Universe*, (Addison-Wesley, Reading, MA, 1990).
2. M. Stevenson, *Z. Phys.* *C* 24 (1984) 87; *Phys. Rev.* **D32** (1985) 1389; N. Banerjee and S. Mallik, *Phys. Rev.* **D43** (1991) 3368.
3. D. Kirzhnits, *Pris'ma Zh. Eksp. Teor. Fiz.* **15** (1972) 374 [JETP Lett. **15** (1972) 529]; D. Kirzhnits and A. Linde, *Phys. Lett.* **42B** (1972) 471.
4. L. Dolan and R. Jackiw, *Phys. Rev.* **D9** (1974) 3320; S. Weinberg, *Phys. Rev.* **D9** (1974) 3357; D. Kirzhnits and A. Linde, Zh. Eksp. Teor. Fiz. **67** (1975) 1263 [Sov. Phys. JETP **40** (1975) 628].
5. J. M. Cornwall, R. Jackiw and E. Tomboulis, *Phys. Rev.* **D10** (1974) 2428.
6. G. Amelino-Camelia and So-Young Pi, *Phys. Rev.* **D47** (1993) 2356.
7. H. Nachbagauer, *Finite temperature effective potential for spontaneously broken $\lambda\Phi^4$ theory*, ENSLAPP-preprint.

THE INFRARED SECTOR OF QCD FROM LATTICE SIMULATIONS *

M. FABER, H. MARKUM, Š. OLEJNÍK, W. SAKULER

Institut für Kernphysik, Technische Universität Wien
A-1040 Vienna, Austria

ABSTRACT

Chiral and topological properties of the QCD vacuum including dynamical fermions were studied on the lattice. Topological charges were computed using two local operators of the charge density and the method of cooling. We observe a suppression of chiral and topological fluctuations in the vicinity of static quarks and mesons. Our results further indicate that known relations between global chiral and topological quantities hold also locally.

1. Introduction and Theory

The physical mechanism of quark confinement in QCD is still not completely understood. To get further insight into the confinement mechanism we investigate chiral and topological properties of the full QCD vacuum including sea quarks from first principles on a fourdimensional hypercubic euclidean lattice. There exists a deep relationship between chirality and topology given by the Atiyah-Singer index theorem and a chiral Ward identity which yields a linear relation between the topological susceptibility and the chiral condensate.[1] It is instructive to study physical quantities also locally around color sources. In this work we analyze chiral and topological fluctuations around static mesons at finite temperature by computing correlation functions between Polyakov loops and the chiral condensate and the topological charge density, respectively.

Global topological properties of a gauge-field configuration are characterized by the topological charge $Q = \int d^4x\, q(x)$ which takes integer values in the continuum and is the integral over the topological charge density:

$$q(x) = \frac{g^2}{64\pi^2} \epsilon^{\mu\nu\rho\sigma} F^a_{\mu\nu}(x) F^a_{\rho\sigma}(x). \tag{1}$$

The formulation of QCD on a lattice and its Monte Carlo simulation are powerful tools for studying nonperturbative aspects of the theory. A straightforward way of defining the topological charge of a lattice configuration is to choose some local lattice operator of topological charge density $q_L(x)$ and define the lattice topological charge

*Supported in part by "Fonds zur Förderung der wissenschaftlichen Forschung" under Contract No. P9428-PHY

Q_L as a sum of such a density over all lattice sites:

$$Q_L = \sum_x q_L(x) \ . \tag{2}$$

However, there is no unique way of defining a lattice operator that converges, in the naive continuum limit, to the continuum topological charge density. We have chosen two commonly used definitions:

$$q_L(x) = -\frac{1}{2^4 32\pi^2} \sum_{\mu\nu\rho\sigma=\pm1}^{\pm4} \tilde{\epsilon}_{\mu\nu\rho\sigma}$$
$$Tr\left[U_\mu(x)U_\nu(x+\mu)U_\mu^\dagger(x+\nu)U_\nu^\dagger(x)U_\rho(x)U_\sigma(x+\rho)U_\rho^\dagger(x+\sigma)U_\sigma^\dagger(x)\right] \tag{3}$$

("plaquette" definition[2,3]), where $U_\mu(x)$ denotes a lattice gauge field element describing the color parallel transport from x to $x+\mu$, and

$$q_L(x) = -\frac{1}{2^4 32\pi^2} \sum_{\mu\nu\rho\sigma=\pm1}^{\pm4} \tilde{\epsilon}_{\mu\nu\rho\sigma} Tr\left[U_\mu(x)U_\nu(x+\mu)U_\rho(x+\mu+\nu)\right.$$
$$\left.U_\sigma(x+\mu+\nu+\rho)U_\mu^\dagger(x+\nu+\rho+\sigma)U_\nu^\dagger(x+\rho+\sigma)U_\rho^\dagger(x+\sigma)U_\sigma^\dagger(x)\right] \tag{4}$$

("hypercube" definition, proposed by Di Vecchia *et al.*[2]). These choices were motivated by our aim of investigating the distributions of the topological charge density *locally* around static quarks and antiquarks.

An apparent drawback of both these definitions is the fact that at finite inverse gluon coupling $\beta = 6/g^2$ and finite lattice spacing a neither of them yields integer topological charges. This is known to be due to renormalization factors relating the continuum and lattice operators of topological charge.[4,5] The relation between the lattice and continuum topological charge density can be written as

$$q_L(x) = a^4 Z(\beta) q(x) + O(a^6), \tag{5}$$

where the multiplicative renormalization constant $Z(\beta)$ which depends also on the definition of the charge approaches 1 for $\beta \to \infty$, but is rather small for typical values of β used in Monte Carlo simulations. A simple procedure that enables one to get rid of the renormalization constants, while preserving physical information contained in lattice configurations, is the *cooling* method. The idea of cooling relies on the fact known from the continuum theory that local changes of a field configuration are unable to change its global topological properties. Technically, cooling is subsequent minimizing of the action with respect to all individual links. It reduces quantum fluctuations, and therefore suppresses differences between different definitions of the topological charge density. If done carefully, one expects the essential features of the confinement mechanism remain unchanged.

A static quark is represented by the Polyakov loop $L(\vec{r})$ which describes the propagation of a charge with infinite mass

$$L(\vec{r}) = \frac{1}{3} Tr \prod_{n=1}^{N_t} U_4(\vec{r}, na). \tag{6}$$

To obtain the local distribution of the chiral condensate $\bar{\psi}\psi(r)$ and the topological charge density $q(r)$ around single quarks and around a meson with quark-antiquark separation d we calculate the correlation functions

$$\langle L(0)\bar{\psi}\psi(r)\rangle,\ \langle L(0)q^2(r)\rangle \quad \text{and} \quad \langle L(0)L^\dagger(d)q^2(r)\rangle \tag{7}$$

and normalize them to the corresponding vacuum expectation values. We take the square of the topological charge density in Eq. (7). The net topological charge of the vacuum is zero, therefore it is of no use to calculate correlations with $q(x)$. The square of the density is then the simplest choice leading to nontrivial correlations.

Our simulations were performed on an $8^3 \times 4$ lattice with periodic boundary conditions using the Metropolis algorithm. The chiral and topological properties were studied in full QCD at inverse gluon coupling $\beta = 5.2$ employing the $SU(3)$ Wilson action for the gluons and the Kogut-Susskind action with flavor number $n_f = 3$ and bare quark mass $ma = 0.1$ for the fermions. We made 10^5 iterations and measured our observables after every 50th iteration. Each of these 2000 configurations was subjected to 50 cooling steps, using the cooling method of Ref. 6.

2. Results and Discussion

In Fig. 1a,b the chiral condensate and the square of the topological charge density (for both definitions) around a single static quark are shown after 0, 5, and 10 cooling steps, respectively. For both quantities a suppression around the color source is clearly visible and becomes more pronounced with increasing number of cooling steps. Especially for the topological charge density without cooling practically no effect is visible, but cooling soon uncovers a suppression of the density. Results from both definitions of the density agree after a few cooling steps. Here one should stress that this agreement is far from trivial. This shows that not only global charges differ by multiplicative renormalization constants approaching 1 during cooling, but the same remains true also locally. One should note that the charge densities in Fig. 1b were calculated at different distances for both definitions. This is due to the fact that the "plaquette" definition gives the topological charge density in a lattice site, while the "hypercube" one rather in the center of the hypercube. The agreement is nevertheless perfect.

The decrease of the chiral condensate in the vicinity of a static quark reflects a local restoration of chiral symmetry. The suppression of both the chiral condensate and the topological charge density indicates a local change of the vacuum structure around external sources from a nonperturbative into a perturbative one. This is in agreement with basic ideas in bag models where it is assumed that hadrons are bubbles of perturbative medium around their constituents, stabilized by the pressure of the nonperturbative medium outside. The qualitatively equal behavior of both quantities further indicates also a local relationship between topological and chiral observables besides the known global relation between the topological susceptibility

Chiral and Topological Densities around Static Quark

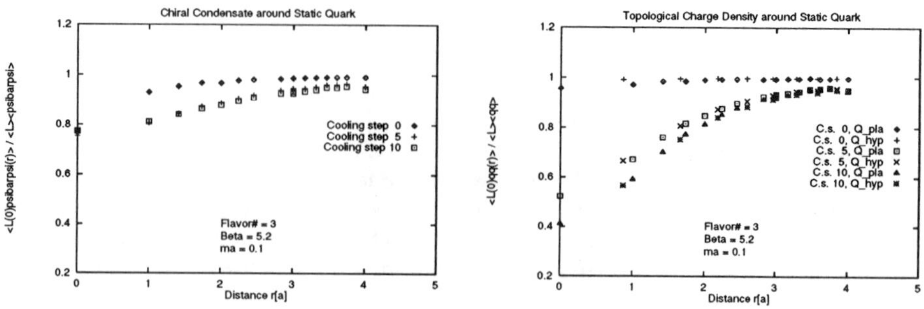

(a) (b)

Fig. 1: Chiral condensate (a) and squared topological charge density (b) around a single color source. The suppression of both quantities becomes more pronounced with increasing number of cooling steps. This effect reflects the local change of the vacuum structure from a nonperturbative into a perturbative one as suggested in phenomenological models.

Topological Charge Densities around Static Meson

$$\frac{\langle L(0)L^{\dagger}(d)q^{2}(r)\rangle}{\langle L(0)L^{\dagger}(d)\rangle\langle q^{2}\rangle}$$

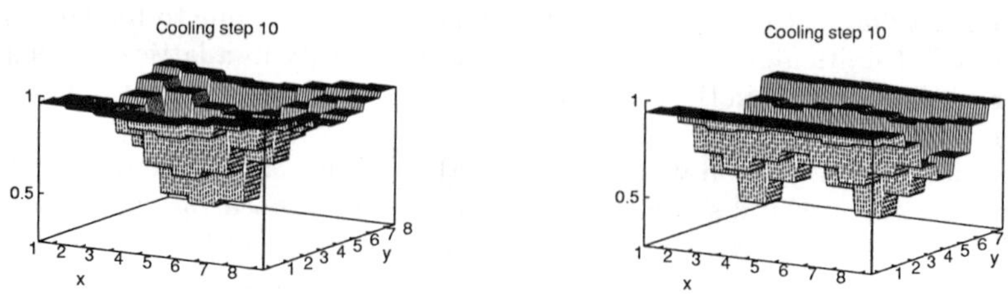

$d = 2$ $d = 4$

Fig. 2: Squared topological charge density around a static quark-antiquark pair at $q\bar{q}$-distances $d = 2, 4$ for 10 cooling steps. At $d = 4$ the suppression is peaked near the position of the color sources indicating the breaking of the flux tube due to hadronization.

χ and the chiral condensate $\langle\bar{\psi}\psi\rangle$ given by

$$\chi = \frac{\langle Q^2\rangle}{V} = \frac{m}{n_f}\langle\bar{\psi}\psi\rangle \ . \tag{8}$$

In Fig. 2 distributions of the topological charge density (squared) around a meson (normalized to $\langle L(0)L^\dagger(d)\rangle\langle q^2\rangle\rangle$) are presented at quark-antiquark separation $d = 2$ and $d = 4$ for 10 cooling steps. The suppression around the external sources is again clearly visible. At $q\bar{q}$-distance $d = 4$ the effect is peaked near the position of the static quarks. Comparing with results[7] of previous calculations for pure QCD where we observed a valley of reduced topological charge reflecting the formation of a flux tube between the quark-antiquark pair, in the presence of sea quarks we recognize the string breaking due to hadronization.

To summarize, our results show that topological objects at least reflect the confinement mechanism of QCD. We observe flux tube formation in pure QCD and string breaking when dynamical fermions are switched on. Both the topological charge density and the chiral density are suppressed in the vicinity of static quarks which describes the local change from a nonperturbative into a perturbative vacuum. The data give evidence that the relationship between topological and chiral quantities holds also locally. Further investigations are necessary to clarify the role of topological charges with respect to confinement. For that reason we plan to study the connection between topological charges and color magnetic monopoles which are topological objects as well and are promising candidates for describing the confinement mechanism.

3. References

1. R. J. Crewther, *Phys. Lett.* **B70** (1977) 349.
2. P. Di Vecchia *et al.*, *Nucl. Phys.* **B192** (1981) 392;
   Phys. Lett. **B108** (1982) 323.
3. K. Ishikawa *et al.*, *Phys. Lett.* **B128** (1983) 309;
   N. V. Makhaldiani and M. Müller-Preußker, *JETP Lett.* **37** (1983) 523.
4. M. Campostrini, A. Di Giacomo and H. Panagopoulos,
   Phys. Lett. **B212** (1988) 206.
5. M. Campostrini *et al.*, *Nucl. Phys.* **B329** (1990) 683.
6. M. Teper, *Phys. Lett.* **B162** (1985) 357; **B171** (1986) 81, 86;
   J. Hoek, M. Teper and J. Waterhouse, *Nucl. Phys.* **B288** (1987) 589.
7. M. Faber *et al.*, *Nucl. Phys.* **B** (Proc. Suppl.) **34** (1994) 167;
   Phys. Lett. **B334** (1994) 145.

Exact $m_{quark} \neq 0$ Condensates in $QCD_{1+1}(N_C \to \infty)$

M.Burkardt
Institute for Nuclear Theory, Univ. of Washington, Seattle, WA 98195

In the limit of an infinite number of colors, we derive an analytic expression for the quark condensate in QCD_{1+1} as a function of the quark mass and the gauge coupling constant. For zero quark mass, a nonvanishing quark condensate is obtained. Nevertheless, we prove that there is no phase transition as a function of the quark mass. It is furthermore shown that the expansion of $\langle 0|\bar{\psi}\psi|0\rangle$ in the gauge coupling has zero radius of convergence but that the perturbation series is Borel summable with finite radius of convergence. The nonanalytic behavior $\langle 0|\bar{\psi}\psi|0\rangle \overset{m_q\to 0}{\sim} -N_C\sqrt{G^2}$ can only be obtained by summing the perturbation series to infinite order.

1. Introduction

What is interesting about the quark condensate in QCD_{1+1}? Zhitnitsky [1], using QCD-sum rule techniques, derived an exact result for the condensate in the limit of an infinite number of colors* and in the limit $m_q \to 0$

$$\langle 0|\bar{\psi}\psi|0\rangle\Big|_{m_q=0} = -\frac{N_C}{\sqrt{12}}\sqrt{\frac{g^2 C_F}{\pi}}, \tag{1}$$

where $C_F = \frac{N_C^2-1}{2N_C}$ and $G^2 \equiv \frac{g^2 C_F}{\pi}$ is held fixed as $N_C \to \infty$. This result is remarkable in several respects: Firstly, $\langle 0|\bar{\psi}\psi|0\rangle\Big|_{m_q=0}$ is nonvanishing, indicating spontaneous breakdown of chiral symmetry. Secondly, the condensate is nonanalytic in the coupling constant G^2, thus indicating nonperturbative effects: although it seems natural from dimensional analysis that $\langle 0|\bar{\psi}\psi|0\rangle \propto \sqrt{G^2}$ for small quark masses, it is impossible to obtain such a behavior in perturbation theory where one can only generate terms $\propto G^{2n}$. Thirdly, one may suspect that there is a phase transition in QCD_{1+1}.

Since the coupling constant g in QCD_{1+1} carries the dimension of a mass, the theory is super-renormalizable and the scale is set both by the coupling and by the mass. In practice this implies that $\langle 0|\bar{\psi}\psi|0\rangle$ can (up to some dimensionful overall factor) depend on G^2 only through the combination $\alpha \equiv G^2/m_q^2$. Therefore, in order to address the abovementioned issues of nonperturbative effects and a possible phase transition, it is necessary to consider nonzero quark masses.

2. $\langle 0|\bar{\psi}\psi|0\rangle$ from sum rules

For nonzero quark masses, the vacuum expectation value of the scalar density diverges

*Note that Coleman's theorem [2] prohibits spontaneous breakdown of chiral symmetry for any finite number of colors, since this is a $1+1$-dimensional model.

already for free fields

$$\langle 0|\bar{\psi}\psi|0\rangle = \frac{N_C}{2\pi} m_q \log \frac{\Lambda^2}{m_q^2}. \tag{2}$$

However, due to the mild UV behavior in $1+1$ dimensions (QCD_{1+1} is super-renormalizable) it is sufficient to subtract the free field expectation value (i.e. to "normal order") to render $\langle 0|\bar{\psi}\psi|0\rangle$ finite. This motivates the definition

$$\langle 0|\bar{\psi}\psi|0\rangle\Big|_{ren} \equiv \langle 0|\bar{\psi}\psi|0\rangle - \langle 0|\bar{\psi}\psi|0\rangle\Big|_{g=0}. \tag{3}$$

The condensate itself can be evaluated using current algebra

$$\begin{aligned} 0 &= \lim_{q\to 0} iq^\mu \int d^2x e^{iqx} \langle 0|T\left(\bar{\psi}\gamma_\mu\gamma_5\psi(x)\bar{\psi}i\gamma_5\psi(0)\right)|0\rangle \\ &= -\langle 0|\bar{\psi}\psi|0\rangle - 2m_q \int d^2x \langle 0|T\left(\bar{\psi}i\gamma_5\psi(x)\bar{\psi}i\gamma_5\psi(0)\right)|0\rangle. \end{aligned} \tag{4}$$

Upon inserting a complete set of meson states† one thus obtains

$$\langle 0|\bar{\psi}\psi|0\rangle = -m_q \sum_n \frac{f_P^2(n)}{M_n^2}, \tag{5}$$

where $f_P(n) \equiv \langle 0|\bar{\psi}i\gamma_5\psi|n\rangle = \sqrt{\frac{N_C}{\pi}} \int_0^1 dx \frac{m_q/2}{x(1-x)}\phi_n(x)$ and the wavefunctions ϕ_n and invariant masses M_n^2 are obtained from solving 't Hooft's bound state equation for mesons in QCD_{1+1}

$$M_n^2\phi_n(x) = \frac{m_q^2}{x(1-x)}\phi_n(x) + G^2 \int_0^1 dy \frac{\phi_n(x)-\phi_n(y)}{(x-y)^2}. \tag{6}$$

The variable x corresponds to the light-front momentum fraction carried by the quark in the meson. Note that 't Hooft's equation was been derived using light-front quantization — we will return to this point below.

In the limit of highly excited mesons, the masses and coupling constants scale: $M_n^2 \stackrel{n\to\infty}{\longrightarrow} n\pi^2G^2$, $f_P(n) \stackrel{n\to\infty}{\longrightarrow} \sqrt{N_C\pi G^2}$ [3] and thus the sum in Eq.(5) diverges logarithmically. Of course this only reflects the free field divergence (2). In order to regularize Eq.(5) in a gauge invariant way we introduce an invariant mass cutoff and obtain

$$\langle 0|\bar{\psi}\psi|0\rangle\Big|_{ren} = -m_q \lim_{\Lambda\to\infty}\left[\sum_n \frac{f_P^2(n)}{M_n^2\left(1+M_n^2/\Lambda^2\right)} - \text{“}g=0\text{''}\right]. \tag{7}$$

Eq.(7) can be used to calculate $\langle 0|\bar{\psi}\psi|0\rangle\Big|_{ren}$ numerically with high precision.

In order to generate an exact result we will use a trick and replace the sum in Eq.(7), where both small and large n contribute, by a sum which is dominated by

†Because we are working at leading order in $1/N_C$, the sum over one meson states saturates the operator product in Eq.(4).

large n only. Due to lack of space, only the basic ideas of the derivation will be sketched here — a more detailed discussion of the limit $\Lambda \to \infty$ can be found in Ref.[4]. For this purpose, let us consider [3]

$$G(x,\Lambda) \equiv \sqrt{\frac{N_C}{\pi}} \sum_n \phi_n(x) \left(\frac{1}{M_n^2} - \frac{1}{M_n^2+\Lambda^2} \right) f_P(n). \tag{8}$$

Obviously this "Green's function" can be used to compute the condensate, via

$$\langle 0|\bar{\psi}\psi|0\rangle\Big|_{ren} = -m_q^2 \lim_{\Lambda\to\infty} \int_0^1 \frac{dx}{x} \left[G(x,\Lambda) - G(x,\Lambda)|_{g=0} \right]. \tag{9}$$

From the equation of motion (6) one can show that $f_P(n) = \frac{M_n^2}{2m_q}\sqrt{\frac{N_C}{\pi}} \int_0^1 dy \phi_n(y)$. If one furthermore invokes completeness of 't Hooft's wavefunctions, i.e. $\sum_n \phi_n(x)\phi_n(y) = \delta(x-y)$ one can simplify the first term in $G(x,\Lambda)$, yielding $\sqrt{\frac{N_C}{\pi}} \sum_n \phi_n(x) f_P(n)/M_n^2 = \frac{N_C}{\pi}\frac{1}{2m_q}$ — independent of x and the coupling constant. This term thus drops out completely when we subtract the free field Green's function.

The crucial point is that, for $\Lambda^2 \to \infty$, the remaining term in $G(x,\Lambda)$ is dominated by the $n \to \infty$: each individual meson yields a negligible contribution $\propto (M_n^2 + \Lambda^2)^{-1}$ when we send the cutoff to infinity, and a nonzero result arises only from the summation over infinitely many highly excited meson states. We have thus succeeded in converting the *low energy sum rule* (7) into a *high energy sum rule* and we can now make use of the abovementioned scaling properties of meson masses M_n^2 and coupling constants $f_P(n)$ as well as of the wavefunction itself $\phi_n(x) \stackrel{n\to\infty}{\longrightarrow} \Phi(M_n^2 x)$. The scaled wavefunction $\Phi(z) \equiv \lim_{n\to\infty} \phi_n(z/M_n^2)$ satisfies the integral equation [3]

$$\Phi(z) = \frac{m_q^2}{z}\Phi(z) + G^2 \int_0^\infty dy \frac{\Phi(z)-\Phi(y)}{(x-y)^2}. \tag{10}$$

In terms of these scaled quantities we thus find

$$\langle 0|\bar{\psi}\psi|0\rangle\Big|_{ren} = \frac{N_C}{\pi} m_q^2 \int_0^1 \frac{dx}{x} \left[\sqrt{G^2} \sum_{n=1,3,..} \frac{\Phi(n\pi^2 G^2 x)}{n\pi^2 G^2 + \Lambda^2} - \text{“}g=0\text{''} \right] \tag{11}$$

Upon performing the substitution $z = n\pi^2 G^2 x$, replacing $\sum_{n=1,3,..} \to (2\pi^2 G^2 x)^{-1} \int_0^\infty dz$ ‡ and performing the x-integral one ends up with

$$\langle 0|\bar{\psi}\psi|0\rangle\Big|_{ren} = \frac{N_C}{\pi} m_q^2 \left[\frac{1}{\sqrt{G^2}} \int_0^\infty \frac{dz \log z}{z} \Phi(z) - \text{“}g=0\text{''} \right]. \tag{12}$$

Note that $\frac{1}{\sqrt{G^2}} \int_0^\infty \frac{dz}{z}\Phi(z) = \frac{\pi}{m_q}$ [5] is independent of G^2 and it does not matter that the argument of the logarithm is dimensionful, since a free theory subtraction is

‡This is exact for $\Lambda \to \infty$ since the series receives nonvanishing contributions only from the $n \to \infty$ region.

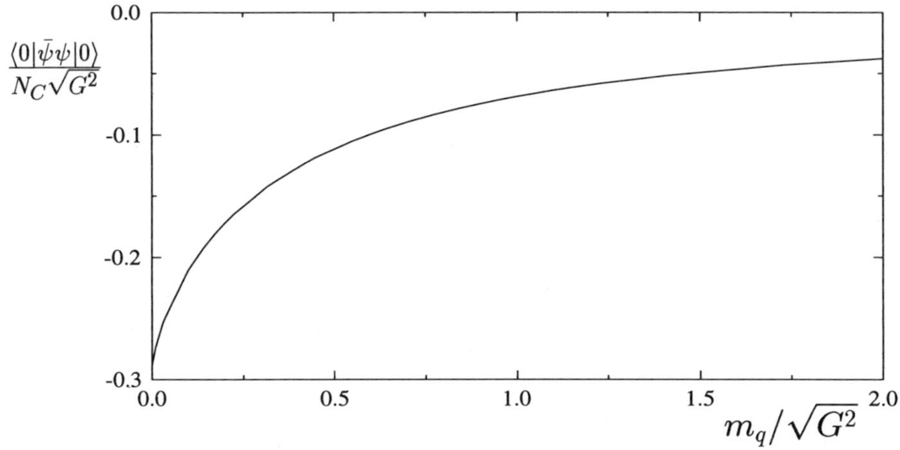

Figure 1: Renormalized quark condensate as a function of the quark mass. Both in units of the effective coupling $G^2 = g^2 C_F/\pi$. Note the absence of singularities.

performed. While 't Hooft's equation cannot been solved exactly, the scaling equation can be solved analytically [5]. The lengthy expression for $\Phi(z)$ will not be displayed here and we merely give the result for the renormalized condensate

$$\langle 0|\bar{\psi}\psi|0\rangle\Big|_{ren} = \frac{m_q N_C}{2\pi}\left\{\log(\pi\alpha) - 1 - \gamma_E + \left(1-\frac{1}{\alpha}\right)[(1-\alpha)I(\alpha) - \log 4]\right\}, \quad (13)$$

where $\alpha = G^2/m_q^2$, $\gamma_E = .5772..$ is Euler's constant and

$$I(\alpha) = \int_0^\infty \frac{dy}{y^2}\frac{1 - \frac{y}{\sinh y \cosh y}}{[\alpha(y\coth y - 1) + 1]}. \quad (14)$$

This result is exact for $N_C \to \infty$ and *all* quark masses. In the limit $\alpha \to \infty$ one recovers Zhitnitsky's result Eq.(1). Furthermore, one can verify that the exact result coincides with the numerical evaluation of Eq.(7). In the limit $\alpha \to 0$ the condensate vanishes, which is not surprising since we have subtracted the free field result.

3. Discussion

We have started from 't Hooft's equation which is based on light-front quantization. The light-front vacuum is trivial, i.e. identical to the Fock space vacuum [6]. Nevertheless, using current algebra and sum rule techniques, we obtained a nonzero result for the quark condensates. The result we obtained agrees with numerical calculations using equal time quantization (see Refs.[6, 7] for $m_q = 0$ and Ref.[8] for the general case).

The exact result which we have obtained (14) is an analytic function of α in the complex plane cut along the negative real axis — i.e. there is *no* phase transition. An asymptotic expansion in powers of α yields

$$\frac{2\pi}{m_q N_C} \left. \langle 0|\bar{\psi}\psi|0\rangle \right|_{ren} = \sum_{\nu=1}^{\infty} c_\nu \alpha^\nu , \tag{15}$$

where the coefficients show factorial growth

$$c_\nu \stackrel{\nu\to\infty}{\sim} (-1)^\nu e^{-2} 2^{1-\nu} (\nu - 1)! \quad , \tag{16}$$

i.e., the asymptotic expansion for $\left. \langle 0|\bar{\psi}\psi|0\rangle \right|_{ren}$ is only Borel summable and the Borel series has a finite radius of convergence. Applying the inverse Borel transfer to the Borel summed series one recovers the exact result which reflects the absence of terms like $e^{-\frac{1}{\alpha}}$. I have compared the first three terms in the asymptotic expansion with the perturbative (Feynman diagrams) expansion and found agreement. Nevertheless Eq.(14) is a completely nonperturbative result, because one has to sum up *all* terms in the perturbation series before one obtains the right scaling behavior (1) for small quark mass (large α). It is also *not* sufficient to keep only the asymptotic behavior of the series: it is easy to write down an expression which has the same asymptotic coefficients for large ν but does not yield the desired $\sqrt{\alpha}$ behavior for $m_q \to 0$.

4. References

[1] A. Zhitnitsky, *Phys. Lett.* **165B** (1985) 405.
[2] S. Coleman, *Comm. Math. Phys.* **31** (1973) 259.
[3] M. B. Einhorn, *Phys. Rev.* **D14** (1976) 3451.
[4] M. Burkardt, to be submitted to Phys. Rev. D.
[5] R. C. Brower, W. L. Spence and J. H. Weis, *Phys. Rev.* **D19** (1979) 3024.
[6] F. Lenz, S. Levit, M. Thies and K. Yazaki, *Ann. Phys. (N.Y.)* **208** (1991) 1.
[7] M. Li, *Phys. Rev.* **D34** (1986) 3888.
[8] M. Thies, *private communications.*

THE FREE ENERGY OF HOT QED AT THREE AND A HALF LOOPS

Rajesh R. Parwani*
Service de Physique Theorique, CE-Saclay
F-91191, Gif-sur-Yvette, France

and

Claudio Corianò†
High Energy Physics Division, Argonne National Laboratory
9700 South Cass, Il 60439, USA.

ABSTRACT

The computation of order e^4 and e^5 contributions to the pressure of massless quantum electrodynamics at a temperature T is overviewed.

The thermodynamic properties of a QED plasma may be determined from the partition function which can be obtained perturbatively using techniques borrowed from field theory at zero temperature (T). Indeed, the simplest approach to calculating the partition function, or the free energy, is to use the imaginary-time formalism whereby the Feynman rules are as at $T = 0$ but the energies take on discrete Matsubara values.

However a naive application of these $T = 0$ like Feynman rules soon leads to the appearance of power-like infrared (IR) singularities in diagrams. When these IR singularities from an infinite set of diagrams are resummed, one obtains an expansion in $\sqrt{e^2}$ rather than e^2, where e is the QED coupling. The best-known example of this phenomenon is the e^3 plasmon correction to the free energy of QED first found by Gell-Mann and Brueckner [1] in the nonrelativistic context and later calculated relativistically [2]. Physically, the appearance of these IR singularities and the consequent breakdown of the naive perturbative expansion is due to the Debye screening of electric fields in a plasma. Unlike at $T = 0$, the particles in the plasma are not free (modulo ultraviolet renormalisations) but perpertually under the influence of the surrounding particles.

In the language of field theory, the Debye screening manifests itself by the non-vanishing limit $\Pi_{00}(p_0 = 0, \vec{p} \to 0)$ of the electric polarization operator. To lowest order in the coupling, and in the limit of high temperature, the electric screening mass-squared is given by a one-loop calculation : $m^2 = \Pi^1(p_0 = 0, \vec{p} \to 0) = e^2T^2/3$.

*E-mail: parwani@wasa.saclay.cea.fr
†E-mail: coriano@hep.anl.gov

The presence of the large scale T means that the "loop correction" need not be small compared to the bare propagator. In imaginary time, this is the case for the zero mode (Matsubara frequency) of the electric propagator at small three momentum ($|\vec{p}| \sim eT$). (By contrast, static magnetic fields are unscreened in a QED plasma.)

In general there are three ways that one can account for large corrections, such as Debye screening, in order to restore the perturbative expansion :

(i) Continue with bare Feynman rules and resum by hand dangerous subsets of diagrams. This procedure is possible in simple cases but has the disadvantage that one must carefully identify the relevant diagrams, account for symmetry factors and prevent overcounting; or

(ii) Begin with the non-perturbative skeleton-expansion and truncate down. This is safe from the point of view of symmetry factors and overcounting but is practical only in simple cases; or

(iii) Re-organise the bare Lagrangian by adding and subtracting the dominant non-negligible effects (termed "hard thermal loops" by Braaten and Pisarski [3]). This method has the virtue that one continues with Feynman perturbation theory but with new effective propagators and vertices.

Usually, methods (i) and (ii) are computationally efficient only for static (zero external energy) Greens functions for which all the power counting analysis can be done in imaginary-time. For a non-static Greens function one first requires its physical definition in real time (either by analytic continuation from imaginary time or through the real-time formalism) and then the method advocated in Ref.[3] is probably the most efficient.

Recently we computed the order e^4 (3-loop) contribution to the free energy density of massless QED at temperature T, going beyond the e^3 term known for many years [2]. The method used was (i) with dimensional regularisation being extensively employed to regulate various singularities appearing at intermediate stages of the calculation. From the technical point of view the fourth order calculation required the evaluation of some complicated overlapping three-loop integrals that did not appear in a similar three-loop calculation in ϕ^4 theory [4].

Following the order e^4 calculation in Ref.[5], the e^5 ($3\frac{1}{2}$ loop) result was also obtained [6]. This latter term may be viewed as a correction to the three-loop result as a consequence of Debye screening, just as the e^3 term is a similar correction to the two-loop (order e^2) result. Computationally, the order e^3 calculation is *simpler* than the order e^2 calculation because only the zero mode of the photon is involved and the loop integral along that line becomes three-dimensional. Similarly, the e^5 piece was easier to obtain than the e^4 piece, the sum of complicated diagrams factorizing themselves into a product of simple one-loop integrals. The e^5 contribution was also reconsidered from the point of view of method (ii) in Ref.[7].

Remarkably, because the odd terms (e^3, e^5) are simple to calculate, one is able to write a general identity [6]. Consider the gauge-invariant contribution to the pressure of *massless* QED, at temperature T, at order $e^{2n}, (n \geq 1)$, coming from diagrams with one-fermion loop. Call this contribution P_{2n}^{1F}. Then the order e^{2n+3} piece is obtained by dressing the photons of P_{2n}^{1F} and is given by

$$P_{2n+3}^{1F} = \frac{e^3 T^2 N^{1/2}}{8\pi\sqrt{3}} \frac{\partial^2 P_{2n}^{1F}}{\partial \mu^2}|_{\mu=0} \quad , n \geq 1 . \tag{1}$$

On the right-hand-side of Ref.(1), μ is the chemical potential and N is the number of massless electron flavours. The generalisation of Eq.(1) to massive electrons at non-zero chemical potential is given in Ref.[7].

Thus at least in this one case the "IR problem" of perturbation theory at non-zero temperature has turned out to be a bonus in giving us a cute relation like Eq.(1). Unfortunately the relation is not quite useful since, as yet, we do not have a simple painless algorithm to get the even pieces.

Note added in proof : At time of writing (August 1994), the three-loop free-energy of hot Yang-Mills theory has been calculated [8].

Acknowledgements

R.P thanks Profs. H. Fried and A. White for the oppurtunity to present these "hot" results at this workshop.

References

1. M. Gell-Mann and K.A. Brueckner, *Phys. Rev.* **106** (1957) 364;

2. I.A. Akhiezer and S.V. Peletminskii, *Sov. Phys. JETP* **11**, (1960) 1316.

3. E. Braaten and R. D. Pisarski, *Nucl. Phys.* **B337** (1990) 569.

4. J. Frenkel, A.V. Saa and J.C. Taylor, *Phys. Rev.* **D46** (1992) 3670.

5. C. Corianò and R. Parwani, preprint ANL-HEP-PR-94-02, SPhT/94-054.

6. R. Parwani, preprint SPhT/94-065.

7. R. Parwani and C. Corianò, preprint SPhT/94-086, ANL-HEP-PR-94-32.

8. P. Arnold and C. Zhai, Seattle preprint UW/PT-94-03.

NEW TRUTHS ABOUT OLD FRIENDS

H. M. FRIED
Physics Department, Brown University, Box 1843
Providence, Rhode Island 02912, USA

The Old Friends of the title are the causal Green's functions of potential theory (PTh) which play a fundamental role in quantum field theory :(QFTh):$< x|G_c[A]|y >$, or $< p|G_c[A]|p' >$, describing the propagation of a particle in an arbitrary, specified, external field $A(z)$, from the space–time point x_μ to y_μ, or from 4–momentum p_μ to p'_μ. The New Truths are two important properties of these functions discovered within the past year, in collaboration with Y. Gabellini (Nice)[1] and B. H. J. McKellar (Melbourne).[2] These properties are important because – as Schwinger[3] taught us, long ago – all of the n–point, or correlation functions of QFTh can be expressed as functional integrals (FIs) over symmetrized products of these $G_c[A]$, multiplied by the exponential of the determinantal, closed–loop functional $L[A] = L\{G_c[A]\}$; in QED, for example $L[A] = \text{Tr}\ \ln[1 + ig\,\gamma \cdot AS_c]$. As Fradkin[4] taught us, also long ago, these Green's functions can be given an exact and explicit representation in terms of a FI corresponding to fluctuations of a ficticious 4–velocity $v_\mu(s)$, where s denotes the proper time of the representation.

Any representation involving a FI can always be written in terms of an infinite number of quadratures, that is, an infinite number of ordinary integrations; and this can be done in many different but equivalent ways. For the past 30 years, a basic procedural question – rarely posed and never answered – has been how to approximate such representations for $G_c[A]$ in a sensible, useful, and non–perturbative way. An answer to that question can now be given – the first Truth, above – and when extended to vector interactions, such as QED or QCD, there appears in these representations – our second Truth – the possibility of a quantum version of the chaotic behavior known to appear in a variety of nonlinear, classical systems. These representations are sufficiently robust to permit one to understand the additional effects of including the full, relativistic, radiative corrections corresponding to quantum fluctuations of the fields about the PTh $A(z)$; and one can now determine whether such corrections enhance, leave unchanged, or suppress the quantum chaos of potential theory.

Imagine that we are dealing with the simplest scalar interaction between scalar fields. The first property then states that it is possible to write for $G_c[A]$ a unique, non–perturbative and time–reversal–invariant (TRI) representation, in terms of a finite number N of quadratures, such that the order–of–magnitude of the errors of this truncation are known in advance, decreasing roughly as N^{-1}. When this construction is extended to vector interactions, independently of the complexity of the finite–N truncation, one finds a new feature in which a central function of the representation is defined in terms of "characteristic" solutions to "integral maps", analogous to those used to describe Navier–Stokes (NS) fluids and other classical systems.[5] One can now see when and why certain PTh vector interactions can lead to chaos, especially in a semi–classical particle limit; and one can follow the course of such chaos when complete radiative corrections are introduced.

Space–time limitations permit only the barest sketch of these properties, as now quoted below. For a scalar situation, an exact solution to the Green's function satisfying

$$\left[m^2 - \partial_x^2 + gA(x)\right]\, G_c(x,y|A) = \delta^4(x-y)$$

can be written, most simply in momentum space, as

$$\begin{aligned} < p|G_c[A]|p' > = i \int_0^\infty ds\, e^{-is(m^2+p^2)} \int d^4z\, e^{-iz\cdot(p-p')+\frac{is}{4}(p-p')^2} \\ \cdot \prod_N{}' \frac{(-i)^2}{(2\pi)^4} \int d^4P_N \int d^4Q_N\, e^{\frac{i}{2}(P_N^2+Q_N^2)} \\ \cdot\, e^{-ig\int_0^s ds' A(z+s'(p+p')-\sum_N{}' R_N(s'))}, \end{aligned} \tag{1}$$

where P_N, Q_N are dimensionless 4–vectors, the primes on $\sum_N{}'$ and $\prod_N{}'$ indicate a sum and product over odd, positive integers N, and $R_N(s') = \frac{2\sqrt{s}}{\pi N}\left[P_N \cos(\frac{N\pi s'}{s}) + Q_N \sin(\frac{N\pi s'}{s})\right]$. Note that if any of the P_N, Q_N are dropped from the argument of A, the normalization integrals over that P_N, Q_N pair produce a factor of unity. Eq. (1) is exact. Its approximation to finite N corresponds directly to the truncation of the sum $\frac{8}{\pi^2}\sum_N{}' N^{-2}\sin^2(\frac{N\pi x}{2})$, which exactly represents in the interval $0 \le x \le 1$ the quantity: x; as illustrated in Fig.1, this approximation becomes better for larger values of N. This is the only form of approximation made in the entire analysis; an estimate of its relative error may be made by comparing the area under the straight–line $f(x) = x$ with the finite–N truncations described above. In the context of a non–trivial but soluble PTh example, one finds that such estimates of relative errors are actually upper bounds.[1] It should also be remarked that any and every finite–N truncation of (1) will satisfy TRI symmetry; and that machine evaluation may be necessary for these truncated integrals.

For interaction with a vector potential $A_\mu(z)$, a similar but profoundly different representation obtains. For clarity, consider the QED fermion Green's function which satisfies

$$[m + \gamma \cdot (\partial_x - igA(x))]\, G_c(x,y|A) = \delta^4(x-y)\,.$$

The corresponding representation for this $G_c[A]$ is given by

$$\begin{aligned}< p|G_c[A]|p > = i \int_0^\infty ds\, e^{-ism^2}\, [m - \gamma \cdot (p - \Omega(s))]\; e^{-i\int_0^s d\tau [p-\Omega(\tau)]^2} \\ \cdot \prod_N{}' \frac{(-i)^2}{(2\pi)^4} \int d^4P_N \int d^4Q_N\, e^{\frac{i}{2}(P_N^2+Q_N^2)} \int d^4z\, e^{-iz\cdot(p\cdot p')} \\ e^{-\int_0^s d\tau \partial_\mu^y \Omega_\mu(\tau)} \cdot \left(e^{g\int_0^s d\tau\sigma\cdot F(\zeta(\tau)-2\int_0^\tau d\tau'\Omega(\tau'))}\right)_+ ,\end{aligned} \tag{2}$$

where $\zeta(\tau) = z + \tau(p+p') - \sum_N{}' R_N(\tau)$, the last factor of (2) is ordered in τ, and the 4–vector Ω_μ is a solution of the map

$$\Omega_\mu(\tau) = gA_\mu\Big(\zeta(\tau) - 2\int_0^\tau \Omega(\tau')d\tau'\Big)\,. \tag{3}$$

If there is more than one root, or solution of (3), then (2) must be summed over all such Ω. With the ansatz $\Omega_\mu(\tau) = dx_\mu(\tau)/d\tau$, and a simplified form of $\zeta(\tau)$, it is clear that strong non–linearities of (3) can lead to behavior which has heretofore been the proper subject of classical dynamical systems.[5] It should be remarked that essentially the same analysis can be performed for the gauge–invariant portion of $G_c[A]$, showing that the possibility of chaos in PTh QED is not a gauge–dependent construct.[2]

Now it is possible to pose and answer the question of what happens to such PTh quantum chaos when the full, radiative corrections of QFTh are included; you have been such a kind and attentive audience, at the very end of this last evening session, that I must tell you the answer. For clarity, denote the previous PTh $A_\mu(z)$ by $A_\mu^{ext}(z)$, and to every such quantity add a field A_μ which is now subject to QFTh fluctuations. The result of such a computation, for every correlation function, will be to remove the equation which lead to the map of (3); the result is certainly complicated, too much so to be calculated exactly; but it is not chaos, or at least not the quantum chaos which results from (3). What happens is that the full radiative corrections – and we are here talking about non–perturbative calculations, as were necessary in order to see the original chaos – provide a measure of what has been called in another context[6] "induced decoherence", effectively replacing a delta–functional which generates the map of (3), by multi-Gaussian FIs which contain no such requirement.

The result of this relatively simple computation thus provides a long–overdue justification for the neglect of chaotic effects in high–energy physics. Essentially all other fields of mathematical science have, in the last two decades, been subject to the conceptual and practical revolutions forced upon them by uncontestable chaotic behavior. Only HE physics has remained outside the chaotic sphere of influence, probably because of its great reliance upon perturbation (and eikonal) approximations, which mask the appearance of chaos. PTh quantum chaos has been noted and discussed;[7,8] and with these new representations may be given a formulation directly in terms of Green's functions. Yet, remarkably, when the same representation is used in the calculation of the full, QFTh correlation functions, all reference to such chaos disappears. One has, at the very least, a justification for the continued state of innocence, and of the good fortune, of HE physics.

References

1. H. M. Fried and Y. Gabellini, Brown-HET-919 and Brown-HET-933; *Phys. Rev.* **D** (to be published).
2. H. M. Fried, Y. Gabellini and B. H. J. McKellar, "Exact and Approximate Fermion Green's Functions in QED and QCD", Brown-HET-959.
3. J. Schwinger, Proc. Nat'l Acad. Sciences (USA) **37** (1961) 452; Harvard Lectures (1954); Stanford Lectures (1956). From a slightly different point of view, much of this material may be found in his book "Particles, Sources, and Fields", Addison-Wesley (1970). See also H. M. Fried, "Functional Methods and Models in Quantum Field Theory", The MIT Press, Cambridge (1972).
4. E. S. Fradkin, *Nucl. Phys.* **76** (1966) 588; Derivations and application to a variety of interactions may be found in H. M. Fried, "Functional Methods and Eikonal Models", Editions Frontiéres, Gif-sur-Yvette, France (1990).
5. See, for example, R. C. Hilborn, "Chaos and Nonlinear Dynamics", Oxford University Press (1994).
6. H-T Elze, CERN-TH preprints 7297/94 and 7131/93; W. Zurek and J. P. Paz, Los Alamos preprint LA-UR 94-927; and other references quoted therin.
7. See for example, J. B. Delos, S. K. Knudsen and D. W. Noid, *Phys. Rev. A* **30** (1984) 1208.

Figure 1

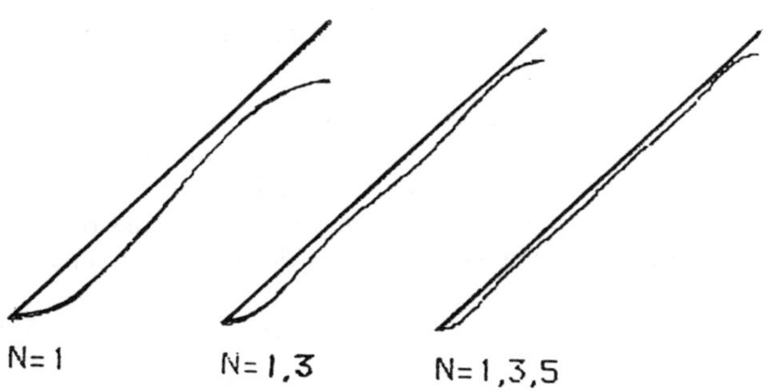

A plot of $(8/\pi^2)\sum_N' N^{-2}\sin^2(N\pi x/2)$ vs x, for $N = 1, N = 1$ and $3, N = 1,3$ and 5; and a comparison with x, the exact value of this sum over all odd N.

A QUANTUM GRAVITATIONAL MECHANISM FOR EXITING INFLATION

RICHARD P. WOODARD
Department of Physics, University of Florida
Gainesville, FL 32611 USA

ABSTRACT

I discuss the proposal by Nick Tsamis and myself that quantum gravity may mediate a naturally slow exit from inflation. The emphasis is on our recently completed two loop computation of the effective Hubble constant. When this result is properly checked and independently confirmed it will provide a proof that the mechanism functions in perturbation theory.

1. Inflation Is Great — But How Do You Stop It?

This is a report on my work with Professor Nicholas Tsamis of the University of Crete on quantum gravity with a non-zero cosmological constant.[1–5] Because this theory differs so radically from conventional quantum gravity — in which the cosmological constant is fine tuned to zero — we distinguish it with the name, "quantum cosmological gravity," or QCG for short. The Lagrangian of QCG is:

$$\mathcal{L} = \frac{1}{16\pi G}\Big(R - 2\Lambda\Big)\sqrt{-g} + \text{counterterms} \tag{1}$$

Although the correct perturbative formulation of QCG was itself a worthy problem, our solution of which[1,2,4,5] will stand even if it lacks applications, the primary interest that QCG holds for Nick and me is the possibility that it can explain a curious hierarchy known as, "the problem of the cosmological constant."[6] To understand this let me review the parameters of QCG, their values and how they affect the universe around us.

The parameter G is Newton's constant, and it controls the strength with which gravity couples to matter. The loop counting parameter of quantum gravity (conventional or cosmological) is $\kappa^2 \equiv 16\pi G$ in the usual unit system for which $\hbar$ and c are one. Depending upon how one deploys factors of $\hbar$ and c the parameter κ can be variously regarded as a length, a time, or an inverse mass:

$$\kappa \sim 10^{-33}\text{cm} \sim 10^{-43}\text{s} \sim \Big(10^{19}\text{GeV}\Big)^{-1} \tag{2}$$

These extreme values are responsible for the ubiquitous belief that quantum gravity is negligible at any scale we can access experimentally.

The parameter Λ is called the cosmological constant, and it controls the global expansion or contraction of the universe. To be more precise, consider the invariant element, $ds^2 \equiv g_{\mu\nu}dx^\mu dx^\nu$ — the quantity which gives the physical distance between infinitesimally close coordinate points. For a homogeneous, isotropic and spatially flat universe the invariant element can be expressed in a co-moving coordinate system for which the "time" coordinate really measures physical time durations at fixed spatial position:

$$ds^2 = -dt^2 + a^2(t)d\vec{x} \cdot d\vec{x} \tag{3}$$

Classical gravity predicts that if the universe is empty and Λ is positive then the scale factor is $a(t) = e^{Ht}$, where $H^2 \equiv \frac{1}{3}\Lambda$. Even if the universe isn't initially homogeneous and isotropic, and even if it contains a finite density of normal matter, the late time behavior still tends locally to this form. This sort of cosmological evolution is known as *inflation* because of its similarity to the history of wages and prices during the last Democratic Administration.[7]

Inflation means that the physical distance between observers at rest on fixed spatial coordinates is actually increasing exponentially. Consider observers A, B, C, and so on, all initially at rest with respect to each other and strung out at equal coordinate separations along the x axis. If A sees his physical distance from B grow, and B sees *his* physical distance from A and C grow, then A must obviously see his physical distance from C grow even faster. The effect builds up so rapidly that it isn't possible even for light waves to cross the gap between observers after the physical distance separating them reaches the critical value of H^{-1}. This separation is known as the *Hubble radius*, and causality obviously prevents any physical effect from propagating beyond it.

Inflation is very unlike what we observe today. In fact one can derive an impressive bound on H by simply noting that humans are capable of seeing their feet. Observations of the wider universe improve this by many orders of magnitude:[8]

$$H_{\text{today}} \lesssim 10^{-60}\kappa^{-1} \tag{4}$$

The problem of the cosmological constant is why the dimensionless product of the two gravitational parameters should be so small. There is no larger hierarchy in fundamental physics. By comparison, the discrepancy between the electroweak scale and the conjectured scale of grand unification is less than one fourth as large.

The current smallness of H is even harder to understand if one accepts the view that the very early universe actually underwent a period of inflation somewhere between the electroweak and the GUT scales:*

$$10^{-34}\kappa^{-1} \lesssim H_{\text{inflation}} \lesssim 10^{-6}\kappa^{-1} \tag{5}$$

* Most inflationary cosmologists prefer the larger value. The electroweak limit arises because the known physics below this scale does not violate CP strongly enough to explain the observed asymmetry between baryons and anti-baryons.

The best evidence for this is the observed homogeneity and isotropy of today's universe. The cosmic microwave background radiation from different parts of the sky is seen to have the same temperature to within about one part in a hundred thousand.[9] Without inflation one must accept the coincidental establishment of regions of nearly identical thermal equilibria which are only now coming into causal contact with one another. Inflation evades this problem by blowing the observed universe up from a single region so tiny that causal processes could have established an initial thermal equilibrium across it. Inflation would also disperse unwanted relics such as primordial black holes and magnetic monopoles, and it would explain why the currently observed energy density is so near to the critical density needed for closing the universe.*

The main disadvantage of inflation is that one must somehow get the effective Hubble constant to fall from the relatively large value (5) it must have had during inflation to the extremely small value (4) it has now. Whatever mechanism is invoked to perform this task must be powerful because it has to slow the expansion rate by at least 26 — and probably by more than 54 — orders of magnitude. On the other hand, this powerful mechanism must operate on a very slow time scale because a minimum of something under 60 e-foldings of inflation is needed to explain the observed degree of isotropy in the cosmic microwave background. Finally, this powerful but very slow mechanism must not embarrass low energy phenomenology with observable but hitherto unseen light quanta.

2. Our Proposal

It is almost irresistible to suspect that the mechanism we seek is an infrared effect mediated by very weakly coupled massless quanta. The strength of this effect would derive from the coherent and causal superposition of weak interactions occurring throughout the past light cone as it extends back to the onset of inflation. The long time lag between the beginning of inflation and its effective extinction would derive from the need for the invariant volume of the past light cone to become enormously large in order to overcome the natural weakness of the interaction. Only massless quanta could mediate interactions coherently over such a large region. Since we have presumably observed the long range forces associated with all massless quanta, it is highly desirable to seek the conjectured mechanism in one of the conventional models of low energy phenomenology. Finally, the model we seek must be capable of distinguishing between flat space and a gravitational background of the form (3) with a time varying scale factor. The precise formulation of this requirement is that the desired model must violate conformal invariance.

The conditions stated above are necessary for a plausible mechanism but they are obviously not sufficient. Happily they do narrow the field to a unique contender:

* This particular explanation may actually be too good because inflation predicts $\rho = \rho_{\text{crit}}$ whereas a variety of different observations seem to give $\rho < \rho_{\text{crit}}$ by one to two orders of magnitude. The matter is still hotly debated.

QCG. Proving — even at the level of perturbation theory — that QCG actually manifests the desired mechanism has consumed the better part of my research effort and Nick's over the past three years. Since I have explained our older work at some length in another conference report[10] I will confine myself here to the barest summary of previous conclusions:

(1) *QCG can be used reliably as a quantum field theory provided the phenomenological focus is on the far infrared.*[3,5] Although QCG is not perturbatively renormalizable, one can of course absorb ultraviolet divergences by adding ever higher local counterterms. This gives reliable predictions for the infrared because the physics of large spatial and temporal separations is dominated by non-local and ultraviolet finite terms which are unaffected by local counterterms. This is a standard idea behind the effective field theory of massless particles. It is why Bloch and Nordsieck were able to solve the infrared problem of QED[11] long before anyone understood renormalization. It is why Weinberg was able to do the same for conventional quantum gravity,[12] and why his result will stand no matter what the correct fundamental theory of gravity turns out to be. And it is why the late Gerald Feinberg and our own Joe Sucher were able to use Fermi theory to compute the long range force due to neutrino exchange.[13]

(2) *QCG corrections become strong at late times in an inflating universe.*[3–5] There are two reasons for this:

   a) QCG allows massless gravitons to self-interact through a coupling of dimension three; and

   b) The inflationary expansion of physical distances induces a corresponding redshift in physical momenta, and so populates the infrared ever more heavily.

   Physicists are very familiar with the notion of the infrared being dominated by massless particles with self-interactions of the lowest possible dimension. One might expect the massless gravitons of QCG to be even more potent in the infrared than the massless gluons of QCD with their dimension four self-interactions. In fact we see a secular effect at late times rather than a static long range force because infrared phenomena in QCG are so powerful that they engender a breakdown of the background. Note as well that even though the momentum redshift is diluted by the decoupling of wavelengths as they pass the Hubble radius, there is still a powerful secular effect due to the ever increasing number of modes whose physical wavelengths are *just reaching* the Hubble radius.

(3) *QCG dominates the physics of late time inflation with respect to any matter theory whose phenomenological viability does not require fine tuning.*[3,5] This is because all the other known massless particles are conformally invariant on the classical level. They therefore cannot distinguish locally between de Sitter space and flat space, and their self-interactions have dimension four.

(4) *If one assumes that the natural vacuum of QCG suffers only perturbatively small corrections then asymptotic graviton scattering amplitudes are infrared divergent even at tree order.*[4] This is a trivial consequence of the fact that QCG vertices grow at late times in a way that is not compensated by either external wave-functions or propagators. This result is sometimes put down as an obvious error owing to the curious fact that the rate of infrared *divergence* in the cut-off time happens to be much stronger than the rate at which the expectation value of the background can possibly *grow* with causal time evolution. Note first that there is nothing inconsistent about this provided both effects become non-perturbatively large at late times. Note as well that although expectation values must be used to follow time evolution, in-out amplitudes provide a perfectly good check on the consistency of any guess one may care to make for this time evolution. For example, our result debunks the oft-made suggestion that the background suffers only perturbatively small corrections at late times.

(5) *Modulo some tensor algebra, QCG corrections act to slow the inflationary expansion rate by an amount which becomes non-perturbatively large at late times.*[5]

3. Our Two Loop Computation

This summer's work has been devoted to removing the qualification, "modulo some tensor algebra," from the fifth conclusion stated above. Starting from a homogeneous and isotropic free vacuum on $T^3 \times R$, we follow the evolution of the expectation value of the invariant element in co-moving coordinates:

$$\left\langle 0 \middle| g_{\mu\nu}(t,\vec{x})\, dx^\mu dx^\nu \middle| 0 \right\rangle = -dt^2 + a^2(t)\, d\vec{x} \cdot d\vec{x} \tag{6}$$

Our choice of state and our organization of perturbation theory enforces the initial condition $a(0) = 1$; we impose $\dot{a}(0) = H$ as a renormalization condition. The subsequent time evolution of $a(t)$ is determined by the dynamics of QCG. From this time evolution we define the (generally coordinate invariant) effective Hubble constant as:

$$H_{\text{eff}}(t) \equiv \frac{d\ln(a)}{dt} \tag{7}$$

I am happy to report the following result — so new that it has not yet appeared elsewhere[14] — from a two loop calculation that consumed most of this summer:

$$H_{\text{eff}}(t) = H\left\{1 - 2\left(\frac{\kappa H}{\pi}\right)^4 \Big[(Ht)^3 + \text{subdominant}\Big] + O\left(\kappa^6 H^6\right)\right\} \tag{8}$$

What we actually computed was the amputated 1-point function for a conformally rescaled version of the graviton field in conformal coordinates. We then attached the external propagator, undid the conformal rescaling, and transformed the result back to co-moving coordinates. The calculation was done using a modified version of the diagrammatic formalism developed by Schwinger for computing expectation values[15]. This technique is manifestly causal in that the only net contributions come from interaction vertices which reside on or within the past light cone of the observer who measures the amputated field. Our modification of Schwinger's formalism is infrared finite because the observer's past light cone has a finite invariant volume, although this invariant volume grows as we observe at later and later times. This growth, and the fact that graviton propagators do not suppress large loops, is the origin of the secular effect so prominent in our result (8).

The ultraviolet regularization we used corresponds to an exponential cutoff upon coordinate momentum modes. Our interaction vertices[1] and propagators[4,5] have been reported elsewhere. The tensor algebra was done using the program FeynCalc[16] within Mathematica on a SUN SPARC station with 8MB of random access memory. The various scalar loop integrals were performed analytically up to a single conformal time integration and then subjected to an asymptotic expansion. The leading contributions were summed using a computer.

Two of the three powers of Ht in the result (8) derive from the fact that at two loops one can have two interaction vertices integrated over the observer's past light cone, the invariant volume of which grows linearly with his co-moving time. The final power of Ht has its physical origin in the increasing vacuum correlation imposed by inflation and evident in the so-called, "logarithm piece" of the bare propagator.[4,5] This enhancement enters (8) through whichever of the propagators happens to connect the earlier of the two interaction vertices to the observation point.

I should conclude this section on a note of caution. This is only the second two loop calculation ever done in quantum gravity, and it is the first one for which the ultraviolet finite part of anything was computed. Such a difficult and complicated piece of work will not be completely accepted — and should not be — until it has been subjected to the most rigorous possible checks, and has also been independently confirmed. Although we checked many things we have not yet had the time — nor will we for many months — to verify that the one loop position space self energy obeys the appropriate BRS Ward identity and that its flat space limit agrees with the momentum space result obtained by Capper.[17]

4. The Physics Behind Our Result

Counting the coupling constants at ℓ loops is simple enough, but it is not so easy to infer the powers of Ht — as witness the fact that we have gotten it wrong twice already![3,5] At $\ell = 3$ there will be up to four free interaction vertices, which should each contribute a factor of Ht as they explore the order Ht invariant volume of the observer's past light cone almost without suppression. It is possible to show as well that the seven propagators can contribute up to two undifferentiated logarithms, each of which should give another factor of Ht. This makes the leading three loop term of order $(H\kappa)^6 \times (Ht)^6$. We therefore conclude that perturbation theory must break down long before the secular dependence at two loops can overwhelm the naturally small prefactor of $(H\kappa)^4 \lesssim 10^{-24}$. In fact the two loop term would go non-perturbative at $Ht \gtrsim 10^8$, whereas the three loop term becomes large when $Ht \gtrsim 10^6$. Our best guess for the leading term at ℓ loops is:

$$\left(\kappa H\right)^{2\ell}\left(Ht\right)^{3\ell-3} \tag{9}$$

This implies that perturbation theory breaks down after:

$$N = Ht \gtrsim 10^4 \tag{10}$$

inflationary e-foldings. Of course this would also be a rough estimate for the lifetime of inflation.

A potentially troublesome point in this analysis is that perturbation theory breaks down when all of the leading terms are still quite small:

$$\left(\kappa H\right)^{2\ell}\left(Ht\right)^{3\ell-3} \longrightarrow (\kappa H)^2 \lesssim 10^{-6} \tag{11}$$

We therefore require some evidence that the contribution at each order is really such as to decrease the expansion rate. The desired argument is supplied by Feynman's tree theorem,[18] a procedure which serves to decompose loops of locally interacting quanta into sums of tree diagrams. When applied to any of the loops which go to make up the amputated 1-point function we obtain a sum of terms whose physical interpretation is the effect on the background of various amounts of gravitational radiation. In fact, the dominant diagrams at ℓ loops decompose to give ℓ-graviton trees.

To understand the significance of this representation first consider the effect of having an initial graviton of wave length λ. At first there is little because the spatial fluctuation of the graviton couples weakly to the zero mode background. However, it is the physical wavelength that matters and this gets expanded by a factor of $a(t)$ as the universe evolves. Eventually the graviton is so spread out that it has a strong coupling to the background. Except, of course that causality prevents it from doing much for times after $a(t)\lambda > H^{-1}$. The strongest effect is therefore at that time when the physical wave length is comparable to the Hubble radius. Note that this effect is to *resist* the inflationary expansion by virtue of the gravitational self-attraction of the wave. The sign is unambiguous because gravitation is universally attractive.

Now consider the classical effect of having some initial radiation in an inflating universe. If the initial geometry is non-singular then there must be a rapid fall-off in the short wavelength modes. Suppose the characteristic wave length of the cutoff is λ. What we see is some effect on the background until t such that $a(t)\lambda \gtrsim H^{-1}$, after which there will be nothing much. At late times the universe inflates without disturbance because the maximum wave length has been expanded past the causal horizon.

The crucial difference between the classical process we have just considered and the representation given by the tree theorem is that *the latter involves sums over gravitons whose initial spatial momenta are arbitrarily large.* The quantum effect never cuts off because we never run out of wave lengths which are just redshifting to the Hubble radius. In fact the number of these *increases* with time as $a^2(t)$ by a simple phase space effect.

Note that this argument does more than we bargained for. Not only do we see that each order acts to slow inflation — because gravitation is universally attractive — we can also understand why relaxation is an inherently quantum mechanism. One finds no large classical effect because all well defined initial configurations must be cut off at small wave lengths, and when the smallest wave length has expanded beyond the Hubble radius there can be no more effect by causality. Quantum gravity is just the opposite because the uncertainty principle *forbids* us from constraining both the initial value and the initial time derivative of each mode to vanish. This is what endows quantum cosmological gravity with the unending supply of ultraviolet modes which the locally de Sitter background eventually redshifts to the infrared.

5. The Implications Of Our Mechanism

If Nick and I are right then the bare cosmological constant is not unreasonably small. Further, there is no need for fine tuning of scalar potentials. Our view is that inflation happened in the early universe for no other reason than that the bare cosmological constant is large and positive. The current observed universe derived from a patch of about one Hubble volume which began inflation when its local temperature dropped below the scale $M = \sqrt{H/\kappa}$.

The reason our universe is not observed to be inflating today — at least not rapidly — is that infrared processes in QCG tend to screen the cosmological constant. We saw in the previous section that any vacuum loop contribution to the background can be represented as a sum of tree corrections due to the presence of some initial gravitational radiation. It is the self-gravitation of this radiation which resists inflation. This is why the contributions at all orders have the same sign. The irresistible force of the effect derives from the fact that it scales as powers of the invariant volume of the past light cone, which grows without bound as evolution continues. The slowness of the effect derives from the fact that only the causal and coherent superposition of interactions over an enormous invariant volume can overcome the natural weakness of the gravitational interaction.

All experimentally confirmed matter quanta are either massive or else conformally invariant at the classical level. This means that they give only negligibly small corrections to the classical geometry of an inflating universe. Although certain conjectured light matter quanta may be competitive with QCG for a time, in the end only the graviton is left. This means that QCG makes *unique* predictions for relaxation. Unfortunately, the most interesting predictions of QCG are not easy to obtain because perturbation theory breaks down before a significant amount of relaxation has occurred. In particular, the end of rapid inflation should tell us what reheating temperature was reached; and the last sixty e-foldings govern the magnitude and spectrum of observable density perturbations.

Finally, we would like to suggest two more points of potential experimental contact. The first arises from the fact that the universe should still be experiencing a residual expansion due to its not completely screened bare cosmological constant. It is conceivable that a modified form of perturbation theory could be used to predict this residual expansion rate. There is some evidence that a suitably time varying effective cosmological constant can explain the apparent discrepancy between the age of the oldest galaxies, as inferred by models of stellar evolution, and the age of the universe as inferred by the currently measured Hubble constant.[19]

The second point of potential experimental contact derives from electroweak and QCD chiral symmetry breaking phase transitions. If these occurred after inflation and relaxation at some high scale then they should have induced an effectively negative cosmological constant. We do not understand the process by which deflation is relaxed — or even if it is relaxed — but one would think that the time scale of this gravitational process should be longer than that of either phase transition. If so then there would have been substantial periods of deflation, which would almost inevitably have affected low scale baryogenesis and nucleosynthesis.

6. Acknowledgements

This work was partially supported by DOE contract DE-FG05-86-ER40272 and by NATO collaborative Research Grant CRG-910627. I wish also to acknowledge the hospitality of the University of Crete during visits in this and each of the three previous years. I am especially grateful to Professor Theodore Tzouros for the generous loan of his SUN SPARC station.

7. References

1. N. C. Tsamis and R. P. Woodard, *Commun. Math. Phys.* **162** (1994) 217.
2. N. C. Tsamis and R. P. Woodard, *Phys. Lett.* **B292** (1992) 269.
3. N. C. Tsamis and R. P. Woodard, *Phys. Lett.* **301** (1993) 351.
4. N. C. Tsamis and R. P. Woodard, "The Physical Basis For Infrared Divergences In Inflationary Quantum Gravity," University of Florida preprint UFIFT-HEP-93-17, July 1993.
5. N. C. Tsamis and R. P. Woodard, "Strong Infrared Effects In Quantum Gravity," University of Florida preprint UFIFT-HEP-92-24, February 1994. (To appear in *Annals of Physics.*)
6. S. Weinberg, *Rev. Mod. Phys.* **61** (1989) 1;
7. A. H. Guth, *Phys. Rev.* **D23** (1981) 347;

A. D. Linde, *Phys. Lett.* **B108** (1982) 389;

A. Albrecht and P. J. Steinhardt, *Phys. Rev. Lett.* **48** (1982) 1220.

8. A. Sandage, *Observatory* **88** (1968) 91.
9. E. W. Kolb and M. S. Turner, *The Early Universe* (Addison-Wesley, Redwood City, CA, 1990).
10. R. P. Woodard, "Results From Quantum Cosmological Gravity," University of Florida preprint UFIFT-HEP-94-6, May 19994. (To appear in *Proceedings of the International Symposium on Unified Symmetry in the Small and in the Large*, 1/27-30/94, Miami, Florida, Plenum Press, New York, 1994.)
11. F. Bloch and H. Nordsieck, *Phys. Rev.* **52** (1937) 54.
12. S. Weinberg, *Phys. Rev.* **B138** (1968) 988.
13. G. Feinberg and J. Sucher, *Phys. Rev.* **166** (1968) 1638.
14. N. C. Tsamis and R. P. Woodard, preprint in preparation.
15. J. Schwinger, *J. Math. Phys.* **2** (1961) 407;

    J. Schwinger, *Particles, Sources and Fields* (Addison-Wesley, Reading, MA, 1970);

    R. D. Jordan, *Phys. Rev.* **D33** (1986) 444.
16. R. Mertig, "Guide to FeynCalc 1.0," University of Würzburg preprint, March 1992.
17. D. M. Capper, *J. Phys.* **A13** (1980) 199;

    D. M. Capper, private communication, July 29, 1982.
18. R. P. Feynman, *Acta Phys. Pol.* **24** (1963) 697;

    R. P. Feynman, in *Magic Without Magic*, ed. J. Klauder (Freeman, New York, 1972), p. 355.
19. T. S. Olson and T. F. Jordan, *Phys. Rev.* **D35** (1987) 3258;

    P. J. E. Peebles and B. Ratra, *Astrophys. J. Lett.* **325** (1988) L17.

STRANGE PARTICLES AND QUARK-GLUON PLASMA

JOHANN RAFELSKI
Department of Physics, University of Arizona, Tucson, AZ 85721

and

JEAN LETESSIER and AHMED TOUNSI
Laboratoire de Physique Théorique et Hautes Energies
Université Paris 7, Tour 24, 2 Pl. Jussieu, F-75251 Cedex 05, France

ABSTRACT

We show that the experimental results obtained in 200 GeV A S→A CERN–SPS collisions suggest that the observed strange particle flow is compatible with a rapidly hadronizing QGP state. The quantities we explore are the strange quark chemical potential μ_s, the strangeness occupancy factor γ_s, the absolute particle yields of multi-strange hadrons and the anomalous high specific entropy S/B. We describe a method of analysis of the data and we develop a reaction scenario. Using QGP equations of state we obtain full compatibility with the experimental results and make predictions for the forthcoming Pb→Pb collisions.

1. Heavy Ion Collisions and Vacuum Structure

Relativistic heavy ion experimental programs at the AGS accelerator at Brookhaven and at the SPS accelerator at CERN, begun in 1986–87, motivated by the desire to recreate conditions akin to those prevailing in the first moments of the Early Universe in the laboratory, and the expectations that a new form of matter, the quark-gluon plasma (QGP) can be explored. Intuitively, it was clear that the higher the energy of the accelerator, the greater the energy density that could be created and hence the more extreme conditions of matter and earlier the time of the Beginning of our Universe one expects to be able to study. This lead on to the idea to build RHIC, the 100 GeV collider presently under construction at Brookhaven and, in turn, to develop a heavy ion collision project ALICE in the proposed Large Hadron Collider (LHC).

At the base of this research program is the hope that large nuclei such as Pb are pieces of nuclear matter large enough to allow us to study the phases and generally hot hadronic matter in its great simplicity arising in statistical systems. In that resides also the great similarity to the situation present in early Universe. Another reason why heaviest nuclei, rather than leptons or protons, are required in these efforts is the recognition that as hadronic particles are heated and compressed one reaches the point of possible phase transition into the deconfined phase. Phase transitions cannot develop in any finite systems, and it can only be hoped that the number of accessible degrees of freedom excited in collisions of heavy nuclei is already sufficiently large to allow us to study the properties of such as phase transition using finite nuclei as projectiles and targets.

High energy nuclear collisions are the only known tools suitable to study extended regions of space containing a locally modified vacuum state: two extended objects, the two colliding nuclei, could produce an extended space region with high energy and particle densities. This volume of space-time devoid of the structured, true vacuum, may turn into the new, perturbative phase. This discovery of the QCD-vacuum 'melting' and the study of the properties of the perturbative and true vacuum is, in our view, the primary fundamental theoretical objective of the high energetic nucleus-nucleus collision experiments. Inside the domain of perturbative vacuum, at sufficiently high excitation the assembled gas of quarks and gluons is subject to the QCD perturbative interactions characterized by the (running) coupling constant α_s. Relative smallness of $\alpha_s/\pi \leq 0.2$ permits us to study the quark matter in a first estimate of its properties, as if it were an ideal gas of quarks and gluons confined to some (large) volume: this is "quark-gluon plasma" (QGP).

In the collision of S with Au nuclei at SPS record temperatures (particle spectra inverse slopes) of the magnitude of 230 MeV have been found. A number of potential signatures of QGP have been studied. We will not discuss here in detail the general situation, rather we will survey our recent progress involving understanding of the strangeness flavor production. While light u, d quarks are brought into the reaction by the colliding nuclei, the strange quarks s are produced, as theoretical calculations suggest, by glue–glue collisions in the QGP phase; their high initial density and considerable final abundance promise to lead to rather characteristic final states. The good news is that some recent results obtained at CERN with 200 GeV A projectiles support the contention that strangeness is a very useful signature of the early hadronic phase.

2. Strangeness

Strangeness production processes constitute a bottleneck in (chemical) equilibration of strongly interacting matter. Enhanced production of strangeness flavor by glue based processes in QGP will lead in the end to enhanced strange particle abundances, and specifically of strange anti-baryons:[1] if the transition from the deconfined state to the confined hadronic gas occurs rapidly, then the high density of strangeness in the early phase will lead to highly anomalous yields of multiply strange particles. The (enhanced) production of (multi)strange anti-baryons is considered to be more specifically related to deconfinement than are multi-strange mesons, because of a large difference in chemical equilibration time scales and chemical properties in the deconfined quark-gluon and hadronic-gas (HG) phases. Furthermore, if QGP is not formed, its absence reveals itself by a relative suppression of strange compared to non-strange particle production, with yields similar to those seen in p–p collisions.

Because of the intrinsic difficulties related to detection of strange particles only recently results have been presented which allow a test of these ideas. Several heavy-ion experiments have reported a strong enhancement of all strange particle yields relative to proton-proton and proton-nucleus collisions, increasing with the centrality of the collision. The enhancement was seen particularly clearly in 200 GeV A collisions of sulphur with various targets in the strange and multi-strange baryon and anti-baryon channels at central rapidity.[2–4] The presently available analysis of all these results is based on a relatively simple, but powerful framework of a thermal fireball model. The particular virtue of this approach is that the spectra and particle abundances can be described in terms of only a few parameters. It presumes that the strong interactions allow to achieve local thermodynamical equilibrium, a fact which is very much in experimental evidence, but which is far from being understood.

The key issue which we will address here is the question to what extent current data can distinguish between the formation of a fireball consisting of a dense and very hot hadron (resonance) gas and the creation of a transient deconfined state such as the quark-gluon plasma (QGP) which subsequently hadronizes. If this final state hadronization involves a full re-equilibration, then the memory of the QGP would be lost. Today, based on diverse results obtained at 200 GeV A it appears that we have a very favorable situation with the observable particles very much representing the expected properties of the primordial QGP state.

2.1. Non-equilibrium Features in the Final State

In order to determine particle abundances, certain non-equilibrium features in the chemical composition of the fireball and of the hadronization process have to be considered. In particular, we cannot expect, unless very special circumstances prevail, that the strangeness flavor has fully saturated the available phase-space. This necessitates the introduction of a strangeness saturation parameter $0 < \gamma_s \leq 1$ which allows to parametrize the effects of incomplete chemical equilibration of the strangeness sector.[5] For a QGP state with its fast gluonic strangeness production even the natural short fireball lifetime of only a few fm/c should be sufficient to reach values of γ_s close to one.[6] In a purely hadronic fireball with its expected much longer strangeness saturation time scale small values of $\gamma_s \sim 0.1$ (like those extracted from N–N collisions) are much more likely. A measurement of γ_s thus in principle provides important information on the strangeness production time scale and hence a relatively large value of γ_s requires some new physics feature in the structure of the collision fireball.

Hadronization must ensue when considering a scenario in which there was formation of an intermediate deconfined state. It leads to another non-equilibrium feature: we must not take for granted that the relative abundances of mesons and baryons, which are produced by quite different mechanisms, obey the laws of relative chemical equilibrium. Processes which convert mesons into baryon-anti-baryon pairs are generally relatively slow. We therefore must envision that the abundance of mesons and baryons is different from the equilibrium values, the difference being characterized by a concentration factor C_i. But should we introduce a different value of C_i for each particle species, this would de-facto destroy the simplicity of the fireball model and would make a full chemo-kinetic description of the collision process necessary. However, on physical grounds one can argue convincingly that there is relative equilibrium between the abundances of baryons and, similarly, mesons. This can be traced back to the OZI rule of hadronic cross-sections: processes described by diagrams in which all quark lines are 'flowing' and no quarks are produced or destroyed are most probable — these dominant processes conserve the number of baryons and mesons separately.

2.2. Sizing up the Final State

The distribution of the final state strange particles can be obtained by considering the Laplace transform of the phase space density, which leads to a partition function like expression $\mathcal{Z}_{\rm s}$, but the weight of individual components is accounted for by the coefficients $\gamma_{\rm s}$, $C_i^{\rm s}$ as described above:

$$\ln \mathcal{Z}_{\rm s} = \frac{VT^3}{2\pi^2}\Big\{(\lambda_{\rm s}\lambda_{\rm q}^{-1}+\lambda_{\rm s}^{-1}\lambda_{\rm q})\gamma_{\rm s}C_{\rm M}^{\rm s}F_K+(\lambda_{\rm s}\lambda_{\rm q}^{2}+\lambda_{\rm s}^{-1}\lambda_{\rm q}^{-2})\gamma_{\rm s}C_{\rm B}^{\rm s}F_Y$$
$$+(\lambda_{\rm s}^{2}\lambda_{\rm q}+\lambda_{\rm s}^{-2}\lambda_{\rm q}^{-1})\gamma_{\rm s}^2C_{\rm B}^{\rm s}F_\Xi+(\lambda_{\rm s}^{3}+\lambda_{\rm s}^{-3})\gamma_{\rm s}^3C_{\rm B}^{\rm s}F_\Omega\Big\}\ , \qquad (1)$$

where the kaon (K), hyperon (Y), cascade (Ξ) and omega (Ω) degrees of freedom are included successively. In the resonance sums all known strange hadrons are counted — we have included in our studies kaons up to 1.78 GeV, hyperons up to 1.94 GeV, cascades up to 1.95 GeV and also the omega resonance at 2.25 GeV. The phase space factors F_i of the strange particles are explicitly

$$F_K = \sum_j g_{K_j}W(m_{K_j}/T)\ , \qquad F_Y=\sum_j g_{Y_j}W(m_{Y_j}/T)\ ,$$
$$F_\Xi = \sum_j g_{\Xi_j}W(m_{\Xi_j}/T)\ , \qquad F_\Omega=\sum_j g_{\Omega_j}W(m_{\Omega_j}/T)\ , \qquad (2)$$

with $W(x)=x^2K_2(x)$, where K_2 is the modified Bessel function, arising as an integral over the free Boltzmann particle phase space.

The abundance factors $\gamma_{\rm s}$, $C_i^{\rm s}$ *correct* the final state freeze-out particle numbers as compared to the yield expected if there were full chemical equilibrium in the total strange quark abundance yield (factor $\gamma_{\rm s}$) and (strange) meson and baryon yield (factors $C_i^{\rm s}$). Therefore we have in the above Eq. (1) the usual abundance factors given in terms of the fugacities, related to the chemical potentials by $\lambda_i = e^{\mu_i/T}$. The fugacity of each hadronic species is simply the product of the valance quark fugacities, *viz.* $\lambda_{\rm p}=\lambda_{\rm u}^2\lambda_{\rm d}$, $\lambda_{\rm K^+}=\lambda_{\rm u}\lambda_{\bar{\rm s}}$, etc. Three fugacities are introduced since the flavors u, d, and s are separately conserved on the time scale of hadronic collisions and can only be produced or annihilated in particle-antiparticle pairs. Furthermore, since the factor $\gamma_{\rm s}$ is introduced to account for the deviation from the full phase space occupancy, (which is assumed implicitly for u, d flavors) the chemical potentials for particles and anti-particles are opposite to each other, implying for the respective fugacities

$$\lambda_{\bar{i}} = \lambda_i^{-1}\ . \qquad (3)$$

2.3. Analysis of Strange Particle Spectra

Our study of the experimental results differs fundamentally from the naive simulations that can be carried out with particle cascade type generators, containing specific models of the reaction and input cross sections. In our approach we seek to obtain by judicious choice of experimental observables the values of the different key parameters which determine the particle freeze-out conditions, as they appear in Eq. (1). This avoids pitfalls such as sensitivity of the experimental results to the exact (trigger) conditions of the experiment, and allows to identify the characteristic features of importance in determining the physical conditions prevailing in the particle source. On the other hand, our approach makes only good sense if all particles of relevance show features characteristic of a local, thermally equilibrated source, and if indeed strange particles originate in a central source (fireball). We recall here that data on the rapidity dependence of strange particles[3,4] gives clear indication that the production of strange particles in nuclear S→Pb collisions is concentrated in the central rapidity domain, which supports the notion of a central fireball reaction picture also for the 200 GeV A S→A collisions (A=100 – 200).

It turns out that independent of the kinematic behavior of the source regarding the longitudinal and collective transverse flow the shape of the particle spectra is dependent only on the transverse particle mass $m^i_\perp = \sqrt{m_i^2 + p_\perp^2}$. Thus, studying the ratios of *spectra* of particles with those of their anti-particles within overlapping regions of $m_\perp$ at some fixed rapidity, the spectral shapes and many other statistical factors cancel leading to quantities which are functions of the equilibrium fugacities only.[5] It was shown[7] that this statement remains approximately valid when feed-down by resonance decays is included. One has specifically:

$$R_\Xi = \frac{\overline{\Xi^-}}{\Xi^-} = \frac{\lambda_d^{-1}\lambda_s^{-2}}{\lambda_d\lambda_s^2}\,, \qquad R_\Lambda = \frac{\overline{\Lambda}}{\Lambda} = \frac{\lambda_d^{-1}\lambda_u^{-1}\lambda_s^{-1}}{\lambda_d\lambda_u\lambda_s}\,. \tag{4}$$

These ratios can easily be related to each other, in a way which shows the Dependance on $\mu_q = (\mu_u + \mu_d)/2$, μ_s and the respective isospin asymmetry $\delta\mu = \mu_d - \mu_u$:

$$R_\Lambda R_\Xi^{-2} = e^{6\mu_s/T} \cdot e^{2\delta\mu/T}\,, \qquad R_\Xi R_\Lambda^{-2} = e^{6\mu_q/T} \cdot e^{-\delta\mu/T}\,. \tag{5}$$

Eq. (5) is valid irrespective of the state of the source at freeze-out (HG or QGP). Once the left hand side is known experimentally, it determines rather accurately the values of μ_q, μ_s which enter on the right hand side with a dominating factor 6, while the (small) isospin asymmetry $\delta\mu$ plays only a minor role. Thus it is possible to determine the chemical potentials with considerable precision in spite of relatively large experimental errors on the measured values of R_Λ, R_Ξ.

Similar considerations allow to determine the strangeness saturation factor. A complete cancellation of the diverse statistical and freeze-out factors occurs when we form the product of the abundances of baryons and anti-baryons. We further need the cancellation of the spectral shape factors which results taking two different particle kinds, *e.g.* we consider (ignoring at first resonance feed-down):

$$\gamma_s^2 \simeq \frac{\Xi^-}{\Lambda} \cdot \frac{\overline{\Xi^-}}{\overline{\Lambda}} \simeq \frac{\Lambda}{p} \cdot \frac{\overline{\Lambda}}{\overline{p}} \simeq \frac{\Omega^-}{2\Xi^-} \cdot \frac{\overline{\Omega^-}}{2\overline{\Xi^-}}, \tag{6}$$

where in the last relation the factors 2 in the denominator correct for the spin-3/2 nature of the Ω.

While the initial analysis of the data carried out 4 years ago[5] was made just in these qualitative terms, the recent efforts are incorporating in numerical studies considerably more detail, including flow deformation of the spectra, cascading of resonances and other phenomena.[7] The vanishing strange quark potential $\mu_s \simeq 0$ is a striking result of different analysis of all today available data (WA85,[7] NA35[8] and NA36[4]). Furthermore, one obtains near saturation of the strange particle phase space $0.5 < \gamma_s < 1$, with th differences indicative of approach to equilibrium as the size of the system increases.[9]

This common results are found despite considerable difference in other details of the experimental results. The greatest difference between NA36[4] and WA85[7,5] analysis result is seen in the extracted value of λ_q. As we just shall see, this could be result of sensitivity of this parameter to the details of the experimental conditions, a point to which we shall return momentarily.

2.4. Strangeness Conservation

The exact balance between s and $\bar{s}$ quarks in hadronic interactions requires non-trivial relations between the parameters characterizing the final state. These are in general difficult to satisfy and the resulting particle distributions are constrained such that they differ greatly e.g. if the source is a rapidly disintegrating QGP or equilibrated HG phase. These two alternatives can be distinguished in particular by the value of the strange quark chemical potential μ_s:

1. In a strangeness neutral QGP fireball μ_s is always exactly zero, independent of prevailing temperature and baryon density, since both s and $\bar{s}$ quarks have the same phase-space size.

2. In any state consisting of locally confined hadronic clusters μ_s is generally different from zero at finite baryon density, in order to correct the asymmetry introduced in the phase-space size by a finite baryon content.

At non-zero baryon density, i.e. for $\mu_B \neq 0$, there is just one (or perhaps at most a few) special value $T(\mu_B)$ for which $\langle s \rangle = \langle \bar{s} \rangle$ at $\mu_s^{HG} = 0$. For the case of a conventional (Hagedorn type) HG these values have been studied carefully.[7,10] In general, for the final state described by Eq. (1) and for the case $\lambda_s = 1$ as determined by the experimental analysis the condition of strangeness conservation takes a simple analytical form:

$$\mu_q^0 = T \cosh^{-1}\left(R_C^s \frac{F_K}{2F_Y} - \gamma_s \frac{F_\Xi}{F_Y} \right), \tag{7}$$

where $R^s_C = C^s_M/C^s_B$. There is at most one real solution, only when the argument on the right hand side in Eq. (7) is greater than unity. It turns out that this result, Eq. (7), is a very sensitive function of the temperature for the observed values of T. Thus it is conceivable that different experiments with nearly the same conditions (of e.g. trigger, rapidity window, etc) obtain similar values of T and $\lambda_s \simeq 1$ but a quite different value of μ_q (i.e. λ_q). Since as we saw above the value of λ_q controls in a sensitive way the particle ratios, such as $\overline{\Lambda}/\Lambda$, these can easily vary from experiment to experiment, while the results regarding the other parameters (μ_s, γ_s, T would be much more in agreement even if some conditions vary.

2.5. Hadronic Gas Source?

To disprove the ad-hoc case of the equilibrated HG we take $R^s_C = 1$, and determine quantitatively in how far can the strangeness conservation constraint be reconciled with the present data. In Fig. 1 the solid line is just $3\mu^0_q = \mu^0_B(T; R^s_C = 1, \gamma_s = 0.7)$, see Eq. (7). The dashed and, respectively, dotted curves are the solutions of the strangeness conservation condition for the experimentally permitted range of $\lambda_s = 0.98$ and, respectively, $= 1.08$ (the mean experimental value is at $\lambda_s = 1.03$,[7]). The two straight dissecting curves correspond to the experimental constraint for WA85 data, $\lambda_q = 1.48 \pm 0.05$ (the disagreement of the HG model is even greater for the NA36 results, which, alas, have greater errors). The hatched area is the region of μ_B and T compatible with a hadronic gas source in which meson to baryon abundance equilibrium is reached. As noted before, in the here interesting domain of values of $100 < \mu_B < 400$ MeV there is great sensitivity of the allowed values of μ_B to the value of T.

The $m_\perp$-spectra of the strange particles used in the multiplicity analysis indicate a common temperature $T = 232 \pm 5$ MeV. We stress that this temperature is not necessarily the freeze-out temperature of final state particles, it combines the freeze-out temperature with the phenomenon of flow, and in first approximation one can argue that it is the flow blue-shifted freeze out value (see Eq. (11) below). Furthermore, as numerical studies show, the value seen at the relatively high $m_\perp \simeq$ 1–2 GeV is near or just below the value of the temperature reached before collective flow sets in. Taking this value as the lower limit on the initial temperature we show in Fig. 1 the experimental cross, 3 s.d. to the right of the range found to be consistent with the experiment.

It is at this point relevant to note that in an initial analysis of the first data, the "experimental" point coincided with the hatched area.[11,12] The increased precision, more recent data lead in the interim to a shift of the experimental point to the right by about 1 s.d.; the theory was also further developed and includes today all known hadronic resonances. This has moved the solid line ($\lambda_s = 1$) in Fig. 1 by about 1 s.d. to the left, to its current location. Finally, the disagreement between HG-theory and experiment is enhanced by another 1 s.d. since previously we had $\lambda_s > 0.93$ (within 1 s.d.) and now it is $\lambda_s > 0.98$ (dashed line in Fig. 1), which moves the error boundary by one further s.d. to the left, to its current location in Fig. 1. Thus we can infer

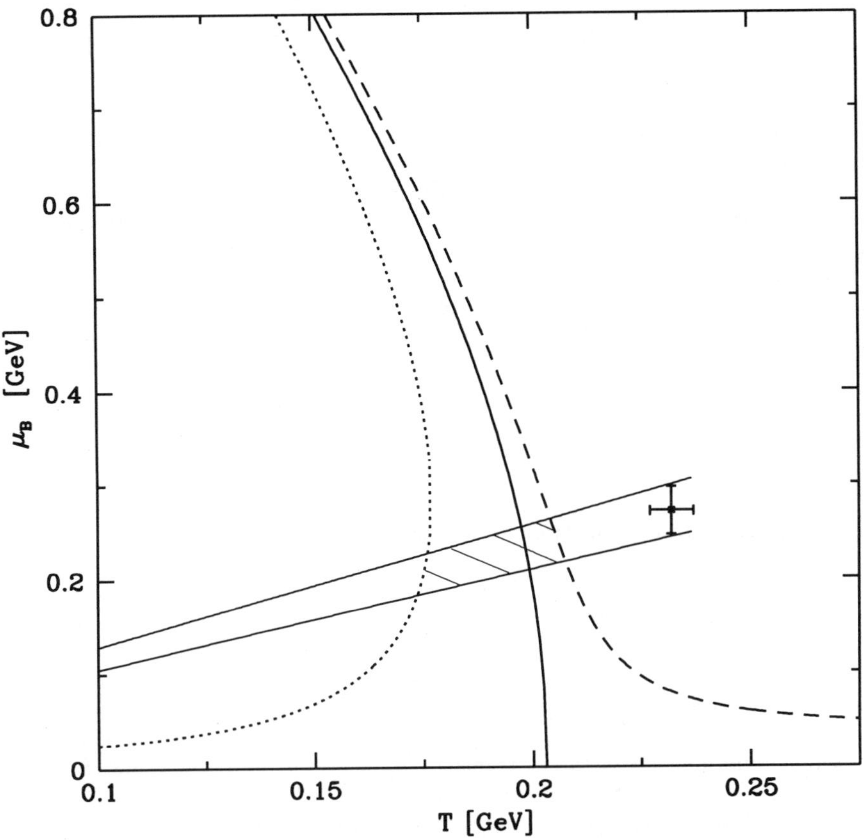

Figure 1: Strangeness conservation constraint in the μ_B–T_f plane. Solid line: $\lambda_s = 1$, dashed curve: $\lambda_s = 0.98$, dotted line: $\lambda_s = 1.08$ (all with $R^s_C = 1$, $\gamma_s = 0.7$). Horizontal intersecting lines correspond to $\lambda_q = 1.48 \pm 0.05$. Hatched area is the region of agreement (within 1 s.d.) of the strangeness conservation condition with the observed values $\lambda_s = 1.03 \pm 0.05$ and $\lambda_q = 1.48 \pm 0.05$. The experimental cross is set at the maximal possible temperature, $T_f = 0.232 \pm 0.005$ GeV, consistent with $m_\perp$-spectra.

that a strangeness neutral HG source, even allowing for off-equilibrium strangeness saturation $\gamma_s = 0.7$, is in disagreement with experimental results.

In order to move the strangeness conservation constraint line to the 'experimental' point, we need the value of $R^s_C = 1.45$. What this means is that if the primordial source were to freeze-out without any collective flow, the relative abundance of strange mesons compared to the equilibrium value should be enhanced by a factor 1.45, or that the relative abundance of strange baryons compared to the chemical equilibrium should be suppressed by a factor 0.69, or both at suitable intermediate value. If, however, we are persuaded that the freeze-out temperature is around 150 MeV, which necessitates presence of a freeze-out flow velocity at 0.41c, then it turns out that $R^{s}_{C} = 0.39$ and accordingly this implies opposite meson-baryon relative enhancement/suppression trends than just described. In any case, it is apparent that the hypothesis only recently seriously entertained[13] of a fireball of hot hadronic gas is incompatible with these results. Also, a HG source appears to have another problem when compared to the data, namely that it does not reproduce the large measured charged particle multiplicities. At fixed (e.g. measured) temperature and chemical potential the specific (per baryon) entropy is a characteristic quantity which distinguishes a HG from a deconfined state of quarks and gluons.[11,14] The fact that the entropy content in the QGP phase is greater than in HG due to the excitation of the abundant gluon degrees of freedom is suggestive that this may be the state formed in the collision. As we shall now show, a quantitative study supports strongly this claim, and leads to a quantitative explanation of the values of the freeze-out parameters obtained from the strange particle data analysis.

3. Formation and Evolution of the Quark–Gluon Plasma

Our objective is to understand if the thermal parameters we determined are indeed consistent with the hypothesis of QGP formation. We need to develop a qualitative understanding of the initial conditions and the evolution of the fireball. The initial conditions are determined by the 'stopping' parameters: the momentum stopping fraction $\eta_{\rm p}$ implies that only a fraction $0 \le \eta_{\rm p} \le 1$ of the incident CM momentum can be used by a particle incident to exercise a force, the balance remains in longitudinal, un-stopped motion. For a target transparent to the incoming flow, i.e. for $\eta_{\rm p} = 0$ there would obviously be no compression force exercised in the collision. We hope that $\eta_{\rm p} \simeq \eta_{\rm E}$, the latter being defined by studying the fraction of the original beam energy found in the transverse direction to the beam motion.[15] In the S→W/Pb reactions at 200 GeV A, the energy stopping is $\eta_{\rm E} \simeq 0.5$, and we therefore will assume a similar value for our parameter $\eta_{\rm p}$. There is a further stopping parameter, $\eta_{\rm B}$ which describes the fraction of baryon number deposited on average in the central interaction region. While in principle $\eta_{\rm B} \neq \eta_{\rm E}$, at the here interesting range of energies it appears that both these parameters are similar. At lower energies we will move towards full stopping, that is $\eta_i = 1$. We also anticipate that $\eta_i = 1$ when we study collisions of largest nuclei such as Pb→Pb.

3.1. Dynamical Pressure

We consider quantitatively the balance between the dynamical compression exercised by the in-flowing matter and the resisting thermal pressure of the constituents of the central fireball.[16] The pressure due to kinetic motion follows from well-established principles, and can be directly inferred, e.g. from Eq. (15) of Ref. [17] for the energy-momentum tensor:

$$T^{ij}(x) = \int p^i u^j f(x,p)\mathrm{d}^3p\,, \quad i,j = 1,2,3\,, \tag{8}$$

where $u^j = p^j/E$. We take for the phase-space distribution of incident particles $f(x,p) = \rho_0(x)\delta^3(\vec{p}-\vec{p}_{\rm CM})$. To obtain the pressure exercised by the flow of matter, we consider the pressure component T^{jj}, with j being the direction of $\vec{v}_{\rm CM}$. The result is:

$$P_{\rm dyn} = \eta_{\rm p}\frac{p_{\rm CM}^2}{E_{\rm CM}}\rho_0\,, \tag{9}$$

where $\eta_{\rm p}$ is momentum stopping fraction. Here it is understood that the energy $E_{\rm CM}$ and momentum $P_{\rm CM}$ is given in the CM frame, and per nucleon.

It is interesting to note that using the normal nuclear density, $\rho_0 = 0.16\,{\rm fm}^{-3}$ and full stopping one finds the bench mark value $P_{\rm dyn} = 0.5$ GeV/fm^3 (respectively 1 GeV/fm^3) at $\gamma_{\rm CM} = 3.4$ (respectively 6.4). In a symmetric collision this corresponds to an incident projectile at 23.6 GeV/c^2 A (respectively = 86.3 GeV/c^2 A).

3.2. Initial Conditions

The QGP picture of the reaction requires that the internal Fermi pressure of quark and gluon constituents is balanced by the external pressure. To determine the initial fireball conditions in the collision, we demand that the internal thermal (and Fermi) pressure $P_{\rm th}$ be balanced by the dynamical and vacuum pressure. We thus obtain the condition:

$$P_{\rm th}(T,\ldots) = P_{\rm dyn} + P_{\rm vac} = \eta_{\rm p}\frac{E_{\rm CM}^2 - m^2}{E_{\rm CM}}\rho_0 + \mathcal{B}\,. \tag{10}$$

The QGP equations of state will determine the left-hand side of Eq. (10). The calculations are carried out[16] using ideal quantum gases for quarks and gluons, allowing for the suppression of the occupancy of gluons G and all quarks by the factors γ_i, $i = q,s,G$. QCD interactions are accounted for incorporating thermal particle masses $m_i^2 = m_0^2 + (cT)^2$ and introducing the quantum degeneracy factors corrected for interactions using the perturbative thermal QCD corrections, according to the prescription of Ref.[9]; In the expression for thermal masses in principle $c^2 \propto \alpha_{\rm s}$, but given the current uncertainty regarding the value of the coefficient c a constant value $c \sim 2$ is used, arising for $\alpha_s \sim 1$ in the current formulas. We take $m_{\rm q}^0 = 5$ MeV, $m_{\rm s}^0 = 160$ MeV, and $m_{\rm G}^0 = 0$. For the QCD coupling, we take $\alpha_s = 0.6$, and the vacuum pressure is $\mathcal{B}$=0.1 GeV/fm^3. Numerical integration of the Bose/Fermi distributions for quarks/gluons including these effects allows us to obtain any physical property of the initial QGP state.

For a given set of three chemical equilibration parameters $\gamma_{\rm G}$, $\gamma_{\rm q}$, $\gamma_{\rm s}$ there are in principle two unknown statistical parameters, T and λ_q, which are further determined after we impose on these equations of state two additional constraints: in our case we prescribe the total pressure $P = P_{\rm dyn} + P_{\rm vac}$ assuming a reasonable value of stopping fractions of momentum. Furthermore assuming that the stopping of energy and baryon number is similar we have that the energy per baryon corresponds to the energy in the beam E/B. For a given set of γ_i, and η_i all parameters are fixed.

3.3. Time Evolution

To describe the time evolution till freeze-out we need to consider in more detail the different stages of the nuclear collision. The particles observed in the collision are emanating at some low, freeze-out temperature $T_{\rm f}$. It cannot be assumed that strange and not strange particles are freezing out exactly in same condition and hence it is convenient to introduce the strange-particle freeze-out temperature $T_{\rm s}$ and in view of the smaller strange particle cross sections very probably $T_{\rm f} \leq T_{\rm s}$. Both values could be much smaller than the temperature T_0 visible in the transverse particle spectra with the collective flow velocity $v_{\rm f}, v_{\rm s}$ superposes a Doppler blue shift with:

$$T_{\rm f} = \sqrt{\frac{1-v_{\rm f}}{1+v_{\rm f}}} T_0 \,, \qquad T_{\rm s} = \sqrt{\frac{1-v_{\rm s}}{1+v_{\rm s}}} T_0 \,. \tag{11}$$

It appears that $T_0 \simeq 232$ MeV for both strange and non-strange-particles at high $m_\perp$, where other effects that distort the spectra (e.g. resonance disintegration) become negligible. This observed temperature T_0 is representative of the conditions of the fireball prior to its expansion, after the initial compression, but after strangeness flavor was produced.

Two other temperatures are also important here: $T_{\rm ch}$ is the temperature reached in the system after the chemical equilibration of light quarks and gluons, but prior to the significant saturation of the strangeness phase-space, the production of strangeness being slower than of light quarks. The temperature $T_{\rm th}$ is the temperature at which we can hypothesize that thermal motion is established already in the initial phase, while the valence quarks, gluons and quark–anti-quark pairs produced are still far from chemical equilibrium. We thus have the following temperature hierarchy. Here the time increases from left to right while temperature decreases:

$$T_{\rm th} > T_{\rm ch} > T_0 \geq T_{\rm s} \geq T_{\rm f} \tag{12}$$

The first inequality is related to cooling during the approach to chemical equilibrium (number equilibrium) of $q, \bar{q}, G$,[18] the second one is due to the cooling while strangeness is produced and reaches its chemical equilibrium, and the last two arise from expansion in which thermal energy is transferred to the collective flow.

This hierarchy of thermal conditions in the fireball can be seen for the cases of particular interest in table 1. In the first column we show the assumed values of the chemical occupancies corresponding to the different stages described. In the second, we list the properties we shall present explicitly (when possible these are given per baryon, with ρ/ρ_0 being the baryon density in units of the equilibrium nuclear density). In 3rd, 4th and 5th columns we give our results for the three cases

Table 1: Conditions in different collision systems at different evolution stages.

Phase-space occupancy	$<s-\bar{s}>=0$, $\lambda_s \equiv 1$	E/B [GeV] 2.6, $\eta=1$, Au–Au	8.8, $\eta=0.5$, S–Pb	8.6, $\eta=1$, Pb–Pb
	T_{th} [GeV]	0.260	0.410	0.471
$\gamma_q=0.2$	λ_q	9.95	1.78	2.00
	n_G/B	0.20	1.55	1.25
	n_q/B	3.00	5.12	3.77
$\gamma_G=0.2$	$n_{\bar{q}}/B$	0.00	2.12	0.77
	$n_{\bar{s}}/B$	0.02	0.16	0.13
	P_{th} [GeV/fm^3]	0.46	0.79	1.46
$\gamma_s=0.03$	ρ/ρ_0	3.34	1.70	3.18
	S/B	11.8	40.0	33.4
	T_{ch} [GeV]	0.212	0.280	0.324
$\gamma_q=1$	λ_q	4.16	1.49	1.61
	n_G/B	0.56	2.50	2.08
	n_q/B	3.11	5.16	4.62
$\gamma_g=1$	$n_{\bar{q}}/B$	0.11	2.16	1.62
	$n_{\bar{s}}/B$	0.05	0.25	0.21
	P_{ch} [GeV/fm^3]	0.46	0.79	1.46
$\gamma_s=0.15$	ρ/ρ_0	3.35	1.80	3.19
	S/B	12.3	41.8	34.9
		$\gamma_s=1$	$\gamma_s=0.8$	$\gamma_s=1$
	T_0 [GeV]	0.190	0.233	0.270
$\gamma_q=1$	λ_q	4.16	1.49	1.61
	n_G/B	0.56	2.50	2.09
$\gamma_G=1$	n_q/B	3.11	5.12	4.60
	$n_{\bar{q}}/B$	0.11	2.12	1.60
$\gamma_s=0.8$	$n_{\bar{s}}/B$	0.28	1.27	1.07
or	P_0 [GeV/fm^3]	0.33	0.47	0.84
$\gamma_s=1$	ρ/ρ_0	2.41	1.05	1.81
	S/B	14.1	49.5	41.7

of specific interest: first for the Si→Au and Au→Au collisions at $E_{CM}=2.6$ GeV, $\eta_i=1$ (corresponding to the condition at BNL–AGS with 11–15 GeV A projectiles), then for the case of S→W/Pb at $E_{CM}=8.8$ GeV, $\eta_i=0.5$ (corresponding to S projectile penetrating a tube of matter in a central impact on a heavy target at 200 GeV A) and finally Pb→Pb collisions at $E_{CM}=8.6$ GeV, $\eta_i=1$ (corresponding to Pb projectiles at 157 GeV A colliding at small impact parameter with Pb target).

The first results, top segment of table 1, depend on how exactly we define the earliest 'th' period, it begins when we can apply statistical methods and speak of temperature. The value of the temperature $T_{\rm th}$ shown there is a direct consequence of the number of gluons and light anti-quarks assumed here to be present when first thermalization is completed (note that we always make sure that for each stopped baryon there are three quarks per baryon in the fireball). Next, in the second segment of the table comes the epoch, beginning with the temperature $T_{\rm ch}$, when the fireball reaches chemical equilibrium for light quarks and gluons, the penetration of nuclei ends, and the expansion and equilibration of strangeness abundance commences. At this point in time the condition of the fireball is fully determined by the values of the stopping parameters. In the last, bottom segment, we consider the properties after strangeness has been produced, for two slightly different values of $\gamma_{\rm s}$. As suggested by the S$\to$W results,[7] we allow a slight deviation from the equilibrium in this case. We recall that in the transverse spectra the temperature T_0 is visible at high $m_\perp$ and we see that the statistical properties determined here are in remarkable agreement with the values extracted from experiments: $T_0 = 232 \pm 5$ MeV, $\lambda_{\rm q} = 1.48 \pm 0.05$.[7]

We note in table 1 the subtle differences between the S$\to$Pb and Pb$\to$Pb collisions (both at similar CERN energies). We predict a rise in temperature and $\lambda_{\rm q}$ — the observable temperature rises from $T_0 = 233$ MeV to $T_0 = 270$ MeV, which should indeed be easily visible in the high $m_\perp$-spectra. Because of the considerable increase in baryon density (from $1.8\rho_0$ to $3.2\rho_0$) we see a noticeable drop in specific entropy. This is intuitively correct, since the energy per baryon is similar; the greater energy content per final-state particle at higher temperature thus entails a smaller number of particles per baryon, and hence a smaller entropy content per baryon. The specific yield of strange-particles is also expected to drop, but it remains relatively high as the number of $\bar{s}$ is greater than the number of $\bar{u}$ or $\bar{d}$ (note that $\bar{q} = \bar{u} + \bar{d}$, and naturally $\bar{s} = s$). Thus the ratios of particles such as $\overline{\Xi^-}/\overline{\Lambda}$, which are considered as a test for the QGP phase,[1] will increase by 35–50% as we move to the Pb$\to$Pb system from S$\to$Pb. Interestingly, this ratio will modestly increase while collision energy decreases, as long as the QGP phase is formed, due to an increase in the value of $\lambda_{\rm q}$, which we predict. A sudden decrease should follow, when $E_{\rm CM}$ drops below the threshold for the formation of QGP.

When looking at the Au–Au results at the lowest (AGS) energy considered here we note a relatively large expected value of $\lambda_{\rm q}$. Thus a simple probe for QGP in this system is the measurement of the ratios of particles such as $\overline{\Lambda}/\bar{p} \simeq \overline{\Xi^-}/\overline{\Lambda} \propto \gamma_{\rm s}\lambda_{\rm q}$, which should thus considerably exceed conventional physics based expectations. However, note that our earlier analysis of the Si–Au results[19] suggested that, should QGP be formed in the 14 GeV A interactions, its disintegration would require a considerable re-equilibration in the final state, since $\lambda_{\rm s} \simeq 1.7$ was found in the final state, and not $\lambda_{\rm s} \simeq 1$ as expected in rapid QGP dissociation and therefore the enhancement will be not as huge as could be otherwise expected. The necessary presence of the intermediate re-equilibration era at these low energies complicates the arguments for, or against, possible formation of QGP .

The entropy content shown at the bottom of table 1, which determines the final-particle multiplicity,[14,7] evolves very little and is to all purposes already present at the initial stage, when quarks and gluons are still far from abundance equilibrium.[18] Practically all the rise we present is due to the formation of the strange flavor against the background of the thermalized and expanding quark–gluon fireball. We have an enormous amount of strangeness, as the tables show, and hence the fact that 20% of entropy is due to strangeness production is not surprising.

4. Final Remarks

We believe strongly, having failed to identify unexpected entropy generating mechanisms, that as the dense matter system evolves towards the final freeze-out conditions, its entropy content remains in essence unchanged. Thus, irrespective of the values of freeze-out temperatures T_s and T_f at which the strange and, respectively, non-strange particles decouple, the specific entropy content of the hadronic system is expected to remain the same. The fact that final hadronic multiplicity is determined in the initial instants of the collision makes numerical cascade type model calculations very difficult since they require the assumption of initial conditions, and thus effectively, assumption of the final state multiplicities.

Interpreting particle abundances in terms of chemical potentials and temperature of a source, and assuming an equation of state, the entropy content can be calculated. For the S→W collisions at 200 GeV A we see in table 1 that we should expect 50 units of entropy per baryon in the final state, as was determined from analysis of the data.[14,7] In a confined hadronic state, with properties described by a mixture of hadronic resonances (Hagedorn gas), we find an entropy content that is 50% of a QGP source at these given thermo-chemical conditions.

We have in hand a full and consistent model of the formation and evolution of QGP, which results in properties of the freeze-out state as needed to fully account for the production of different strange baryons and anti-baryons measured in recent experiments, and of the total particle multiplicity (entropy). We have demonstrated that the competing models using conventional particles miss key properties of the final state in both these respects and hence they aught now to be definitively abandoned. The QGP alternative lends itself as a natural and indeed expected explanation of the data, but we cannot at this stage exclude the possibility that someone will propose a detailed model which will be similarly successfully. Therefore, our QGP picture of the reaction needs to be confirmed and we propose that this is done studying the forthcoming Pb→Pb 160 GeV A collisions. Here we stress without further discussion that certain easily accessible variations of triggers and eventually, beam energy will lead to variations of the data which are particularly characteristic for the QGP state. For example, we can argue that decrease of the beam energy will entail an increase of the ratio of $\overline{\Xi}/\overline{\Lambda}$ as long as the deconfined state is formed. If and when the energy is reduced below the deconfinement limit, this ratio can be expected to become rather rapidly small, $\simeq 0.05$. Such a threshold behavior coupled with counter-intuitive response to the energy variation would provide a very convincing confirmation of the QGP formation hypothesis.

5. Acknowledgements

J. R. was partially supported by the US-DOE grant DE-FG02-92ER40733. He thanks his co-authors for their very warm hospitality in Paris.

6. References

1. J. Rafelski, *Phys. Rep.* **88** (1982) 331.
2. S. Abatzis *et al.* (WA85 collaboration), *Phys. Lett.* **B270** (1991) 123; **B259** (1991) 508; **B316** (1993) 615. D. Evans et al. (WA85 collaboration), *Nucl. Phys.* **A566** (1994) 225c.
3. T. Alber *et al*, The NA35 collaboration, *Strange Particle Production in Nuclear Collisions at 200 GeV per Nucleon Z. Phys.* **C** in press, Univ. Frankfurt preprint, IKF-HENPG/1-94, April 1994.
4. E. Andersen *et al.*, NA36 Collaboration, *Phys. Lett.* **B316** (1993) 603. *Nucl. Phys.* **A566** (1994) 217c; *Phys. Lett.* **B327** (1994) 433.
5. J. Rafelski, *Phys. Lett.* **B262** (1991) 333; *Nucl. Phys.* **A544** (1992) 279c.
6. J. Rafelski and B. Müller, *Phys. Rev. Lett.* **48** (1982) 1066.
7. J. Letessier, A. Tounsi, U. Heinz, J. Sollfrank and J. Rafelski, *Strangeness Conservation in hot nuclear fireballs*, Paris preprint LPTHE/92–27R, *submitted to Phys. Rev. D.*
8. J. Sollfrank, M. Gaździcki, U. Heinz and J. Rafelski, *Z. Physik* **C61** (1994) 659.
9. J. Letessier, J. Rafelski and A. Tounsi, *Phys. Lett.* **B323** (1994) 393.
10. J. Letessier, J. Rafelski and A. Tounsi, *Phys. Lett.* **B321** (1994) 394.
11. J. Letessier, A. Tounsi and J. Rafelski, *Phys. Lett.* **B292** (1992) 417.
12. J. Cleymans and H. Satz, *Z. Phys.* **C57** (1993) 135.
13. J. Cleymans, K. Redlich, H. Satz, and E. Suhonen, *Z. Phys.* **C58** (1993) 347.
14. J. Letessier, A. Tounsi, U. Heinz, J. Sollfrank and J. Rafelski, *Phys. Rev. Lett.* **70** (1993) 3530.
15. G.W. London, *Result from the CERN Pilot Study of Ultra-Relativistic Nucleus-Nucleus Interaction*, in: *proceedings of the VIII International Conference on Physics in Collision*, Capri, 1988, P. Strolin, ed., Editions Frontières, Paris, 1989, 397.
    I. Otterlund, *Physics of Relativistic Nuclear Collisions*, in: *Particle Production in Highly Excited Matter*, H. H. Gutbrod and J. Rafelski, eds, NATO Physics series Vol. **B 303**, Plenum Press, New York, 1993, 57.
    S.P. Sorensen et al. *Nuclear Stopping Power*, in: *Proc. XXI Int. Symposium on Multiparticle Dynamics 1991*, World Scientific, Singapore, 1992.
16. J. Letessier, J. Rafelski and A. Tounsi, *Phys. Lett.* **B333**, (1994) 484.
17. S.R. de Groot, W.A. van Leeuwen, Ch.G. van Weert, *Relativistic Kinetic Theory,* North Holland, Amsterdam, 1980.
18. J. Letessier, J. Rafelski and A. Tounsi, *Phys. Rev.* **C50** (1994) 406.
19. J. Letessier, J. Rafelski and A. Tounsi, *Phys. Lett.* **B328** (1994) 499.

ON THE SELF-COHERENCE OF THE HADRON HIGH-ENERGY TOTAL CROSS-SECTION DATA

Basarab NICOLESCU

Division de Physique Théorique, Institut de Physique Nucléaire, 91406 Orsay Cedex and LPTPE, Université P. & M. Curie, 4 Place Jussieu, 75252 Paris Cedex 05 (France)*

ABSTRACT — We perform an esentially model-independent study of the internal coherence between high-energy total cross-section data by using classes of functions satisfying general principles. The recent CDF data are included in our analysis. Our study is practically independent of the ρ-parameter values and is fit-independent. We conclude that the $\ln^2 s$ increase of total cross-sections at high energies is clearly favoured when compared with the other possible behaviours.

Talk at the Workshop on Quantum Infrared Physics, The American University of Paris, 6-10 June 1994

---

* *Unité de Recherche des Universités Paris 11 et Paris 6 Associée au CNRS*

In this talk, based on work done in collaboration with P. Gauron, we question the *coherence* between high-energy σ_T data we compared with general theoretical requirements.

In order to perform this study we will not resort to any fitology. Fitology - the science of fits, based on χ^2 minimization - has, of course, many virtues but has also a major defect : the *details* of minimization are sometimes crucial in getting the main physical conclusions.

We consider classes of functions respecting the general principles and modelling in fact an infinite number of possible scattering amplitudes. We then inject as little as possible experimental input in order to extract as much as possible theoretical predictions. We finally compare the theoretical numbers with the experimental bounds and suggest that there are possible internal self-contradictions of the high energy σ_T data (i.e. ISR, CERN $S\bar{p}pS$, Tevatron and cosmic-ray data).

We first consider the following class of functions supposed to be valid for both pp and $\bar{p}p$ scattering (no-Odderon case) :

$$\sigma_T = C_1 + C_2 \ln^\gamma \left(\frac{s}{s_+} \right) \tag{1}$$

where γ is left free (of course, the Froissart-Martin bound [1] requires $\gamma \leq 2$).

We are namely interested in the *increase* of σ_T in the range (s_1, s_2), with $s_2 > s_1$:

$$\Delta\sigma(\gamma) \Big|_{s_1}^{s_2} = C_2 \left[\ln^\gamma \left(\frac{s_2}{s_+} \right) - \ln^\gamma \left(\frac{s_1}{s_+} \right) \right], \tag{2}$$

the parameter C_1 being so eliminated from our study.

Let us fix for the moment the scale factor s_+ to be

$$s_+ = 1 \text{ GeV}^2 \tag{3}$$

in order to get the Amaldi form*[2], popular amongst experimentalists and recently used in fits performed by the UA4/2 collaboration [3]. Of course, there is no physical

* The familiar Regge terms, present in the Amaldi form, are practically negligible in our energy range.

reason to fix s_+ to such a value. Moreover, from the mathematical point a view such a procedure is simply wrong : the 3-parameter form of eq. (2) cannot be reduced to a 2-parameter form without an arbitrary supplementary condition such as (3). However let us accept for the moment the condition (3) in order to compare with the results of ref. [3]. The case of arbitrary s_+ will be considered later.

The procedure is the following : 1) we first take as experimental input the increase of σ_T between CERN $S\bar{p}pS$ and Tevatron energies ; 2) we then compute, at arbitrary fixed γ, the value of C_2 extracted from eq. (2) ; 3) we finally compute $\Delta\sigma(\gamma)\Big|_{s_1}^{s_2}$ by using eq. (2), for the cases ($\sqrt{s_1} =$ 62.3 GeV, $\sqrt{s_2} =$ 541 GeV) and ($\sqrt{s_1} = 1.8$ TeV, $\sqrt{s_2} = 30$ TeV) and compare with the experimental data.

It has to be noted that, in the CDF case[4], the values of ρ and σ_T are not given separately, only the value $(1+\rho^2)\sigma_T$ being measured. We will assume for the moment $\rho = 0.135$ in agreement with the UA4/2 [5] and E710 [6] values. This corresponds to $\sigma_T = 62.2 \pm 1.5$ mb at CERN $S\bar{p}pS$ energy $\sqrt{s} = 541$ GeV [3].

In fact our study has the advantage of being practically *independent* of the assumed value of ρ. Of course, at fixed s, the value of σ_T is sensitive to the assumed value of ρ, when the experimental $(1+\rho^2)\sigma_T$ value is imposed : for example, when ρ increases from 0.135 to 0.2, the σ_T decreases by 1.3 mb at $\sqrt{s} = 541$ GeV and by 1.7 mb at $\sqrt{s} = 1.8$ TeV. However this variation at fixed s is practically cancelled when the difference of σ_T between s_1 and s_2 is considered : when ρ varies, $\sigma_T(s_1)$ and $\sigma_T(s_2)$ *both* decrease (or increase) by almost the same amount. This is in fact the reason why we chose to study $\Delta\sigma(\gamma)\Big|_{s_1}^{s_2}$ and not $\sigma_T(s_1)$ and $\sigma_T(s_2)$ themselves.

There are two experimental inputs at fixed γ : $\Delta\sigma^{\exp}(\gamma)\Big|_{UA4/2}^{CDF}$ and $\Delta\sigma^{\exp}(\gamma)\Big|_{UA4/2}^{E710}$, corresponding to the two different experimental results at $\sqrt{s} = 1.8$ TeV [4,6]. The results of our procedure are shown in figs. 1 and 2, where only the range $0.5 \leq \gamma \leq 2.5$ is explicitely exhibited. In these figures the lower band corresponds to the allowed experimental increase of σ_T between $\sqrt{s} = 62.3$ GeV [7] and $\sqrt{s} = 541$ GeV [3], while the upper band is, in fact, the *lower bound* on the increase of σ_T between $\sqrt{s} = 1.8$

TeV and $\sqrt{s} = 30$ TeV, the AKENO best fit [8] being considered in the cosmic-ray range of energies. The central line in the upper band is affected by an indicative error of ± 10 mb, typical for the cosmic-ray energy range.

From the fig. 1 it is seen that, in the case of the $\Delta\sigma^{\exp}(\gamma)\Big|_{UA4/2}^{CDF}$ input, the theoretical points corresponding to the increase of σ_T between ISR and CERN $S\bar{p}pS$ energies marginally meet the allowed experimental range only for values of γ corresponding to the violation of the Froissart-Martin bound : $2 \lesssim \gamma \lesssim 2.5$. The lower bound on the increase of σ_T between Tevatron (TEV) and cosmic-ray (CR) energies is respected for all γ values in the considered range (0.5, 2.5).

The situation in fig. 2, corresponding to the $\Delta\sigma^{\exp}(\gamma)\Big|_{UA4/2}^{E710}$ input, is somewhat reversed. The theoretical points corresponding to $\Delta\sigma(\gamma)\Big|_{ISR}^{CERN\ S\bar{p}pS}$ meet the allowed experimental strip for all values of γ, with a slight preference for the range $0.7 \lesssim \gamma \lesssim 1.4$ (to which the magical value $\gamma = 1$ prefered in ref. 9 belongs). However the theoretical points are systematically lower than the lower experimental bound on $\Delta\sigma(\gamma)\Big|_{TEV}^{CR}$.

It is now interesting to compare our results with the detailed 8-parameter fit of σ_T in the energy range $5 \lesssim s \lesssim 541$ GeV performed by the UA4/2 collaboration [3]. The relevant parameter in the context of our discussion is γ. The value of γ extracted from the fit of ref. [3] is 2.25 affected by a positive error of 0.35 and by a negative error of 0.31. This result is perfectly compatible with our result shown in fig. 1. (It is amusing to note that, in ref. [3], the predicted value of σ_T at $\sqrt{s} = 1.8$ TeV is 76.5 ± 2.3 mb, in between the E710 and CDF experimental values : $\simeq 72$ mb and $\simeq 80$ mb, respectively). However, the violation of the Froissart-Martin bound is just a *mathematical artefact* of the arbitrary condition (3), as we will show in the following.

Let us now study the effect of ruling out the condition (3), which obviously introduces (see eq. 1) a strong and artificial correlation between the parameters γ and C_2 and therefore leads to doubtful predictions at high energies. The above procedure can be easily extended in the case of arbitrary s_+ by systematically exploring, for fixed γ, a given range of s_+, say $10 \leq s_+ \leq 500\ GeV^2$. It has to be noted that $\gamma = 1$ is a special case : the quantity $\Delta\sigma(\gamma=1)\Big|_{s_1}^{s_2}$ defined by eq. (2) is obviously independent of s_+. However, for $\gamma \neq 1$, the dependence on s_+ has interesting consequences, especially when γ is far from the special case $\gamma = 1$:

1. For the first input $\Delta\sigma^{\exp}(\gamma)\Big|_{UA4/2}^{CDF}$, the scale factor s_+ *improves* the situation described in fig. 1. Namely, for $\gamma > 1$ the quantity $\Delta\sigma(\gamma)\Big|_{ISR}^{CERN\ Sp\bar{p}S}$ decreases and for $\gamma \simeq 2$ the theoretical points meet the experimental allowed region (while $\Delta\sigma(\gamma)|_{TEV}^{CR}$ increases, the agreement with the experimental bound being kept). For example, for

$$\gamma = 2 \qquad \text{and} \qquad s_+ = 240 \text{ GeV}^2 \tag{4}$$

we get

$$C_2 = 0.47 \pm 0.07 \text{ mb} \tag{5}$$

and, therefore,

$$\Delta\sigma(\gamma=2)\Big|_{ISR}^{CERN\ Sp\bar{p}S} = 20.08 \pm 2.97 \text{ mb} \tag{6}$$

and

$$\Delta\sigma(\gamma=2)\Big|_{TEV}^{CR} = 65.13 \pm 9.65 \text{ mb}, \tag{7}$$

in agreement with the experimental constraints (see fig. 1).

2. Paradoxically enough, for the second input $\Delta\sigma^{\exp}(\gamma)\Big|_{UA4/2}^{E710}$, the scale factor s_+ *worsens* the situation described in fig. 2 : $\Delta\sigma(\gamma)\Big|_{ISR}^{CERN\ Sp\bar{p}S}$ increases for $\gamma < 1$ and decreases for $\gamma > 1$ and meets the experimental band for γ-values around 1. The agreement of $\Delta\sigma(\gamma)\Big|_{TEV}^{CR}$ with the experimental bound is somewhat improved for $2 \lesssim \gamma \lesssim 2.5$ but, for this range of γ, the agreement of $\Delta\sigma(\gamma)\Big|_{ISR}^{CERN\ Sp\bar{p}S}$ with the experimental constraint is destroyed.

In conclusion, our study, involving the general class of functions (1-2), singles out the CDF data (when compared with the E710 data) and the physical value $\gamma = 2$. The agreement with the Froissart-Martin bound, in the CDF case, is restored by considering the scale factor $s_+ \neq 1$ GeV^2. In this context, the CDF data appears as one of the most significant experimental results obtained in the last years.

The class of functions proposed by Bourrely and Martin [10] leads to similar conclusions [11].

Our main conclusion - the fact that the $\ln^2 s$ increase of total cross-sections at high energies is clearly favoured - is strenghten by the impressive agreement between CDF [4] and UA4 σ_T - data [3,5] at the same energy $\sqrt{s} = 541$ GeV.

The publication of the luminosity dependent σ_T obtained in the already performed UA4/2 experiment would bring interesting information for the clarification of the present situation. Without waiting for the important (but far away) experimental results from RHIC and LHC machines, a new high-precision E-710 type experiment at the Tevatron is badly needed. The theoretical interest of such experiments is clear : striking trends of a spectacular *simplification* appear already in the *maximum complexity* region of hadron physics - the elastic and diffractive scattering.

REFERENCES

[1] M. Froissart, Phys. Rev. 123 (1961) 1053;
A. Martin, Phys. Rev. 129 (1963) 1432; Nuovo Cimento 42A (1966) 930.

[2] U. Amaldi et al., Phys. Lett. B66 (1977) 390.

[3] UA4/2 Collab., C. Augier et al., Phys. Lett. B315 (1993) 503.

[4] CDF Collab., F. Abe et al., FERMILAB-Pub-93/234-E preprint, submitted to Phys. Rev D.

[5] UA4/2 Collab., C. Augier et al., Phys. Lett. B316 (1993) 448.

[6] E710 Collab., N.A. Amos et al., Phys. Lett. B243 (1990) 158; Phys. Rev. Lett. 68 (1992) 2433;
S. Sadr,in Proc. of the Vth International Conference on the Elastic and Diffractive Scattering (Vth Blois Workshop), Providence, USA., June 1993,

World Scientific (1994), edited by H.M. Fried, K. Kang and C. -I Tan, pp. 59-63.

[7] R210 Collab., G. Carboni et al., Nucl. Phys. B254 (1985) 697.
R211 Collab., N.A. Amos et al., Nucl. Phys. B262 (1985) 689.

[8] M. Honda et al., Phys. Rev. Lett. 70 (1993) 525.

[9] M.M. Block and A.R. White, Phys. Lett. B273 (1991) 145.

[10] C. Bourrely and A. Martin, Proc. of the ECFA-CERN Workshop, CERN, Service d'Information Scientifique (1984), p. 233.

[11] P. Gauron and B. Nicolescu, Phys. Lett. B (in press).

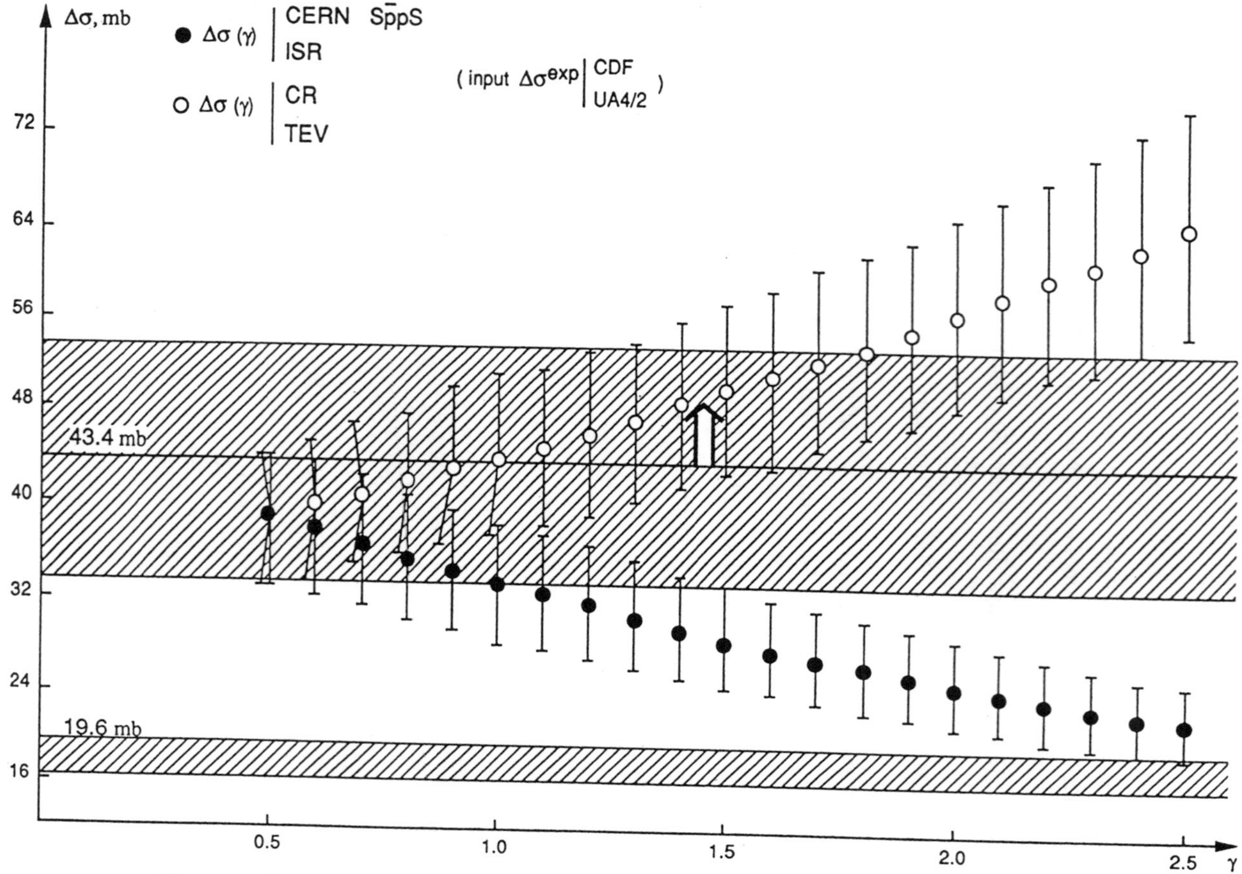

Fig. 1 The increase $\Delta\sigma$ of σ_T in terms of γ when the CDF data [4] are used as input at $\sqrt{s} = 1.8$ TeV.

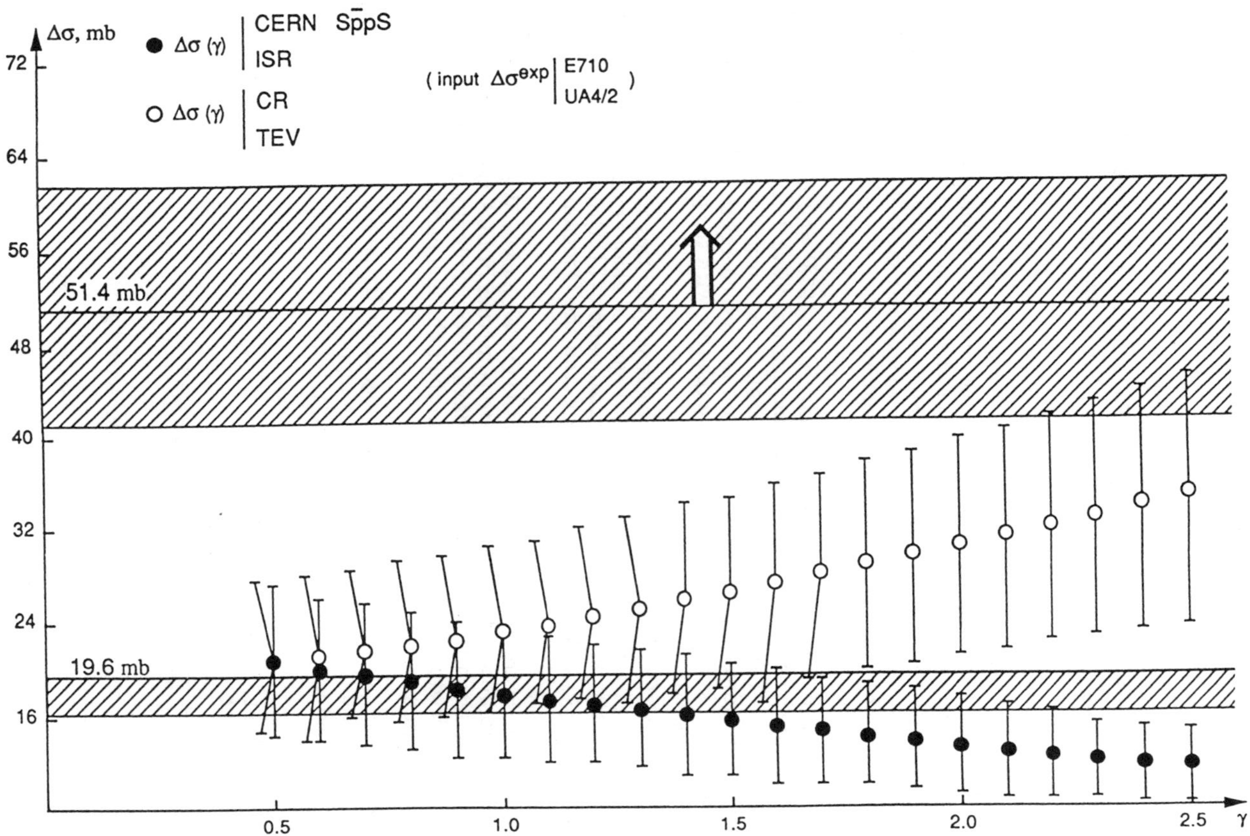

Fig. 2 The increase $\Delta\sigma$ of σ_T in terms of γ when the E710 data [6] are used as input at $\sqrt{s} = 1.8$ TeV.

THE ODDERON INTERCEPT IN PERTURBATIVE QCD

P. GAURON

*Division de Physique Théorique *, IPN, 91406 Orsay Cedex and*

LPTPE, Université Pierre et Marie Curie, 4 Place Jussieu 75252 Paris Cedex 05, France

ABSTRACT

We construct, in the framework of QCD, the conformally invariant functional whose maximal value gives the J-plane location of the leading singularity of the t-channel partial waves in LLA for diagrams with n reggeized gluons. By using a variational method with conformal techniques we show that the Odderon J-plane singularity in the LLA approximation of QCD lies above 1.

The concept of the Odderon was introduced in 1973[1] and corresponds to a singularity situated near $J = 1$ in the t-channel partial wave $f_-^J(t)$ with the quantum numbers $P = -1$ and $C = -1$. The Odderon is responsible for the differences between hadron-hadron and hadron-antihadron at high energies[2]. Various models investigated this possibility near $t = 0$ [3] but the relation with the perturbative QCD is not yet established.

In pertubative QCD the dominant rightmost singularities of the partial waves in the J-plane come from diagrams with several reggeized gluons exchanged between the particles[4]. In the leading logarithmic approximation (LLA)the Pomeron arises as a compound state of two reggeized gluons[4,5] and the Odderon consists of three gluons[6].

In this work we are mainly interested in the Odderon case. We use the conformally invariant approach developed in ref. 5 to construct in LLA the functional $\mathcal{J}$ depending on Odderon wave functions. Its maximal value obtained by variational method gives the leading J-plane singularity in the t-channel partial waves with the Odderon quantum numbers[7].

To begin with, let us write down the homogeneous Bethe-Salpeter equation in LLA for the wave function $\chi_\omega(\vec{\rho}_1, \ldots, \vec{\rho}_n)$ of the compound state of n reggeized gluons in the impact-parameter space

$$\Omega \prod_{i=1}^{n} |\partial_i|^2 \chi_\omega = \sum_{r=1}^{n} \sum_{s=1}^{n} \prod_{i \neq r,s} |\partial_i|^2 R^{rs}(-T_r^a T_s^a) \chi_\omega \qquad (r \neq s), \tag{1a}$$

$$\partial_i = \frac{\partial}{\partial \rho_i}, \quad \partial_i^* = \frac{\partial}{\partial \rho_i^*}, \quad \Omega = \frac{8\pi^2}{g^2}\omega, \; \omega = J - 1 \tag{1b}$$

T_r^a, T_s^a $(a = 1, \ldots, 8)$ act on colour indices of interacting gluons r,s.

---

* Unité de Recherche des Universités Paris 11 et Paris 6 Associée au CNRS

R^{rs} is an integral operator describing pair-like gluon interactions

$$R^{rs}\chi_\omega(\vec{\rho}_1,...,\vec{\rho}_n) = \frac{1}{2\pi}\int d^2\rho_{r'} \left\{ \partial_s\partial_r^* \frac{\theta\left(\left|\frac{\rho_{rr'}}{\rho_{rs}}\right| - \varepsilon\right)}{|\rho_{rr'}|^2} \partial_s^*\partial_{r'} + \partial_s^*\partial_r \frac{\theta\left(\left|\frac{\rho_{rr'}}{\rho_{rs}}\right| - \varepsilon\right)}{|\rho_{rr'}|^2} \partial_s\partial_{r'}^* \right\}$$

$$\cdot\, \chi_\omega(\vec{\rho}_1,...,\vec{\rho}_{r-1},\vec{\rho}_{r'},\vec{\rho}_{r+1},...,\vec{\rho}_n) - ln(\frac{1}{\varepsilon^2})|\partial_r|^2|\partial_s|^2\chi_\omega(\vec{\rho}_1,...,\vec{\rho}_n), \tag{2}$$

where θ is the step function, $\rho_{rs} = \rho_r - \rho_s$ and the last term $\sim \ln(\frac{1}{\epsilon^2})$, $\epsilon \to +0$ removes the ultraviolet divergence in the integral when $\rho_{rr'} \to 0$. Note that $R^{rs} \neq R^{sr}$.

The kernels R^{rs} and R^{sr} describe the contribution of the graphs a) and b) involving two reggeized gluons:

a) The first graph is

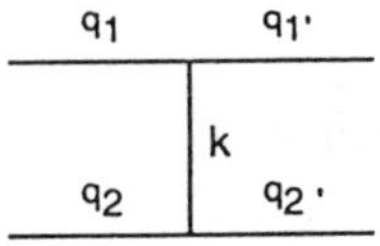

$$\sim \frac{1}{k_\perp^2}(q_1^2 q_{2'}^2 + q_{1'}^2 q_2^2) - (q_1 + q_2^2)^2 \sim q_1 q_2^* \frac{1}{k_\perp^2} q_{1'}^* q_{2'} + h.c.$$

and by Fourier transform this gives terms like

$$\partial_2^*\partial_1 \ln|\rho_{12}|^2 \partial_1^*\partial_2 \quad + h.c.$$

Such terms are not explicit in the kernel (2). They come from the step function. If instead $\theta\left(\left|\frac{\rho_{rr'}}{\rho_{rs}}\right| - \varepsilon\right)\frac{1}{|\rho_{rr'}|^2}$ we write $\theta(|\rho_{rr'}| - \varepsilon)\frac{1}{|\rho_{rr'}|^2}$, we have an additional term

$$\int_\varepsilon^{\varepsilon\rho_{rs}} d^2\rho_{r'} \frac{1}{|\rho_{rr'}|^2} \sim \ln|\rho_{rs}|^2 \,.$$

b) The second graph is

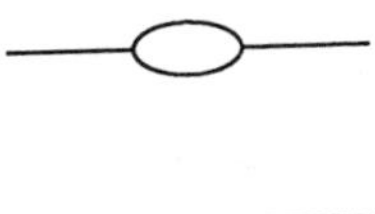

$$\sim \ln(k^2\varepsilon^2),$$

which by Fourier transform gives $\dfrac{\theta(|\rho_{rr'}| - \varepsilon)}{|\rho_{rr'}|^2}$.

The upper boundary of the spectrum of eigenvalues Ω of eq. (1) determines the position ω_{max} of the leading singularity of the corresponding solution f_ω of the inhomogeneous Bethe-Salpeter equation.

Eq. (1) can be derived from the variational principle for the functional

$$\mathcal{J}^{rs} = \int \prod_i d^2\rho_i \left[(\prod_{k\neq r,s} \partial_k^*)\chi_\omega^* \right] R^{rs}(-T_r^a T_s^a) \left[(\prod_{k\neq r,s} \partial_k)\chi_\omega \right] , \tag{3}$$

where the sum over colour indices c_i of gluons is implied.

The trial function χ_ω must satisfy the normalization condition

$$\int \prod_i d^2\rho_i \left| (\prod_k \partial_k)\chi_\omega \right|^2 = 1 \tag{4}$$

and Bose-symmetry properties

$$\chi_\omega^{c_1,\ldots,c_n}(\vec{\rho}_1,\ldots,\vec{\rho}_n) = \chi_\omega^{c_1,\ldots,c_n}(\vec{\rho}_1,\ldots,\vec{\rho}_n)\Big|_{\substack{c_i \rightleftarrows c_j \\ \rho_i \rightleftarrows \rho_j}} \tag{5}$$

Only colourless composite states symmetric in colour indices c_i are considered

$$\sum_r T_r^a \chi_\omega = 0, \qquad \chi_\omega^{c_1\ldots c_n} = \chi_\omega^{c_1\ldots c_n}\Big|_{c_i \rightleftarrows c_j}$$

This is important for the cancellation of the infrared singularities.

The eigenvalues Ω of eq. (1) are expressed through the stationary values $\tilde{\mathcal{J}}$ of functional $\mathcal{J}$

$$\Omega = n(n-1)\tilde{\mathcal{J}}\ , \tag{6}$$

where $n(n-1)$ is the total number of the kernels. Note that Ω does not depend on the indices r and s. The position of the leading singularity of the n-reggeon contribution is $\omega^{(n)} = (g^2/8\pi^2)\ max\ \Omega$.

It can be shown that these equations are invariant under the Moebius transform $\rho_i \to (a\rho_i + b)/(c\rho_i + d)$. Therefore χ_ω are eigenfunctions of the Casimir operators of the conformal group and can be classified with respect to their eigenvalues $m(m+1)$ and $\tilde{m}(\tilde{m}+1)$ with the conformal weights

$$m = \frac{1}{2} + i\nu + \frac{n}{2} \quad , \qquad \tilde{m} = \frac{1}{2} + i\nu - \frac{n}{2}\ , \tag{7}$$

where integer n is the conformal spin.

Let us consider the case $n = 3$ describing in LLA the Odderon.

Then due to conformal invariance $\chi_\omega^{m,\tilde{m};c_1c_2c_3}(\rho_1,\rho_2,\rho_3;\rho_o)$ takes the form

$$\chi_\omega^{m,\tilde{m};c_1c_2c_3}(\rho_1,\rho_2,\rho_3;\rho_o) = d^{c_1c_2c_3}\left(\frac{\rho_{12}\rho_{13}\rho_{23}}{\rho_{10}^2\rho_{20}^2\rho_{30}^2}\right)^{\frac{m}{3}}\left(\frac{\rho_{12}^*\rho_{13}^*\rho_{23}^*}{\rho_{10}^{*2}\rho_{20}^{*2}\rho_{30}^{*2}}\right)^{\frac{\tilde{m}}{3}} \cdot \psi^{m,\tilde{m}}(x,x^*)\ , \tag{8}$$

where $\rho_i(i=1,2,3)$ are the complex coordinates of the gluons in the impact-parameter space and ρ_0 is the coordinate of the Odderon with quantum numbers m and $\tilde{m}$. $d^{c_1c_2c_3}$ is the unique totally symmetric tensor corresponding to the d-couplings of gluons in the colourless state.

The function $\psi^{m,\tilde{m}}$ depends on only one complex variable

$$x = \frac{\rho_{12}\rho_{30}}{\rho_{10}\rho_{32}} \tag{9}$$

Further the colour structure simplifies by substituting $-T_r^a T_s^a \to 3/2$. Therefore in the odderon case we obtain from eqs. (1) and (6)

$$\omega^{Odd} = \frac{g^2}{8\pi^2}\Omega = \frac{g^2}{8\pi^2} \cdot 9\tilde{S} \tag{10}$$

where $\tilde{S}$ is the maximal value of the functional S obtained by the insertion of ansatz (8) into eq. (3).

$$S = \frac{1}{2\pi}\int d^2x d^2x' \Bigg\{ \alpha^*(x,x^*) \frac{\theta\left(\left|\frac{x-x'}{x}\right| - \epsilon\right)}{|x-x'|^2} \alpha(x',x'^*)$$
$$+ \beta^*(x,x^*) \frac{\theta\left(\left|\frac{x-x'}{x}\right| - \epsilon\right)}{|x-x'|^2} \beta(x',x'^*) - 2\pi \ln(\frac{1}{\epsilon^2})\delta^2(x-x')|\gamma(x,x^*)|^2 \Bigg\} \tag{11}$$

$\alpha(x,x^*), \beta(x,x^*)$ and $\gamma(x,x^*)$ correspond to the derivatives $\partial_1^*\partial_2\partial_3\chi_\omega$, $\partial_1\partial_2^*\partial_3\chi_\omega$ and $\partial_1\partial_2\partial_3\chi_\omega$ which appear in formulae (2) and (3) :

$$\begin{aligned} \alpha(x,x^*) &\equiv B_{m,\tilde{m}}(x,x^*)Z(x,x^*) \\ \beta(x,x^*) &\equiv B_{\tilde{m},m}(x^*,x)Z(x,x^*) \\ |\gamma(x,x^*)|^2 &= \frac{1}{2}[|\alpha(x,x^*)|^2 + |\beta(x,x^*)|^2] \end{aligned} \tag{12}$$

where

$$B_{m,\tilde{m}}(x,x^*) = [m + (1-x)\partial]\partial^*[-\tilde{m} + x^*\partial^*] \tag{13}$$

$$Z(x,x^*) \equiv [x(1-x)]^{\frac{m}{3}}[x^*(1-x^*)]^{\frac{\tilde{m}}{3}}\psi^{m,\tilde{m}}(x,x^*). \tag{14}$$

$\psi^{m,\tilde{m}}(x,x^*)$ satisfies the normalization condition

$$\int d^2x|\gamma(x,x^*)|^2 = 1 \ , \tag{15}$$

where $\gamma(x,x^*)$ is determined in terms of $\psi^{m,\tilde{m}}(x,x^*)$ by eqs. (12)-(14), and the modular invariance properties

$$\psi^{m,\tilde{m}}(x,x^*) = e^{i\pi(m-\tilde{m})/3}\psi^{m,\tilde{m}}\left(-\frac{x}{1-x}, -\frac{x^*}{1-x^*}\right),$$
$$\psi^{m,\tilde{m}}(x,x^*) = e^{i\pi(m-\tilde{m})/3}\psi^{m,\tilde{m}}\left(\frac{1}{x}, \frac{1}{x^*}\right). \tag{16}$$

For convenience of numerical computation[8], we perform the Fourier transform of the function α, β and make an integration by part which allows to solve the problem of the derivatives.

We choose the simplest trial function in agreement with the conformal invariance and the correct small-x asymptotics:

$$\psi_{m,\tilde{m}}(x,x^*) = C\Big\{[a(x,x^*)]^{\frac{1}{3}(m+\tilde{m})} +$$
$$+ C_1[a(x,x^*)]^{1-\frac{1}{6}(m+\tilde{m})}[C_2 - 2\ln a(x,x^*)]\Big\} \tag{17}$$

$$a(x,x^*) = \frac{|x|^2|1-x|^2}{(|x|^2+1)(|1-x|^2+1)(|x|^2+|1-x|^2)} \ . \tag{18}$$

The rightmost singularity in the J-plane is obtained for $\nu = n = 0$, i.e. $m = \tilde{m} = 1/2$. The constant C is fixed by the normalization condition. The constants C_1 and C_2 are determined from the condition of maximality of the functional S. The numerical value of $\widetilde{S}$ is approximately $\widetilde{S} \simeq 0.25$ and therefore from eq. (10) we get

$$\alpha_{Odd} \geq 1 + 0.28 \frac{g^2}{\pi^2} \quad . \tag{19}$$

Thus the Odderon intercept is bigger than 1.

In conclusion, we have demonstrated, within the LLA approximation, that the Odderon intercept is bigger than 1. For the first time it is shown that in perturbative QCD the differences between hadron-hadron and hadron-antihadron grow with energy.

Experimentally there is already an experimental evidence for a difference between pp and $\bar{p}p$ interactions at $\sqrt{s} = 53$ GeV and at relatively large t². The Odderon effects can be checked experimentally at HERA where it is possible to select hard processes with the exchange of Reggeons with Odderon quantum numbers only. In the future it could be checked at LHC and SSC provided that both pp and $\bar{p}p$ options are adopted.

REFERENCES

1. L.Łukaszuk and B. Nicolescu, *Nuovo Cimento Lett.*, **8** (1973) 405.
2. A. Breakstone et al., *Phys. Rev. Lett.* **54** (1985) 2180;
   S. Erhan et al., *Phys. Lett.* **B152** (1985) 131.
3. K. Kang and B. Nicolescu, Phys. Rev. D11 (1975) 2461;
   D. Joynson, E. Leader, C. Lopez and B. Nicolescu, *Nuov. Cim.* **30A** (1975) 345;
   P. Gauron, E. Leader and B. Nicolescu, *Nucl. Phys.* **B299** (1985) 640;
   A. Donnachie and P.V. Landshoff, *Nucl. Phys.* **B244** (1984) 322;
   L.L. Jenkovszky, E.S. Martynov and B.V. Struminsky, *Phys. Lett.* **B249** (1990) 535.
4. E. Kuraev, L. Lipatov and V. Fadin, *Sov. Phys. JETP* **44** (1976) 433; **45** (1977) 199.
5. Ya. Balitski and L. Lipatov, *Sov. J. Nucl. Phys.* **28** (1978) 822;
   L. Lipatov, *Sov. Phys. JETP* **63(5)** (1986) 904;
   For a review, see L. Lipatov, in Perturbative QCD, ed. A.H. Mueller (World Scientific, Singapore, 1989).
6. J. Bartels, *Nucl. Phys.* **B175** (1980) 365;
   J. Kwiecinski and M. Praszalowicz, *Phys. Lett.* **B94** (1980) 413.
7. P. Gauron, L.N. Lipatov and B. Nicolescu, *Phys. Lett.* **B260** (1991) 407.
8. P. Gauron, L.N. Lipatov and B. Nicolescu, *Phys. Lett.* **B304** (1993) 334.

Generalized Theory of Gravitation

J. W. Moffat
Department of Physics
University of Toronto
Toronto, Ontario M5S 1A7
Canada

Abstract

A generalized gravitational theory is presented based on a non-Riemannian geometry with a nonsymmetric metric $g_{\mu\nu}$.

e-mail: moffat@physics.utoronto.ca

1. Introduction

A nonsymmetric gravitational theory (NGT), based on the decompositions of the metric $g_{\mu\nu}$ and the connection $\Gamma^{\lambda}_{\mu\nu}$:

$$g_{(\mu\nu)} = \frac{1}{2}(g_{\mu\nu} + g_{\nu\mu}), \quad g_{[\mu\nu]} = \frac{1}{2}(g_{\mu\nu} - g_{\nu\mu}), \tag{1.1}$$

and

$$\Gamma^{\lambda}_{\mu\nu} = \Gamma^{\lambda}_{(\mu\nu)} + \Gamma^{\lambda}_{[\mu\nu]}, \tag{1.2}$$

is presented[1]. The theory is free of ghosts, tachyons and higher-order poles in the propagator in the linear approximation. An expansion of $g_{\mu\nu}$ about an arbitrary Einstein background metric also yields field equations to first order in $g_{[\mu\nu]}$, which are free of couplings to unphysical (negative energy) modes; the solutions of the field equations have consistent asymptotic boundary conditions.

In view of the difficulty in obtaining physically consistent geometrical generalizations of Einstein gravitational theory (EGT), it is interesting that such a consistent theory can be formulated[2].

A static spherically symmetric solution of the NGT field equations in the short-range approximation is everywhere regular and does not contain a black hole event horizon[3].

2. Nonsymmetric Gravitational Theory

The non-Riemannian geometry is based on the nonsymmetric field structure with a nonsymmetric $g_{\mu\nu}$ and affine connection $\Gamma^\lambda_{\mu\nu}$, defined in Eqs.(1.1) and (1.2)[4]. The contravariant tensor $g^{\mu\nu}$ is defined in terms of the equation:

$$g^{\mu\nu}g_{\sigma\nu}=g^{\nu\mu}g_{\nu\sigma}=\delta^\mu_\sigma. \tag{2.1}$$

The Lagrangian density is given by

$$\mathcal{L}_{NGT}=\mathcal{L}_R+\mathcal{L}_M, \tag{2.2}$$

where

$$\mathcal{L}_R=\mathbf{g}^{\mu\nu}R_{\mu\nu}(W)-2\lambda\sqrt{-g}-\frac{1}{4}\mu^2\mathbf{g}^{\mu\nu}g_{[\nu\mu]}-\frac{1}{6}g^{\mu\nu}W_\mu W_\nu, \tag{2.3}$$

where λ is the cosmological constant and μ^2 is an additional cosmological constant associated with $g_{[\mu\nu]}$. Moreover, $\mathcal{L}_M$ is the matter Lagrangian density ($G=c=1$):

$$\mathcal{L}_M=-8\pi g^{\mu\nu}\mathbf{T}_{\mu\nu}. \tag{2.4}$$

Here, $\mathbf{g}^{\mu\nu}=\sqrt{-g}g^{\mu\nu}$ and $R_{\mu\nu}(W)$ is the NGT contracted curvature tensor:

$$R_{\mu\nu}(W)=W^\beta_{\mu\nu,\beta}-\frac{1}{2}(W^\beta_{\mu\beta,\nu}+W^\beta_{\nu\beta,\mu})-W^\beta_{\alpha\nu}W^\alpha_{\mu\beta}+W^\beta_{\alpha\beta}W^\alpha_{\mu\nu}, \tag{2.5}$$

defined in terms of the unconstrained nonsymmetric connection:

$$W^\lambda_{\mu\nu}=\Gamma^\lambda_{\mu\nu}-\frac{2}{3}\delta^\lambda_\mu W_\nu, \tag{2.6}$$

where

$$W_\mu=\frac{1}{2}(W^\lambda_{\mu\lambda}-W^\lambda_{\lambda\mu}). \tag{2.7}$$

Eq.(2.6) leads to the result:

$$\Gamma_\mu = \Gamma^\lambda_{[\mu\lambda]} = 0. \tag{2.8}$$

The NGT contracted curvature tensor can be written as

$$R_{\mu\nu}(W) = R_{\mu\nu}(\Gamma) + \frac{2}{3}W_{[\mu,\nu]}, \tag{2.9}$$

where $R_{\mu\nu}(\Gamma)$ is defined by

$$R_{\mu\nu}(\Gamma) = \Gamma^\beta_{\mu\nu,\beta} - \frac{1}{2}\left(\Gamma^\beta_{(\mu\beta),\nu} + \Gamma^\beta_{(\nu\beta),\mu}\right) - \Gamma^\beta_{\alpha\nu}\Gamma^\alpha_{\mu\beta} + \Gamma^\beta_{(\alpha\beta)}\Gamma^\alpha_{\mu\nu}. \tag{2.10}$$

The field equations in the presence of matter sources are given by:

$$G_{\mu\nu}(W) + \lambda g_{\mu\nu} + \frac{1}{4}\mu^2 C_{\mu\nu} - \frac{1}{6}(P_{\mu\nu} - \frac{1}{2}g_{\mu\nu}P) = 8\pi T_{\mu\nu}, \tag{2.11}$$

$$\mathbf{g}^{[\mu\nu]}{}_{,\nu} = -\frac{1}{2}\mathbf{g}^{(\mu\beta)}W_\beta, \tag{2.12}$$

$$\mathbf{g}^{\mu\nu}{}_{,\sigma} + \mathbf{g}^{\rho\nu}W^\mu_{\rho\sigma} + g^{\mu\rho}W^\nu_{\sigma\rho} - \mathbf{g}^{\mu\nu}W^\rho_{\sigma\rho} + \frac{2}{3}\delta^\nu_\sigma\mathbf{g}^{\mu\rho}W^\beta_{[\rho\beta]}$$

$$+\frac{1}{6}(\mathbf{g}^{(\mu\beta)}W_\beta\delta^\nu_\sigma - \mathbf{g}^{(\nu\beta)}W_\beta\delta^\mu_\sigma) = 0. \tag{2.13}$$

Here, we have

$$G_{\mu\nu} = R_{\mu\nu} - \frac{1}{2}g_{\mu\nu}R, \tag{2.14}$$

$$C_{\mu\nu} = g_{[\mu\nu]} + \frac{1}{2}g_{\mu\nu}g^{[\sigma\rho]}g_{[\rho\sigma]} + g^{[\sigma\rho]}g_{\mu\sigma}g_{\rho\nu}, \tag{2.15}$$

$$P_{\mu\nu} = W_\mu W_\nu, \tag{2.16}$$

and $P = g^{\mu\nu}P_{\mu\nu} = g^{(\mu\nu)}W_\mu W_\nu$. We have not included a phenomenological vector density source, $\mathbf{S}^\mu$, in Eq.(2.12), although such an option is possible.

In empty space, the field equations (2.11) become:

$$R_{\mu\nu}(\Gamma) = \frac{2}{3}W_{[\nu,\mu]} + \lambda g_{\mu\nu} - \frac{1}{4}\mu^2(C_{\mu\nu} - \frac{1}{2}g_{\mu\nu}C) + \frac{1}{6}P_{\mu\nu}. \tag{2.17}$$

The generalized Bianchi identities:

$$[\mathbf{g}^{\alpha\nu}G_{\rho\nu}(\Gamma)+\mathbf{g}^{\nu\alpha}G_{\nu\rho}(\Gamma)]_{,\alpha}+g^{\mu\nu}{}_{,\rho}\mathbf{G}_{\mu\nu}=0, \tag{2.18}$$

give rise to the matter response equations:

$$g_{\mu\rho}\mathbf{T}^{*\mu\nu}{}_{,\nu}+g_{\rho\mu}\mathbf{T}^{*\nu\mu}{}_{,\nu}+(g_{\mu\rho,\nu}+g_{\rho\nu,\mu}-g_{\mu\nu,\rho})\mathbf{T}^{*\mu\nu}+\frac{1}{12\pi}\mathbf{g}^{(\alpha\nu)}W_\alpha W_{[\nu,\rho]}=0, \tag{2.19}$$

where

$$T^{*\mu\nu}=T^{\mu\nu}-\frac{1}{32\pi}\mu^2C^{\mu\nu}+\frac{1}{48\pi}(P^{\mu\nu}-\frac{1}{2}g^{\mu\nu}P). \tag{2.20}$$

In general, the last term in Eq.(2.3) is of the form, $\frac{1}{2}\sigma g^{\mu\nu}W_\mu W_\nu$. The coupling constant σ is chosen to be $\sigma=-\frac{1}{3}$, so that the theory yields consistent ghost and tachyon free perturbative solutions to the field equations.

3. Linear Approximation

Let us assume that $\lambda=0$ and expand $g_{\mu\nu}$ about Minkowski spacetime:

$$g_{\mu\nu}=\eta_{\mu\nu}+{}^{(1)}h_{\mu\nu}+..., \tag{3.1}$$

where $\eta_{\mu\nu}$ is the Minkowski metric tensor: $\eta_{\mu\nu}=\mathrm{diag}(-1,-1,-1,+1)$. We also expand $\Gamma^\lambda_{\mu\nu}$ and $W^\lambda_{\mu\nu}$:

$$\Gamma^\lambda_{\mu\nu}={}^{(1)}\Gamma^\lambda_{\mu\nu}+{}^{(2)}\Gamma^\lambda_{\mu\nu}+...,$$
$$W_\mu={}^{(1)}W_\mu+{}^{(2)}W_\mu+\dots. \tag{3.2}$$

Let us adopt the notation: $\psi_{\mu\nu}=h_{[\mu\nu]}$. To first order of approximation, Eq.(2.12) gives

$$\psi_\mu=-\frac{1}{2}W_\mu, \tag{3.3}$$

where for convenience W_μ denotes ${}^{(1)}W_\mu$. Moreover,

$$\psi_\mu=\psi_{\mu\beta}{}^{,\beta}=\eta^{\beta\sigma}\psi_{\mu\beta,\sigma}. \tag{3.4}$$

¿From Eq.(2.13), we obtain the result to first order:

$${}^{(1)}\Gamma^\lambda_{\mu\nu}=\frac{1}{2}\eta^{\lambda\sigma}({}^{(1)}h_{\sigma\nu,\mu}+{}^{(1)}h_{\mu\sigma,\nu}-{}^{(1)}h_{\nu\mu,\sigma})+\frac{1}{6}(\delta^\lambda_\nu W_\mu-\delta^\lambda_\mu W_\nu). \tag{3.5}$$

The antisymmetric and symmetric field equations derived from Eq.(2.11) decouple to lowest order; the symmetric equations are the usual Einstein field equations in the linear approximation. The skew equations are given by

$$(\Box + \mu^2)\psi_{\mu\nu} = J_{\mu\nu}, \tag{3.6}$$

where

$$J_{\mu\nu} = 16\pi(T_{[\mu\nu]} + \frac{2}{\mu^2}T_{[[\mu\sigma],\nu]}{}^{,\sigma}). \tag{3.7}$$

We have

$$W_\mu = -\frac{32\pi}{\mu^2}T_{[\mu\nu]}{}^{,\nu}, \tag{3.8}$$

and from Eq.(3.3) we get

$$\psi_\mu = \frac{16\pi}{\mu^2}T_{[\mu\nu]}{}^{,\nu}. \tag{3.9}$$

In the wave-zone, $T_{\mu\nu} = 0$, and Eqs.(3.6) and (3.9) become

$$(\Box + \mu^2)\psi_{\mu\nu} = 0, \tag{3.10}$$

$$\psi_\mu = 0. \tag{3.11}$$

These equations can be obtained from the Lagrangian:

$$\mathcal{L}_\psi = \frac{1}{4}\psi^2_{\mu\nu,\lambda} - \frac{1}{2}\psi^2_\mu - \frac{1}{4}\mu^2\psi^2_{\mu\nu}. \tag{3.12}$$

The $\psi_{\mu\nu}$ has the spin decomposition:

$$\psi_{\mu\nu} = 1_b \oplus 1_e, \tag{3.13}$$

where 1_b and 1_e denote the magnetic and electric vectors, respectively. Only the magnetic vector propagates corresponding to a massive spin 1^+ pseudovector field with the propagator:

$$\Pi = \frac{P^1_b}{k^2}, \tag{3.14}$$

where P^1_b is the magnetic projection operator defined by

$$P^1_b = \frac{1}{2}(\theta_{\mu\rho}\theta_{\nu\sigma} - \theta_{\mu\sigma}\theta_{\nu\rho}), \tag{3.15}$$

$$\theta_{\mu\nu} = \delta_{\mu\nu} - \frac{k_\mu k_\nu}{k^2}, \tag{3.16}$$

and k_μ denotes the momentum four vector. The Lagrangian (3.12) is free of ghosts, tachyons and higher-order poles[5] and the Hamiltonian is positive and bounded from below.

The physical requirement that the weak field linear approximation of the NGT field equations is free of ghosts, tachyons and higher-order poles in the propagator is satisfied by the version of NGT presented here. This overcomes the problem of obtaining a consistent geometrical NGT, in which the physical vacuum in the theory is stable.

As shown by van Nieuwenhuizen[5], in flat space there exist only two physically consistent Lagrangians for the antisymmetric potential field $\psi_{\mu\nu}$, which are free of ghosts, tachyons and higher-order poles in the propagator, namely, the massless gauge invariant theory, and the massive Proca-type theory with the Lagrangian (3.12). Since NGT does not possess a massless gauge invariance, then the only other possibility is that it should reduce to the physical, massive Proca-type theory. We have, in fact, now discovered a fully geometrical NGT scheme that fulfils this requirement.

4. Expansion of the Field Equations Around a Curved Background

Let us now consider the expansion of the field equations around an arbitrary Einstein background metric. We shall introduce the notation: $g_{[\mu\nu]} = a_{\mu\nu}$. We have

$$g_{\mu\nu} = g_{S\mu\nu} + {}^{(1)}g_{\mu\nu} + \dots, \quad \Gamma^\lambda_{\mu\nu} = \Gamma^\lambda_{S\mu\nu} + {}^{(1)}\Gamma^\lambda_{\mu\nu} + \dots, \quad W_\mu = {}^{(1)}W_\mu + {}^{(2)}W_\mu + \dots, \tag{4.1}$$

where $g_{S\mu\nu}$ and $\Gamma^\lambda_{S\mu\nu}$ denote the Einstein background metric tensor and connection, respectively, and $g_S^{\lambda\alpha} g_{S\beta\lambda} = \delta^\alpha_\beta$. We also introduce the notation for the background Riemann tensor:

$$B^\sigma{}_{\mu\nu\rho} = \Gamma^\sigma_{S\mu\nu,\rho} - \Gamma^\sigma_{S\mu\rho,\nu} - \Gamma^\sigma_{S\alpha\nu}\Gamma^\alpha_{S\mu\rho} + \Gamma^\sigma_{S\alpha\rho}\Gamma^\alpha_{S\mu\nu}. \tag{4.2}$$

By performing the contraction on the suffixes σ and ρ, we get the Ricci tensor, $B_{\mu\nu} = B^\alpha{}_{\mu\nu\alpha}$. We get to first order in $a_{\mu\nu}$ and W_μ (we denote ${}^{(1)}W_\mu$ by W_μ and ${}^{(1)}a_{\mu\nu}$ by $a_{\mu\nu}$) the field equations:

$$B_{\mu\nu} = 0, \tag{4.3}$$

$$\nabla^\sigma a_{\mu\sigma} = \frac{1}{2\mu^2}\left[\nabla^\nu\left(4g_S^{\lambda\sigma}g_S^{\alpha\beta}B_{\alpha\nu\lambda\mu}a_{\sigma\beta} - 2(Ba)_{\mu\nu}\right)\right]. \tag{4.4}$$

$$(\Box_S + \mu^2)a_{\mu\nu} = M_{\mu\nu}, \tag{4.5}$$

where

$$M_{\mu\nu} = 2g_S^{\lambda\sigma}g_S^{\alpha\beta}B_{\alpha\nu\lambda\mu}a_{\sigma\beta} + \frac{1}{\mu^2}\nabla_{[\nu}\nabla^\rho\left[4g_S^{\lambda\sigma}g_S^{\alpha\beta}B_{\alpha\rho\lambda\mu]}a_{\sigma\beta} - 2(Ba)_{\mu]\rho}\right], \tag{4.6}$$

and $\Box_S = \nabla^\sigma\nabla_\sigma, \nabla^\sigma = g_S^{\sigma\alpha}\nabla_\alpha$. Also, $(Ba)_{\mu\nu}$ denotes additional terms involving products of the Riemann tensor B and $a_{\mu\nu}$.

As before, W_μ does not propagate and there is no coupling to unphysical modes through the effective source tensor formed from the Riemann tensor and $a_{\mu\nu}$.

The energy associated with the flux of gravitational waves, calculated in the wave-zone for $r \to \infty$, is positive definite.

5. Static Spherically Symmetric Solution

In the case of a static spherically symmetric field, Papapetrou has derived the canonical form of $g_{\mu\nu}$ in NGT[6]:

$$g_{\mu\nu} = \begin{pmatrix} -\alpha & 0 & 0 & w \\ 0 & -\beta & f\sin\theta & 0 \\ 0 & -f\sin\theta & -\beta\sin^2\theta & 0 \\ -w & 0 & 0 & \gamma \end{pmatrix}, \tag{5.1}$$

where α, β, γ and w are functions of r. For the theory in which there is no NGT magnetic monopole charge, we have $w = 0$ and only the $g_{[23]}$ component of $g_{[\mu\nu]}$ survives.

The line element for a spherically symmetric body is given by

$$ds^2 = \gamma(r)dt^2 - \alpha(r)dr^2 - \beta(r)(d\theta^2 + \sin^2\theta d\phi^2). \tag{5.2}$$

We have

$$\sqrt{-g} = \sin\theta[\alpha\gamma(\beta^2 + f^2)]^{1/2}. \tag{5.3}$$

The vector W_μ can be determined from Eq.(2.12):

$$W_\mu = -2k_{\mu\rho}\mathbf{g}^{[\rho\sigma]}{}_{,\sigma}, \tag{5.4}$$

where $k_{\mu\nu}$ is defined by $k_{\mu\alpha}g^{(\mu\beta)} = \delta^\beta_\alpha$. For the static spherically symmetric field with $w = 0$ it follows from (5.4) that $W_\mu = 0$.

We choose $\lambda = 0, \beta = r^2$ and demand the boundary conditions:

$$\alpha \to 1, \quad \gamma \to 1, \quad f \to 0, \tag{5.5}$$

as $r \to \infty$. We also assume the short-range approximation for which the μ^2 contributions in the vacuum field equations can be neglected. If $\mu^{-1} >> 2m$, then we can use the static, spherically symmetric Wyman solution[7]:

$$\gamma = \exp(\nu), \tag{5.6}$$

$$\alpha = m^2(\nu')^2\exp(-\nu)(1+s^2)(\cosh(a\nu) - \cos(b\nu))^{-2}, \tag{5.7}$$

$$f = [2m^2\exp(-\nu)(\sinh(a\nu)\sin(b\nu)$$

$$+s(1-\cosh(a\nu)\cos(b\nu))](\cosh(a\nu) - \cos(b\nu))^{-2}, \tag{5.8}$$

where ν is implicitly determined by the equation:

$$\exp(\nu)(\cosh(a\nu) - \cos(b\nu))^2\frac{r^2}{2m^2} = \cosh(a\nu)\cos(b\nu) - 1 + s\sinh(a\nu)\sin(b\nu). \tag{5.9}$$

Here, s is a dimensionless constant of integration.

We find for $2m/r < 1$ and $0 < sm^2/r^2 < 1$ that α and γ take the Schwarzschild form[3]:

$$\gamma = \alpha^{-1} = 1 - \frac{2m}{r}. \tag{5.10}$$

Near $r = 0$ we can develop expansions where $r/m < 1$ and $0 < |s| < 1$. The leading terms are

$$\gamma = \gamma_0 + \frac{\gamma_0(1+\mathcal{O}(s^2))}{2|s|}\left(\frac{r}{m}\right)^2 + \mathcal{O}\left(\left(\frac{r}{m}\right)^4\right), \tag{5.11}$$

$$\alpha = \frac{4\gamma_0(1+\mathcal{O}(s^2))}{s^2}\left(\frac{r}{m}\right)^2 + \mathcal{O}\left(\left(\frac{r}{m}\right)^4\right), \tag{5.12}$$

$$f = m^2\left(4 - \frac{|s|\pi}{2} + s|s| + \mathcal{O}(s^3)\right) + \frac{|s| + s^2\pi/8 + \mathcal{O}(s^3)}{4}r^2 + \mathcal{O}(r^4), \tag{5.13}$$

$$\gamma_0 = \exp\left(-\frac{\pi+2s}{|s|} + \mathcal{O}(s)\right)\dots. \tag{5.14}$$

These solutions clearly illustrate the non-analytic nature of the limit $s \to 0$ in the strong gravitational field regime.

The singularity caused by the vanishing of $\alpha(r)$ at $r = 0$ is a *coordinate* singularity, which can be removed by transforming to another coordinate frame of reference[3]. The curvature invariants do not, of course, contain any coordinate singularities.

The NGT curvature invariants such as the generalized Kretschmann scalar:

$$K = R^{\lambda\mu\nu\rho}R_{\lambda\mu\nu\rho} \tag{5.15}$$

are finite.

The solution is everywhere non-singular and there is no event horizon at $r = 2m$. A black hole is replaced in the theory by a superdense object which can be stable for an arbitrarily large mass[3,8].

6. Conclusions

We have succeeded in deriving a geometrical version of NGT which yields a stable, ghost and tachyon free linear approximation, when the field equations are expanded about Minkowski spacetime or about an arbitrary Einstein background metric.

In the short-range approximation, in which μ^{-1} is large compared to $2m$, a static spherically symmetric solution can be obtained from the field equations which is regular everywhere in spacetime and which does not contain an event horizon at $r = 2m$. The black hole that is predicted by the collapse of a sufficiently massive star in EGT is replaced by a superdense object that is stable for an arbitrarily large mass.

An important result of this study is that a classical theory of gravity can be formulated, which has a physically consistent perturbative expansion for weak fields, and does not have singularities and black holes. It satisfies the standard gravitational experimental tests. Recently, a solution has been published for inhomogeneous gravitational collapse of a matter cloud with a general form of matter, which leads to a naked singularity[9]. The collapse is related to the choice of initial data for the Einstein field equations, and the naked singularity would occur in generic situations based on regular initial conditions satisfying the weak energy condition. This result would lead to the demise of EGT, for it represents a local violation of the Cauchy data for collapse; it would provide a strong motivation for seriously considering a classical gravity theory such as NGT with everywhere regular solutions of the field equations.

Because there is no event horizon in the static spherically symmetric solution, we can resolve the problem of information loss[10] associated with black holes at a classical level.

Acknowledgements

I thank M. Clayton, N. J. Cornish, P. Savaria and I. Yu.Sokolov for helpful and stimulating discussions. I thank the Natural Sciences and Engineering Research Council of Canada for the support of this work.

References

1. For earlier work on NGT, see: J. W. Moffat, Phys. Rev. D **19**, 3554 (1979); J. W. Moffat, J. Math. Phys. **21**, 1978 (1980); J. W. Moffat, Phys. Rev. D **35**, 3733 (1987). For a review of NGT, see: J. W. Moffat, Proceedings of the Banff Summer Institute on Gravitation, Banff, Alberta, August 12-25, 1990, edited by R. Mann and P. Wesson, World Scientific, p.523, 1991.

2. R. B. Mann and J. W. Moffat, J. Phys. A **14**, 2367 (1981); T. Damour, S. Deser and J. McCarthy, Phys. Rev. D **47**, 1541 (1993).

3. N. J. Cornish and J. W. Moffat, Phys. Letts. **336B**, 337 (1994); University of Toronto preprint, UTPT-94-08, 1994, gr-qc/9406007. To be published in J. Math. Phys.; I. Yu.Sokolov, University of Toronto preprint, UTPT-94-23, 1994.

4. A. Einstein, The Meaning of Relativity, Menthuen and Co. London, U.K. fifth edition, Appendix II, 1951. Einstein used the nonsymmetric field structure to derive a unified field theory of gravitation and electromagnetism. The

nonsymmetric field equations cannot produce a consistent physical theory of electromagnetism. We use the nonsymmetric field structure to descibe a non-Riemannian theory of gravitation.

5. P. van Nieuwenhuizen, Nucl. Phys. B **60**, 478 (1973).
6. A. Papapetrou, Proc. Roy. Ir. Acad. Sec. A **52**, 69 (1948).
7. M. Wyman, Can. J. Math. **2**, 427 (1950).
8. N. J. Cornish, University of Toronto preprint, UTPT-94-10, 1994. Revised November, 1994.
9. P. S. Joshi and I. H. Dwivedi, Commun. Math. Phys. **146**, 333 (1992); to appear in Commun. Math. Phys. (1994); P. S. Joshi, Global Aspects in Gravitation and Cosmology, Clarendon Press, Oxford, 1994.
10. S. Hawking, Phys. Rev. D **14**, 2460 (1976); Commun. Math. Phys. **87**, 395 (1982).

Electroweak Physics at LEP*

Kyungsik Kang

Department of Physics, Brown University , Providence, RI, USA†

and

LPTPE, Universite P.& M. Curie, 4 Place Jussieu, 75252 Paris Cedex05, France

and

Sin Kyu Kang

Department of Physics, Seoul National University, Seoul, Korea‡

ABSTRACT

We have examined the evidence for the electroweak radiative corrections in the LEP precision data sets of 1993 and 1994 along with the intriguing possibility that the QED corrections only may be sufficient to fit the data within the framework of the minimal standard model. We find that the situation is very sensitive to the precise value of M_W. The current world average value of M_W and the improved 1994 LEP data strongly favor nonvanishing electroweak radiative corrections, and are consistent with a heavy m_t as reported by CDF but with a heavy Higgs scalar of about 400 GeV. We discuss how future precision measurements of M_W and m_t can provide a decisive test for the standard model with radiative corrections.

*Presented at the Workshop on Quantum Infrared Physics, American University of Paris (Paris, France, 6 - 10 June, 1994)

†Permanent address and supported in part by the USDOE contract DE-FG02-91ER40688-Task A

‡Supported in part by the Basic Science Research Institute Program, Ministry of Education, Project No. BSRI-94-2418, and the Korea Science and Engineering Foundation through SNU CTP

Recently much interests have been paid to the electroweak radiative corrections (EWRC) and precision tests of the standard model (SM) thanks to the accurate data obtained at LEP [1,2]. Numerous articles have appeared on the subject as has been documented in [1 - 3]. The LEP data are generally regarded as the success of the SM and as the evidence for the nonvanishing EWRC [4].

There have been new experimental developments since last year that warrant a renewed examination of the precision tests of the SM, namely, the new measurements of M_W [5], the improved LEP precision data [2], and the evidence of m_t from CDF [6]. We would like to report on the results of the new precision tests of the SM based on these new experimental informations and implication on the Higgs mass range. At the same time we reexamine the intriguing claim made by Novikov, Okun, and Vysotsky [7] based on the 1993 experimental data from LEP that the electroweak parameters as defined in the SM could be explained by the QED Born approximation (QBA) in which $\alpha(M_Z^2)$ is used instead of $\alpha(0)$ and the corresponding redefinition of the weak mixing angle $\sin^2\theta$ instead of $\sin^2\theta_W$ in the tree-level SM within 1 σ level.

The full one-loop EWRC are calculated with the aid of the ZFITTER program [8] modified by the improved QCD correction factor and the χ^2 minimization to the fit. In order to achieve QBA, we neglect the terms of non-photonic and pure weak interaction origin systematically in the program

Since the basic lagrangian contains the bare electric charge e_0, the renormalized physical charge e is fixed by a counter term δe; $e_0 = e + \delta e$. The counter term δe is determined by the condition of the on-shell charge renormalization in the $\overline{MS}$ or on-shell scheme. It is well known that the charge renormalization in the conventional QED fixes the counter term by the renormalized vacuum polarization $\hat{\Pi}^\gamma(0)$ and one can evaluate $\hat{\Pi}^\gamma(q^2) = \hat{\Sigma}^{\gamma\gamma}(q^2)/q^2$ from the photon self energy $\hat{\Sigma}^{\gamma\gamma}(q^2)$, for example, by the dimensional regularization method. This gives at $q^2 = M_Z^2$,

$$\hat{\Pi}^\gamma(M_Z^2) = \sum_f Q_f^2 \frac{\alpha}{3\pi}\left(\frac{5}{3} - \ln\frac{M_Z^2}{m_f^2} + i\pi\right), \tag{1}$$

where Q_f is the charge of the fermion f in the unit of e and α is the hyperfine structure constant $\alpha = \frac{e^2}{4\pi} = 1/137.0359895(61)$.

This gives the total fermionic contribution of $m_f \leq M_Z$ to the real part, $Re\hat{\Pi}^\gamma(M_Z^2) = -0.0602(9)$, so that the "running" charge defined as $e^2(q^2) = \frac{e^2}{1+Re\hat{\Pi}^\gamma(q^2)}$ gives $\alpha(M_Z^2) = 1/128.786$ in the on-shell scheme. The concept of the running charge is, however, scheme dependent [9]: the $\overline{MS}$ fine structure constant at the Z mass scale is given by

$$\hat{\alpha}(M_Z) = \alpha/[1 - \Pi^\gamma(0)|_{\overline{MS}} + 2\tan\theta_W(\Sigma^{\gamma Z}(0)/M_Z^2)_{\overline{MS}}]. \tag{2}$$

so that one can show $\hat{\alpha}(M_Z) = (127.9 \pm 0.1)^{-1}$, which differs by some 0.8 % from the on-shell $\alpha(M_Z^2)$.

The electroweak parameters are evaluated numerically with the hyperfine structure constant α, the four-fermion coupling constant of μ-decay, $G_\mu = 1.16639(2) \times$

10^{-5} GeV^{-2}, and Z-mass, i.e., $M_Z = 91.187(7)$ GeV for the 1993 data fit and 91.1899(44) GeV for the 1994 data fit. Numerical estimate of the full EWRC requires the mass values of the leptons, quarks, Higgs scalar and W-boson besides these quantities. While Z-mass is known to an incredible accuracy from the LEP experiments largely due to the resonant depolarization method, the situation with respect to the W-mass is desired to be improved, i.e., $M_W = 80.22(26)$ GeV [10] and 80.21(16) GeV [5] vs. the CDF measurement $M_W = 79.91(39)$ GeV [11] and 80.38(23) GeV [5].

One has, in the standard model, the on-shell relation $\sin^2\theta_W = 1 - \frac{M_W^2}{M_Z^2}$, and the four-fermion coupling constant G_μ

$$G_\mu = \frac{\pi\alpha}{\sqrt{2}M_W^2}\left(1 - \frac{M_W^2}{M_Z^2}\right)^{-1}(1-\Delta r)^{-1} \tag{3}$$

so that Δr, representing the radiative corrections, is given by

$$\Delta r = 1 - \left(\frac{37.28}{M_W}\right)^2 \frac{1}{1 - M_W^2/M_Z^2}. \tag{4}$$

Notice that the radiative correction Δr is very sensitive to the value of M_W: Mere change in M_W by 0.59% results as much as a 75% change in Δr. Also precise determination of the on-shell value of $\sin^2\theta_W$ can constrain the needed radiative correction and the value of M_W, thus providing another crucial test for the evidence of the EWRC in the standard model.

We have made χ^2-fits to both 1993 and 1994 data sets of the Z-decay parameters measured at LEP and M_W as shown in Tables 1 - 4. In each set, the fit is carried out for both the CDF and world average values of M_W. Details of the analysis can be found elsewhere [12].

The Z-decay parameters are calculated with a modified ZFITTER program, in which the best χ^2 fit search is made with the gluonic coupling constant $\bar{\alpha}_s(M_Z^2) = 0.123 \pm 0.006$ in the improved QCD correction factor [13] $R_{\rm QCD} = 1 + 1.05\frac{\bar{\alpha}_s}{\pi} + 0.9(\pm 0.1)\left(\frac{\bar{\alpha}_s}{\pi}\right)^2 - 13.0\left(\frac{\bar{\alpha}_s}{\pi}\right)^3$ for all quarks. The partial width for $Z \to f\bar{f}$ is given by

$$\Gamma_f = \frac{G_\mu}{\sqrt{2}}\frac{M_Z^3}{24\pi}\beta R_{\rm QED}c_f R_{\rm QCD}(M_Z^2)\left\{[(\bar{v}_f^Z)^2 + (\bar{a}_f^Z)^2]\times\left(1 + 2\frac{m_f^2}{M_Z^2}\right) - 6(\bar{a}_f^Z)^2\frac{m_f^2}{M_Z^2}\right\} \tag{5}$$

where $\beta = \beta(s) = \sqrt{1 - 4m_f^2/s}$ at $s = M_Z^2$, $R_{\rm QED} = 1 + \frac{3}{4}\frac{\alpha}{\pi}Q_f^2$ and the color factor $c_f = 3$ for quarks and 1 for leptons. Here the renormalized vector and axial-vector couplings are defined by $\bar{a}_f^Z = \sqrt{\rho_f^Z}2a_f^Z = \sqrt{\rho_f^Z}2I_3^f$ and $\bar{v}_f^Z = \bar{a}_f^Z[1 - 4|Q_f|\sin^2\theta_W\kappa_f^Z]$ in terms of the familiar notations [8, 9, 14]. Note that $\Delta\alpha$ is contained in the couplings through G_μ and all other non-photonic and pure weak loop corrections are grouped in ρ_f^Z and κ_f^Z. Thus the case of the QBA can be achieved simply by setting ρ_f^Z and κ_f^Z to 1 in the vector and axial-vector couplings.

	Experiment	QBA	Full EW	Full EW	Full EW
m_t (GeV)	150		120	138	158
m_H (GeV)	$60 \le m_H \le 1000$		60	300	1000
M_W (GeV)	79.91 ± 0.39	79.95	80.10	80.10	80.13
Γ_Z (MeV)	2488.0 ± 7.0	2488.4	2489.0	2488.9	2488.8
$\Gamma_{b\bar{b}}$ (MeV)	383.0 ± 6.0	379.4	377.4	376.5	375.4
$\Gamma_{l\bar{l}}$ (MeV)	83.52 ± 0.28	83.47	83.53	83.53	83.63
Γ_{had} (MeV)	1739.9 ± 6.3	1740.3	1738.8	1738.2	1737.7
$R(\Gamma_{b\bar{b}}/\Gamma_{had})$	0.220 ± 0.003	0.218	0.217	0.217	0.216
$R(\Gamma_{had}/\Gamma_{l\bar{l}})$	20.83 ± 0.06	20.85	20.82	20.81	20.78
$\sigma_h^P(nb)$	41.45 ± 0.17	41.41	41.37	41.38	41.40
g_V	-0.0372 ± 0.0024	-0.0372	-0.0341	-0.0334	-0.0334
g_A	-0.4999 ± 0.0009	-0.5000	-0.5003	-0.5005	-0.5006
χ^2		0.9621	4.2664	6.0802	7.7723
$\sin^2\theta_W$	0.2321	0.2314	0.2284	0.2283	0.2278
Δr	0.0623	0.06022	0.05162	0.05131	0.04967

Table 1. Numerical results including full EWRC for nine experimental parameters of the Z-decay and M_W. The results of QBA are shown also for comparison. Each pair of m_t and m_H represents the case of the best χ^2 -fit to the 1993 LEP data and $M_W = 79.91(39)$ GeV.

	Experiment	QBA	Full EW	Full EW	Full EW
m_t (GeV)	150		126	142	160
m_H (GeV)	$60 \le m_H \le 1000$		60	300	1000
M_W (GeV)	80.22 ± 0.26	79.95	80.13	80.13	80.15
Γ_Z (MeV)	2488.0 ± 7.0	2488.4	2490.2	2489.7	2489.3
$\Gamma_{b\bar{b}}$ (MeV)	383.0 ± 6.0	379.4	377.3	376.4	375.3
$\Gamma_{l\bar{l}}$ (MeV)	83.52 ± 0.28	83.47	83.53	83.63	83.63
Γ_{had} (MeV)	1739.9 ± 6.3	1740.3	1739.6	1738.8	1738.1
$R(\Gamma_{b\bar{b}}/\Gamma_{had})$	0.220 ± 0.003	0.218	0.217	0.216	0.216
$R(\Gamma_{had}/\Gamma_{l\bar{l}})$	20.83 ± 0.06	20.85	20.83	20.79	20.78
$\sigma_h^P(nb)$	41.45 ± 0.17	41.41	41.38	41.39	41.40
g_V	-0.0372 ± 0.0024	-0.0372	-0.0344	-0.0337	-0.0335
g_A	-0.4999 ± 0.0009	-0.5000	-0.5004	-0.5006	-0.5007
χ^2		2.0545	4.1346	5.9538	7.5517
$\sin^2\theta_W$	0.2261	0.2314	0.2278	0.2279	0.2275
Δr	0.0448	0.06022	0.04975	0.04991	0.04895

Table 2. The same as Table 1 but for the experimental $M_W = 80.22 \pm 0.26$ GeV.

	Experiment	QBA	Full EW	Full EW	Full EW
m_t (GeV)	$174 \pm 10^{+13}_{-12}$		187	172	155
m_H (GeV)	$60 \le m_H \le 1000$		1000	300	60
M_W (GeV)	80.38 ± 0.23	79.95	80.33	80.32	80.30
Γ_Z (MeV)	2497.1 ± 3.8	2488.7	2496.8	2497.3	2496.7
$\Gamma_{l\bar{l}}$ (MeV)	83.98 ± 0.18	83.49	83.90	83.87	83.80
Γ_{had} (MeV)	1746.0 ± 4.0	1740.5	1743.4	1744.1	1744.2
$R(\Gamma_{b\bar{b}}/\Gamma_{had})$	0.2210 ± 0.0019	0.2180	0.2149	0.2155	0.2160
$R(\Gamma_{had}/\Gamma_{l\bar{l}})$	20.790 ± 0.04	20.847	20.778	20.794	20.813
$\sigma_h^P(nb)$	41.51 ± 0.12	41.41	41.41	41.40	41.39
g_V/g_A	0.0711 ± 0.002	0.0745	0.0711	0.0714	0.0723
χ^2		25.8	11.6	9.99	9.86
$\sin^2\theta_W$	0.2231	0.2314	0.2240	0.2242	0.2245
Δr	0.0355	0.06022	0.03841	0.03913	0.03998

Table 3. Numerical results including full EWRC for seven experimental parameters of the Z-decay and M_W. The case of QBA is shown also for comparison. Each pair of m_t and m_H represents the case of the best χ^2- fit to the 1994 LEP data and $M_W = 80.38(23)$ GeV.

	Experiment	QBA	Full EW	Full EW	Full EW
m_t (GeV)	$174 \pm 10^{+13}_{-12}$		185	174	153
m_H (GeV)	$60 \le m_H \le 1000$		1000	400	60
M_W (GeV)	80.21 ± 0.16	79.95	80.32	80.31	80.29
Γ_Z (MeV)	2497.1 ± 3.8	2488.7	2496.3	2496.8	2496.2
$\Gamma_{l\bar{l}}$ (MeV)	83.98 ± 0.18	83.49	83.88	83.87	83.79
Γ_{had} (MeV)	1746.0 ± 4.0	1740.5	1743.0	1743.6	1743.9
$R(\Gamma_{b\bar{b}}/\Gamma_{had})$	0.2210 ± 0.0019	0.2180	0.2150	0.2154	0.2161
$R(\Gamma_{had}/\Gamma_{l\bar{l}})$	20.790 ± 0.04	20.847	20.779	20.791	20.813
$\sigma_h^P(nb)$	41.51 ± 0.12	41.41	41.41	41.40	41.38
g_V/g_A	0.0711 ± 0.002	0.0745	0.0707	0.0711	0.0720
χ^2		25.8	12.1	10.6	10.1
$\sin^2\theta_W$	0.2263	0.2314	0.2243	0.2244	0.2247
Δr	0.0455	0.06022	0.03925	0.03957	0.04070

Table 4. The same as Table 3 but for the experimental $M_W = 80.21 \pm 0.16$ GeV.

Numerical results for the best χ^2 fit to the 1993 LEP experimental parameters of Z-decay are shown in Tables 1 and 2 for $M_W = 79.91(39)$ GeV and $M_W = 80.22(26)$ GeV respectively as experimental inputs. The results for the best χ^2-fit to the improved 1994 LEP data and M_W are given in Tables 3 and 4 for $M_W = 80.38(23)$ GeV and $M_W = 80.21(16)$ GeV respectively. Also included in the Tables are the results of QBA as well as the output $\sin^2\theta_W$ and Δr for comparison. We see that the contributions of the weak corrections are generally small and in particular for the 1993 data the QBA is close to the experimental values within the uncertainty of the measurements.

The near absence of the pure weak loop contributions to the radiative corrections for the 1993 data is more impressive for $M_W = 79.91$ GeV than for $M_W = 80.22$ GeV. At closer examination, however, the QBA in the latter case over-estimates the radiative corrections and the full one-loop EWRC fair better.

From the global fit to the data with two variables m_t and m_H in the range $60 - 1000$ GeV, we find the best fits to the 1993 data is obtained by $m_t = 142^{-16}_{+18}$ GeV, the central value being the best χ^2 case of $m_H = 300$ GeV in Table 2. The best global fits to the 1993 data give a rather stable output $M_W = 80.13 \pm 0.03$ GeV if the full EWRC are taken into account, which is to be contrasted to the output $M_W = 79.95$ GeV from the QBA, for either experimental M_W value. Also $\sin^2\theta_W = 0.2279 \pm 0.0005$ in the case of the full EWRC is to be compared to $\sin^2\theta_W = 0.2314$ in the case of QBA. While the 1993 world average value of M_W supports strongly for the evidence of the full EWRC in the LEP data, the QBA appears to be in statistically comparable agreement, i.e., within 2σ, with the precisions of the 1993 data. If M_W were to be definitely at around 79.95 GeV with the uncertainty of the 1993 LEP data, then the QED correction would have been all that was observed at LEP and one would have been cultivating the null result of the weak correction to produce the range of t-quark mass as pointed out in [7].

The situation with the χ^2-fit to the improved 1994 LEP data and M_W is significantly different from the case of the 1993 data as one can see from Tables 3 and 4. Not only there is clear evidence for the full EWRC in each of the seven LEP data but also the QBA gives distinctively inferior χ^2 in either case of new M_W. From the best fits to the 1994 data, one gets again a stable output $M_W = 80.31 \pm 0.02$ GeV for m_H in the range of 60 - 1000 GeV. In particular, the CDF m_t value 174 GeV is a possible output solution (in the case $M_W = 80.21(16)$ GeV) but with a m_H about 400 GeV among the many possible combinations of (m_t, m_H) given by the 'Best.fit' curve in Fig. 1. In general the χ^2-value tends to prefer lower m_t and accordingly smaller m_H combination of the curve but any pair of (m_t, m_H) on this curve is statistically comparable to each other. We see from Fig. 1 that the best global fits to the 1994 data are obtained by $m_t = 153 - 185$ GeV for $m_H = 60 - 1000$ GeV.

Fig.2 shows how M_W changes with m_t for fixed m_H from full EWRC where the new world average M_W and CDF m_t are also shown. The central values of the world average M_W and CDF m_t are consistent with a Higgs scalar mass somewhat heavier than 1000 GeV, though $m_H = 200$ GeV is only less than 1.5 σ away. Clearly a better precision measurement of M_W is desired to distinguish different m_H. For example, a change of m_H by 200 GeV, i.e., from 400 GeV to 200 GeV, requires from the best χ^2-fits a change of 9 GeV in m_t, i.e., from 174 GeV to 165 GeV, as one can see from Fig. 1. This in turn requires a precision of 20 MeV or better in M_W from Fig. 2.

In short, we find that the QBA is in agreement with the 1993 data within 2σ level of accuracy but the new world average value of M_W and the improved 1994 LEP data disfavor the QBA and definitely support for the evidence of the nonvanishing weak-loop correction. Furthermore, the CDF m_t is a solution of the minimal χ^2-fit to the 1994 data but then the Higgs scalar mass is *about* 400 GeV. Further precision measurement of M_W can provide a real test of the standard model as it will give a tight constraint for the needed amount of the EWRC and provide a profound implication to the mass of t-quark and Higgs scalar in the context of the standard model. If M_W is determined within 20 MeV uncertainty, Δr within the context of the standard model can distinguish the mass range of the t-quark and Higgs scalar and provide a crucial test for and even the need of new physics beyond the standard model.

References

1. See, for example, L. Rolandi, in: Proc. XXVI ICHEP 1992, CERN-PPE/92 -175 (1992); M. P. Altarelli, talk given at Les Rencontres de Physique de la Vallee d'Aoste (La Thuil, 1993), LNF-93/019(p); and J. Lefranceis, in: Proc. Int. EPS. Conf. H. E. Phys. (Marseille, July 1993).
2. P. Clarke; Y. K. Kim; B. Pictrzyk; P. Siegris; and M. Woods, in: Proc. 29th Rencontres de Moriond (Meribel, 1994); D. Schaile (private communication).
3. F. Dydak, in: Proc. 25 Int. Conf. H. E. Phys., Eds. K. K. Phua, Y. Yamaguchi (World Scientific, Singapore, 1991).
4. W. J. Marciano, Phys. Rev. **D20** (1979) 274; A. Sirlin, Phys. Rev. **D22** (1980) 971; (1984) 89; A. Sirlin and W. J. Marciano, Nucl. Phys. **B189** (1981) 442; and A. Sirlin, NYU-TH-93/11/01. See also W. Hollik, in: *Precision Tests of the Standard Model*, ed. P. Langacker (World Scientific Pub., 1993) and G. Altarelli, in: Proc. Int.EPS. Conf. H. E. Phys. (Marseille, July 1993).
5. D. Saltzberg, Fermilab-Conf-93/355-E, and S. B. Kim (private communication)
6. The most recent m_t lower bound from D0 is 131 GeV, S. Abachi et al, Phys. Rev. Lett. **72** (1994) 2138 while the CDF reported $m_t = 174 \pm 10^{+13}_{-12}$ GeV in F. Abe et al., FERMILAB-PUB-94/097-E, and Phys. Rev. Lett. **73** (1994) 255.
7. V. A. Novikov, L. B. Okun and M. I. Vysotsky, Mod. Phys. Lett. **A8** (1993) 5929. See, however, CERN-TH-7214/94 for more recent analysis.
8. D. Bardin et al., CERN-TH-6443-92 (1992).
9. See, for example, W. Hollik (Ref. 4); CERN Yellow Book CERN 89-08, vol.1, p45 ; and K. Kang, in: Proc. 14th Int. Workshop Weak Interactions and Neutrinos (Seoul, 1993), Brown-HET-931 (1993).

10. Particle Data Group, Review of Particle Properties, Phys. Rev. **D45**, No.11, Part II (1992).

11. CDF Collab., F. Abe et al., Phys. Rev. **D43** (1991) 2070.

12. Kyungsik Kang and Sin Kyu Kang, SNUTP-94-59 (to be published).

13. T. Hebbeker, Aachen preprint PITHA 91-08 (1991); and S. G. Gorishny, A. L. Kataev and S. A. Larin, Phys. Lett. **B259** (1991) 144. See also L. R. Surguladze and M. A. Samuel, Phys. Rev. Lett. **66** (1991) 560.

14. M. Consoli and W. Hollik, in *Z Physics at LEP 1*, Vol. 1, eds. G. Altarelli et al., CERN 89-08 (1989).

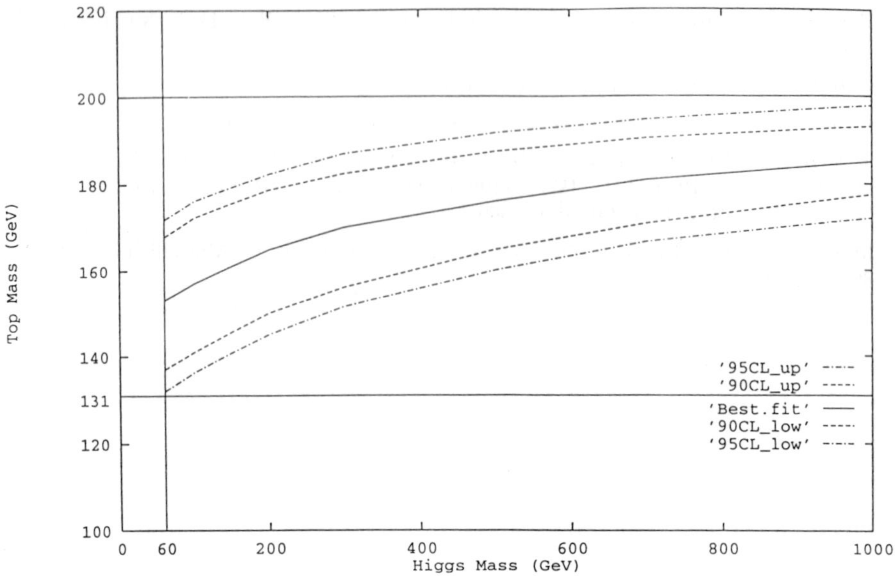

Fig. 1 : The mass ranges of m_t and m_H from the minimal χ^2-fit to the 1994 LEP data and $M_W = 80.21$ GeV.

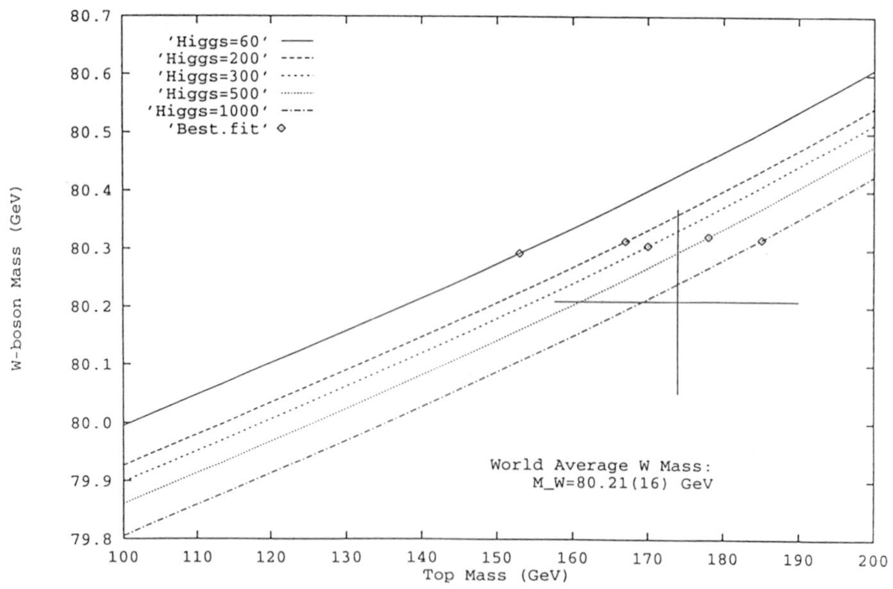

Fig. 2 : M_W versus m_t for fixed values of m_H from the full radiative correction in the standard model. The case of the minimal χ^2-fit to the 1994 LEP data corresponding to the full EWRC in Table 4 are indicated by $\diamond$.

CHAOS IN NONABELIAN GAUGE THEORIES

B. MÜLLER
Department of Physics, Duke University
Durham, NC 27708-0305, USA

ABSTRACT

Recent results revealing the universally chaotic nature of the Yang-Mills fields in the classical limit are discussed. The results enhance our understanding of the basis of statistical models for hadronic reactions and of the mechanisms of thermalization in nonabelian gauge theories.

Multiparticle production and other phenomena of what today is called "soft" hadronic interactions have often been successfully explained in the framework of statistical models[1]. However, it has remained largely obscure why the assumption of a random statistical distribution of final states is warranted in high energy reactions that last not much longer than 1 fm/c (or 3×10^{-24}s), and why are the final states not dominated by coherent quantum states or collective excitations of a small subset of the available hadronic degrees of freedom? Maybe it is good to recall that similar questions posed themselves in the context of Bohr's statistical model of compound nucleus reactions. In this case, the conceptual difficulties were eventually resolved by the insight that the highly excited compound nucleus is a chaotic quantum system[2,3] exhibiting rapid exchange of energy between the accessible degrees of freedom. Here I want to show that the same mechanism may be responsible for the apparent thermalization in high-energy hadron-hadron interactions: nonabelian gauge theories are strongly chaotic.

1. Chaos and Ergodicity

Let us begin with some general remarks about the applicability of statistical concepts to a single system. A dynamical system exhibits *ergodic* behavior, if the time average of an observable A can be replaced by the phase space average

$$\lim_{T\to\infty} \frac{1}{T} \int_0^T dt A(t) = \langle A \rangle \equiv \frac{1}{Z_E} \int d\Gamma_E A(\Gamma_E). \tag{1}$$

$\langle A \rangle$ here denotes the *microcanonical* average, Z_E is the microcanonical partition function, and $d\Gamma_E$ is the phase space measure at constant energy. For systems with very many degrees of freedom it is equivalent to take the *canonical* average

$$\langle A \rangle = \frac{1}{Z(\beta)} \int d\Gamma A(\Gamma) \exp[-\beta E(\Gamma)], \tag{2}$$

where $Z(\beta) = \int d\Gamma \exp[-\beta E(\Gamma)]$ is the canonical partition function and the inverse temperature $\beta = 1/T$ is determined by the condition $E = -\partial(\ln Z)/\partial\beta$.

For practical applications it is crucial to know the time scale on which ergodicity is attained. It can be shown that this time scale is related to the rate h of exponential divergence of neighboring trajectories in phase space; this rate is called the (maximal) Lyapunov exponent[4]. The complete spectrum of Lyapunov exponents is defined as follows. Consider a given trajectory in phase space, $x_\alpha(t)$, where $\alpha = 1, \ldots \nu$ enumerates the degrees of freedom. $x_\alpha(t)$ is a solution of the classical equations of motion of the system. Now take a set of neighboring trajectories:

$$\tilde{x}'^{(i)}_\alpha(t) = x_\alpha(t) + \delta x^{(i)}_\alpha(t), \quad \alpha = 1, \ldots, \nu. \tag{3}$$

For infinitesimal δx_α they are solutions of a second-order linear differential equation of the form

$$D[x_\alpha(t)]\delta x^{(i)}_\alpha(t) = 0. \tag{3}$$

One can then obtain a complete orthogonal set of solutions of this equation; they define ν Lyapunov exponents λ_i according to

$$\lambda_i = \lim_{t\to\infty} \frac{1}{t} \ln \frac{\|\delta x^{(i)}_\alpha(t)\|}{\|\delta x^{(i)}_\alpha(0)\|}, \quad i = 1, \ldots, \nu. \tag{5}$$

In other words, for long times one has the norm of $\delta x^{(i)}_\alpha$ growing (or shrinking) as $\|\delta x^{(i)}_\alpha(t)\| \propto \exp(\lambda_i t)$. One usually assumes the Lyapunov exponents to the ordered in size:

$$\lambda_1 \equiv h \geq \lambda_2 \geq \ldots \geq \lambda_\nu. \tag{6}$$

For conservative (Hamiltonian) systems the Lyapunov exponents occur in pairs of equal size, but opposite sign: $\lambda_i = -\lambda_{N+1-i}$. This is in accordance with Liouville's theorem which states that the volumes in phase space filled by an ensemble remains unchanged with time, implying that there must be a direction of contraction for every direction in which the phase space volume expands. For each conservation law there occur two vanishing Lyapunov exponents; the conservation of energy always ensures the existence of one such pair for a Hamiltonian system. Since the extent of the ensemble rapidly shrinks below any practially achievable resolution in the exponentially contracting directions, the *observable* volume in phase space grows as

$$\overline{\Gamma}(t) \propto \exp\left(t \sum_i \lambda_i \theta(\lambda_i)\right) \equiv \exp(\dot{S}_{\mathrm{KS}} t) \tag{7}$$

where the sum only includes the positive Lyapunov exponents. The exponential growth rate of the observable phase space volume implies a *linear* rate of growth of the observable "coarse-grained" entropy associated with the ensemble. This rate,

$\dot{S}_{\rm KS} = \sum_i \lambda_i \theta(\lambda_i)$, is called the Kolmogorov-Sinai entropy, or short, KS-entropy. Dynamical systems that have a positive KS-entropy everywhere in phase space are called K-systems; they exhibit all the properties required for a statistical description on time scales that are long compared with the ratio between the equilibrium entropy $S_{\rm eq}$ and the KS-entropy, i.e. for times

$$t \gg \tau_s = S_{\rm eq}/\dot{S}_{\rm KS}. \tag{8}$$

An illustration of these properties is provided by the simple dynamical system

$$H(x,y;p_x p_y) = \tfrac{1}{2}\left(p_x^2 + p_y^2 + x^2 y^2\right), \tag{9}$$

which occurs as part of the extreme infrared limit of Yang-Mills fields[5]. The system described by the Hamiltonian (9) has a positive Lyapunov exponent $\lambda \approx 0.4$. Almost all its trajectories are unstable against small perturbations[6] and the analogous quantum system has been shown to exhibit a Wigner distribution of its level spacings[7]. The remarkable ability of this system is to randomize an initially localized phase space distribution. After a rather limited time the phase space distribution is indistinguishable from a microcanonical ensemble.

2. Chaos in Nonabelian Gauge Theories

If we want to apply these concepts to nonabelian gauge theories, we must consider these as classical Hamiltonian systems with many degrees of freedom, and we need a gauge invariant distance measure in the space of field configurations. The first part is easy; the Hamiltonian formulation of lattice gauge theory by Kogut and Susskind can form the basis for a study for the gauge group SU(n) of nonabelian gauge theories as dynamical systems. The lattice Hamiltonian is expressed as (a denotes the lattice spacing)

$$H = \frac{g^2}{2a}\sum_\ell \mathrm{tr}(E_\ell E_\ell^\dagger) + \frac{2}{g^2 a}\sum_p (n - \mathrm{Re}\,\mathrm{tr}\,U_p)\,, \tag{10}$$

where electric field strength E_ℓ and the link variables U_ℓ are defined on the lattice links, and U_p denotes the ordered product of the U_ℓ around an elementary plaquette $p : U_p = U_4^\dagger U_3^\dagger U_2 U_1$. The U_ℓ are elements of the gauge group (SU(3) in the case of QCD) and the E_ℓ are elements of the associated Lie algebra. In the classical limit, the link variables U_ℓ are functions of time, and the electric field variables are given by

$$E_\ell = \frac{a}{ig^2}\dot{U}_\ell U_\ell^\dagger. \tag{11}$$

The Hamiltonian equations then provide a set of coupled equations for the time evolution of $U_\ell(t)$ and $E_\ell(t)$, which can be integrated numerically. We have taken great care to ensure a numerically exact solution. The energy and Gauss' law remain conserved to better than 10^{-6} over the whole course of the numerical integration.

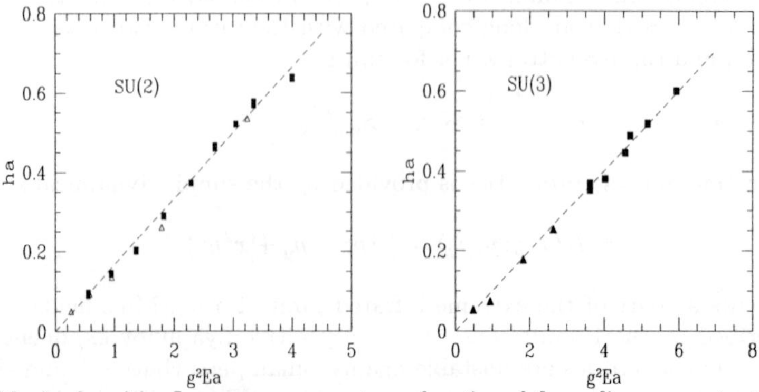

Figure 1. Maximal positive Lyapunov exponent as a function of the scaling parameter ($g^2 E_p a$) for (a) SU(2), (b) SU(3) gauge theory.

An appropriate measure for the distance between two field configurations is[8]

$$D[U_\ell, E_\ell; U'_\ell, E'_\ell] = \frac{1}{2N_p}\left(a^2 \sum_\ell \left|\mathrm{tr}(E_\ell E_\ell^\dagger) - \mathrm{tr}(E'_\ell E'^\dagger_\ell)\right| + \sum_p \left|\mathrm{tr}\; U_p - \mathrm{tr}\; U'_p\right|\right). \tag{12}$$

It is gauge invariant, gives a vanishing distance between gauge equivalent field configurations, and goes over into

$$D[A, E; A', E'] \propto \frac{1}{2V}\int d^3x \left(|E^2 - E'^2| + |B^2 - B'^2|\right) \tag{13}$$

in the continuum limit, measuring the local differences in the electric and magnetic field energy. If one only wants to determine the largest Lyapunov exponent, it is sufficient to consider either the electric or the magnetic contribution to $D[U, U']$.

If one starts from two randomly chosen neighboring field configurations and integrates these in time, one finds that the distance quickly grows exponentially, until it saturates due to the compactness of the space of gauge fields. The growth rate h quickly reaches a constant limit as function of the lattice size N^3, if the energy density is kept fixed by choosing the same average energy per plaquette E_p in each case. It is easy to see[9] that the Hamiltonian (10) exhibits a scaling behavior such that the Lyapunov exponents, if they are universal functions of the average energy density, as expressed by E_p, can only depend on the dimensionless scaling variable $g^2 E_p a$. The nontrivial surprise is that, as shown in Figures 1a,b, the dependence is linear[8,10]

$$ha \equiv \lambda_1 a = b_n g^2 E_p a \quad \text{for SU}(n), \tag{14}$$

where $b_2 \approx \frac{1}{6}$ and $b_3 \approx \frac{1}{10}$. The linear relationship means that the lattice spacing drops out, yielding $h = b_n g^2 E_p$ independent of a. The maximal Lyapunov exponent hence has a well-defined continuum limit.

What about the other Lyapunov exponents? Their calculation for large lattices is prohibitively expensive, as there are in total $6(n^2 - 1)N^3$ degrees of freedom for SU(n) gauge theory on a N^3 lattice, but Gong[11] has evaluated the complete spectrum for SU(2) on lattices of size $N = 1, 2$, and 3. The result again is a surprise: when the Lyapunov exponents are scaled by the maximal one, and are plotted on the interval $[0, 1]$, the spectra for $N = 2$ and $N = 3$ are indistinguishable, and there is only a small difference between $N = 1$ and $N = 2$ (see Figure 2).

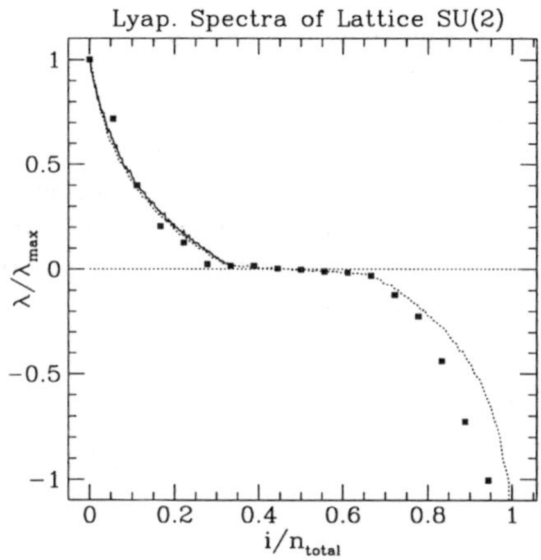

Figure 2. Spectrum of Lyapunov exponents for SU(2) lattice gauge theory. The black dots are for a 1^3, the dotted line is for a 2^3, and the solid line is for a 3^3 lattice. The $18N^3$ exponents are plotted on the fixed interval [0,1] to exhibit the scaling with N.

The Lyapunov spectrum for SU(2) shows three separate components: there are $(6N^3 - 1)$ positive and negative exponents each, and there are $(6N^3 + 2)$ exponents that converge to zero in the limit $t \to \infty$. Their vanishing reflects the existence of $(3N^3 + 1)$ conservation laws: Gauss' law at every lattice point and the overall energy conservation. Since the density of points on the line over the fixed interval [0,1] grows as N^3, this implies that the sum over positive Lyapunov exponents increases like the volume of the lattice, yielding a constant *KS-entropy density* $\dot{\sigma}_{\rm KS}$ in the thermodynamic limit. Gong finds

$$\dot{S}_{\rm KS} = \sum_i \lambda_i \theta(\lambda_i) = c_2 \lambda_1 N^3, \tag{15}$$

which together with (14) and the lattice volume $V = (Na)^3$ yields

$$\dot{\sigma}_{\rm KS} = \frac{1}{V} \dot{S}_{\rm KS} = \frac{1}{3} b_2 c_2 g^2 \varepsilon, \tag{16}$$

where $\varepsilon = 3E_p/a^3$ is the average energy density on the lattice. (Note that there are $3N^3$ plaquettes.) No one has yet calculated the complete Lyapunov spectrum for SU(3), but I expect a similar relationship as (15) to hold in that case, too. The coefficient c_2 is not completely independent of the scaling variable g^2E_pa, but has a value around 2. We will return below to the question how the physically relevant value of g^2E_pa can be chosen.

3. Physics Perspectives: Thermalization Time, Gluon Damping Rate

The instability of all degrees of freedom of the nonabelian gauge field (in the classical limit) leads to a very rapid "thermalization" of the energy density on the lattice. This is illustrated in Figure 3 showing the distribution of magnetic energy on the lattice plaquettes[12]. The initial state was chosen according to a random (not thermal) distribution of lattice link variables with vanishing electric field everywhere. Within two lattice units ($t/a = 2$) the energy has been equilibrated between electric and magnetic fields and, as the exponentially falling distribution shows, has assumed the form of a Gibbs distribution. The time scale for this "thermalization" is in good agreement from the time scale estimated from the inverse of the maximal Lyapunov exponent which is $h \approx 0.6$ in lattice units in SU(3) at this energy density.

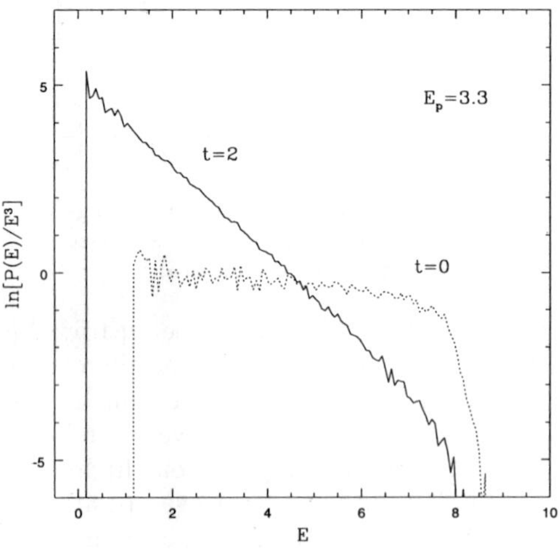

Figure 3. Logarithmic plot of the evolution of the magnetic energy density distribution on the lattice for SU(3) gauge theory. The distribution appears "thermalized" (exponential) after a time $t \approx 2$.

The fact that the energy density thermalizes on a time scale much shorter than that required for the numerical determination of the Lyapunov exponents (typically $t/a = 1000$) allows us to relate the Lyapunov exponents to quantities in the presence of a thermal environment. First, we can replace the average energy per plaquette

E_p in (14) by the "temperature" T, because $E_p = \frac{2}{3}(n^2-1)T$ for the classical, equilibrated SU(n) gauge field. This implies that

$$h \approx \begin{cases} \frac{1}{3}g^2T \approx 0.33g^2T & \text{for SU(2)}, \\ \frac{8}{15}g^2T \approx 0.53g^2T & \text{for SU(3)}. \end{cases} \tag{17}$$

We can use this result to obtain a model independent, nonperturbative estimate of the thermalization time solely due to gauge field dynamics in QCD. To compensate for the lack of "running" of the gauge coupling constant in the context of our classical gauge field calculation, we may use the one-loop result for $g(T)^2$ in (17) to evaluate $\tau_h = h^{-1}$, as shown in Figure 4. Clearly, this time is much smaller than 0.5 fm/c for all relevant temperatures, indicating a very rapid thermalization of the available energy. One should note that the Lyapunov exponents usually approach their asymptotic values from *above*, i.e. the dynamical instabilities are actually greater before the energy has been completely thermalized. This indicates

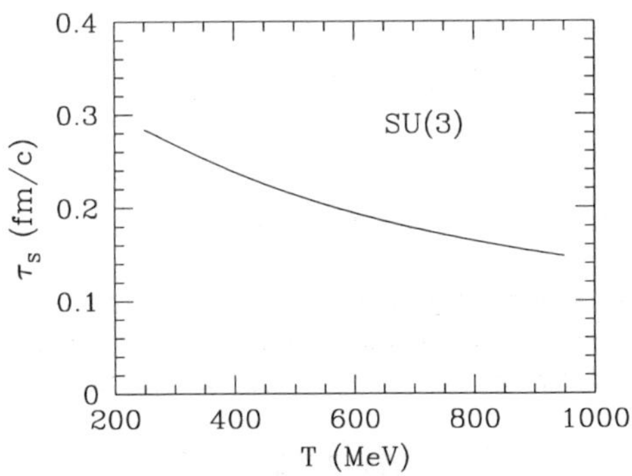

Figure 4. "Thermalization" time scale in SU(3) gauge theory as defined by the inverse of the maximal Lyapunov exponent (17), using $g(T)^2 = 16\pi^2/[11\ln(\pi T/\Lambda)^2]$ with Λ=200 MeV.

indicates that thermalization of field configurations far away from equilibrium may proceed even more rapidly.

It first appeared as a remarkable coincidence that the maximal Lyapunov exponents for SU(2) and SU(3) agree within numerical errors with the analytically calculated damping rate of a nonabelian plasmon at rest[13]:

$$\Gamma = 2\gamma(0) = 6.635\frac{n}{24\pi}g^2T \approx \begin{cases} 0.35g^2T & \text{for SU(2)}, \\ 0.53g^2T & \text{for SU(3)}. \end{cases} \tag{18}$$

The reason for the factor 2 is that the plasmon pole is usually parametrized as $\omega_p(k) = \omega^*(k) - i\gamma(k)$, so that the energy density of the soft plasmon mode falls off as $\exp(-2\gamma(0)t)$. The observation that the Lyapunov exponent is numerically evaluated in the vicinity of a "thermalized" field configuration over time scales that are much longer than those of thermal fluctuations allows us to establish a connection between these two quantities. According to (5) the Lyapunov exponents are determined from the long time behavior of solutions of the linearized equation for the fluctuation $a_\mu(x,t)$ around an exact solution $A_\mu(x,t)$ of the Yang-Mills equations. In the continuum limit this equation reads:

$$\left(g_{\mu\nu}D(A)^2 - D_\mu(A)D_\nu(A)\right) a^\nu(x,t) - 2i[F_{\mu\nu}, a^\nu(x,t)] = 0 \tag{19}$$

where $D_\mu(A)$ denotes the gauge covariant derivative and $[F_{\mu\nu}, a^\nu]$ denotes the Lie algebra commutator. Equation (19) is the usual starting point for quantization in a background field, but we will *not* impose here the background gauge constraint $D_\nu a^\nu = 0$. The initial value problem for (19) can be solved by means of the retarded propagator in the background field:

$$a_\mu(x,t) = \int d^3x' \Delta^{\mu\nu}_{\text{ret}}(x,t;x',0|A)a_\nu(x',0). \tag{20}$$

$\Delta^{\mu\nu}_{\text{ret}}$ has the formal representation as difference between the causal and the anti-causal propagator:

$$\Delta^{\mu\nu}_{\text{ret}}(A) = i\theta(t)\left[D^{\mu\nu}_{\text{F}}(A) - D^{\mu\nu^*}_{\text{F}}(A)\right] \tag{21}$$

Now recall that the maximal Lyapunov exponent is obtained from the long time average of the growth rate of $a^\nu(x,t)$. Assuming ergodicity, we may therefore replace the long-time average of the propagators by the canonical, i.e. thermal, average:

$$\overline{\Delta}^{\mu\nu}_{\text{ret}} = i\theta(t)\left[D^{\mu\nu}_{\text{F},T} - D^{\mu\nu^*}_{\text{F},T}\right] \tag{22}$$

where $D^{\mu\nu}_{\text{F},T}$ denotes the *exact* finite temperature Feynman propagator of the gauge field. Note that the causal Feynman propagator describes damped fluctuations, the anti-causal propagator $D^{\mu\nu}_{\text{F}}$ describes exponentially growing perturbations. The only remaining obstacle before establishing the equivalence of h and $2\gamma(0)$ is that the thermal average in (22) should be calculated for a *classical* emsemble of gauge fields, whereas the usual perturbative approach to $D^{\mu\nu}_{\text{F},T}$ is based on the quantum mechanical ensemble. However, this difference turns out to be irrelevant for the plasmon damping rate $\gamma(0)$, although it has a large effect on the effective plasmon mass $\omega^*(0)$. This is not entirely fortuitous, because the damping rate is given by tree diagrams, such as Compton scattering and bremsstrahlung, that have an exact low-energy classical limit[12,13]. We can therefore identify the maximal Lyapunov exponent with (twice) the damping rate of the most unstable mode in a thermal nonabelian plasma, which turns out to be a plasmon at rest[14].

4. Conclusions and Outlook

The short thermalization time scales of less than 1 fm/c found in our studies of the time evolution of classical nonabelian gauge fields show why Hagedorn was right thirty years ago with his assumption that final states in "soft" strong interaction physics are populated statistically. The reason for the success of these classical studies is that the dynamical instabilities in thermal gauge theories are of order g^2T, which is a classical inverse length or time scale that does not involve $\hbar$.

It would be interesting to see whether other quantities of order g^2T, such as the thermal magnetic screening mass on the "spatial" string tension, can also be calculated in the framework of classical Yang-Mills theory. This consideration leads into the problem of deriving an effective quasi-classical theory for thermal Yang-Mills theories at the length scale $(g^2T)^{-1}$ that consistently incorporates quantum effects from shorter distances in the form of transport coefficients. Presumably such an effective theory will contain a gauge invariant mass term of order gT (as in the Taylor-Wong action[15]) and a Langevin noise term describing the fluctuations due to interactions with hard thermal modes.

Another interesting problem concerns the application of real-time evolution of gauge fields to processes far off equilibrium as they occur in the earliest stage of hadron-hadron or nucleus-nucleus interactions. We have recently studied the instability of the superposition of two counter-propagating plane waves, i.e. of a standing abelian plane wave, in SU(2) Yang-Mills theory[16]. Here one finds that the Lyapunov exponent is proportional to the *amplitude* of the wave, not to the energy as it is the case in random fields. Once the coherent wave is only slightly perturbed it decays rapidly, exciting modes of all wavelengths, and quickly generates a thermal energy spectrum. The evolution of more realistic initial configurations, such as the interaction between nonabelian wave packets, is presently under investigation.

5. Acknowledgements

I thank T. S. Biró, C. Gong, S. G. Matinyan and A. Trayanov for their enthusiastic help in unraveling the intricacies of chaotic dynamics in gauge theories. I would also like to thank D. Egolf, H. B. Nielsen, S. E. Rugh, and G. K. Savvidy for illuminating discussions. This work was supported in part by the U. S. Department of Energy (grant DE-FG05-90ER40592) and the North Carolina Supercomputing Program.

6. References

1. R. Hagedorn, *Suppl. Nuovo Cimento* **3** (1965) 147.
2. J. J. M. Verbaarschot, H. A. Weidenmüller, and M. R. Zirnbauer, *Phys. Rep.* **129** (1985) 367; see also: C. Mahaux and H. A. Weidenmüller, *Ann. Rev. Nucl. Part. Science* **29** (1979) 1.
3. G. E. Mitchell, E. G. Bilpuch, P. M. Endt, J. F. Shriner, and T. von Egidy, *Nucl. Instr. Meth.* **B46/47** (1991) 446.

4. see e.g. A. J. Lichtenberg and M. A. Liebermann, *Regular and Stochastic Motion*, (Springer-Verlag, New York, 1983).
5. S. G. Matinyan, G. K. Savvidy, and N. G. Ter-Arutyunyan-Savvidy, *Zh. Eksp. Teor. Fiz.* **80** (1981) 830 [*Sov. Phys. JETP* **53** (1981) 421].
6. P. Dahlquist and G. Russberg, *Phys. Rev. Lett.* **65** (1990) 2837.
7. E. Haller, M. Köppel, and L. Cederbaum, *Phys. Rev. Lett.* **52** (1984) 1665.
8. B. Müller and A. Trayanov, *Phys. Rev. Lett.* **68** (1992) 3387.
9. T. S. Biró, C. Gong, B. Müller, and A. Trayanov, *Int. J. Mod. Phys.* **C5** (1994) 113.
10. C. Gong, *Phys. Lett.* **B298** (1993) 257.
11. C. Gong, *Phys. Rev.* **D49** (1994) 2642.
12. C. Gong, Dissertation, Duke University, 1994 (unpublished).
13. E. Braaten and R. D. Pisarski, *Phys. Rev.* **D42** (1990) 2156.
14. T. S. Biró, C. Gong, and B. Müller, to be published.
15. J. C. Taylor and S. Wong, *Nucl. Phys.* **B346** (1990) 115.
16. C. Gong, S. G. Matinyan, B. Müller, and A. Trayanov, *Phys. Rev.* **D49** (1994) R607.

SOFT PHOTON PRODUCTION
IN QUARK-GLUON PLASMA

E. PETITGIRARD
Laboratoire de physique théorique ENSLAPP
LAPP, BP 110, 74941 Annecy-le-Vieux Cedex, France

ABSTRACT

We calculate the production rate of soft photons from a quark-gluon plasma using real time formalism and Braaten-Pisarski's perturbative resummation method of high temperature QCD. A logarithmically divergent result is found for photon energies of order gT, where g is the QCD coupling constant and T the plasma temperature.

1. Introduction

Many studies are devoted to promising signatures of the expected quark-gluon plasma in high-energy heavy ion collisions. In particular, electromagnetic signals, such as photons and dileptons, can reveal information about this phase of hot and dense matter[1]. If they are produced in the early thermal stage of the process, they are not affected by the subsequent hadronization of the system: as they interact only electromagnetically, they usually pass through the surrounding matter without interacting. In the present paper we are concerned with real photons. Therefore, their mean free path l_γ has to be much larger than the transverse size L of the collision volume created in the nuclear collision. Since only photons with wave length less than L are radiated[2], the size of the plasma satisfies the constraint:

$$\frac{2\pi}{E} < L < l_\gamma, \tag{1}$$

with the photon energy E. The order of magnitude of the photon's mean free path can be estimate with the Compton cross section, responsible for the photon absorption. Introducing the couplings of QED and QCD, respectively $\alpha = e^2/4\pi$ and $\alpha_s = g^2/4\pi$, we have $\sigma_{Compton} \sim \alpha\alpha_s/ET$. Since the mean free path is given by $l_\gamma \simeq 1/n_q\sigma_{Compton}$ with $n_q \sim O(T^3)$ the quark density, it turns out that $l_\gamma \simeq E/\alpha\alpha_s T^2$. It is for soft photon energies $E \sim O(gT)$ that the dimension L of the plasma is the most constrained. Eq. (1) becomes:

$$O(\frac{1}{g}) < LT < O(\frac{1}{\alpha g}), \tag{2}$$

which reads $1fm < L < 100fm$ for typical values of $T \sim 400MeV$ and $\alpha_s \sim 0.25$. This order of magnitude is compatible with the expected dimensions of a realistic plasma produced in colliders. From the theoretical problem's point of view , Eq. (2) means that the quark-gluon plasma can be treated as a thermodynamic system in equilibrium with respect to QCD, emitting non-thermalized photons.

According to the results given by Weldon[3] and by Kobes and Semenoff[4], the thermal production rates (which represent the number of reactions producing photons or dileptons per unit time per unit volume) are calculated from the absorptive parts of two-point Green's functions. Since we are interested in the emission rate of real non-thermalized photons, the differential rate is related with the on-shell photon polarization tensor by

$$E\frac{dW^{\gamma}}{d^3q} = \frac{1}{(2\pi)^3} n_B(E) \mathrm{Im}\Pi^{\mu}_{\mu}(E,\vec{q}), \tag{3}$$

where the Bose-Einstein factor $n_B(E) = 1/(\exp(E/T) - 1)$ reflects the thermal equilibrium of the QCD system.

For dileptons or heavy photon production, the quark mass singularities, due to massless exchanged quarks, cancel at least to two-loop order of the perturbation theory[5]. On the contrary, the production of hard real photons involves basic QCD processes, like the Compton scattering ($qg \to q\gamma$, $\bar{q}g \to \bar{q}\gamma$) and the anihilation ($q\bar{q} \to g\gamma$), which lead to logarithmic divergent results in leading order. These mass singularities should be screened by thermal effects in order to have infrared-safe calculations of the production rate[6,7].

Pisarski and Braaten[8], as well as Frenkel and Taylor[9], proposed a resummation technique of diagrams (called hard thermal loops) for investigating the screening effect in hot gauge theories. When aplying the resummed perturbative expansion, two energy scales, the hard one of $O(T)$ and the soft one of $O(gT)$ have clearly to be distinguished. For soft energies and momenta, effective propagators and vertices are required.

In the following, we want to check if the resummation program advocated by Braaten and Pisarski succeeds in screening mass singularities in the calculation of real photons production rates, in the cases of hard as well as soft photons.

2. Hard real photon production

When the external photon is hard, its energy is of the order of the temperature of the heat bath. As it was mentioned above, the discontinuity of the polarization tensor is related to the basic processes, Compton and anihilation processes. The kinematical region of interest is the soft contribution, when the effective quark-loop diagram of Fig.1, with one soft (dressed) and one hard (bare) quark line, is considered. Propagation of soft momenta usually gives rise to infrared divergences in loops, but the main property of the effective propagator is that it is screened on the momentum scale of order gT for space-like momenta, due to the mechanism of

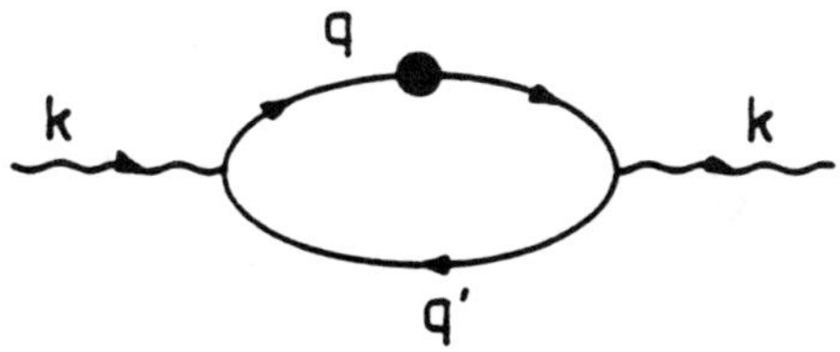

Fig. 1. One-loop diagram for the production of a real photon, with the effective (dressed) quark propagator.

Landau damping. The cut propagator of the soft exchanged quark involves

$$\mathrm{Re}\left(\frac{1}{\not{Q} - \Sigma(q_0, q) + i\epsilon}\right) = \sum_{s=\pm1} \frac{1}{2}(\gamma_0 - s\vec{\gamma})\mathrm{Im}\frac{1}{D_s(q_0, q)}, \tag{4}$$

where one has resummed an infinite set of diagrams contributing to the same order as the Born term. The functions $D_\pm$ for small space-like momenta in the static limit $q_0 = 0$, $q \to 0$ yield

$$\mathrm{Im}\frac{1}{D_s(q_0, q)} \simeq \frac{q}{q^2 + m_f^2(4 + \pi^2)/8}. \tag{5}$$

The fermion mass m_f, with $m_f^2 = 2\pi\alpha_s T^2/3$, acts as an infrared cut-off, and will give rise to a $\ln(1/\alpha_s)$ term in the final result, which is a reminiscence of the logarithmic divergence.

For the hard contribution, the amplitudes of the basic processes are calculated using first order perturbation theory. The final result for the hard photon rate is obtained[6,7] by adding the soft and hard contributions:

$$E\frac{dW^\gamma}{d^3q} = \Sigma e_q^2 \frac{\alpha\alpha_s}{2\pi^2} T^2 e^{-E/T} \ln\left(\frac{cE}{\alpha_s T}\right), \tag{6}$$

with $c \simeq 0.23$, when $E > T$. Thus the Braaten-Pisarski's method works well in removing the infrared divergences in this case.

3. Soft photon production

When the photon energy is of order gT[10,11], the one-loop diagram using effective propagators and vertices are necessary (Fig.2). The tadpole diagram (Fig.2b) vanishes because it is traceless in its indices[8], $\Pi^\mu_{\mu,tadpole} = 0$. Therefore the self-energy like diagram (Fig.2a) with both effective propagators and vertices gives the dominant contribution.

As the internal quark lines are resummed, we expect that screening occurs as

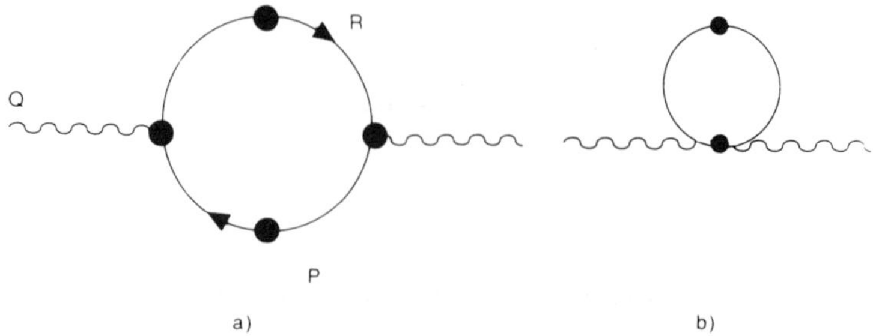

Fig. 2. The soft fermion loop contribution to the soft real photon production rate.

for the hard photons. But the introduction of effective vertices will lead to non-vanishing collinear singularities. This can be seen from the expression of the effective quark-photon vertex:

$$ {}^{*}V^{\mu}_{\alpha\beta\delta}(P,Q,R) = \gamma^{\mu} + m_f^2 \int \frac{d\Omega}{4\pi} \frac{L^{\mu} \not{L}}{(L.P + i\epsilon_{\alpha})(L.R + i\epsilon_{\delta})}, \tag{7} $$

where L is a light-like 4-momentum, $L^2 = 0$, P and $R = P + Q$ the quark momenta with $Q = (q, \vec{q})$ the photon vector. Here we have used a particular reformulation of the real time formalism, the retarded/ advanced formalism[12]. The prescriptions $i\epsilon_{\alpha}$, $i\epsilon_{\delta}$ are no more the Feynman prescriptions, but the retarded or advanced ones. This formalism has the advantage to exhibit more clearly the analytic structure of Green's functions, and the correspondence between the real time and the imaginary time correlation functions[11,12]. The mass singularity arises when both quark momenta P and R are space-like, i.e. from:

$$ \mathbf{P}\frac{1}{L.R}\delta(L.P) = \mathbf{P}\frac{1}{L.Q}\delta(L.P), \tag{8} $$

when the real photon is collinear with the bare gluon, carrying the momentum L. It is worth noting that the divergence is related to the vanishing mass of the photon. If Q^2 were slightly off-shell, the pole in Eq. (8) would never be reached in the time-like case or would be defined through a principal value prescription in the space-like case. The collinear divergences will arise from the discontinuity of the three diagrams shown in Fig.3, where the configurations of the effective vertices of Fig.2a are represented. A partial cancellation between the divergent terms of these three diagrams occurs. Indeed, when calculating the residue at the poles, we can use the Ward identity:

$$ Q^{*}_{\mu} V^{\mu}_{\alpha\beta\delta}(P,Q,R) = \frac{1}{q}\left(\Sigma_{\alpha}(P) - \Sigma_{\delta}(R)\right), \tag{9} $$

where Σ is the hard thermal loop part of the self-energy, related to the inverse of the effective quark propagator. Without going into details, the Ward identity

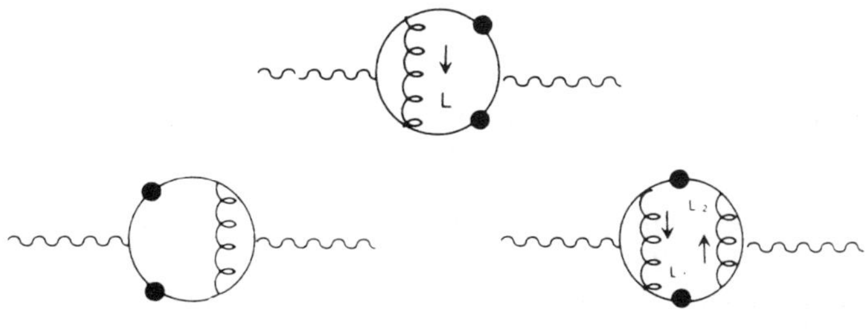

Fig. 3. Diagrams from the soft fermion loop contribution, displaying the structure of the effective vertices.

allows the compensation of the singular single-vertex terms with a divergent part of the double-vertex diagram. However, there still remains a collinear divergent piece. Indeed, in $D = 4 + 2\epsilon$ dimensions, the leading singular contribution to the soft photon production rate reads:

$$E\frac{dW^{\gamma}}{d^3q} = \frac{1}{\epsilon}\Sigma e_q^2 \frac{\alpha\alpha_s}{2\pi^2} T^2 n_F(E) \left(\frac{m_f}{E}\right)^2 \ln\left(\frac{1}{\alpha_s}\right), \tag{10}$$

which shows that the Braaten-Pisarski's resummation does not succeed in removing the divergences for the soft real production rate[10,11].

4. Conclusion

It is interesting and somewhat paradoxical that the divergence in the calculation of the soft photon production rate is a property of the resummed theory which tells us to use the effective propagators and vertices rather than the bare ones. All the more, since, for scales of $O(gT)$ to T, resummation turns out to be successful in screening fermionic and bosonic mass singularities. The production of real photons with energies of $O(gT)$ is, however, one exception.

There are several ways the divergent result Eq. (10) could be regularized. One of the possibilities is to include corrections to the hard internal fermion lines. But the crucial problem is that going beyond Braaten-pisarski's resummation in a consistent way, by keeping gauge invariance for physical quantities, is not actually an easy task. This is in the spirit of several attempts to define the damping rate of a hard fermion[13].

5. Acknowledgements

I kindly thank P. Aurenche, R. Baier, R. Kobes, S. Peigné and D. Schiff for discussions.

References

1. For a recent review: P.V. Ruuskanen, in *Particle Production In Highly Excited Matter* (eds. H.H. Gutbrod and J. Rafelski, Proc. of ASI, Il Ciocco, Lucca, Italy, 1993).
2. H.A. Weldon, *Phys. Rev.* **D44** (1991) 3955.
3. H.A. Weldon, *Phys. Rev.* **D28** (1983) 2007.
4. R.L. Kobes and G.W. Semenoff, *Nucl. Phys.* **B260** (1985) 714; *Nucl. Phys.* **B272** (1986) 329.
5. For a review and references: M. Le Bellac, *Lectures given at the XXXth Schladming Winter School* (Schladming, Austria, March 1991).
6. J.I. Kapusta, P. Lichard and D. Seibert, *Phys. Rev.* **D44** (1991) 2774.
7. R. Baier, H. Nakkagawa, A. Niégawa and K. Redlich, *Z. Phys.* **C53** (1992) 433.
8. E. Braaten and R.D. Pisarski, *Nucl. Phys.* **B337** (1990) 569, *Nucl. Phys.* **B339** (1990) 310.
9. J. Frenkel and J.C. Taylor, *Nucl. Phys.* **B334** (1990) 199.
10. R. Baier, S. Peigné and D. Schiff, *Orsay preprint* LPTHE-Orsay 93-46, BI-TP-93-55 (1993).
11. P. Aurenche, T. Becherrawy and E. Petitgirard, *ENSLAPP preprint* ENSLAPP-A-452/93, NSF-ITP-93-155 (1993).
12. P. Aurenche and T. Becherrawy, *Nucl. Phys.* **B379** (1992) 259; C.M.A. van Eijck and Ch.G. van Weert, *Phys. Lett.* **B278** (1992) 305.
13. R. Baier, *These proceedings and references therein.*

Many-Body Structure in Hamiltonian QCD

M.K.Mak and D.Schütte
Institut für Theoretische Kernphysik der Universität Bonn
Nußallee 14 – 16, D-5300 Bonn, Germany

Abstract

The problems connected with the application of many-body techniques to the lattice QCD Hamiltonian are investigated. Various approaches to the pure Yang Mills case (Kogut-Susskind framework) including the exp S method of Coester and Kümmel are compared yielding encouraging results. We then discuss the generalization of this exp S method to QCD including quarks.

1 Introduction

The investigation of the eigenvalue problem of the QCD Hamiltonian is considered to be an alternative to standard lattice Monte Carto QCD giving possibly new insight into the structure of such non-abelian gauge theories. The basic problem of such a treatment of QCD is to find a combination of a *consistent renormalization* and a *controlable many-body approach.*

Since cutoff procedures are in general only consistent with renormalization if one adds to the Hamiltonian an infinite number of counter terms wich are difficult to control[?], at present the following two alternatives seem to be the most promosing strategies:

1) *The approach of Lüscher - van Baal[?]:* Here, the many body approach is chosen as a model space method (folded diagram method) enabling the perturbative computation of a model space Hamiltonian whose coefficients can be renormalized using dimensional regularization. The diagonalization of the model space Hamiltonian is done by non-perturbative methods allowing especially the incorporation of topological symmetries.

At present, the limitation of the mehod is a restriction to "intermediate volumes" (box with dimension $\leq 1fm$) because the "Θ-symmetry" is still missing in the formulation. Attempt to improve this situation are discussed by P. van Baal in a contribution to this conference.

2) *The lattice Hamiltonian approach:* Here, the same regularization is used as in standard lattice QCD, and the same consistency checks with

dimensional regularization exist as in the 4-dimensional case[?, ?]. Consequently, the usual scaling test for the continuum limit holds, especially dimensionless quantities should become independent of the coupling g for $g \to 0$. In this sense, renormalization is under control in this framework.

At present, the limitation of this approach arises from the fact that all approximation techniques - defined in analogy to standard many-body theory - are most reliable in the strong coupling limit ($g \to \infty$) and that extrapolations to $g \to 0$ have to surmount a "critical coupling" (usually around $g = 1$) where perturbation theory breaks down. A signature of this structure is e.g. the well-known roughening transition for the string tension.

Within this talk, we will first give a survey on the situation for the lattice Hamiltonian approach for the pure Yang Mills theory. After that, we will discuss QCD with quarks. We will only deal with 3+1 dimensional SU(3) gauge theories.

2 Kogut-Susskind Theory

Excluding quarks, QCD is given by the Kogut-Susskind theory[?]. In this case, the many-body states are "top" wave functions depending on N link variables $\Psi(U_1, ..U_N)$ where the quantities U_l ($l = 1, .., N$) are elements of the gauge group $SU(3)$ and N is the number of oriented links in a three-dimensional cubic lattice.

The Kogut-Susskind Hamiltonian [?] is

$$H_{KS} = \frac{g^2}{2\epsilon} E_{la} E_{la} - \frac{1}{\epsilon g^2} \sum_p (\chi_p + \chi_{-p} - 6) \ , \tag{1}$$

where g is the coupling constant, ϵ denotes the lattice constant and a, l and p are colour, link and plaquette labels, respectively. Summation over repeated indices is always assumed. The colour-magnetic flux operator χ_p is given by

$$\chi_p := \text{tr}\ U_{l_1} U_{l_2} U_{l_3} U_{l_4} \tag{2}$$

with $l_1, \ldots, l_4 \in p$. The colour-electric field operators $E_{\alpha a}$ and the link operators U_β are conjugate quantum operators obeying the commutation relations

$$[E_{la}, U_{l'}] = \delta_{ll'} \lambda^a U_l \ , \tag{3}$$

where the generators λ^a are normalized according to tr $\lambda^a\lambda^b = \delta^{ab}/2$.

The problem is to determine the spectrum of the Hamiltonian H in the space of gauge invariant states obeying Gauss' law

$$\begin{aligned} \Psi(U_1^g, .., U_N^g) &= \Psi(U_1, .., U_N) \\ U_l^g &= g(\mathbf{x}_l)U_l g^{-1}(\mathbf{y}_l) \end{aligned} \tag{4}$$

where $g(\mathbf{x})$ is an arbitrary, time independent gauge transformation and the lattice points $\mathbf{x}_l$ and $\mathbf{y}_l$ are connected by the link l.

Different many body approaches have been used to approximate the low lying spectrum of H_{KS}:

1) *Padé extrapolation of strong coupling expansion*[?]

Here, the starting point is the knowlegde of the result of standard strong coupling perturbation theory for the desired observable. In order to overcome the roughening transition point, Padé and Shafer techniques are employed[?].

2) *exp -tH method*[?]

Using the fact that the operator exp -tH projects on eigenstates for large values of the parameter t, D. Horn et al. have developed a method by computing matrix elements of exp -tH as polynomials and performing the extrapolation to large t by Padé techniques.

3) *Variational method*[?]

In principle, the Ritz variational method yields a rather simple and safe approximation method. With the standard exp S ansatz, however, (see below) no analytical evaluation of energy expectation values is possible, but "Hamiltonian" Monte Carlo techniques can be employed[?].

4) *exp S and shifted coupled cluster method*[?]

An important result of perturbation theory - valid beyond the applicability of the expansion - is that the ground state has the structure

$$|\psi\rangle = \mathrm{e}^S\,|0\rangle \tag{5}$$

where $|0\rangle$ is the strong coupling ground state and S is a linked operator. An approximation may then be defined by making a restricted ansatz for S - consistent with Gauss' law - and by determining the coefficients in the ansatz by a suitable truncation of the Schrödinger equation written in the form

$$\mathrm{e}^{-S}H\mathrm{e}^S\,|0\rangle = E\,|0\rangle \tag{6}$$

(The commutator expansion of (6) terminates with the third term!). The "art" of this exp S method is the suitable choice of an ansatz for S and of a truncation for (6).

Llewelling Smith and Watson by their choice of S try to sum up all short range correlations. A shifted coupled cluster method is then employed to rewrite eq. (6) (and a corresponding generalization for excited states) such that higher order correlations are consistently neglegted[?]

5) *The method of Coester, Kümmel and Emrich*[?]

In view of the relative success of the first two approaches which are based on perturbation theory, an alternative approximate exp S method is to use the perturbative structure of S as a guiding principle. For the standard many-body case, this method has been systemized by Coester and Kümmel[?] for the ground state and by Emrich[?] for excited states. A generalization for the Kogut-Susskind theory has been worked out[?] and results in lowest (Brueckner-type) order have been obtained.

In figure 1, a comparison is made for the predictions of the various many body approaches for the dimensionless quantity

$$R := M(0^{++})/\sqrt{\sigma} \tag{7}$$

where $M(0^{++})$ is the smallest glueball mass and σ is the string tension. (For the methods 3) and 4), no calculations for this ratio exist.)

Compared to a recent standard lattice Monte Carlo computation[?] we see that all approximation schemes predict the correct order of magnitude for the ratio R for $g \simeq 1$. The scaling behaviour (R should be independent of g for $g \to 0$) is not so satisfactory, especially the exp S method breaks down for small values of g in the lowest (Brückner type) order. The other approximations are carried to up to 6-th[?] and 12-th[?] order, respectively.

We consider these results as encouraging enough to warrant an investigation of QCD including quarks.

3 Inclusion of Quarks

The lattice QCD Hamiltonian with quarks - using staggered fermions[?] without mass term - has the structure

$$\begin{aligned} H &= H_{KS} + H_F \\ H_F &= h_{\alpha\beta}(U) a^\dagger_\alpha a_\beta \end{aligned} \tag{8}$$

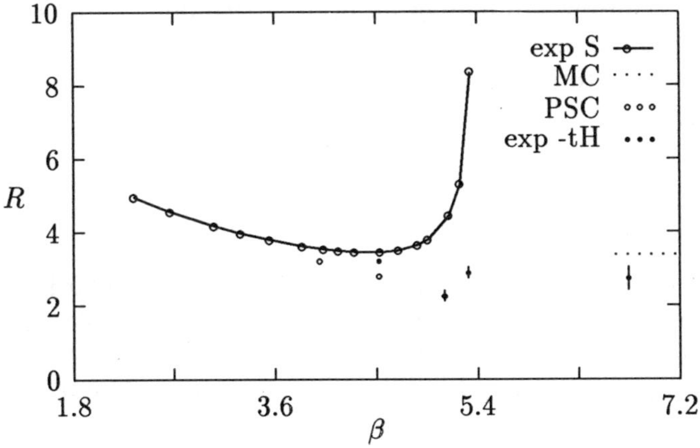

Figure 1: The ratio R as a function of $\beta = 6/g^2$ for different many-body approaches: Padé-Shafer improved strong coupling expansion (PSC)[?], exp -tH method[?] (both with error bars marking the uncertainty of the extrapolation to large β) and lowest order Coester-Kümmel method (exp S)[?]. The lattice Monte Carlo result (MC) is from ref.[?].

Here, $\alpha = (\mathbf{x}_\alpha, c_\alpha)$ abbreviates lattice point and color quantum numbers ($c = 1,.,3$). The matrix elements $h_{\alpha\beta}$ depend on the link variable (U_l): if $(\mathbf{x}_\alpha, \mathbf{y}_\beta) = l$ the $h_{\alpha\beta}$ are chosen to make H_F gauge invariant and $h_{\alpha\beta} = 0$ if $(\mathbf{x}_\alpha, \mathbf{y}_\beta)$ is not a link. Due to Fermion doubling, the theory generates two (up and down) Dirac spinor fields in the continuum limit if the $h_{\alpha\beta}$ are suitably chosen[?].

Only few calculations for the hadron sprectrum with this Hamiltonian exist: Except for "ancient" strong coupling expansion attempts, only a recent exp -tH work of Horn and Schreiber[?] is known. Here we want to discuss some problems connected with the extension of the exp S method for the treatment of H for the vacuum.

Since the Fermion number is conserved, the vacuum state is given by the ground state of H for the M-particle problem if M is related to the number of lattice points by $2M = 3N_p$. A convenient (in the strong coupling sense) "unperturbed" choice ϕ_0 of the vacuum is a Slater determinant where all color quantum numbers of all even lattice points are occupied[?]. This "occupied Dirac sea" is gauge invariant.

A first step to improve ϕ_0 within the exp S method would be an ansatz of the type

$$\begin{aligned} \phi &= e^{S_F}\phi_0 \\ S_F &= S_{Ab}a_A^\dagger a_b \end{aligned} \tag{9}$$

where A (b) denote odd (even) lattice points. In standard many-body theory, optimizing the coefficients S_{Ab} by the Ritz principle would amount to a Hartree Fock (HF) theory[?], in field theoretical models involving a four-point interaction it would correspond to the Nambu Jona-Lasignio theory. A generalization to QCD should be, therefore, interesting.

A literal application of the HF theory, however, is not possible because the state Φ is not gauge invariant if the coefficients S_{Ab} are constant. In order to guarantee Gauss' law, the S_{Ab} have to depend suitably on the link variable related to a string between the lattice points $\mathbf{x}_A$ and $\mathbf{x}_b$.

A formulation of such a "gauge invariant HF theory" is possible[?]. Whereas the Fermionic part of the scalar products can be worked out analytically, the gluonic part - involving an N_p-fold Haar measure integral - can only be approximated by a suitable cluster expansion. Details will be given elsewhere[?], here we will only discuss some general features of this "QCD HF theory":

In principle, it provides a generalization of Nambu Jona-Lasignio-type calculations to QCD, where *renormalization is completely under control.* This should allow a refined investigation of chiral symmetry breaking. Due to the Fermion doubling problem[?], however, the regularized QCD described by H exhibits only a discrete remnant of chiral symmetry. Consequently, one does not get the standard structure of chiral symmetry breaking, e.g. the pion does not become a Goldstone boson[?]. Possibly, using Wilson Fermions or including four flavours (which corresponds to standard 4-dimensional lattice QCD) may improve this situation.

Before one can hope to decrible hadrons realistically one has, of course, to refine the ansatz for the vacuum $\phi = exp\ S\ \phi_0$ (and for excited states) to include higher order terms in the operator S: Pure gluonic contributions as in (5) would amount to a quenched approximation, generalization of this have to take into account "two-particle two-hole" terms, etc.[?]. A detailed investigation of this problem has to be the subject of future investigations.

Financial support by the Deutsche Forschungsgemeinschaft is gratefully acknowlegded.

References

[1] K.G. Wilson et al., Phys. Rev. D**49** (1994) 6720

[2] M. Lüscher, Nucl. Phys. B**219** (1983) 233

P. van Baal and J. Koller, Ann. of Phys. **174** (1987) 299; Nucl. Phys. B**303** (1988) 1

[3] A. Hasenfratz and P. Hasenfratz, Phys. Lett. B**93** (1980) 165

[4] B. Diekmann, M. Langer and D. Schütte, Nucl. Phys. B (Proc. Suppl.)**30** (1993) 932

[5] J. Kogut , L. Susskind ; Phys.Rev. D**11** (1975) 395

[6] C.J. Hamer, A.C. Irving and T.E. Preece; Nucl. Phys. B **270** [FS16] (1986) 553

[7] C.P. van den Doel and D. Horn; Phys. Rev. D**35** (1987) 2824

[8] S.A. Chin, C. Long and D. Robson; Phys.Rev. D**37** (1988) 3001 and 3006

[9] C.H. Llewellyn Smith and N.J. Watson, Phys. Letters B **302** (1993) 463

[10] M.K. Mak; *Anwendung einer Clusterentwicklung auf Gittereichtheorien*; Dissertation; Bonn 1993 (unpublished)

[11] F. Coester and H. Kümmel; Nucl.Phys. **17** (1960) 477

[12] K. Emrich; Nucl.Phys. **A351** (1981) 379

[13] G.S. Bali et al., Phys. Lett. B**309** (1993) 378

[14] L. Susskind, Phys. Rev. D**16** (1977) 3031

[15] D. Horn and D. Schreiber; Phys. Rev. D**47** (1993) 2081

[16] P. Ring and P. Schuck, The Nuclear Many-Body Problem, Springer Verlag 1980

[17] D. Schütte, A. Spitz and V. Zota, work in preparation

LIST OF PARTICIPANTS

Dr. Abdellatif Abada
Universite de Nantes

Prof. Gregory D. Adkins
Franklin and Marshall College

Dr. Michael Albrow
FERMILAB

Dr. Tanguy Altherr
CERN

Dr. Patrick Aurenche
LAPP-Annecy

Dr. Rudolf Baier
Universitat Bielefeld

Prof. Marshall Baker
University of Washington

Prof. James S. Ball
University of Utah

Dr. Jean Paul Blaizot
Saclay

Dr. Vincenzo Branchina
CRN-Strasbourg

Dr. Conrad Burden
Australian Natl. Univ.

Dr. Matthias Burkardt
University of Washington

Prof. P. Carruthers
University of Arizona

Prof. Leon Cooper
Brown University

Prof. A. Donnachie
University of Manchester

Prof. Michael I. Eides
Konstantinov Nuc. Phys. Inst.

Dr. Hans-Thomas Elze
CERN

Dr. Albert Cogollos Ferrando
Universitad de Valencia

Prof. H. M. Fried
Brown University

Dr. Santiago Garcia
Brown University

Dr. Pierre Gauron
IPN-Orsay

Prof. Howard Grotch
Penn State University

Prof. Francoise Guerin
Universite de Nice

Prof. Gerald Guralnik
Brown University

Dr. Suzhou Huang
University of Washington

Prof. Munir Islam
University of Connecticut

Prof. Kenneth Johnson
MIT

Dr. Dimitri Kalafatis
University of Manchester

Prof. Kyungsik Kang
Brown University

Dr. Marc Knecht
INP-Orsay

Dr. Randy Kobes
LAPP-Annecy

Dr David A. Kosower
Saclay

Dr. Gabor Kunstatter
University of Winnipeg

Prof. P. Landshoff
DAMTP

Prof. Michel LeBellac
Universite de Nice

Dr. P. Levai
Budapest

Dr. R. Levi
Boston University

Prof. Maurice Levy
LPTHE-VI

Dr. Senben Liao
Duke University

Dr. Marcello Lissia
INFN

Prof. K. T. Mahanthappa
University of Colorado

Prof. Uri Maor
Tel-Aviv University

Dr. Harold Markum
Technische Univ.,Wien

Dr. Andre Martin
CERN

Dr. C. Martin
INP-Orsay

Prof. John Moffat
University of Toronoto

Dr. Emil Mottola
Los Alamos National Lab.

Dr. Bechir Moussallam
IPN-Orsay

Prof. Berndt Mueller
Duke University

Dr. Herbert Nachbagauer
LAPP - Annecy

Dr. Basarab Nicolescu
Universite de Paris VI

Prof. David Owen
Ben-Gurion University

Dr. Rajesh R. Parwani
Saclay

Dr. Emmanuel Petitgirard
LAPP - Annecy

Dr. Bernard Pire
Ecole Polytechnique

Dr. Janos Polonyi
CRN- Strasbourg

Prof. J. Rafelski
LPTHE - Paris VII

Prof. John Ralston
University of Kansas

Dr. Anton Rebhan
DESY

Prof. H. Reinhardt
Universitat Tubingen

Dr. Patrick Reynaud
Universite de Nice

Dr. Craig Roberts
Argonne National Lab.

Dr. V. E. Rochev
HEP - Protvino

Prof. Douglas A. Ross
University of Southampton

Dr. Serge Rudaz
University of Minnesota

Prof. David S. Saxon
UCLA

Dr. M. Schaden
New York University

Dr. Bernd Schreiber
MIT

Dr. D. Schutte
Universitat Bonn

Dr. Erhard Seiler
Max Planck Institut

Prof. J. Ely Shrauner
Washington University

Prof. Edward Shuryak
SUNY at Stony Brook

Dr. Jacques Soffer
CNRS - Marseille

Dr. J. Stern
IPN-Orsay

Dr. Dieter Stoll
University of Tokyo

Prof. Joseph Sucher
University of Maryland

Prof. U. Sukhatme
Univ. of Illinois at Chicago

Prof. Chung-I Tan
Brown University

Prof. Pierre van Baal
University of Leiden

Dr. D. Vautherin
IPN-Orsay

Prof. R. Vinh Mau
Universite de Paris VI

Dr. Alan White
Argonne National Lab

Prof. Richard Woodard
University of Florida

Prof. Tai T. Wu
Harvard University

Dr. Ken Yee
Louisiana State University

Dr. Jean Zinn-Justin
Saclay

Prof. Dan Zwanziger
New York University

Directions in High Energy Physics — Vol. 14

PRECISION TESTS OF THE STANDARD ELECTROWEAK MODEL

Paul Langacker [illegible]

[illegible]

World Scientific